Emerging Research in the Analysis and Modeling of Gene Regulatory Networks

Ivan V. Ivanov
Texas A&M University, USA

Xiaoning Qian
Texas A&M University, USA

Ranadip Pal
Texas Tech University, USA

A volume in the Advances in Medical Technologies and Clinical Practice (AMTCP) Book Series

Medical Information Science
REFERENCE
An Imprint of IGI Global

Published in the United States of America by
Medical Information Science Reference (an imprint of IGI Global)
701 E. Chocolate Avenue
Hershey PA 17033
Tel: 717-533-8845
Fax: 717-533-8661
E-mail: cust@igi-global.com
Web site: http://www.igi-global.com

Library of Congress Cataloging-in-Publication Data

Names: Ivanov, Ivan V., 1962- | Qian, Xiaoning, 1975- | Pal, Ranadip, 1980-
Title: Emerging research in the analysis and modeling of gene regulatory networks / Ivan V. Ivanov, Xiaoning Qian, and Ranadip Pal, editors.
Description: Hershey, PA : Medical Information Science Reference, [2016] | Includes bibliographical references and index.
Identifiers: LCCN 2016005966| ISBN 9781522503538 (hardcover) | ISBN 9781522503545 (ebook)
Subjects: LCSH: Genetic regulation--Computer simulation. | Gene regulatory networks.
Classification: LCC QH450 .E44 2016 | DDC 572.8/65--dc23 LC record available at https://lccn.loc.gov/2016005966

This book is published in the IGI Global book series Advances in Medical Technologies and Clinical Practice (AMTCP) (ISSN: 2327-9354; eISSN: 2327-9370)

British Cataloguing in Publication Data
A Cataloguing in Publication record for this book is available from the British Library.

All work contributed to this book is new, previously-unpublished material. The views expressed in this book are those of the authors, but not necessarily of the publisher.

Advances in Medical Technologies and Clinical Practice (AMTCP) Book Series

ISSN: 2327-9354
EISSN: 2327-9370

Mission

Medical technological innovation continues to provide avenues of research for faster and safer diagnosis and treatments for patients. Practitioners must stay up to date with these latest advancements to provide the best care for nursing and clinical practices.

The **Advances in Medical Technologies and Clinical Practice (AMTCP) Book Series** brings together the most recent research on the latest technology used in areas of nursing informatics, clinical technology, biomedicine, diagnostic technologies, and more. Researchers, students, and practitioners in this field will benefit from this fundamental coverage on the use of technology in clinical practices.

Coverage

- Medical Imaging
- Nursing Informatics
- Clinical High-Performance Computing
- Neural Engineering
- E-Health
- Nutrition
- Clinical Studies
- Biometrics
- Clinical Data Mining
- Clinical Nutrition

IGI Global is currently accepting manuscripts for publication within this series. To submit a proposal for a volume in this series, please contact our Acquisition Editors at Acquisitions@igi-global.com or visit: http://www.igi-global.com/publish/.

The Advances in Medical Technologies and Clinical Practice (AMTCP) Book Series (ISSN 2327-9354) is published by IGI Global, 701 E. Chocolate Avenue, Hershey, PA 17033-1240, USA, www.igi-global.com. This series is composed of titles available for purchase individually; each title is edited to be contextually exclusive from any other title within the series. For pricing and ordering information please visit http://www.igi-global.com/book-series/advances-medical-technologies-clinical-practice/73682. Postmaster: Send all address changes to above address.

Titles in this Series

For a list of additional titles in this series, please visit: www.igi-global.com

Applied Case Studies and Solutions in Molecular Docking-Based Drug Design
Siavoush Dastmalchi (Tabriz University of Medical Sciences, Iran) Maryam Hamzeh-Mivehroud (Tabriz University of Medical Sciences, Iran) and Babak Sokouti (Tabriz University of Medical Sciences, Iran)
Medical Information Science Reference • copyright 2016 • 367pp • H/C (ISBN: 9781522503620) • US $225.00 (our price)

Methods and Algorithms for Molecular Docking-Based Drug Design and Discovery
Siavoush Dastmalchi (Tabriz University of Medical Sciences, Iran) Maryam Hamzeh-Mivehroud (Tabriz University of Medical Sciences, Iran) and Babak Sokouti (Tabriz University of Medical Sciences, Iran)
Medical Information Science Reference • copyright 2016 • 456pp • H/C (ISBN: 9781522501152) • US $235.00 (our price)

Advancing Pharmaceutical Processes and Tools for Improved Health Outcomes
Tagelsir Mohamed Gasmelseid (International University of Africa, Sudan)
Medical Information Science Reference • copyright 2016 • 388pp • H/C (ISBN: 9781522502487) • US $185.00 (our price)

Classification and Clustering in Biomedical Signal Processing
Nilanjan Dey (Techno India College of Technology, India) and Amira Ashour (Tanta University, Egypt)
Medical Information Science Reference • copyright 2016 • 463pp • H/C (ISBN: 9781522501404) • US $225.00 (our price)

Virtual Reality Enhanced Robotic Systems for Disability Rehabilitation
Fei Hu (University of Alabama, USA) Jiang Lu (University of Houston -Clear Lake, USA) and Ting Zhang (University of Alabama, USA)
Medical Information Science Reference • copyright 2016 • 383pp • H/C (ISBN: 9781466697409) • US $210.00 (our price)

Software Innovations in Clinical Drug Development and Safety
Partha Chakraborty (Cognizant Technology Solutions, India) and Amit Nagal (GVK Bioscience, India)
Medical Information Science Reference • copyright 2016 • 305pp • H/C (ISBN: 9781466687264) • US $225.00 (our price)

External Advisory Board

List of Reviewers

Table of Contents

Detailed Table of Contents

The inference of Gene Regulatory Networks (GRNs) is a very challenging problem which has attracted increasing attention since the development of high-throughput sequencing and gene expression measurement technologies. Many models and algorithms have been developed to identify GRNs using mainly gene expression profile as data source. As the gene expression data usually has limited number of samples and inherent noise, the integration of gene expression with several other sources of information can be vital for accurately inferring GRNs. For instance, some prior information about the overall topological structure of the GRN can guide inference techniques toward better results. In addition to gene expression data, recently biological information from heterogeneous data sources have been integrated by GRN inference methods as well. The objective of this chapter is to present an overview of GRN inference models and techniques with focus on incorporation of prior information such as, global and local topological features and integration of several heterogeneous data sources.

Genetic Regulatory Networks (GRNs) represent the interconnections between genomic entities that govern the regulation of gene expression. GRNs have been represented by various types of mathematical models that capture different aspects of the biological system. This chapter discusses the relationships among the most commonly used GRN models that can enable effective integration of diverse types of sub-models. A detailed model in the form of stochastic master equation is described, followed by it coarse-scale and deterministic approximations in the form of Probabilistic Boolean Networks and Ordinary Differential Equation models respectively.

Stochasticity plays important roles in many biological networks. A fundamental framework for studying the full stochasticity is the Discrete Chemical Master Equation (dCME). Under this framework, the combination of copy numbers of molecular species defines the microstate of the molecular interactions in the network. The probability distribution over these microstates provide a full description of the properties of a stochastic molecular network. However, it is challenging to solve a dCME. In this chapter, we will first discuss how to derive approximation methods including Fokker-Planck equation and the chemical Langevin equation from the dCME. We also discuss the widely used stochastic simulation method. After that, we focus on the direct solutions to the dCME. We first discuss the Finite State Projection (FSP) method, and then introduce the recently developed finite buffer method (fb-dCME) for directly solving both steady state and time-evolving probability landscape of dCME. We show the advantages of the fb-dCME method using two realistic gene regulatory networks.

In order to derive system-based methods to control dynamic behavior of biological systems of interest for future gene-based intervention therapeutics, two basic categories of intervention strategies have been studied based on the Markov chain theory and Markov decision processes: Structural intervention by function perturbation and external control based on state perturbation. The chapter reviews the existing network analysis and control methods in these two categories and discusses their extensions for more robust and clinically relevant intervention strategies considering collateral damages from intervention.

A central problem in translational medicine is to provide a framework for deriving and studying effective intervention methods to elicit desired steady-state behavior for a gene regulatory network of interest with Markovian dynamics. Heretofore, two rather different external control approaches have been taken. The first optimizes a subjectively defined cost function while modeling treatment constraints; therefore, desirable shift of the steady-state mass is a by-product. The second approach, on the other hand, focuses solely on the steady-state behavior of the network and provides the maximal shift achievable. Although both approaches are optimal with respect to their objectives, the choice of which to use depends on the treatment goals.

The human genome has been proposed to contain numerous regions called "consensus sequences" that are recognized by DNA-binding proteins required for gene expression. Results obtained with a computer simulation showed that: (1) all random sequences consisting of less than 15 base pairs were present in the human genome, and (2) consensus sequences reported to date were found in an artificial genome created randomly by a computer. These results suggest that conserved sequences consisting of more than 15 base pairs are required for accurate gene expression. No consensus sequence consisting of more than 15 base pairs has been identified to date. Thus, the co-association of multiple proteins bound side by side is required for appropriate gene expression because the total number of conserved sequences can exceed 15, whereas the binding of a single protein has no physiological role because its consensus sequence is present randomly in the human genome.

To ensure totipotency of new zygotes, immediately after fertilization, two gametes activate genetic and epigenetic processes that highly interplay and lead to sequential waves of gene expression. Since the first report of fertilization a hundred years ago, a long-standing question to what factors govern the dramatic transcriptional changes remains elusive. Recently, significant advances in biological experiment technology and computational methodology have taken place in understanding of the regulatory dynamics during fertilization. This chapter provides an overview of the recent progress in characterizing mammalian fertilization, and focuses on the computational methods applicable to the inference of regulation for understanding early embryo development. In particular, this chapter introduces the linear regression modeling and log-linear graphical modeling to identify potential key regulators in the higher-order conditional distribution where statistical paradox often occurs.

Computational prediction of cis-regulatory elements for a set of co-expressed genes based on sequence analysis provides an overwhelming volume of potential transcription factor binding sites. It presents a challenge to prioritize a set of functional transcription factors and their binding sites on the regulatory regions of the target genes that are relevant to the gene expression study. A novel approach based on the use of lasso multinomial regression models is proposed to address this problem. We examine the ability of the lasso models using a time-course microarray data obtained from a comprehensive study of gene expression profiles in skin and mucosal in mouse over all stages of wound healing.

To the development of effective cancer drug, it is necessary to, first, identify drugs and their possible combinations that could exert desired control over the type of cancer being considered; second, have a drug testing method that allows one to assess the variety of responses that can be provoked by drugs. To facilitate such an experiment-modeling-experiment cycle for drug development, a method based on the dynamical systems of pathways is presented. It involves a three-state experimental design: (1) formulate an oncologic pathway model of relevant cancer; (2) perturb the pathways with the drugs of known effects on components of the pathways of interest; and (3) measure process activity indicators at various points on cell populations. To evaluate the drug response in a high-throughput manner, a green fluorescent protein reporter-based technology has been developed. The authors apply the dynamical approach to several issues in the context of colon cancer cell lines.

Preface

The intention and the result of a scientific inquiry is to obtain an understanding and a control of some part of the Universe -Arturo Rosenblueth and Norbert Wiener

While the advances in technology can help in obtaining more and better quality data about the cellular processes and signals, the design and practical applications of predictive computational models of genomic regulation continue to face major challenges. This edited book draws attention to several of them and outlines important future research directions.

One can think of a *Gene Regulatory Network (GRN)* as a network of relations among strands of DNA (genes) and the regulatory activities associated with those genes (Dougherty & Braga-Neto, 2006). This general definition allows for many mathematical (usually dynamical) systems to be called GRNs. The goodness of each such model is evaluated using several important criteria: the level of description of the biochemical reactions involved, complexity of the model, model parameter estimation, and its predictive power. There have been many attempts to model the structure and dynamical behavior of GRNs, ranging from deterministic with discrete state and time space to fully stochastic with continuous state and time space (de Jong, 2002). The so called 'central dogma' of molecular biology (Crick, 1970) implies that genes communicate via the proteins they encode. All of the stages of protein production, transcription and translation, are controlled by a multitude of biochemical reactions, and are influenced by both internal and external to the cell factors. This perspective suggests that the expression of a given gene *i*, i.e. the quantity of either protein or messenger RNA, should be considered as a random function $X_i(t)$ of the cell's internal and external environments. Thus, if one wants to study the dynamical behavior of a GRN, one must design a mathematical model for the gene-expression vector $\boldsymbol{X}(t) = (X_1(t), X_2(t), \ldots, X_n(t))$ for the n genes that are selected to constitute the network. The stochastic master differential equation model appears to provide the most detailed description of the dynamics of $\boldsymbol{X}(t)$. In principle, it could include all of the information about the biochemical processes involved in gene regulation. At the same time, the estimation of its parameters cannot be done without large amount of reliable time-series data. As a result, one is often forced to take a more pragmatic approach and look for simpler models for the dynamics of the gene-expression vector. One of the most extreme simplifications is the *Boolean Network* (BN) model, originally proposed by Kauffman (1969). The BN model had been successfully used in physics before attracting the attention of the biology community. The initial application of the model was to study the evolution of ensembles of networks that were restricted to a specific type of fitness landscape. The BN model is based on the observation that during the regulation of its functional states the cell often exhibits switch-like behavior. Work using the NCI 60 Anti-Cancer Drug Screen has demonstrated that Boolean logic type interactions can be detected in gene expression data (Pal, Ivanov, & Dougherty, 2005). While there are instances in gene regulation where the Boolean logic is the appropriate level of description of the interactions – for instance, when transcription factors have to form a complex that binds to the *cis*-regulatory DNA to activate transcription, one should keep in mind that coarse discrete models cannot capture the details of the biochemical reactions involved in those processes, and models like the Chemical Master Equation (CME) should be used, instead.

In recent years, it has become increasingly clear that systems biology in general and genomics in particular, underscores the role played by stochastic nonlinear dynamical systems which in turn, precludes uncritical reductionist realism. Moreover, systems biology and genomic regulation cannot be considered without the appreciation of their relation to translational science. *Translational science* transforms a mathematical model, whose purpose is to provide a predictive conceptualization of some portion of the physical world, into an intervention (action) strategy, again in the physical world. This ultimately leads to an appropriate framework addressing the specific challenges associated with the goals for controlling cellular regulatory activities. Thus, a properly designed translational mathematical system guides:

1. The scientist in building a model applicable to the specific problem;
2. The engineer in studying costs and benefits of actions;
3. The technologist in devising devices or treatments.

This book provides a compilation of recent and emerging research topics that address these important points. The first chapter concerns a broad overview of different GRN models with emphasis on model inference and prior knowledge incorporation. The problem of prior knowledge incorporation into the inference procedures for GRNs is a fundamental one in the field of genomic signal processing. The modern high-throughput technologies for acquiring genomic data and the intrinsic high-dimensional space of variables involved in genomic regulation leave one with a virtually unbounded model space where it is often insufficient to configure mathematical theories in the absence of prior knowledge. Therefore, a scientist must come to the table with sufficient knowledge of the problem to formulate a small class of models for which it remains only necessary to utilize data to estimate some set of parameters to instantiate the model or the model class. This is imperative for the typically ill-posed GRN inference problem.

The second and third chapters of the book discuss the relationships between various flavors of GRN models and the basic Chemical Master Equation (CME) model. The topics of model approximations, their properties, advantages and disadvantages as well as the means to infer them from data, including rare-event sampling are presented in detail. Several examples of applications to real biological GRNs are provided to illustrate the general discussion.

The reader can notice that there are clear overlaps in the topics discussed in the first three chapters. However, the editorial team decided not to attempt to reduce those because on one hand, each one of these chapters presents the authors' unique perspective of the modeling issues and on the other hand, that allows for preserving the consistency within each separate chapter. In addition, the reader might find it useful and easier to read self-contained chapters.

The fourth and fifth chapters focus on the second major translational point: the control of the model dynamics and the associated costs and benefits of the control actions. *Structural stability* is one of the central concepts in the theory of dynamical systems. It describes persistent behavior that cannot be destroyed by small changes to the system. As real GRNs are capable of maintaining metabolic homeostasis and stable developmental program in the face of a changing environment, they certainly possess structural stability. These two chapters discuss these important topics in the setting of Markovian GRNs. While this specific class of models is clearly a simplification of the detailed stochastic differential equation GRN model, it is rich enough to model such important notions as cellular states (Huang, 1999, 2001) and cellular types (Kauffman, 1993). Moreover, the stochastic nature of the Markovian GRNs allows for incorporation of uncertainty on various levels. First, due to biological variability, gene expression is inherently stochastic. Second, the complex measurement process, e.g. the microarray or next-generation sequencing signal acquisition and processing, create experimental noise that has to be taken into account during the inference of the model. All of these combined with the presence of latent or unobservable variables such as proteins or environmental conditions present one with the problem to infer deterministic predictor functions under uncertainty which is one of the main reasons why the Markovian GRN models have attracted significant attention among the research community. As the main objective of GRN modeling is to predict the dynamical behavior of the underlying real regulatory system and ultimately provide the effective intervention strategies for prevention and control of disease that is associated with altered genomic regulation, the control policies are often evaluated by their effects on the steady-state probability distribution of the model.

Chapter 6 presents a convincing case why one cannot expect to successfully use uncritical reductionist realism when modeling real gene regulatory interactions. The author argues convincingly why multiple regulators are required for the transcription of each gene. The implications of the presented evolutionary and computational considerations are clear: modeling of gene regulation leads, with some notable exceptions, to highly complex models. It is important to point out that the term *complexity* is overloaded with many different meanings depending on the field of study. Chapter 6 emphasizes that GRNs are composed of many parts that interact with each other and those interactions involve feedback loops and are stochastic from the observer's point of view. Thus, GRN satisfy the definition of a complex system and this observation brings up the need for complexity-reducing strategies. One of the aspects of such strategies in this setting is to find subsystems with a fewer number of interacting genes and with simpler rules of interactions between them that preserve, to a degree, critical properties of the whole system. Unfortunately, this important topic – how to optimally compress a large and complex GRN model remained outside the scope of this book.

Chapter 7 considers a specific application of GRN modeling – deciphering of the regulatory machinery in early embryos using whole-transcriptome data. This particular application of GRNs is especially important in terms of translational science given the well documented influence of epigenetic and environmental factors on the regulatory processes guiding the developmental stages starting from an embryo to an adult organism.

Chapter 8 complements the discussion presented in the previous two chapters and proposes a framework for prioritizing the transcription factor binding sites for multiple co-expressed gene sets. The framework uses Lasso multinomial regression modeling and ultimately leads to constraints that reduce the size of the search space for the reverse engineering of GRNs.

Chapter 9 provides a description of an impressive combination of specialized genomic, image acquisition and processing devices and techniques that allow for observing and computational modeling of drug efficacy on specific cell regulatory pathways over time and for the same cell population. The presented technology and modeling approaches underscore the importance of close collaborative interactions between scientists and engineers of various backgrounds. The discussion in the chapter serves as an example of a properly functioning relationship, where the scientist does not hand the engineer or statistician a set of data and asks them to find something in it; instead, assuming a translational goal, the enterprise is guided by that goal and it leads to a carefully designed experiment. Moreover, the authors present a convincing case for the need of time series data on the same cell population or even individual cells in that population if one would hope for a GRN model with sufficient predictive capability.

Predictive computational modeling of genomic regulation has the potential to unravel the mechanisms of cellular functions from a systemic perspective. There are several important goals of the modeling process:

- To infer optimal models from data using prior knowledge as a constraint in the optimization process.
- To characterize the dynamical behavior of the real GRN in terms of long-run probability distributions of the model and the structure of its attractors and their basins.
- To investigate how structure determines dynamics and the restrictions imposed by a specific dynamical behavior on the structure of the GRN.
- To characterize classes of intervention strategies for control of the dynamical behavior of the system.

We hope that this edited book will provide the reader with a good mix of theoretical and practical topics representing the goals of mathematical and computational modeling of genomic regulation in the context of translational science. The book could serve as a guide to some of the important and highly relevant current and emerging research trends in the analysis and modeling of GRNs.

REFERENCES

Crick, F. (1970). Central dogma of molecular biology. *Nature*, *227*(5258), 561–563. doi:10.1038/227561a0 PMID:4913914

de Jong, H. (2002). Modeling and simulation of genetic regulatory systems: A literature review. *Journal of Computational Biology*, *9*(1), 67–103. doi:10.1089/10665270252833208 PMID:11911796

Dougherty, E. R., & Braga-Neto, U. (2006). Epistemology of computational biology: Mathematical models and experimental prediction as the basis of their validity. *Journal of Computational Biology*, *14*(1), 65–90.

Huang, S. (1999). Gene expression profiling, genetic networks and cellular states: An integrating concept for tumorigenesis and drug discovery. *Molecular Medicine (Cambridge, Mass.)*, *77*(6), 469–480. PMID:10475062

Huang, S. (2001). Genomics, complexity and drug discovery: Insights from Boolean network models of cellular regulation. *Pharmacomgenomics*, *2*(3), 203–222. doi:10.1517/14622416.2.3.203 PMID:11535110

Kauffman, S. (1969). Metabolic stability and epigenesis in randomly generated genetic nets. *Journal of Theoretical Biology*, *22*(3), 437–367. doi:10.1016/0022-5193(69)90015-0 PMID:5803332

Kauffman, S. (1993). *The origins of order: Self-organization and selection in evolution*. Oxford University Press.

Pal, R., Ivanov, I., & Dougherty, E. R. (2005). Generating Boolean networks with a prescribed attractor structure. *Bioinformatics (Oxford, England)*, *21*(21), 4021–4025. doi:10.1093/bioinformatics/bti664 PMID:16150807

Chapter 1
Inference of Gene Regulatory Networks by Topological Prior Information and Data Integration

David Correa Martins Jr.
Federal University of ABC (UFABC), Brazil

Fabricio Martins Lopes
Federal University of Technology – Paraná (UTFPR), Brazil

Shubhra Sankar Ray
Indian Statistical Institute, India

ABSTRACT

The inference of Gene Regulatory Networks (GRNs) is a very challenging problem which has attracted increasing attention since the development of high-throughput sequencing and gene expression measurement technologies. Many models and algorithms have been developed to identify GRNs using mainly gene expression profile as data source. As the gene expression data usually has limited number of samples and inherent noise, the integration of gene expression with several other sources of information can be vital for accurately inferring GRNs. For instance, some prior information about the overall topological structure of the GRN can guide inference

DOI: 10.4018/978-1-5225-0353-8.ch001

techniques toward better results. In addition to gene expression data, recently biological information from heterogeneous data sources have been integrated by GRN inference methods as well. The objective of this chapter is to present an overview of GRN inference models and techniques with focus on incorporation of prior information such as, global and local topological features and integration of several heterogeneous data sources.

1. INTRODUCTION

Systems Biology is an interdisciplinary research field that aims at the study of complex interactions occurring in living organisms (Snoep & Westerhoff, 2005). Research in this field focuses on the study of biological processes such as cell cycles and the conditions for the origin of certain diseases. The ultimate goal of these studies is to help the development of new treatments and drugs against diseases, biofuel production techniques, among many other applications.

The genome of an organism has a central role in the control of cell processes such as cell response to environmental stimuli, cell differentiation in its respective functional groups, DNA replication for cell division, and many others. An organism can be seen as a network of molecules connected by biochemical reactions (Voet, Voet & Pratt, 2005). Proteins synthesized from genes may work as transcription factors which bind to regulatory sites of other genes, such as enzymes which catalyze metabolic reactions or components of signal transduction pathways. Such regulatory mechanism forms a complex system of sending and receiving signals (RNAs) which can be investigated to identify the control mechanisms of the cell and the relationships among various biological entities like genes, RNAs and proteins. However, there is still much to be discovered about the functional relationships of control mechanisms, e.g., transcription levels and proteins, in the regulatory system (Barabasi,2002; Fall, Marland, Wagner & Tyson, 2002; Shmulevich & Dougherty, 2007).

With few exceptions, all cells of an organism contain the same genetic material, although cells of different tissues are functionally different. The cell function is partially determined and controlled by gene expression profiles. With the aim of understanding how genes are involved in control of intra and inter cell processes, the scope of the molecular biology studies needs to be enlarged to include not only the discovery of nucleotide sequences that codes for proteins, but also the unraveling of the regulatory systems which determine what genes are expressed, when, where, and to how much extent (Snoep & Westerhoff, 2005). The explanation of these regulatory networks functioning, by means of sending and receiving signals, is currently one of the main objectives of the systems biology studies.

One of the most challenging research problems of Systems Biology is the inference (or reverse-engineering) of gene regulatory networks (GRNs) from expression profiles (Werhli, Grzegorczyk & Husmeier, 2006; Marbach et al, 2012). This research issue became important after the development of high-throughput technologies for extraction of gene expressions such as DNA microarrays (Schena, Shalon, Davis & Brown, 1995) or SAGE (Velculescu, Zhang, Vogelstein, & Kinzler, 1995), and more recently RNA-Seq (Wang, Gerstein, & Snyder, 2009). The importance of GRN reconstruction can be seen through initiatives taken for this purpose such as DREAM (Dialogue for Reverse Engineering Assessments and Methods) (Marbach et al, 2012). The inference problem involves discovery of complex regulatory relationships among biological molecules which can describe not only diverse biological functions, but also the dynamics of molecular activities. Once the network is recovered, intervention studies can be conducted to control the dynamics of the biological systems aiming to prevent or treat diseases (Shmulevich & Dougherty, 2007).

In general, it is not possible to recover GRNs very accurately based only on gene expression profiles for several reasons, including the presence of significant noise in the data, limited number of samples and large dimensionality. Also, GRN inference is considered an ill-posed problem, meaning that many networks may be able to explain the data in hand. Besides, the lack of information about the biological organism and the high complexity of the networks are additional challenges involved in GRN inference. From the computational point of view, this problem is NP-hard, requiring the development of approximation algorithms and high performance computing (including parallelization) techniques.

To infer, analyze and compare the interrelationship between genes with adequate precision is an open research problem. In this regard, the integration of mathematical models with several types of molecular information can be crucial for GRN inference and discovery of biological knowledge as well as their characterization (synthesis) (Ray, Bandyopadhyay, & Pal, 2009; Hecker, Lambeck, Toepfer, van Someren, & Guthke, 2009; Ristevski, 2013; Lopes, Ray, Hashimoto, & Cesar Jr., 2014b). Some widely studied available sources of biological data are Gene Ontology (GO) (Ashburner et al, 2000), GenBank (Benson, Karsch Mizrachi, Lipman, Ostell, & Wheeler, 2008), KEGG (Kyoto Encyclopedia of Genes and Genomes) (Kanehisa, Goto, Furumichi, Tanabe, & Hirakawa, 2010), The Arabidopsis Information Resource (TAIR) (D'Angelo, Kilian, Kudla, Batistic, & Weinl, 2009), Munich Information Center for Protein Sequences (MIPS) (Mewes, et al, 2011), Protein-Protein Interaction databases (Prasad et al, 2009; Licata et al, 2012; Orchard et al, 2014), to cite but a few. Another important trend is the inclusion of global and local topological information to improve GRN inference methods (Hecker et al, 2009; Lopes, Martins Jr, Barrera, Cesar Jr, 2014a).

The objective of this chapter is to present an overview of GRN inference techniques with focus on incorporation of prior information such as global and local topological features and integration of several heterogeneous data sources. The chapter will cover modeling, inference, validation and computational efficiency of these techniques, specially giving emphasis into the models and approaches proposed by the authors of this chapter such as Probabilistic Gene Networks (Section 2.2.4.3), SFFS-MCE (Secion 2.3.4.2), SFFS-BA (Section 3.3), Biological Score and Entropic Biological Score (Section 5). Section 2 presents the fundamentals of modeling and inference of gene regulatory networks from gene expression data, discusses the main models and inference methods proposed in the literature, including Probabilistic Gene Networks model and SFFS-MCE method. Section 3 presents some basic principles for including global topological information to help the gene regulatory networks inference process, including the discussion of SFFS-BA method. Section 4 provides principles behind integration of several biological sources to reconstruct gene networks. Finally, Section 5 briefly discusses the methods Biological Score and Entropic Biological Score, which infer gene regulatory networks by integrating several biological data sources.

2. BACKGROUND

Cells define their specific behaviors according to the genes expressed by them. The transcription is the first step to convert the stored information in DNA to proteins. The transcription process is regulated by a control network which coordinates the cell activity (Shmulevich & Dougherty, 2007). Genes express a certain mRNA concentration from which the proteins will be synthesized in ribosomes. Some of these proteins are enzymes that catalyze metabolic reactions vital for the cell maintenance and signalling. Other proteins interact with the DNA to regulate the RNA synthesis of genes.

The microarray technology (Schena, 1995) was developed twenty years ago, but it is still the most popular technique for extraction of gene transcription signals. Such method uses biochemical process that measures the expression of thousands of genes simultaneously. More recently, next generation sequencing techniques have allowed the development of more advanced methods to measure gene expression, such as RNA-Seq (Wang et al, 2009). These technologies allow, for the first time, the study of genes in an integrated and systematic way, since it is known that genes do not act alone, but rather jointly with other genes, proteins and other molecules.

Gene regulatory networks (GRN) inference is fundamentally involved in discovering which genes indirectly interact with each other (and how are these interactions) by means of their mRNA expressions and, ultimately, by their proteins. The study of the dynamics of biological processes requires computational approaches to model the fast state transitions of molecules. The information about the dynamics, combined with computational analysis, can lead to a new stage of discovery about the understanding of the gene regulatory process. In particular, the gene expression analysis allows to identify the activation and inhibition of genes while the cell responds rapidly to these events in the cellular microenvironment (Shav Tal, Singer, & Darzacq, 2004). In this context, mathematical and computational modeling of biological processes may provide a possibility to improve the understanding of molecular organization, inter-dependence between molecules (or genes) and their regulatory interrelationships.

2.1. GRN Inference Challenges

The GRN inference is considered an ill-posed problem in the sense that, for a given dataset of gene expression profiles, there are many (if not infinity) networks capable to generate such dataset. Such problem is hardened by a typically limited number of samples, a huge dimensionality (number of variables, i.e., genes) and presence of noise (Shmulevich & Dougherty, 2007). The performance of the GRNs inference methods depends on the interrelationship between sample sizes, number of features and the complexity of the inference method. For example, a naive GRNs inference method requires the number of observations (e.g., microarrays) to be an exponential function of the number of genes (Jain, Duin & Mao, 2000). This phenomenon is known as the ``curse of dimensionality'' in the field of pattern recognition. In this regard, novel computational and statistical techniques are required to infer relationships among the variables and to alleviate the intrinsic errors, committed in the presence of small number of samples with huge dimensionality.

In general, it is not possible to recover the GRNs very accurately due to aforementioned reasons combined with the lack of information about the biological organism and the high complexity of the networks. Part of this complexity is due to the intrinsically multivariate prediction phenomenon (Martins Jr, Braga Neto, Hashimoto, Dougherty, & Bittner, 2008; Martins Jr, Oliveira, Braga Neto, Hashimoto, & Cesar Jr, 2013), also known as synergy (Anastassiou, 2007) or combinatorial regulation (Marbach et al, 2010) in the context of GRNs. A set of genes X is considered as intrinsically multivariate predictive with regard to a given target gene Y if Y has its behavior (expression signal) strongly predicted by the expression signals of X and, at the same time, weakly predicted by any proper subset of X (Martins Jr et al, 2008; Martins Jr et al, 2013). The multivariate nature of the relationship of some predictor variables with the target variable is one of the main causes of the occurrence of the well known nesting effect (Martins Jr et al, 2008; Martins Jr et al, 2013). The nesting effect occurs when an inserted feature in the solution subset may never be removed from it by a suboptimal feature selection, even if such feature does not belong to the optimal solution. Analogously, a feature removed from the solution subset may never be inserted into it, even if such feature belongs to the optimal solution (Pudil, Novovicová, & Kittler, 1994).

Thus, to infer, analyze and compare the interrelationship between genes with adequate precision is an open research problem. Although the integration of biological information with mathematical models is critically important in discovering and characterizing novel biological knowledge, it is restricted by the prior biological information of each gene/biological entity involved in the process. Moreover, the integration of multiple data types, associated with the local and global properties of the biological organisms, is also important for analysis of topological features of connectivity between genes, effective prediction of GRNs, and their functions (Troyanskaya, 2005; Vidal, 2005; Aittokallio & Schwikowski, 2006; Lopes et al, 2014a).

Another issue is the validation of the inferred networks which requires prior knowledge about the real gene connections and their functional relations. In the validation process, it is possible to use computational gene network models (Mendes, Sha, & Ye, 2003; Marbach, Schaffter, Mattiussi, & Floreano, 2009; Lopes, Cesar Jr, & Costa, 2011a), for which the mechanisms are known, or to choose a biological model as a gold standard network.

2.2. GRN Modeling

The current technologies for gene expression measurement, such as microarrays and RNA-seq allow the study of aspects of GRNs in a global fashion. Such experiments extract the states of a regulatory system of a certain organism under a condition of interest, generating a set of expression concentrations. Such data are compared with predictions performed by mathematical models constructed from some biological knowledge about the regulatory system. These models are constantly corrected and enhanced to make the predictions reliable with respect to the reality represented by the experimental data. Figure 1 presents a fluxogram which illustrates this process.

The GRN modelling is an area in constant evolution. This section presents a brief overview of the main GRN models based mainly on surveys from De Jong, 2002; Schlitt & Brazma, 2007; Shmulevich & Dougherty, 2007; Kelemen, Abraham, & Chen, 2008; Hecker, 2009; Shmulevich & Aitchison, 2009; Penfold & Wild, 2013; Ristevski, 2013.

2.2.1 List of Parts

A list of parts is a collection and description of the components involved in a network of a biological system (Schlitt & Brazma, 2007). Such elements include transcrip-

Figure 1. Fluxogram illustrating the GRN analysis process. Rectangles represent procedures, triangles represent information sources and arrows represent information flux. Figure adapted from (De Jong, 2002).

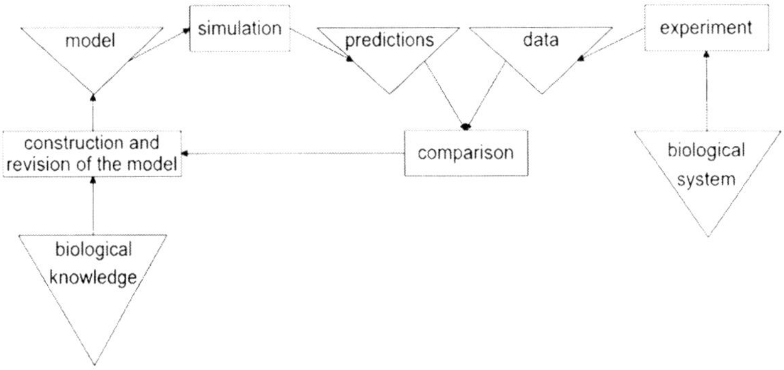

tion factors, promoters and binding sites for transcription factors. The building of a list of parts is the first step in the development of a complex model. Such lists are resulting from a genomic sequencing and annotation project in which the DNA of an organism is determined and a large part of genes and proteins are identified. These lists can be represented as a database of regulatory elements or functions performed by genes. Once the transcription factors and their respective binding sites are known, the topology of a regulatory network can be derived.

2.2.2 Topological Models

Topological models describe the connection between the elements composing the list of parts. They can be represented by graphs whose nodes correspond to genes and edges (or arcs) representing the regulatory interactions. Graph representations have been used for a myriad of biological datasets, including gene expression networks and protein-protein interaction networks (Albert & Barabási, 2002; da-Silva & Stumpf, 2005).

The representation of regulatory networks by means of graphs can be generalized in several ways (De Jong, 2002). Nodes and arcs can be labeled to provide information about the functionality of each gene and the type of interactions existing among the genes. A simple example of gene regulatory network can be viewed in Figure 2 which illustrates 3 nodes (genes) and arcs indicating activation (→) or inhibition (-|).

There are many graph operations that can be performed to obtain biologically relevant predictions about regulatory systems (De Jong, 2002). For instance, a search for paths between two genes can provide clues about network redundancy. Characteristics of global connectivity of a network, such as the average and the distribution

Figure 2. Directed graph representing a gene regulatory network. The arcs indicated by → and -| represent activation and inhibition interactions, respectively

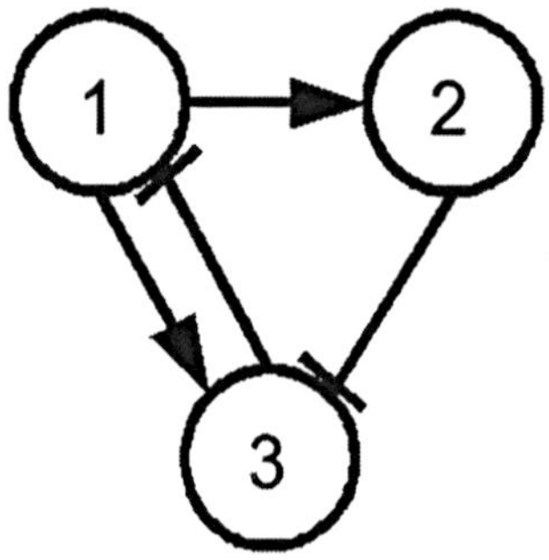

of the number of regulator per gene, may indicate the network complexity. A strongly connected subgraph that is loosely connected with other subgraphs may be a functional module of the regulatory system which can be separately analysed. Largely connected genes (hubs) normally provide tolerance to random faults occurring in elements of the network (Albert, Jeong, & Barabási, 2001). Repetition of local structures (motifs), frequently occurring in real data but almost absent in random networks, can be examined. For example, cycles may indicate feedback relationships important to the homeostasis and differentiation (Thomas, 1998; Milo, 2002). Finally, comparison of regulatory networks from distinct organisms can be conducted to search for evolutionary features preserved in both species (De Jong, 2002).

2.2.3. Logic Control Models

Once the network topology is known, the next step is the study of the interaction rules among the network elements. For instance, considering a promoter consisting of only one binding site to a transcription factor, it is important to know whether such factor is either an activator or inhibitor. If several transcription factors bind to a promoter, the knowledge of the functioning of each factor and the way they interact is important. Figure 3 illustrates a Boolean logic circuit among three input genes which provides as a output the result of the fourth gene. Some studies show that some promoters present multivariate behaviors approximated by Boolean functions (AND, OR, NOT and combinations of these logics) (Kauffman, 1993; Shmulevich, Dougherty, Kim, & Zhang, 2002; Martins-Jr et al, 2008). In certain cases the interactions are even more complex (Louis & Becskei 2002).

Boolean functions, decision trees, relevance networks, Bayesian networks and linear functions are often used to describe the network logics. These models can be

Figure 3. Logic Boolean circuit with three input genes producing the value of the fourth gene as a result (adapted from (Schlitt & Brazma, 2007). The fourth gene is activated only when genes 1 and 2 are activated and gene 3 is inhibited.

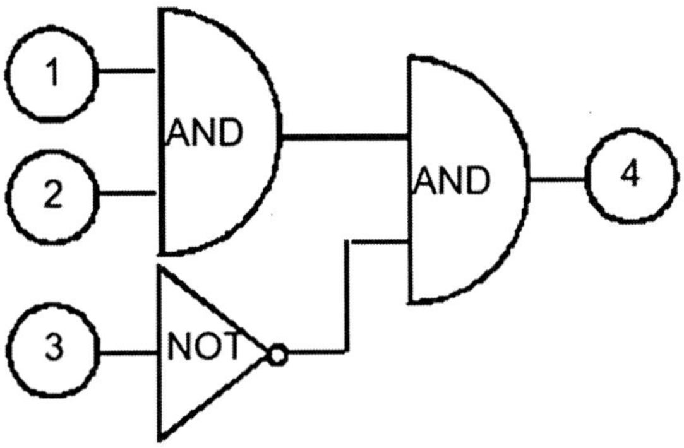

classified in two major categories: discrete or continuous (D'haeseleer, Liang, & Somogyi, 2000). Discrete functions are based on the hypothesis that a gene can be in a finite number of states. In the simplest case, only two states describe the state of a gene: expressed and not expressed. In this way, Boolean functions can describe the interactions among genes. Such functions are simplifications of gene networks, since gene expression data are in continuous domain (concentration of molecules change in a smooth fashion). Nevertheless, Boolean networks can be a good starting point to model gene networks since they are suitable to capture the global behavior of the biological system, require less data to be estimated, and the implementation, simulation and analysis are straightforward (Kauffman,1969; Hecker, 2009).

Continuous functions use real values to represent the gene activity. One of the widely used approaches in this category is the linear function which considers the target gene as a linear combination of its predictors, each one with a different weight and a signal that can be positive or negative. The network topology determines which predictors possess weight equal to zero for a given target (those genes without arcs going to the target). Unlike Boolean functions, linear functions cannot model a situation in which a gene can exert both activation and inhibition to another gene depending on the state of other genes (non-linear relationships). Non-linear relations are better modeled by Boolean functions.

Decision trees are models that allow to predict the behavior of a particular gene (tree leaf) based on the expression of other genes (internal nodes of the tree) (Soinov, Krestyaninova, & Brazma, 2003). This approach works with both discrete and continuous variables.

Relevance networks approaches consider the similarity between gene pairs in the continuous domain (Butte & Kohane, 2000; Kelemen et al, 2008). To define the network topology, all gene pairs are compared using similarity metrics such as correlation coefficients or mutual information. An approach in this direction is available in ARACNe (Algorithm for the Reconstruction of Accurate Cellular Networks) (Margolin et al, 2006). Such approaches cannot capture multivariate (N-to-1) relationships, since they capture only pairs (1-to-1).

Bayesian Networks provide a probabilistic approach to model gene regulatory networks (Friedman, Linial, Nachman, & Pe'er, 2000; Pe'er, Regev, Elidan, & Friedman, 2001; Pournara & Wernisch, 2004). These networks are represented by a directed acyclic graph where each node represents a gene and each arc represents a dependency. For each gene expression level *i* denoted by X_i, there exists a conditional distribution of probabilities $P(X_i | \rho(X_i))$ where $\rho(X_i)$ corresponds to the set of direct predictors of *i*. An important property is that the joint probability distribution $P(X)$ of all nodes of a Bayesian network is given by the product of the conditional probabilities of each gene given its predictors (Markov Hypothesis), i.e.,

$$P(X) = \prod_{i=1}^{n} P(X_i | \rho(X_i))$$

Figure 4 illustrates an example of Bayesian network with all associated conditional probabilities and the mathematical expression which provides the joint probability distribution of the network.

Bayesian network approaches are attractive to model GRNs for its solid statistical basis, which allows to deal with stochastic aspects of gene expression and noisy measures in a natural way (De Jong, 2002). However, this model does not take into account dynamic aspects. Such disadvantage can be circumvented through a generalization of this model (Dynamic Bayesian Networks) which allows feedback relationships (loops) to be modelled (Perrin et al, 2003).

2.2.4. Dynamic Models

The knowledge of the list of parts of a network, as well as its topology and the control logics, is required to expand the model and to capture dynamic changes along time (Schlitt & Brazma, 2007). Typically, dynamic models take into account a relatively small number of genes. The objective of these models is to describe and simulate the dynamic behavior of the system state and to predict the network response to several stimuli and environmental changes.

Figure 4. Bayesian network with all assigned conditional probabilities and the corresponding joint probability of X

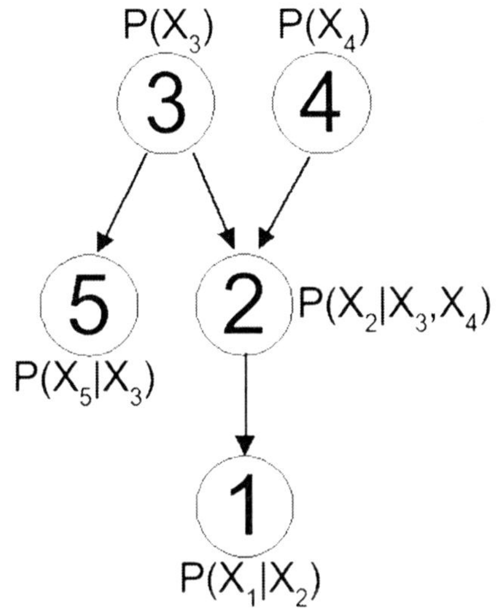

$$P(\mathbf{X}) = P(X_2|X_3,X_4)\,P(X_5|X_3)\,P(X_1|X_2)\,P(X_3)\,P(X_4)$$

Several dynamic models are proposed in the last decade. Greller and Somogyi (2002) suggested dualities to categorize these models. Some dualities are highlighted below.

- Static (without time-series information) versus dynamic (with time-series information);
- Qualitative (discrete) versus quantitative (continuous) measures (e.g. Boolean or abundances);
- Deterministic versus probabilistic state transitions (e.g. differential equations or Markov chains);
- Linear versus non-linear interactions among elements and state transition rules;

In this section, the main dynamic models used in GRNs will be discussed.

2.2.4.1 Boolean Networks

The Boolean Networks model has been employed in GRNs since 1969 (Kauffman, 1969). Boolean networks are based on the hypothesis where the values "active/

inhibited" of the elements can describe important aspects of the gene regulation. In this model, the network state is represented by a sequence of n bits (0s and 1s) indicating which genes are expressed or not in a given time instant. As the time proceeds, the network transits along the state space as illustrated in Figure 5. For a network of n genes, there are 2^n possible states. Some states may never be reached. Other states, called attractors, can be periodically reached depending on the initial condition of the system. Figure 5(D) illustrates two attractors: one of them is composed by only one state (0,0,1), while the other is composed by two alternating states (1,0,1) and (0,1,0).

Attractors are stationary states of a dynamic system which captures the long term behavior of the system (Shmulevich & Dougherty, 2007). According to Kauffman, attractors can be viewed as cell types (Kauffman, 1969). Different cells are characterized by their recurrent gene expression patterns, which correspond to the attractors. Real GRNs are largely stable in presence of perturbations on the expression of one or several genes caused by external factors. Considering the Boolean

Figure 5. Example of a Boolean network consisting of 3 genes. (A) State transition logic rules; (B) Graph of dependencies (topology); (C) Table listing all possible states at time t and their respective state transitions at time t+1; (D) State transition diagram. Adapted from (Schlitt & Brazma, 2007)

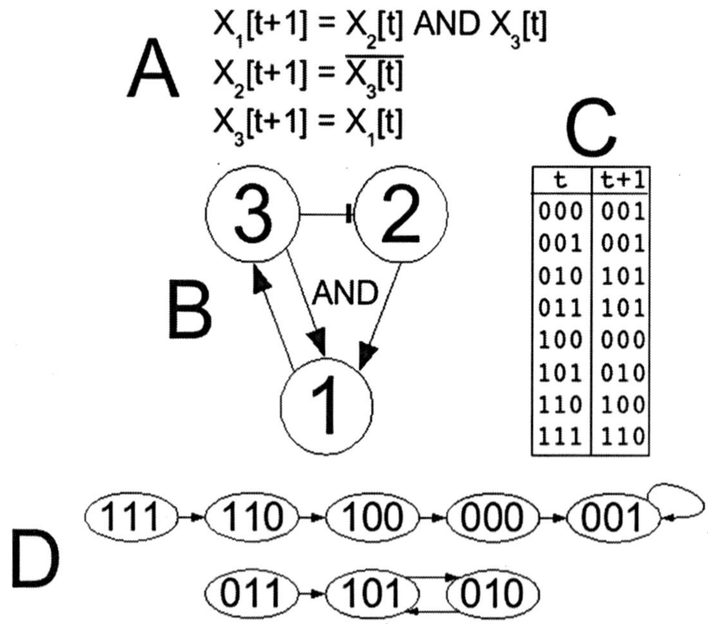

networks formalism, this means that when a small number of genes are perturbed, their values (states) may change, but the network states usually remain in the same attractor basin, which leads to the same attractor (Shmulevich & Dougherty, 2007). In this sense, attractors of large attractor basins lead to strong stability of the system (Li, Long, Lu, Ouyang, & Tang, 2004; Zhang et al, 2006). Such stability of regulatory networks in live organisms allows cells to maintain its normal functional state in the organism even when exposed to some degree of external perturbations.

Generalized Logical Networks are a generalization of the Boolean network model in which the elements can have an arbitrary number of values and the state transitions occur asynchronously (Thomas, Thieffry, & Kaufman, 1995).

2.2.4.2. Probabilistic Boolean Networks (PBN)

Probabilistic Boolean Networks is a stochastic model of Boolean Networks where the binary expression value of each gene at a certain timepoint is determined by a set of Boolean functions of some genes at the previous timepoint. Here, each function is applied probabilistically (Shmulevich et al, 2002). Normally, the quasi-determinism, inherent to GRNs, is modelled by assigning probability close to 1 to one of the functions, while the functions with small probability play the role of perturbations or changes of biological context (Dougherty, Brun, Trent, & Bittner, 2007). As a consequence, Boolean networks represent a particular type of probabilistic Boolean networks in which each gene has only one predictor function.

2.2.4.3. Probabilistic Gene Networks (PGN)

Probabilistic gene networks (PGN) follows the feature selection principle. For each target gene, a search for the subset is performed which best describes the target behavior according to the expression signals. Barrera et al (2007) discussed this approach in analyzing the dynamical expression signals of the Plasmodium falciparum (one of the malaria agents) and provided interesting biological results. Such approach assumes that the temporal samples follow a first order Markov Chain in which each target gene in a given timepoint depends only on its predictor values in the previous time instant. The transition function is homogeneous (it does not change along time), almost deterministic (from any state, the system has a preferential state to go) and conditionally independent (i.e., the value of a given gene is dependent only on its predictors, following the Markov hypothesis). These assumptions are important simplifications to deal with the limited number of samples typically available in real data.

Conceptually, PGN is a restricted type of PBN when the variables are binary. PGN assumes that a target gene presents several different predictor functions like PBN, although the difference is that all these functions necessarily present the same set of predictor genes as inputs in PGN. Another important restriction is the quasi-determinism assumed by PGN. Finally, PGN model assumes that genes can have a variable number of discrete values. For example, Barrera et al (2007) considered three possible states for each gene: -1 (underexpressed), 0 (normally expressed), +1 (overexpressed).

2.2.4.4. Dynamic Bayesian Networks

Bayesian networks are inherent static, as they do not model cycles in the network topology. In contrast, Dynamic Bayesian Networks (DBN) is a model that considers time-dependency, thus allowing cycles (Perrin et al, 2003). Shmulevich et al. (2002) showed that PBN is a type of DBN which considers only binary values. Similar to Bayesian networks, DBN can consider a variable number of discrete states for each gene.

2.2.4.5. Differential Equations

Boolean networks model and its derivatives can reveal important global properties of the network, but they are insufficient to capture certain aspects of the dynamics (Schlitt & Brazma, 2007). On the other hand, differential equations models allow more detailed descriptions of the network dynamics by modelling concentration alterations along time (Chen, He, & Church, 1999; Maki, Tominaga, Okamoto, Watanabe, & Eguchi, 2001). Basically, the regulatory network is modelled by equations of the form:

$$\frac{dx_i}{dt} = F_i(x_1, x_2, ..., x_n), i \in \{1, 2, ..., n\}, F_i : R^n \to R$$

where F_i is a logic control function that can be linear or non-linear (generally non-linear (Shmulevich & Dougherty, 2007; De Jong, 2002)).

A disadvantage in using differential equations in the context of GRN inference is the requirement of a large number of parameters to be estimated. Additionally, a high temporal resolution dataset (many timepoints) is required. However, datasets with large number of temporal gene expression samples is normally very difficult to obtain. This is the main reason by which differential equation models are applied only to very small networks with few components. Another difficulty is that the experimental data in question are very susceptible to noises, which may deny the appropriate quantitative modelling. Prior biological knowledge and simplifications can help this type of modelling (Kelemen et al, 2008).

2.2.4.6. Stochastic Differential Equations

Similar to differential equations, stochastic differential equations try to capture quantitative molecular details. However, this model tries to incorporate stochasticity to also capture fluctuations on the molecular level, which are important for many biological processes (Raser & O'Shea, 2005; Shmulevich & Aitchson, 2009). Besides, the inclusion of stochasticity is general and does not require specific knowledge about the stochasticity nature. There are simulation methods that accurately simulate the system dynamics of reactions of biochemical molecules, although their computational costs strongly increases with the number of chemical variables considered (Cao & Samuels, 2009; Shmulevich & Aitchson, 2009). On the other hand, there are methods which incorporate stochasticity on the deterministic differential equations, which perform much better in terms of computational complexity than simulation methods, specially for a large number of chemical molecules (Manninen, Linne, & Ruohonen, 2006). Such methods can introduce stochasticity into reaction rates, rate constants and concentrations. The main disadvantage of this model is the requirement of prior knowledge of concentrations and kinetic parameters, as happens for quantitative models in general.

2.2.4.7. Chemical Master Equations

In this model, chemical reactions are modeled as a Markov chain process where the set of states is discrete and the time is continuous (Hegland, Burden, Santoso, MacNamara, & Boooth, 2007; Smadbeck & Kaznessis, 2014). Each element of the state corresponds to the number of molecules of a given type. When a chemical reaction occurs, a transition happens from a given state to another one. There are several types of reactions including covalent reactions, bindings, conformational changes, transcriptional elongation events, among others. This model presents an important disadvantage due to the involvement of a large number of interacting substances, suffering from the curse of dimensionality. To address such problem, stochastic approximations have been proposed. Even so, this model is very computationally intensive, where even a moderate number of substances (~ 100) cannot be feasibly modelled.

2.2.4.8. Petri Nets

Petri nets are an extension of graph models which allows simple quantitative representation of dynamic processes (Pinney, Westhead, & McConkey, 2003; Hardy & Robillard, 2004). Usually Petri Nets are directed graphs consisting of arcs and two different types of nodes: position and transition nodes. The arcs connect only position nodes to transition nodes and vice-versa. The dynamic aspect is introduced by tokens, which can be contained by position nodes. Each arc has a value indicating the number of tokens required to perform the transition through it. The transition nodes determine the change rate along the way. Petri nets are particularly useful to model metabolic reactions, since a detailed information about the reaction rates is not needed. This is an advantage because such rates are usually unknown, although such information is important to completely describe certain metabolic networks (Schlitt & Brazma, 2007).

2.2.4.9. Hybrid Models

Biological systems present continuous and discrete aspects (Schlitt & Brazma, 2007). In general, concentrations are continuous values whereas binding of a transcription factor to DNA is expressed as a discrete event (on or off). However, the boundary between the continuous and discrete aspects depends on the detail level of the adopted model. Hybrid models are developed to describe both aspects (Akutsu, Miyano, & Kuhara, 2000).

There exist extensions of the Petri nets model which allow to include knowledge about the reactions dynamics. One of these extensions include stochastic time delays of the transitions (Goss & Peccoud, 1998) . In these networks, the trigger of transition nodes depends not only on the number of tokens, but also on a stochastic component. Another type of extension is the Hybrid Functional Petri Nets (HFPN) model where the position and transition nodes are continuous (Matsuno, Inouye, Okitsu, Fujii, & Miyano, 2006). The problem with using HPFN is that it cannot utilize (lost) the main advantage of conventional Petri nets over differential equations (independence on reaction rates information). On the other hand, HFPN allows a mixture of continuous and discrete nodes, which enables the inclusion of information about known reaction rates only (Schlitt & Brazma, 2007).

2.2.4.10. Finite State Linear Model

Finite State Linear Model (FSLM) combines the simplicity and the small computational cost of Boolean networks with a continuous representation of the element values (Ruklisa, Brazma, & Viksna, 2005). In this model, the gene activity is represented by two states (overexpressed and underexpressed) but the expression levels are represented by real values. The time is continuous and the concentration rate change is determined by the network state. This model can represent a good degree of complexity of the networks, although it is not appropriate to model relatively large regulatory networks due to lack of detailed biological information (Schlitt & Brazma, 2007).

2.3. Brief Overview of GRN Inference Methods based on Gene Expression Data

The GRN inference from gene expression data, also known as reverse engineering, is not an easy task due to the huge volume of variables (genes) and the small number of available samples (experiments). It also involves the large complexity of the biological networks, representing an important challenge in bioinformatics and computational biology (Hovatta et al, 2005). The literature related to GRN identification is vast and growing rapidly which reflects the importance of this research field. Some reviews about this topic can be found in D'haeseleer et al, 2000; De Jong, 2002; Styczynski & Stephanopoulos, 2005; Schllit & Brazma, 2007; Markowetz & Spang, 2007; Karlebach & Shamir, 2008; Hecker et al, 2009; Xiao, 2009; de Smet & Marchal, 2010; Marbach et al., 2010; Lu & Zhou, 2011 and Penfold & Wild, 2013. The objective of this section is to present a brief overview of some GRN inference methods based on the aforementioned surveys.

2.3.1. REVEAL

The strategy adopted by REVEAL (REVerse Engineering ALgorithm) (Liang, Fuhrman, & Somogyi, 1998; Kelemen et al, 2008) is to infer regulatory interactions between genes through mutual information of the joint probability distribution of a target, at a given time instant, and its predictors at the previous instant of time. In this way, REVEAL estimates the relationship between genes based on mutual information of the temporal measurements of gene expression (e.g. temporal microarrays).

Given a set of temporal expressions and a target gene Y, REVEAL applies an exhaustive search for a given subset of genes X, such that I(X,Y) = H(X), where H is the Shannon's entropy of X. This equality means that the behavior of Y is fully predicted by X. A parameter k indicating The maximum size of the subsets to be searched is required. The value of this parameter needs to be small (at most 3), otherwise the exhaustive search for subsets of size k becomes computationally unfeasible. In addition, another importance to constrain the k values to be small is due to the curse of dimensionality (Theodoridis & Koutroumbas, 2008), where larger k lead to poorer statistical estimation of the joint probability distributions.

2.3.2. Signal Perturbation Approaches (Predictor and Chooser)

The methods Predictors and Chooser, developed by Ideker et al., identify GRNs models as Boolean Networks by applying successive perturbations to genes with the most uncertain connections and analyzing the alterations in the system behavior (Ideker, Thorsson, & Karp, 2000). One of the steps of these methods is the Hitting Set Problem (HSP) solving, which is NP-hard (Garey & Johnson, 1999). Recently, Graphical Processing Unit (GPU) implementations of HSP is developed to make HSP feasible for larger number of Boolean clauses and higher dimensionalities (Steinbach & Posthoff, 2012; Santos et al, 2015), which are intrinsic characteristics of GRNs.

2.3.3. Network Relevance Approaches

Approaches based on relevance networks (Butte, & Kohane, 2000) look for similarity or dissimilarity between gene pairs. Two steps are required for network inference: (i) all gene pairs are compared using some similarity or dissimilarity metric (in general, correlation coefficient and mutual information are used (Kelemen et al, 2008)); (ii) the full set of comparisons between pairs is filtered to determine relevant connections, typically done by using a threshold. The selection of these relevant connections represents the discovery of the edges between genes in the inferred network.

2.3.3.1. ARACNE

ARACNE (Algorithm for the Reconstruction of Accurate Cellular NEtworks) (Margolin et al, 2006) is an algorithm based on relevance networks which uses mutual information to identify the regulatory relationships between genes. Once such relationships are identified, ARACNE applies a filter to eliminate edges representing indirect relationships between genes. This is performed by using the data processing inequality, which states that if tree random variables X, Y, and Z depend on each other in a linear fashion $X \rightarrow Y \rightarrow Z$, then the mutual information between X and Z is less than or equal to the minimum of the mutual informations between X and Y, and between Y and Z (Kelemen et al, 2008). In this way, each triplet of genes is analysed with the aim to eliminate indirect relationships.

2.3.3.2. CLR

CLR (Context Likelihood of Relatedness) algorithm (Faith et al, 2007) assigns an edge between two genes if some similarity metric is larger than a given threshold. Instead of considering the mutual information between two genes directly, as ARACNE does, CLR calculates a measure based on the distribution of mutual information values. This metric is given by $z_{ij} = \sqrt{z_i^2 + z_j^2}$, where

$$z_i = \max_i(0, \frac{I(X_i, X_j) - \mu_i}{\sigma_i}).$$

Here, I is the mutual information, μ_i is the mean and σ_i is the standard deviation of the empirical distribution of the mutual information values between genes X_i and X_j.

2.3.4. Feature Selection Based Approaches

Feature selection based approaches searches for the subset of genes which best predicts the behavior of a considered target gene according to the expression data, a search algorithm and a relevance criterion (criterion function). This process is applied to all genes considered as target (all genes or a reduced list of genes of interest).

2.3.4.1. MRNET

MRNET method applies feature selection guided by maximum relevance / minimum redundancy criterion to infer GRNs (Peng, Long, & Ding, 2005). For each target gene Y, a search is performed to determine the variable X_i which presents the largest mutual information with respect to the chosen target gene $I(X_i,Y)$. The second chosen gene X_j is the one with the largest mutual information with respect to the target gene $I(X_j,Y)$ and at the same time it has the least mutual information with the gene already chosen as predictor $I(X_i,X_j)$, generating a predictor set Z. In the next steps, this set is updated by inclusion of new genes X_k which maximize the difference u_k-r_k, where u_k represents the relevance given by the mutual information $I(X_k,Y)$, and r_k represents the redundancy given by:

$$r_k = \frac{1}{|Z|} \sum_{X_j \in Z} I(X_k, X_j)$$

2.3.4.2. SFFS-MCE

The method SFFS-MCE (Sequential Floating Forward Search - Mean Conditional Entropy) considers the regulatory interactions represented by the PGN model (see Section 2.2.4.3) to infer GRNs (Barrera et al, 2007; Lopes, Martins Jr, & Cesar Jr, 2008). The inference process is initiated by choosing one gene of interest (target). As the search space is usually very large, an exhaustive search for predictors of a target gene is unfeasible, unless the number of predictors that each target can possess is limited. This is similar to what is done by REVEAL (Section 2.3.1). To circumvent this problem, SFFS-MCE applies SFFS, a sub-optimal feature selection with good cost-benefit in terms of computational complexity and quality of the solution returned (Pudil et al, 1994). The objective is to find a good subset of predictors for the chosen target within a small computation time. The criterion function is based on penalized mean conditional entropy, which needs to be minimized to achieve the best subsets. This criterion is, in essence, based on mutual information with a factor that penalizes joint probability distributions in which the number of poorly observed predictor states is large (Martins Jr, Cesar Jr, & Barrera, 2006; Barrera et al, 2007; Lopes et al., 2008).

Feature selection algorithms like SFFS are suboptimal in the sense that the optimum solution (i.e., the best predictor set according to the criterion function adopted) is not guaranteed. This is due to the presence of intrinsically multivariate predictive relationships in GRNs (Martins-Jr et al, 2008), resulting from the nesting effect (Pudil et al, 1994) (see Section 2.1). Although SFFS was designed to alleviate the nesting effect problem, it is not completely removed. In this regard, several improvements on the search algorithm and/or the criterion function of the SFFS-MCE are developed (Lopes, Martins Jr, Barrera, & Cesar Jr, 2010; Lopes, de Oliveira, & Cesar Jr, 2011b; Borelli, Camargo, Martins Jr, & Rozante, 2013; Montoya Cubas, Martins Jr, Santos, & Barrera, 2014; Lopes et al, 2014a; Lopes et al, 2014b; Vicente & Lopes, 2014). In particular, the method SFFS-BA (Lopes et al, 2014a), which will be described and discussed in Section 3.3, utilizes the SFFS algorithm to obtain networks with scale-free properties. Finally, a method, called Entropic Biological Score (Lopes et al, 2014b), is discussed in Section 4 which creates a criterion function based on mean conditional entropy computed from gene expression data and also incorporates several other sources of biological information.

2.3.5. Genetic Algorithms (GA)

Evolutionary algorithms is a branch of the artificial intelligence where computational models, mimicking the natural process of evolution, are proposed to solve combinatorial optimization problems. In this context, the genetic algorithms (GA) utilize concepts of the natural selection theory such as inheritance, selection, crossover and mutation to solve combinatorial optimazation problems until local optimal or global optimal solutions are achieved (Holland, 1992; Mitchell, 1996).

The genetic algorithms framework has been employed to infer gene regulatory networks (GRNs) in several investigations (Larrañaga, Poza, Yurramendi, Murga, & Kuijpers, 1996; Shin & Iba, 2003; Mamakou, Sirakoulis, Andreadis, & Karafyllidis, 2005; Swain, Hunniford, Dubitsky, Mandel, & Palfreyman, 2005; Knabe, Wegner, Nehaniv, & Schilstra, 2010; Davidson, 2010; Mendoza, Lopes & Bazzan, 2012; Jimenez, Martins Jr, & Santos, 2014). In particular, Davidson (2010) presented a discussion about genetic algorithms for Bayesian networks inference through topological sort of genes (nodes) and developed a hybrid genetic algorithm, personalized for this purpose. Mendoza et al (2012) proposed a genetic algorithm for inference of GRN, modeled by Boolean networks from simulated and real data. In this work, the topologies are computed by integer numbers representing the predictor indices for each gene. Here, it is necessary to determine a *priori* the maximum number of predictors per gene as the random initial population is restricted to one predictor per gene. The fitness function is based on Tsallis entropy (Lopes et al, 2011b) with a penalty factor that needs to be appropriately defined to obtain networks with a good balance between complexity (number of edges in the network) and consistency with the temporal expression data.

In general, the GA methods developed for GRN inference codify the structure of the network as a whole, i.e., each chromosome represents a network, and the initial population consists of random networks generation. On the other hand, a recent work proposed by Jimenez et al. (2014) presents the application of one genetic algorithm per gene to seek for the best subset of predictors for each gene. In this way, a chromosome codifies a subset of predictor genes, specific for a given gene. Besides, this work suggests the generation of initial populations based on an inference method (for instance, exhaustive search by mutual information with a predefined small number of predictors per gene as proposed by Barrera et al. (2007).

2.3.6. Ensemble (Wisdom of Crowds)

The DREAM (Dialogue for Reverse Engineering Assessments and Methods) project is an initiative which is conducted by a community of researchers to provide important systems biology and translational medical challenges and to invite participants to propose solutions (Marbach et al, 2010; Marbach et al, 2012). The main idea behind this project is to combine methods proposed by the community to generate a "wisdom of the crowd" to solve intricate challenges. Marbach et al. observed that GRN inference methods usually fail to accurately infer multiple regulatory inputs (combinatorial or intrinsically multivariate regulation) of genes (Marbach et al, 2010) by assessing several methods that try to solve this problem. Based on this observation, Marbach and colleagues compared 35 methods, submitted to DREAM5 (fifth edition) challenge, with community-based consensus networks obtained by ensembling those 35 methods. The results revealed that only the consensus networks are robust across species and datasets, achieving best overall performances (Marbach et al, 2012). Those 35 methods are classified in five broad categories (regression, mutual information, correlation, Bayesian and other) and strengths and weaknesses of each category are highlighted.

Although the authors conclude that the ensemble of networks is the best approach in general, it still does not accurately infer GRNs in all situations. This is due to the very known inherent limitations of gene expression data involved with the fact that mRNA levels of transcription factors and their targets not always tend to be correlated. The authors suggest that the accurate inference of eukaryotic regulatory networks can be achieved only by integrating other sources of biological information. This is precisely the key point of the present chapter. In Section 3, a method that integrates prior global topological information (scale free topology) to the SFFS-MCE (Lopes et al, 2008), called SFFS-BA (Lopes et al, 2014a), is discussed. Finally, Section 4 discusses about a method, called Entropic Biological Score (Lopes et al, 2014b), that integrates several sources of biological information to the MCE (mean conditional entropy) criterion function.

3. GRN INFERENCE BY INCORPORATING TOPOLOGICAL INFORMATION

3.1. Introduction

The biological data integration and mathematical models can be crucial for the discovery of biological knowledge, as well as their characterization (synthesis). Among the mathematical models, the FunNet (Prifti, Zucker, Clement, & Henegar, 2008) is an integrative tool for exploring interactions in gene expression data sets. Another class of biological data can be obtained in the form of images, which try to represent in a pictorial way the cells and molecules in various biological phenomena of interest. Currently there are some biological imaging databases such as Allen Brain Atlas (Lein et al, 2007), Cell Centered Database (Martone et al, 2002) and FlyEx Database (Poustelnikova, Pisarev, Blagov, Samsonova, & Reinitz, 2004).

There are several recent initiatives where biological information is incorporated in the inference/prediction methods. One kind of initiative is the use of known information of biological entities, available in public databases, into the clustering process (exploratory analysis / unsupervised classification), resulting in more biologically meaningful clusters (Cui, Wang, Chen, Busov, & Wei, 2010; De-haan et al, 2010; Macintyre, Bailey, Gustafsson, Haviv, & Kowalczyk, 2010).

Another initiative is also to use biological information for the discovery of transcriptional regulation relationships, i.e., to infer GRNs (Ernst et al, 2008; Seok, Kaushal, Davis, & Xiao, 2010; Werhli & Husmeier, 2008). A variety of biological data integration techniques for GRNs inference are described in (Troyanskaya, 2005; Karlebach & Shamir, 2008; Baumbach, Tauch, A., & Rahmann, 2009; Hecker et al, 2009). Although the integration of biological information with mathematical models is critically important in discovering and characterizing novel biological knowledge, it is restricted by the prior biological information of each gene/biological entity involved in the process.

Another important aspect is the integration of multiple data types associated with the local and global properties of the biological organisms such as the analysis of topological features of connectivity between genes. This can be decisive for the effective prediction of GRNs and their functions in face of the known limitations (Troyanskaya, 2005; Vidal, 2005; Aittokallio & Schwikowski, 2006).

3.2. Local and Global Properties of GRNs

The analysis of the local and global biological network properties and its application on the inference process is very recent and has been very promising (Karlebach & Shamir, 2008; Klamt, Saez Rodriguez, J., & Gilles, 2007; Lacroix, Cottret, Thébault,

& Sagot, 2008; Lenas, Moos, & Luyten, 2009; Przytycka & Kim, 2010; Lopes et al, 2014a; Vicente & Lopes, 2014). For example, the application of local and global structures of biological organisms on the inference methods may occur in the form of similarities of connected network modules (Ulitsky & Shamir, 2007), structural and graph-theoretic interpretation of the network components (Kuchaiev, Milenkovic, Memisevic, Hayes, & Przulj, 2010; Narasimhan, Rengaswamy, & Vadigepalli, 2009), sparseness of the network (Andrecut, Huang, & Kauffman, 2008; Andrecut & Kauffman, 2008; Christley, Nie, & Xie, 2009), gene network motifs (Ott, Hansen, Kim, & Miyano, 2005) and search for cliques in network graphs (Yu, Paccanaro, Trifonov, & Gerstein, 2006). More importantly, biological networks and particularly GRNs are known not only to be sparse, but also organized, as nodes belong to different classes of connectivity (Charbonnier, Chiquet, & Ambroise, 2010).

The information about network topology can help the investigation of biological process by adopting the complex network theory and its properties (Albert & Barabasi, 2002; Costa, Rodrigues, & Cristino, 2008; Newman, 2003; Strogatz, 2001). It is known that many kinds of relationships can be successfully described by complex networks. In particular, the complex networks theory describes various types of network topologies, each one with well defined properties. It has been broadly applied to characterize biological processes and gene relationships. Some biological networks present the scale-free property in which many nodes have a low degree and a few of them have a high degree (hubs). Here, the degree distribution of scale-free networks is approximated by a power-law distribution (Albert, 2005; Barabasi, 2009; Barabasi & Oltvai, 2004; Costa et al, 2008; Farkas et al, 2003; Guelzim, Bottani, Bourgine, & Képès, 2002; Jeong, Tombor, Albert, Oltvai, & Barabási, 2000). In general, topological patterns and its structural analysis is one of the most promising topics in the analysis of complex networks (Alon, 2007; Goemann, Wingender, & Potapov, 2009), and particularly the application of structural properties of the network can be a very valuable prior information for GRNs inference methods.

3.3. SFFS-BA

Recently, an approach by using the scale-free topological information for GRNs inference, called Sequential Forward Floating Selection - Barabási-Albert (SFFS-BA), is proposed by Lopes et al., 2014a. This new feature selection algorithm is guided by a criterion function applied to the GRNs inference from expression data, which assumes that the structure of GRNs follows a scale-free topology. Therefore, the scale-free topology is used as a prior information by the search algorithm for the GRNs inference.

Considering the limitations of data, knowledge about biological organisms, presence of noise in the expression measures and the complexity of biological networks, the main objective of the SFFS-BA method is to show the importance of considering a priori information in order to improve the inference process and make it more suitable for a class of problems, like the scale-free networks in this work.

The SFFS-BA approach is based on the Probabilistic Gene Networks (PGNs) (Barrera et al, 2007), which are established on some axioms. In the SFFS-BA search algorithm, it is assumed that the axioms for PGNs are already defined and two new assumptions are made:

- For each target gene, adding a new predictor in the subset of predictors, which has a cardinality greater than 1, there must be a gain of information in predicting the target gene if the chosen predictor is a part of true subset of predictors for the target gene. If the information gain with the inclusion of a new predictor in its subset of predictors is very poor then the predictor cannot be part of the subset of the true predictors or there is not enough data to perform the prediction. In both the cases, the inclusion of the predictor in the subset of predictors should be avoided.
- The network topology follows a power law of connections distribution, which means it is a scale-free network.

By considering PGN model and these two new assumptions, the main contribution of this method is to include structural information as a priori knowledge, and therefore it can perform a search in a reduced search space and achieve better results. The heuristic is based on the idea that genes without predictive genes tend to show a random behavior, while genes with real predictors tend to show a more ordered behavior. Thus, it is possible to expect that a gene with no real predictors in a GRN presents small variation in the criterion function, when trying to identify a possible predictor for it. On the other hand, when a target gene has a real predictor, it is expected a different behavior, especially when including a real predictor in its subset of predictors. In other words, given a target gene with no real predictor, small changes are expected in the criterion function value when new predictors are included in its subset of inferred predictors, as shown in Figure 6.

On the other hand, when a target gene has a real predictor it is expected to show a different behavior, especially when a real predictor is included in its subset of inferred predictors (true positives) as shown in Figure 6 (Normal and IMP - intrinsically multivariate prediction (Martins Jr et al, 2008)).

Given the first new assumption, the SFFS-BA algorithm performs a search for all individual predictors (best and worst) based on the SFFS-MR[1] algorithm (Lopes et al, 2010), considering all the genes of the network. In the second stage, the search is performed again for all the genes, but the algorithm chooses target genes that provides

Figure 6. Behavior of the criterion function by including new predictors in the subset of inferred predictors. The objective is to minimize the criterion function value to its optimal value, zero (Reprinted with permission from (Lopes et al., 2014)).

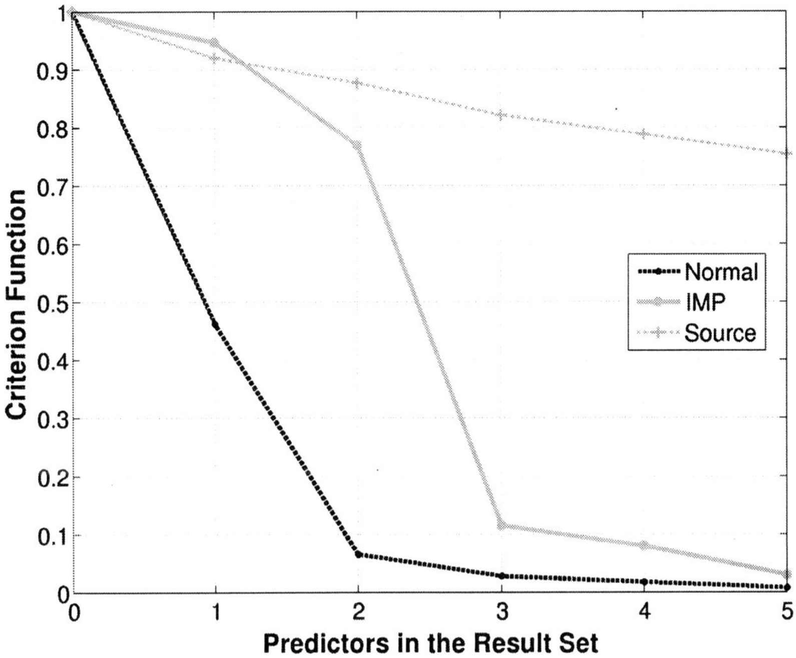

higher gain of prediction (i.e., the value of the criterion function approaches to the goal) when its subset of predictors is incremented. Therefore, when a new predictor is added in its subset of predictors, the target genes with higher gain of prediction have their subsets of predictors preserved for the next iteration. The target genes with little improvement in their criterion function results, i.e., poor predictions when its subset of predictors is increased, are not considered in the next iterations of the algorithm. The target genes that show a subset of predictors which reach the optimal value of the criterion function or come very close to, are also not considered in the next iterations of the SFFS-BA algorithm. Here, it is assumed that there is noise in the input data which may cause small errors in the criterion function. Thus, in each iteration the algorithm discards the expansion of subsets of predictors for target genes that achieve the optimal value of the criterion function or get too close, and assigns a priority queue to target genes, according to the gain in criterion function to add a new predictor in its subset of predictors. Therefore, the target genes with higher gain are more likely to be chosen by the algorithm to expand its subsets of predictors in the next iteration of the algorithm.

In this context, the topology of scale-free networks and the assumption of the gain in function criteria while adding new predictors are considered in the SFFS-BA algorithm. This helps to integrate knowledge as a priori information and guide the search process in the inference of GRNs from expression profiles.

The scale-free topology, defined in the second new assumption and characterized by a power law $P(k) \approx k^{-\gamma}$ in its connections distribution, is used by the algorithm as a limiter or "pruning" of the number of target genes that are chosen for the expansion of its subsets of inferred predictors at each iteration. Therefore, the number of target genes, considered for expansion at each iteration of the algorithm, is also governed by a power law.

Algorithms 1 and 2 together (Figures 7 and 8, respectively), present the formalization of the SFFS-BA approach. Algorithm 1 (Figure 7) starts by considering all the available samples of the target gene and its expression values. The initial iteration, i.e., k = 1, is aimed to select the individual predictors for all target genes. In this regard, the SFFS-BA (Algorithm 2 in Figure 8) is applied to identify the best individual inferred predictors for each of the target genes, which are classified according to the adopted criterion function.

An important aspect in the SFFS-BA feature selection method is that the algorithm will explore the search space in steps, i.e., the predictors are chosen itera-

Figure 7. Algorithm 1 - Network Inference
Reprinted with permission from Lopes, Martins Jr and Barrera, Cesar Jr, (2014a)

Algorithm 1 Network-Inference (*targets*, *exps*, γ, Δ)

```
var list exelist
var integer k ← 1, n ← targets.size()
for i = 1 to n do
    exelist.append(targets[i], ∅, 1, 0)
end for
while n > 1 do
    for i = 1 to n do
        [target, psets, cfv, gain] ← exelist.removefirst()
        [newpsets, newcfv, gain] ← SFFS-BA(target,cfv,psets,k,exps,Δ)
        exelist.append(target,newpsets,newcfv,gain)
    end for
    SortPredictorSetsbyGain(exelist)
    n ← n × k^(-γ)
    k ← k + 1
end while
return exelist
```

Figure 8. Algorithm 2 – SFFS-BA
Reprinted with permission from Lopes, Martins Jr and Barrera, Cesar Jr. (2014a)

```
Algorithm 2 SFFS-BA (target,cfv,psets,k,exps,Δ)
if psets = ∅ then
    for predictoridx = 1 to exps.size() do
        psets.append(predictoridx)
    end for
end if
while psets is not empty and psets.first.cardinality ≤ k do
    newpset ← psets.removefirst()
    newcfv ← SFS(target,newpset,k,exps)
    if newcfv < bestcfv and (bestcfv − newcfv) > Δ then
        newcfv ← SBS(target,newpset,exps)
        bestcfv ← newcfv
        bestset ← newpset
    end if
    if newpset.cardinality = 1 then
        psets.append(newpset)
    end if
end while
if k > 1 then
    psets ← bestset
end if
return [psets, bestcfv,(cfv − bestcfv)]
```

tively by considering the cardinality parameter k. Another feature is that for k = 1, the SFFS-BA algorithm returns all subsets of predictors and their criterion function values to explore all individual predictors in the next iteration and to recover the predictors of IMP genes more precisely.

Considering the subsets of inferred predictors with cardinality k > 1, the SFFS-BA algorithm runs by considering only the best subset of inferred predictors for each target gene, assuming that some of the true predictors may already be in the subset of selected predictors. At the end of each iteration of Algorithm 3.1, the target genes are sorted by its prediction gain variation by including a new predictor in its subset of inferred predictors. The number of target genes to be considered in the next iteration of the algorithm is updated according to a power law given by $n \leftarrow n \times k^{-\gamma}$ and the cardinality of the subset of predictors is incremented (k+1). Therefore, in the next iterations (i.e., k > 1) the algorithm will consider only the *n* target genes with higher prediction gains, and will increment the cardinality *k* of its subset of predictors. It is important to note that if a target gene reaches the optimal value of the criterion function, or is close enough, it is not considered to increase the cardinality of its subset of predictors. The search is performed until the stop condition, $n < 1$, is reached.

In summary, the SFFS-BA algorithm differs from SFFS algorithm (Pudil et al, 1994) in operating the combinations of subsets of cardinality $k < 3$ and in the inclusion of a factor that reduces the search space based on the assumption that the expression data (input data) are generated from the genetic interrelationships that originate from a scale-free network. The parameter, *exps,* corresponds to a set of gene expression profiles where the genes are typically arranged in rows and the experiments in columns to form a matrix. The parameter γ is a constant which determines the exponential decay of the number of target genes in the next iteration of the algorithm.

In Lopes et al., 2014, the constants γ and Δ are chosen as 2.5 and 0.05, repectively. The value of the constant γ is chosen as the average value found in the literature for this constant. The work presented in Jeong et al., 2000, pointed out that the scale-free topology is suitable to represent the metabolic network of the *Escherichia coli*, which has a probability distribution on its connections well approximated by a power law with constant $\gamma = 2.2$. In some investigations (Jeong et al, 2001; Boccaletti, Latora, Moreno, Chavez, & Hwang, 2006) the probability distribution of the interactions, for a given yeast protein with *k* other proteins, follows a power law with $\gamma = 2.4$. In general, it can be seen that the distributions available in the literature for the connections in the various networks such as the Internet, human collaboration networks and metabolic networks follow a power law decay of the number of elements while increasing the number of connections, usually in finding the value of the constant (decay exponent) in the range $2 < \gamma < 3$ (Albert & Barabasi, 2002; Albert, 2005).

The work described in Lopes et al., 2014, presents some application and results of the SFFS-BA approach. Finally, it is possible to find a new approach for GRNs inference by considering the small-world network topology in Vicente & Lopes, 2014.

4. FUNCTIONAL DATA SOURCES USED IN GRN INFERENCE

An important trend to overcome the GRN inference issues like accuracy of prediction is to integrate several types of molecular biology data sources (omics) such as genomic (SNPs, CNVs, etc), transcriptomic (gene expression), proteomic (PPI networks and protein similarity relations), metabolomic and epigenomic (Ray et al, 2009, Lopes et al., 2014b). Several researchers (Cui et al, 2010; De Haan et al, 2010; Macintyre et al, 2010) pointed out that true gene relations can be predicted more accurately if a higher number of biological data sources are used in gene inference methods. While biological information is incorporated in gene regulatory networks in (Ernst et al, 2008; Seok et al, 2010) to discover transcriptional regulation relationships, others (Troyanskaya, 2005; Karlebach & Shamir, 2008; Baumbach et al, 2009, Hecker et al, 2009, Ray et al, 2009, Ray, Bandyopadhyay, & Pal; Vicente & Lopes, 2012) integrate different biological data for predicting gene relations. In our recent investigations the following data sources are mainly used for integration and discovering new knowledge in GRN frameworks:

Gene Expression: It is a measure which represents the degree of activity of a gene in a cell during various biological processes like cell division, DNA metabolism etc. The technology that enables to measure this activity of thousands of genes simultaneously is the microarray which contains thousands of spots representing those genes. During microarray preparation mRNAs, representing genes, are extracted from cells and are converted into cDNAs using reverse transcription and colored with a fluorescent dye to measure the intensity level. The intensity level gives the amount of expressed genes as compared to unexpressed ones which are colored with another dye. Similar experiments are performed at various time points with all the genes and the time series data is prepared. For example, the All Yeast data (Eisen, Spellman, Brown, & Botstein, 1998) (http://rana.lbl.gov./EisenData.htm), has 6221 genes and 80 time points and can be used for yeast functional analysis. The similarity between two genes can be extracted by computing Pearson correlation between their gene expression profiles. The value lies within -1 to 1.

Phenotypic Profile: The phenotypic profiles method is first demonstrated by Brown et al (2006) by using genes of budding yeast. The method proceeds by homozygous deletion of nonessential genes which respond to agents like cytotoxic or cytostatic. The function analysis of genes is performed by hierarchical clustering of the quantitative sensitivity profiles of the 4756 strains. From the analysis they observed that 860 un-annotated genes are grouped with known genes in various clustering solutions. This in turn helped to predict the function of unknown genes from the functional enrichment (using p-value) of known genes in a given cluster. This phenotypic profile data can be used in a very similar way as gene expression data to extract similarity between any two genes using Pearson correlation.

KEGG pathway: The Kyoto Encyclopedia of Genes and Genomes (KEGG) database (Kanehisa et al, 2010) is a collection of all pathway profiles of genes for many species. There are more than 200 subprofiles under the main profiles metabolism, genetic information processing, environmental information processing and cellular processes. This pathway information can be converted to numerical information by constructing profile vectors for each gene/protein where each component will be marked with 0 or 1 depending on whether that gene/protein participates in a particular sub pathway profile or not. For a particular species, the participation of a protein in a pathway can be predicted by downloading all the protein sequences, except for that species, corresponding to that pathway from Protein Information Resource (PIR) database and comparing them with the target protein using BLAST (Altschul, Gish, Miller, Myers, & Lipman, 1990). The degree of similarity between two proteins can be predicted by computing the ratio of dot product value and OR value between their pathway profiles.

Protein Similarity through Transitive Homology: At the present scenario, the relations discovered by direct similarity search are already known to the research community and one should not expect anything new from this relations. On the other hand, protein transitive homologues (Ma, Chim, Cai, Szustakowski, & Nirmala, 2005; Park et al, 1998) can reveal the relationship between two proteins which cannot be realized by searching direct sequence similarity/homology. This technique works as follows: first, for a given sequence, its homologues are searched within a database and then within the same database homologues of these homologues are pointed out which serve as transitive homologues of the given sequence. As a typical database one can use the sequences available from the UniProt (Bairoch et al, 2005). Further precautions should be taken in the search process such as a threshold of 200 residues on total length of the protein alignment to avoid homology arising from conserved domains. Further, the similarities arising from the raw BLAST score should be normalized within 0 to 1 to make them usable for computing transitive homology scores. For example, consider that an intermediate sequence Y identifies a transitive homology relation between sequence X and Z. Let us consider that the direct normalized BLAST score similarities B(X,Y), B(Y,Z) and B(X,Z) are 0.7, 0.8 and 0.1, respectively. The direct homology between sequence X and Z is very low and cannot be used for further similarity analysis. But interestingly the transitive homology between X and Z can be computed as $0.7 \times 0.8 = 0.56$ which indicates a strong transitive relation between X and Z. The basic idea behind this relation is that, these two sequences (X and Z) are thought to be a part of the larger sequence Y and they separated with the evolutionary process of the original sequence Y. In many cases the transitive homologues are seen to identify new relations better than the direct homology search (Park et al, 1998).

Protein-Protein Interaction: Protein interactions assemble the molecular machines of the cell and represent the dynamics of virtually all cellular responses (Pawson & Nash, 2003). It takes place by physical contacts between two or more proteins as a result of some biochemical or electrostatic process. The results of such interaction can be observed in the formation of homo-oligomeric or hetero-oligomeric or enzyme-inhibitor or antibody-antigen complexes which can be stable or transient. Interactions can occur between domain-peptide and domain-domain. In the manual curation process the interactions can be of many type such as growth defect, co-crystal structure, co-purification, co-fractionation, phenotypic enhancement and suppression, affinity capture, two-hybrid, synthetic lethality and protein-peptide experiments.

These interactions can be useful in constructing GRNs and protein-protein maps in the cell and are represented by binary values 1 or 0, depending on the availability of interaction. There are several databases like BioGRID (Stark et al, 2006) where known protein-protein interactions are manually curated and preserved. For example, there are more than 0.3 million yeast protein-protein interactions are available in the BioGRID database which are collected from aforementioned interactions.

Comparison Among Data Sources: The aforementioned data sources show certain degree of accuracy for identifying true gene relations and functional predictions. For example, using yeast data sources the degree of accuracy can be measured by positive predictive value (PPV) involving yeast GO-slim process annotations of genes in the Saccharomyces Genome database (SGD) (Dwight et al, 2002). Note that, GO-Slim annotations are available through a webserver at http://www.yeast-genome.org/cgi-bin/GO/goSlimMapper.pl which provides the biological processes, molecular functions, and cellular components of Yeast Saccharomyces cerevisiae genes. At a particular similarity value, extracted from a specific data source, there can be thousands of gene pairs. Using these gene pairs the positive predictive value (PPV) at a particular similarity value can be defined as:

$$PPV = \frac{\text{number of predicted pairs with common yeast GO-slim process annotations}}{\text{total number of predicted pairs}}$$

Figure 9. Comparing the positive predictive values at various similarity values for different data sources using Yeast GO-slim process annotations.

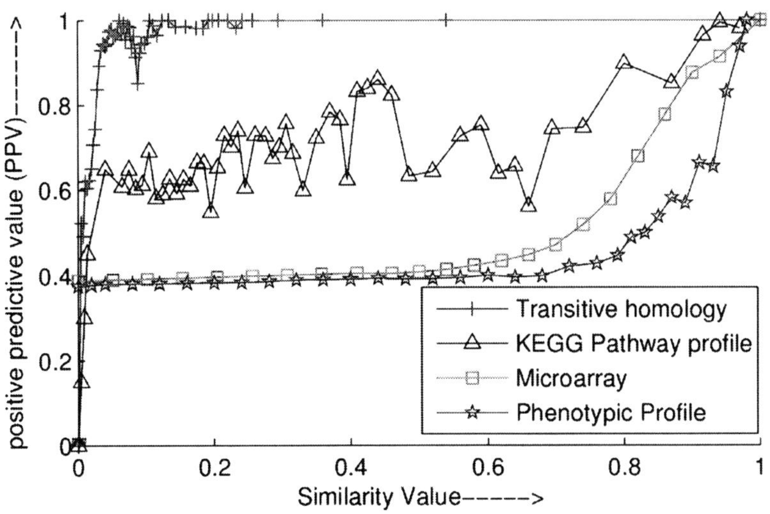

5. DATA SOURCE INTEGRATION IN GRN INFERENCE

Computational methods have begun to forge connections between cellular events and to foster numerous hypotheses on gene functions. While gene expressions and phenotypic profiles can be used for hypothesis generation, a higher degree of specificity is needed for tasks like gene function prediction and gene regulatory network inference. This can be achieved by the integration of several data sources. However a key requirement to integrate heterogeneous data sources is that different data sources should have a unified observational scale such as PPV's as mentioned in the previous section. This allows us to directly compare and integrate the different types of datasets. The PPV reflects the accuracy of data sets, but do not provide any information about relative importance of one data source in presence of the other data sources. Here we describe the method in Ray, Bandyopadhyay, & Pal, 2009 which is focused on integrating information from data sources in a weighted arithmetic mean framework. A `Biological Score' score, called BS is developed, where, PPVs computed from phenotypic similarity (P), gene expression similarity (G), KEGG pathway profile similarity (K), protein similarity through transitive homologue (B), and protein-protein interaction information (I) between two genes X and Y are integrated through weights a, b, c, d, and e in a linear combination style. This score is referred to as Biological Score (BS) and is defined as

$$BS=\frac{a\times P+b\times G+c\times K+d\times B+e\times I}{a+b+c+d+e}$$

where a, b, c, d, and e are varied within range 0 to 100 in steps of 0.1 to find a combination that maximizes the PPV of BS for a user defined number of top gene pairs. Note that, the weights a, b, c, d, and e are assigned to the complete PPV matrices calculated from individual data sources. Intuitively, more important data sources should be assigned larger weights than less important ones, while irrelevant ones should be assigned zero weight. The following steps are used to estimate the weight factors in the BS:

Step 1: All the factors are assigned an initial value of 1.

Step 2: BS values are calculated for all the gene pairs and sorted in descending order to identify the gold standard cut-off value above which the top 26432 gene pairs are available. The gold standard cut-off value and gold standard of top gene pairs are determined from KEGG pathway profiles, which provides 26432 gene pairs with similarity value 1 and constant PPV of 0.81. These gene pairs are the most predictive of all, whereas the PPV of other data sources, as well as gene pairs below top 26432 for KEGG pathway profiles, vary considerably.

Step 3: PPV is calculated for the top 26432 gene pairs.

Step 4: The weight factors are now varied in steps of 0.1 and the steps from 2 to 3 are repeated to find a combination of weights for which the PPV is maximized.

Some typical instances in the exhaustive search process, showing the variation of PPV of BS for different sets of weight coefficients, ranging from 0 to 100, in steps of 1, are shown in Figure 10. Yeast GO-Slim process annotations are used in determining the PPVs.

Once the weights are finalized (optimized) BS can be used for GRN inference. To infer GRN here we discuss the methodology developed in Lopes, Ray, Hashi-

Figure 10. Comparing the values of PPV using BS, by varying weight coefficients of PPV of different data sources for top 24632 gene pairs. When a particular coefficient is varied the others are kept constant at the values shown in the figure.

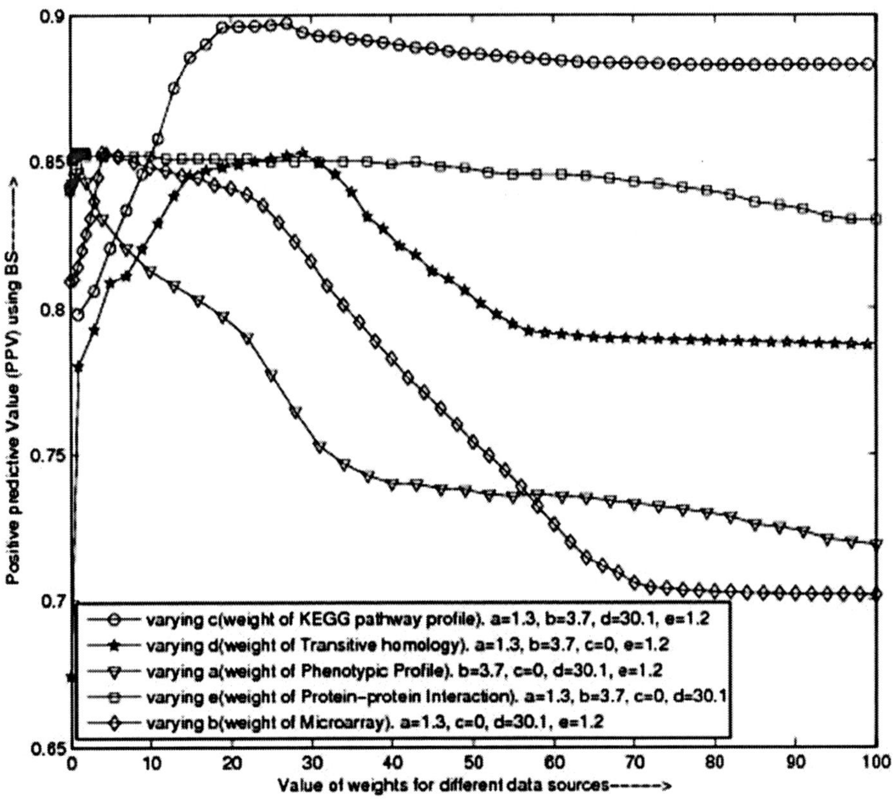

moto, & Cesar Jr., 2014b, called Entropic Biological Score (EBS). It linearly combines the mean conditional entropy (MCE) (Section 2.3.4.2) from expression levels and the aforementioned Biological Score (BS). The process of combining is similar to that of combining multiple data sources in BS. The best results for EBS are obtained by considering the weights as 0.2 and 0.8 for MCE and BS, respectively, where approximately 40% of the inferred connections are found to be correct and significantly better than related methods like MRNET, ARACNE and CLR. The EBS is validated with the Cell Cycle related functional annotation information, available from Munich Information Center for Protein Sequences (MIPS). Some typical results, are shown in Figures 11 and 12 which indicate that expression profile is able to recover some true connections that are not present in biological annotations, thus leading to the possibility of discovering new relations between its genes.

Figure 11. Applying similarity measure, $Similarity(A,B) = \sqrt{PPV(A,B) \times Sensitivity(A,B)}$*, obtained by SFS, SFFS, and SFFS-BA (Lopes et al, 2014a), to infer network edges from expression profiles with different numbers of time points (signal size).*

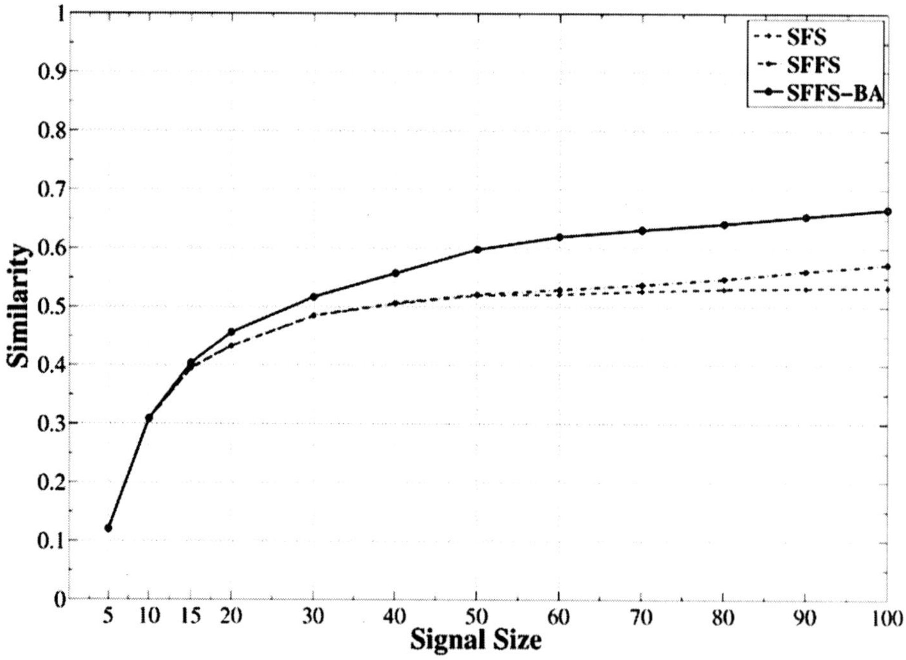

Figure 12. The Positive Predictive Value (PPV) reaches the highest value at w=0.8 for any K. The precision decreases fast when the algorithm does not consider the MCE (w=1.0) showing that both expression and topology are necessary to increase the precision for SFFS-WS (Vicente & Lopes, 2014).

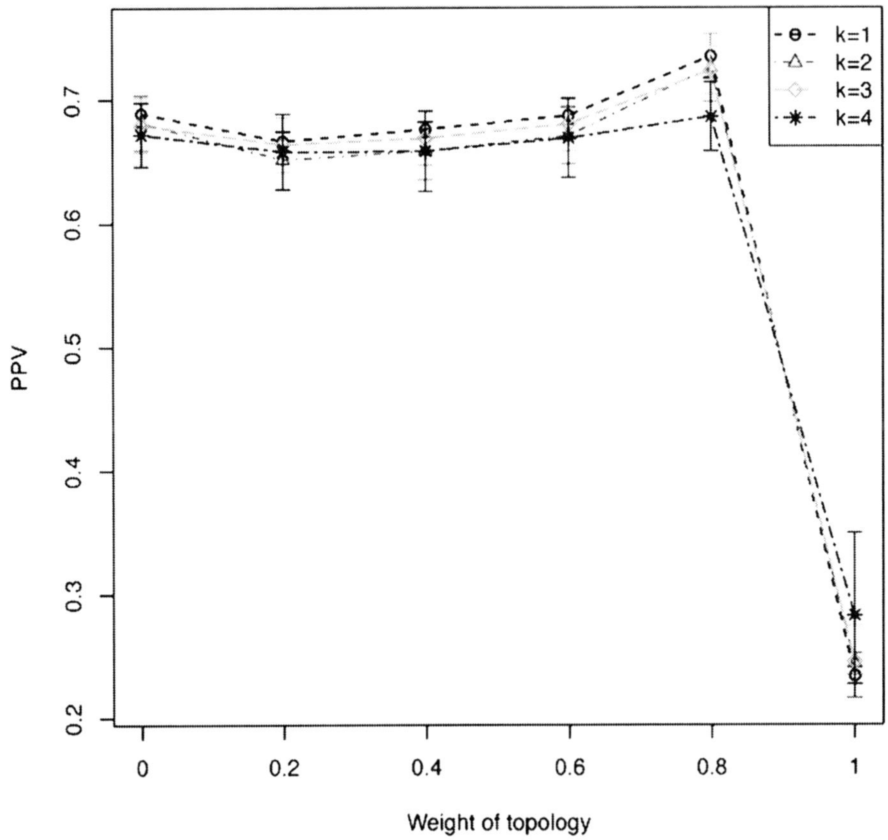

6. CONCLUSION

The living beings respond rapidly to the changes in the internal and external environment by adjusting their gene expression profiles, i.e., their genes regulation. As a result, the gene regulatory networks (GRNs) are capable to adapt with these changes. The regulatory systems can be viewed as a complex network of signals codified in the genetic material which controls the gene expression dynamics. Such dynamics can explain some phenomena like cell differentiation and the development of an organism. However, there is still much to be discovered about the way by which these alterations in the transcripts and proteins production are related and regulated along time. To comprehend this regulatory system is crucial and it is one of the main concerns of the Systems Biology studies.

Although a myriad of models and methods for identification of GRNs based on gene expression profiles has been proposed in the literature, generally the information required to validate the inferred network is incomplete or unknown. To circumvent this problem, a recent trend in the literature is to incorporate other types of information such as global and local topological properties found in biological networks, and to integrate other *omics* data with the traditional transcriptomic data.

Regarding the inclusion of topological properties to improve the GRN inference methods, an interactive and floating search strategy, which includes the hypothesis that GRNs possess scale free topology as prior information in the inference process (Lopes et al, 2014a), is discussed in this chapter in some details. This method is adequate and efficient to infer GRNs from temporal gene expression signals having small number of samples and a huge number of genes. Such method, called SFFS-BA, explores the scale free property by pruning the search space based on the power law function where the number of gene targets, chosen to grow their sets of predictors, decreases at each step according to this function. SFFS-BA provides better results in terms of similarity than SFFS-MCE, which does not include any prior topological information. Besides, a method which includes small world features as prior information, called SFFS-WS (Vicente & Lopes, 2014), is recently developed, showing encouraging results as showed in Figures 11 and 12.

In predicting GRNs, the accuracy of biological data sources as well as the integration of several data sources are also very important (Lopes et al, 2014b). Protein similarity through transitive homology and KEGG pathway profiles should be given more importance than gene expression and phenotypic profiles. In other words, one should design a supervised GRN inference method using multiple heterogeneous data sources in such a way that the system itself should give more importance to data sources with higher positive predictive values (PPVs) by using this information. Further, from the analysis discussed in this chapter, it is clear that the performance of a GRN inference method using single data sources will be in decreasing order for data sources like protein similarity through transitive homology, KEGG pathway profiles, gene expression and phenotypic profiles. Its performance will be much improved if all the data sources are used together where each data source will cover the limitation and low PPV gene relations of other data sources. The challenge to the researchers will be in designing a better system to take care of the aforementioned PPV related factors. Note that, protein-protein interaction in general cannot be used alone to design a GRN as the database does not cover the relations for all possible gene pairs.

REFERENCES

Aittokallio, T., & Schwikowski, B. (2006). Graph-based methods for analysing networks in cell biology. *Briefings in Bioinformatics*, *73*(3), 243–255. doi:10.1093/bib/bbl022 PMID:16880171

Akutsu, T., Miyano, S., & Kuhara, S. (2000). Inferring qualitative relations in genetic networks and metabolic pathways. *Bioinformatics (Oxford, England)*, *16*(8), 727–734. doi:10.1093/bioinformatics/16.8.727 PMID:11099258

Albert, R. (2005). Scale-free networks in cell biology. *Journal of Cell Science*, *118*(21), 4947–4957. doi:10.1242/jcs.02714 PMID:16254242

Albert, R., & Barabasi, A. L. (2002). Statistical mechanics of complex networks. *Reviews of Modern Physics*, *74*(1), 47–97. doi:10.1103/RevModPhys.74.47

Albert, R., Jeong, H., & Barabási, A. L. (2001). Error and attack tolerance of complex networks. *Nature*, *409*(6819), 542. doi:10.1038/35054111 PMID:10935628

Alon, U. (2007). Network motifs: Theory and experimental approaches. *Nature Reviews. Genetics*, *8*(6), 450–461. doi:10.1038/nrg2102 PMID:17510665

Altschul, S. F., Gish, W., Miller, W., Myers, E. W., & Lipman, D. J. (1990). Basic local alignment search tool. *Journal of Molecular Biology*, *215*(3), 403–410. doi:10.1016/S0022-2836(05)80360-2 PMID:2231712

Anastassiou, D. (2007). Computational analysis of the synergy among multiple interacting genes. *Molecular Systems Biology*, *3*, 83. doi:10.1038/msb4100124 PMID:17299419

Andrecut, M., Huang, S., & Kauffman, S. (2008). Heuristic approach to sparse approximation of gene regulatory networks. *Journal of Computational Biology*, *15*(9), 1173–1186. doi:10.1089/cmb.2008.0087 PMID:18844584

Andrecut, M., & Kauffman, S. (2008). On the sparse reconstruction of gene networks. *Journal of Computational Biology*, *15*(1), 21–30. doi:10.1089/cmb.2007.0185 PMID:18257675

Ashburner, M., Ball, C. A., Blake, J. A., Botstein, D., Butler, H., Cherry, J. M., & Sherlock, G. (2000). Gene Ontology: Tool for the unification of biology. *Nature Genetics*, *25*(1), 25–29. doi:10.1038/75556 PMID:10802651

Bairoch, A., Apweiler, R., Wu, C. H., Barker, W. C., Boeckmann, B., Ferro, S., & Yeh, L. S. et al. (2005). The universal protein resource (UniProt). *Nucleic Acids Research*, *33*(Database issue), 154–159. doi:10.1093/nar/gki070 PMID:15608167

Barabasi, A. L. (2002). *Linked: How Everything Is Connected to Everything Else and What It Means*. Perseus.

Barabasi, A. L. (2009). Scale-Free Networks: A Decade and Beyond. *Science*, *325*(5939), 412–413. doi:10.1126/science.1173299 PMID:19628854

Barabasi, A. L., & Oltvai, Z. (2004). Network biology: Understanding the cell's functional organization. *Nature Reviews. Genetics*, *5*(2), 101–113. doi:10.1038/nrg1272 PMID:14735121

Barrera, J., Cesar-Jr, R. M., Martins-Jr, D. C., Vencio, R. Z. N., Merino, E. F., Yamamoto, … Portillo, H. A. (2007). Constructing probabilistic genetic networks of Plasmodium falciparum, from dynamical expression signals of the intraerythrocytic development cycle. In P. McConnell, S. M. Lin, & P. Hurban (Eds.), Methods of Microarray Data Analysis V (pp. 11-26). Springer.

Baumbach, J., Tauch, A., & Rahmann, S. (2009). Towards the integrated analysis, visualization and reconstruction of microbial gene regulatory networks. *Briefings in Bioinformatics*, *10*(1), 75–83. doi:10.1093/bib/bbn055 PMID:19074493

Benson, D. A., Karsch Mizrachi, I., Lipman, D. J., Ostell, J., & Wheeler, L. (2008). GenBank. *Nucleic Acids Research*, *36*(suppl 1), D25–D30. doi:10.1093/nar/gkm929 PMID:18073190

Boccaletti, S., Latora, V., Moreno, Y., Chavez, M., & Hwang, D. U. (2006). Complex networks: Structure and dynamics. *Physics Reports*, *424*(4-5), 4–5, 175–308. doi:10.1016/j.physrep.2005.10.009

Borelli, F. F., Camargo, R. Y., Martins, D. C. Jr, & Rozante, L. C. S. (2013). Gene regulatory networks inference using a multi-GPU exhaustive search algorithm. *BMC Bioinformatics*, *14*(Suppl 18), S5. doi:10.1186/1471-2105-14-S18-S5 PMID:24564268

Brown, J. A., Sherlock, G., Myers, C. L., Burrows, N. M., Deng, C., Wu, H. I., & Brown, J. M. et al. (2006). Global analysis of gene function in yeast by quantitative phenotypic proling. *Molecular Systems Biology*, *2*, 1–9. doi:10.1038/msb4100043 PMID:16738548

Butte, A., & Kohane, I. (2000). Mutual information relevance networks: functional genomic clustering using pairwise entropy measurements. In *Pacific Symposium on Biocomputing*, (pp. 418-429).

Cao, Y., & Samuels, D. C. (2009). Discrete stochastic simulation methods for chemically reacting systems. *Methods in Enzymology, 454*, 115–140. doi:10.1016/S0076-6879(08)03805-6 PMID:19216925

Charbonnier, C., Chiquet, J., & Ambroise, C. (2010). Weighted-lasso for structured network inference from time course data. *Statistical Applications in Genetics and Molecular Biology, 9*(1), 15.

Chen, T., He, H. L., & Church, G. M. (1999). Modeling gene expression with differential equations.*Pacific Symposium on Biocomputing*, (pp. 29-40).

Christley, S., Nie, Q., & Xie, X. (2009). Incorporating existing network information into gene network inference. *PLoS ONE, 4*(8), e6799.

Costa, L. F., Rodrigues, F. A., & Cristino, A. S. (2008). Complex networks: The key to systems biology. *Genetics and Molecular Biology, 31*(3), 591–601. doi:10.1590/S1415-47572008000400001

Cui, X., Wang, T., Chen, H. S., Busov, V., & Wei, H. (2010). Tf-finder: A software package for identifying transcription factors involved in biological processes using microarray data and existing knowledge base. *BMC Bioinformatics, 11*(1), 425. doi:10.1186/1471-2105-11-425 PMID:20704747

D'Angelo, C., Kilian, J., Kudla, J., Batistic, O., & Weinl, S. (2009). *The Arabidopsis Information Resource*. TAIR.

D'haeseleer, P., Liang, S., & Somogyi, R. (2000). Genetic network inference: From co-expression clustering to reverse engineering. *Bioinformatics (Oxford, England), 16*(8), 707–726. doi:10.1093/bioinformatics/16.8.707 PMID:11099257

Da Silva, E., & Stumpf, P. (2005). Complex networks and simple models in biology. *Journal of the Royal Society, 2*(5), 419-430.

Davidson, C. (2010). Identifying gene regulatory networks using evolutionary algorithms. *Journal of Computing Sciences in Colleges, 25*(5), 231–237.

De Haan, J., Piek, E., van Schaik, R., de Vlieg, J., Bauerschmidt, S., Buydens, L., & Wehrens, R. (2010). Integrating gene expression and GO classification for PCA by preclustering. *BMC Bioinformatics, 11*(1), 158. doi:10.1186/1471-2105-11-158 PMID:20346140

De Jong, H. (2002). Modeling and Simulation of Genetic Regulatory Systems: A Literature Review. *Journal of Computational Biology, 9*(1), 67–103. doi:10.1089/10665270252833208 PMID:11911796 de Smet, R., & Marchal, K. (2010). Advantages and limitations of current network inference methods. *Nature Reviews. Microbiology*, *8*(10), 717–729. PMID:20805835

Dougherty, E. R., Brun, M., Trent, J., & Bittner, M. L. (2009). A Conditioning-Based Model of Contextual Regulation. *IEEE/ACM Transactions on Computational Biology and Bioinformatics*, *6*(2), 310–320. doi:10.1109/TCBB.2007.70247 PMID:19407354

Dwight, S. S., Harris, M. A., Dolinski, K., Ball, C. A., Binkley, G., Christie, K. R., & Cherry, J. M. et al. (2002). Saccharomyces genome database (SGD) provides secondary gene annotation using the gene ontology (GO). *Nucleic Acids Research*, *30*(1), 69–72. doi:10.1093/nar/30.1.69 PMID:11752257

Eisen, M. B., Spellman, P. T., Brown, P. O., & Botstein, D. (1998). Cluster analysis and display of genome-wide expression patterns. *Proceedings of the National Academy of Sciences of the United States of America*, *95*(25), 14863–14868. doi:10.1073/pnas.95.25.14863 PMID:9843981

Ernst, J., Beg, Q. K., Kay, K. A., Balazsi, G., Oltvai, Z. N., & Bar-Joseph, Z. (2008). A semi-supervised method for predicting transcription factor-gene interactions in *Escherichia coli*. *PLoS Comput Biol*, *4*(3), e1000044.

Faith, J., Hayete, B., Thaden, J., Mogno, I., Wierzbowski, J., Cottarel, G., & Gardner, T. S. et al. (2007). Large-scale mapping and validation of escherichia coli transcriptional regulation from a compendium of expression proles. *PLoS Biology*, *5*(1), 259–265. doi:10.1371/journal.pbio.0050008 PMID:17214507

Fall, C., Marland, E., Wagner, J., & Tyson, J. (Eds.). (2002). *Computational Cell Biology*. New York: Springer.

Farkas, I. J., Jeong, H., Vicsek, T., Barabási, A.-L., & Oltvai, Z. N. (2003). The topology of the transcription regulatory network in the yeast, saccharomyces cerevisiae. *Physica A: Statistical Mechanics and its Applications*, *318*(3-4), 601–612.

Friedman, N., Linial, M., Nachman, I., & Pe'er, D. (2000). Using Bayesian networks to analyze expression data. *Journal of Computational Biology*, *7*(3-4), 601–620. doi:10.1089/106652700750050961 PMID:11108481

Garey, M. R., & Johnson, D. S. (1979). *Computers and intractability: a guide to the theory of NP-completeness*. San Francisco: WH Freeman & Co.

Goemann, B., Wingender, E., & Potapov, A. (2009). An approach to evaluate the topological significance of motifs and other patterns in regulatory networks. *BMC Systems Biology*, *3*(1), 53. doi:10.1186/1752-0509-3-53 PMID:19454001

Goss, P. J., & Peccoud, J. (1998). Quantitative modeling of stochastic systems in molecular biology by using stochastic Petri nets. *Proceedings of the National Academy of Sciences of the United States of America, 95*(12), 6750–6755. doi:10.1073/pnas.95.12.6750 PMID:9618484

Greller, L. D., & Somogyi, R. (2002). Reverse engineers map the molecular switching yards. *Trends in Biotechnology, 20*(11), 445–447. doi:10.1016/S0167-7799(02)02051-6 PMID:12413811

Guelzim, N., Bottani, S., Bourgine, P., & Képès, F. (2002). Topological and causal structure of the yeast transcriptional regulatory network. *Nature Genetics, 31*(1), 60–63. doi:10.1038/ng873 PMID:11967534

Hardy, S., & Robillard, P. N. (2004). Modeling and simulation of molecular biology systems using petri nets: Modeling goals of various approaches. *Journal of Bioinformatics and Computational Biology, 2*(4), 595–613. doi:10.1142/S0219720004000764 PMID:15617157

Hecker, M., Lambeck, S., Toepfer, S., van Someren, E., & Guthke, R. (2009). Gene regulatory network inference: Data integration in dynamic models-a review. *Bio Systems, 96*(1), 86–103. doi:10.1016/j.biosystems.2008.12.004 PMID:19150482

Hegland, M., Burden, C., Santoso, L., MacNamara, S., & Booth, H. (2007). A solver for the stochastic master equation applied to gene regulatory networks. *Journal of Computational and Applied Mathematics, 205*(2), 708–724. doi:10.1016/j.cam.2006.02.053

Holland, J. H. (1992). *Adaptation in Natural and Artificial Systems*. MIT Press.

Hovatta, I., Kimppa, K., Lehmussola, A., Pasanen, T., Saarela, J., Saarikko, I., … Laine, M. (2005). DNA microarray data analysis (2nd ed.). CSC - Scientific Computing Ltd.

Ideker, T. E., Thorsson, V., & Karp, R. M. (2000). Discovery of regulatory interactions through perturbation: inference and experimental design.*Pacific Symposium on Biocomputing*, (vol. 5, pp. 302-313).

Jain, A. K., Duin, R. P. W., & Mao, J. (2000). Statistical Pattern Recognition: A Review. *IEEE Transactions on Pattern Analysis and Machine Intelligence, 22*(1), 4–37. doi:10.1109/34.824819

Jeong, H., Mason, S. P., Barabási, A. L., & Oltvai, Z. N. (2001). Lethality and centrality in protein networks. *Nature, 411*(6833), 41–42. doi:10.1038/35075138 PMID:11333967

Jeong, H., Tombor, B., Albert, R., Oltvai, Z. N., & Barabási, A.-L. (2000). The large-scale organization of metabolic networks. *Nature*, *407*(6804), 651–654. doi:10.1038/35036627 PMID:11034217

Jimenez, R. D., Martins, D. C. Jr, & Santos, C. S. (2014). Gene networks inference through one genetic algorithm per gene. In *14th IEEE International Conference on Bioinformatics and Bioengineering Conference (BIBE)*. Boca Raton, FL: IEEE. doi:10.1109/BIBE.2014.9

Kanehisa, M., Goto, S., Furumichi, M., Tanabe, M., & Hirakawa, M. (2010). KEGG for representation and analysis of molecular networks involving diseases and drugs. *Nucleic Acids Research*, *38*(suppl 1), D355–D360. doi:10.1093/nar/gkp896 PMID:19880382

Karlebach, G., & Shamir, R. (2008). Modelling and analysis of gene regulatory networks. *Nature Reviews. Molecular Cell Biology*, *9*(10), 770–780. doi:10.1038/nrm2503 PMID:18797474

Kauffman, S. A. (1969). Homeostasis and differentiation in random genetic control networks. *Nature*, *224*(215), 177–178. doi:10.1038/224177a0 PMID:5343519

Kauffman, S. A. (1993). *The Origins of Order*. Oxford University Press.

Kelemen, A., Abraham, A., & Chen, Y. (2008). *Computational Intelligence in Bioinformatics*. Springer. doi:10.1007/978-3-540-76803-6

Klamt, S., Saez Rodriguez, J., & Gilles, E. (2007). Structural and functional analysis of cellular networks with cellnetanalyzer. *BMC Systems Biology*, *1*(1), 2.

Knabe, J. F., Wegner, K., Nehaniv, C. L., & Schilstra, M. J. (2010). Genetic algorithms and their application to in silico evolution of genetic regulatory networks. *Methods in Molecular Biology (Clifton, N.J.)*, *673*, 297–321. doi:10.1007/978-1-60761-842-3_19 PMID:20835807

Kuchaiev, O., Milenkovic, T., Memisevic, V., Hayes, W., & Przulj, N. (2010). Topological network alignment uncovers biological function and phylogeny. *Journal of the Royal Society, Interface*, *7*(50), 1341–1354. doi:10.1098/rsif.2010.0063 PMID:20236959

Lacroix, V., Cottret, L., Thébault, P., & Sagot, M.-F. (2008). An introduction to metabolic networks and their structural analysis. *IEEE/ACM Transactions on Computational Biology and Bioinformatics*, *5*(4), 594–617. doi:10.1109/TCBB.2008.79 PMID:18989046

Larrañaga, P., Poza, M., Yurramendi, Y., Murga, R. H., & Kuijpers, C. M. H. (1996). Structure learning of bayesian networks by genetic algorithms: A performance analysis of control parameters. *IEEE Transactions on Pattern Analysis and Machine Intelligence*, *18*(9), 912–926. doi:10.1109/34.537345

Lein, E. S., Hawrylycz, M. J., Ao, N., Ayres, M., Bensinger, A., Bernard, A., & Jones, A. R. et al. (2007). Genome-wide atlas of gene expression in the adult mouse brain. *Nature*, *445*(7124), 168–176. doi:10.1038/nature05453 PMID:17151600

Lenas, P., Moos, M. Jr, & Luyten, F. P. (2009). Developmental engineering: A new paradigm for the design and manufacturing of cell-based products. part ii. from genes to networks: Tissue engineering from the viewpoint of systems biology and network science. *Tissue Engineering. Part B, Reviews*, *15*(4), 395–422. doi:10.1089/ten.teb.2009.0461 PMID:19589040

Li, F., Long, T., Lu, Y., Ouyang, Q., & Tang, C. (2004). The yeast cell-cycle network is robustly designed. *Proceedings of the National Academy of Sciences of the United States of America*, *101*(14), 4781–4786. doi:10.1073/pnas.0305937101 PMID:15037758

Liang, S., Fuhrman, S., & Somogyi, R. (1998). Reveal: a general reverse engineering algorithm for inference of genetic network architectures. In *Proceedings of Pacific Symposium on Biocomputing*, (pp. 18-29).

Licata, L., Briganti, L., Peluso, D., Perfetto, L., Iannuccelli, M., Galeota, E., & Cesareni, G. et al. (2012). MINT, the molecular interaction database: 2012 update. *Nucleic Acids Research*, *40*(D1), D857–D861. doi:10.1093/nar/gkr930 PMID:22096227

Lopes, F. M., Cesar, R. M. Jr, & Costa, L. F. (2011a). Gene expression complex networks: Synthesis, identification and analysis. *Journal of Computational Biology*, *18*(10), 1353–1367. doi:10.1089/cmb.2010.0118 PMID:21548810

Lopes, F. M., de Oliveira, E. A., & Cesar, R. M. Jr. (2011b). Inference of gene regulatory networks from time series by Tsallis entropy. *BMC Systems Biology*, 5(1), 61. doi:10.1186/1752-0509-5-61 PMID:21545720

Lopes, F. M., Martins, D. C. Jr, & Barrera, J. (2008). Feature selection environment for genomic applications. *BMC Bioinformatics*, *9*(1), 451. doi:10.1186/1471-2105-9-451 PMID:18945362

Lopes, F. M., Martins, D. C., Jr., Barrera, J., & Cesar, R. M., Jr. (2010). SFFS-MR: a floating search strategy for grns inference. In Pattern Recognition in Bioinformatics, Proceedings, (LNCS), (vol. 6282, pp. 407-418). Springer. doi:10.1007/978-3-642-16001-1_35

Lopes, F. M., Martins, D. C. Jr, Barrera, J., & Cesar, R. M. Jr. (2014a). A feature selection technique for inference of graphs from their known topological properties: Revealing scale-free gene regulatory networks. *Information Sciences*, *272*, 1–15. doi:10.1016/j.ins.2014.02.096

Lopes, F. M., Ray, S. S., Hashimoto, R. F., & Cesar, R. M. Jr. (2014b). Entropic biological score: A cell cycle investigation for GRNs inference. *Gene*, *541*(2), 129–137. doi:10.1016/j.gene.2014.03.010 PMID:24631265

Louis, M. & Becskei, A. (2002). Binary and graded responses in gene networks. *Science's Signal Transduction Knowledge Environment*, *2002*(143), PE33.

Lu, L. and Zhou, T. (2011). Link prediction in complex networks: a survey. *Physica A: Statistical Mechanics and its Applications*, *390*(6), 1150-1170.

Ma, Q., Chirn, G. W., Cai, R., Szustakowski, J. D., & Nirmala, N. (2005). Clustering protein sequences with a novel metric transformed from sequence similarity scores and sequence alignments with neural networks. *BMC Bioinformatics*, 6. PMID:16202129

Macintyre, G., Bailey, J., Gustafsson, D., Haviv, I., & Kowalczyk, A. (2010). Using gene ontology annotations in exploratory microarray clustering to understand cancer etiology. *Pattern Recognition Letters*, *31*(14), 2138–2146. doi:10.1016/j.patrec.2010.01.006

Maki, Y., Tominaga, D., Okamoto, M., Watanabe, S., & Eguchi, Y. (2001). Development of a system for the inference of large scale genetic networks.*Pacific Symposium on Biocomputing*, (pp. 446-458).

Mamakou, M. E., Sirakoulis, G. C., Andreadis, I., & Karafyllidis, I. (2005). Adaptive reverse engineering of gene regulatory networks using genetic algorithms. In *IEEE International Conference on EUROCON.*Belgrade: IEEE. doi:10.1109/EURCON.2005.1629947

Manninen, T., Linne, M. L., & Ruohonen, K. (2006). Developing Itô stochastic differential equation models for neuronal signal transduction pathways. *Computational Biology and Chemistry*, *30*(4), 280–291. doi:10.1016/j.compbiolchem.2006.04.002 PMID:16880117

Marbach, D., Costello, J. C., Küffner, R., Vega, N. M., Prill, R. J., Camacho, D. M., & Zimmer, R. et al. (2012). Wisdom of crowds for robust gene network inference. *Nature Methods*, *9*(8), 796–804. doi:10.1038/nmeth.2016 PMID:22796662

Marbach, D., Prill, R. J., Schaffter, T., Mattiussi, C., Floreano, D., & Stolovitzky, G. (2010). Revealing strengths and weaknesses of methods for gene network inference. *Proceedings of the National Academy of Sciences of the United States of America, 107*(14), 6286–6291. doi:10.1073/pnas.0913357107 PMID:20308593

Marbach, D., Schaffter, T., Mattiussi, C., & Floreano, D. (2009). Generating realistic in silico gene networks for performance assessment of reverse engineering methods. *Journal of Computational Biology, 16*(2), 229–239. doi:10.1089/cmb.2008.09TT PMID:19183003

Margolin, A. A., Nemenman, I., Basso, K., Wiggins, C., Stolovitzky, G., Dalla Favera, R., & Califano, A. (2006). Aracne: An algorithm for the reconstruction of gene regulatory networks in a mammalian cellular context. *BMC Bioinformatics, 7*(Suppl 1).

Markowetz, F., & Spang, R. (2007). Inferring cellular networks - a review. *BMC Bioinformatics, 8*(Suppl 6), S5. doi:10.1186/1471-2105-8-S6-S5 PMID:17903286

Martins, D. C. Jr, Braga Neto, U. M., Hashimoto, R. F., Dougherty, E. R., & Bittner, M. L. (2008). Intrinsically Multivariate Predictive Genes. *IEEE Journal of Selected Topics in Signal Processing, 2*(3), 424–439. doi:10.1109/JSTSP.2008.923841

Martins, D. C. Jr, Cesar, R. M. Jr, & Barrera, J. (2006). W-operator window design by minimization of mean conditional entropy. *Pattern Analysis & Applications, 9*(2-3), 139–153. doi:10.1007/s10044-006-0031-0

Martins, D. C. Jr, Oliveira, E. A., Braga Neto, U. M., Hashimoto, R. F., & Cesar, R. M. Jr. (2013). Signal propagation in Bayesian networks and its relationship with intrinsically multivariate predictive variables. *Information Sciences, 225*, 18–34. doi:10.1016/j.ins.2012.10.027

Martone, M. E., Gupta, A., Wong, M., Qian, X., Sosinsky, G., Ludascher, B., & Ellisman, M. H. (2002). A cell-centered database for electron tomographic data. *Journal of Structural Biology, 138*(1-2), 1–2, 145–155. doi:10.1016/S1047-8477(02)00006-0 PMID:12160711

Matsuno, H., Inouye, S. T., Okitsu, Y., Fujii, Y., & Miyano, S. (2006). A new regulatory interaction suggested by simulations for circadian genetic control mechanism in mammals. *Journal of Bioinformatics and Computational Biology, 4*(1), 139–153. doi:10.1142/S021972000600176X PMID:16568547

Mendes, P., Sha, W., & Ye, K. (2003). Artificial gene networks for objective comparison of analysis algorithms. *Bioinformatics (Oxford, England), 19*(Suppl 2), 122–129. doi:10.1093/bioinformatics/btg1069 PMID:14534181

Mendoza, M. R., Lopes, F. M., & Bazzan, A. L. C. (2012). Reverse engineering of grns: An evolutionary approach based on the tsallis entropy. In *Proceedings of the 14th international conference on Genetic and evolutionary computation (GECCO)*. doi:10.1145/2330163.2330190

Mewes, H. W., Ruepp, A., Theis, F., Rattei, T., Walter, M., Frishman, D., & Antonov, A. et al. (2011). MIPS: Curated databases and comprehensive secondary data resources in 2010. *Nucleic Acids Research*, *39*, D220–D224. doi:10.1093/nar/gkq1157 PMID:21109531

Milo, R., Shann-Orr, S., Itzkovitz, S., Kashtan, N., Chklovskii, D., & Alon, U. (2002). Network motifs: Simple building blocks of complex networks. *Science*, *298*(5594), 824–827. doi:10.1126/science.298.5594.824 PMID:12399590

Mitchell, M. (1996). *An introduction to Genetic Algorithms*. Cambridge, MA: MIT Press.

Montoya Cubas, C. F., Martins, D. C. Jr, Santos, C. S., & Barrera, J. (2014). Gene networks inference through linear grouping of variables. In *14th IEEE International Conference on Bioinformatics and Bioengineering Conference (BIBE)*. Boca Raton, FL: IEEE. doi:10.1109/BIBE.2014.10

Narasimhan, S., Rengaswamy, R., & Vadigepalli, R. (2009). Structural properties of gene regulatory networks: Definitions and connections. *IEEE/ACM Transactions on Computational Biology and Bioinformatics*, *6*(1), 158–170. doi:10.1109/TCBB.2007.70231 PMID:19179709

Newman, M. E. J. (2003). The structure and function of complex networks. *SIAM Review*, *45*(2), 167–256. doi:10.1137/S003614450342480

Orchard, S., Ammari, M., Aranda, B., Breuza, L., Briganti, L., Broackes Carter, F., & Hermjakob, H. et al. (2014). The MIntAct project - IntAct as a common curation platform for 11 molecular interaction databases. *Nucleic Acids Research*, *42*(D1), D358–D363. doi:10.1093/nar/gkt1115 PMID:24234451

Ott, S., Hansen, A., Kim, S.-Y., & Miyano, S. (2005). Superiority of network motifs over optimal networks and an application to the revelation of gene network evolution. *Bioinformatics (Oxford, England)*, *21*(2), 227–238. doi:10.1093/bioinformatics/bth484 PMID:15377501

Park, J., Karplus, K., Barrett, C., Hughey, R., Haussler, D., Hubbard, T., & Chothia, C. (1998). Sequence comparisons using multiple sequences detect three times as many remote homologues as pairwise methods. *Journal of Molecular Biology*, *284*(4), 1201–1210. doi:10.1006/jmbi.1998.2221 PMID:9837738

Pawson, T., & Nash, P. (2003). Assembly of cell regulatory systems through protein interaction domains. *Science*, *300*(5618), 445–452. doi:10.1126/science.1083653 PMID:12702867

Pe'er, D., Regev, A., Elidan, G., & Friedman, N. (2001). Inferring subnetworks from perturbed expression profiles. *Bioinformatics (Oxford, England)*, *17*(Suppl 1), S215–S224. doi:10.1093/bioinformatics/17.suppl_1.S215 PMID:11473012

Penfold, C. A., Wild, D. L. (2011). How to infer gene networks from expression profiles, revisited. *Interface Focus*, 1.

Peng, H., Long, F., & Ding, C. (2005). Feature selection based on mutual information: Criteria of max-dependency, max-relevance, and min-redundancy. *IEEE Transactions on Pattern Analysis and Machine Intelligence*, *27*(8), 1226–1238. doi:10.1109/TPAMI.2005.159 PMID:16119262

Perrin, B. E., Ralaivola, L., Mazurie, A., Bottani, S., Mallet, J., & d'Alché Buc, F. (2003). Gene networks inference using dynamic Bayesian networks. *Bioinformatics (Oxford, England)*, *19*(Suppl 2), 138–148. doi:10.1093/bioinformatics/btg1071

Pinney, J. W., Westhead, D. R., & McConkey, G. A. (2003). Petri Net representations in systems biology. *Biochemical Society Transactions*, *31*(6), 1513–1515. doi:10.1042/bst0311513 PMID:14641101

Pournara, I., & Wernisch, L. (2004). Reconstruction of gene networks using Bayesian learning and manipulation experiments. *Bioinformatics (Oxford, England)*, *20*(17), 2934–2942. doi:10.1093/bioinformatics/bth337 PMID:15180938

Poustelnikova, E., Pisarev, A., Blagov, M., Samsonova, M., & Reinitz, J. (2004). A database for management of gene expression data in situ. *Bioinformatics (Oxford, England)*, *20*(14), 2212–2221. doi:10.1093/bioinformatics/bth222 PMID:15059825

Prasad, T. S. K., Goel, R., Kandasamy, K., Keerthikumar, S., Kumar, S., Mathivanan, S., & Pandey, A. (2009). Human Protein Reference Database - 2009 update. *Nucleic Acids Research*, *37*(Database), D767–D772. doi:10.1093/nar/gkn892 PMID:18988627

Prifti, E., Zucker, J., Clement, K., & Henegar, C. (2008). Funnet: An integrative tool for exploring transcriptional interactions. *Bioinformatics (Oxford, England)*, *24*(22), 2636–2638. doi:10.1093/bioinformatics/btn492 PMID:18799481

Przytycka, T., & Kim, Y. A. (2010). Network integration meets network dynamics. *BMC Biology*, *8*(1), 48. doi:10.1186/1741-7007-8-48 PMID:20513250

Pudil, P., Novovicová, J., & Kittler, J. (1994). Floating search methods in feature-selection. *Pattern Recognition Letters, 15*(11), 1119–1125. doi:10.1016/0167-8655(94)90127-9

Raser, J. M., & O'Shea, E. K. (2005). Noise in gene expression: Origins, consequences, control. *Science, 309*(5743), 2010–2013. doi:10.1126/science.1105891 PMID:16179466

Ray, S. S., Bandyopadhyay, S., & Pal, S. (2009). Combining multisource information through functional-annotation-based weighting: Gene function prediction in yeast. *Biomedical Engineering. IEEE Transactions on, 56*(2), 229–236.

Ray, S. S., Bandyopadhyay, S., & Pal, S. (2012). A Weighted power framework for integrating multi-source information: Gene function prediction in yeast. *IEEE Transactions on Bio-Medical Engineering, 59*(4), 1162–1168. doi:10.1109/TBME.2012.2186689 PMID:22318478

Ristevski, B. (2013). A survey of models for inference of gene regulatory networks. *Nonlinear Analysis: Modelling and Control, 18*(4), 444–465.

Ruklisa, D., Brazma, A., & Viksna, J. (2005). Reconstruction of gene regulatory networks under the Finite State Linear Model. *Genome Informatics, 16*(1), 225–236. PMID:16901105

Santos, D. C., Camargo, R. Y., Martins, D. C., Jr., Song, S. W., Rozante, L. C. S., & Borelli, F. F. (2015). A multi-GPU hitting set algorithm for GRNs inference. *15th IEEE/ACM International Symposium on Cluster, Cloud and Grid Computing (CCGrid)*. Shenzhen, China: IEEE/ACM.

Schena, M., Shalon, D., Davis, R. W., & Brown, P. O. (1995). Quantitative monitoring of gene expression patterns with a complementary DNA microarray. *Science, 270*(5235), 467–470. doi:10.1126/science.270.5235.467 PMID:7569999

Schlitt, T., & Brazma, A. (2007). Current approaches to gene regulatory network modelling. *BMC Bioinformatics, 59*(Suppl 6). PMID:17903290

Seok, J., Kaushal, A., Davis, R., & Xiao, W. (2010). Knowledge-based analysis of microarrays for the discovery of transcriptional regulation relationships. *BMC Bioinformatics, 11*(Suppl 1), S8. doi:10.1186/1471-2105-11-S1-S8 PMID:20122245

Shav Tal, Y., Singer, R. H., & Darzacq, X. (2004). Imaging gene expression in single living cells. *Nature Reviews. Molecular Cell Biology, 5*(10), 855–862. doi:10.1038/nrm1494 PMID:15459666

Shin, A., & Iba, H. (2003). Construction of genetic network using evolutionary algorithm and combined fitness function. *Genome Informatics*, *14*, 94–103. PMID:15706524

Shmulevich, I., & Aitchison, J. D. (2009). Deterministic and stochastic models of gene regulatory networks. *Methods in Enzymology*, *467*, 335–356. doi:10.1016/S0076-6879(09)67013-0 PMID:19897099

Shmulevich, I., & Dougherty, E. R. (2007). *Genomic Signal Processing*. Princeton University Press. doi:10.1515/9781400865260

Shmulevich, I., Dougherty, E. R., Kim, S., & Zhang, W. (2002). Probabilistic Boolean networks: A rule-based uncertainty model for gene regulatory networks. *Bioinformatics (Oxford, England)*, *18*(2), 261–274. doi:10.1093/bioinformatics/18.2.261 PMID:11847074

Smadbeck, P., & Kaznessis, Y. N. (2014). Solution of Chemical Master Equations for Nonlinear Stochastic Reaction Networks. *Curr Opin Chem Eng*, *5*, 90–95. doi:10.1016/j.coche.2014.07.001 PMID:25215268

Snoep, J. L., & Westerhoff, H. V. (2005). From isolation to integration, a systems biology approach for building the Silicon Cell. *Topics in Current Genetics*, *13*, 13–30. doi:10.1007/b106456

Soinov, L. A., Krestyaninova, M. A., & Brazma, A. (2003). Towards reconstruction of gene networks from expression data by supervised learning. *Genome Biology*, *4*(1), R6. doi:10.1186/gb-2003-4-1-r6 PMID:12540298

Stark, C., Breitkreutz, B., Reguly, T., Boucher, L., Breitkreutz, A., & Tyers, M. (2006). BioGRID: A general repository for interaction datasets. *Nucleic Acids Research*, *34*(90001), D535–D539. doi:10.1093/nar/gkj109 PMID:16381927

Steinbach, B., & Posthoff, C. (2012). Sources and Obstacles for Parallelization-a Comprehensive Exploration of the Unate Covering Problem Using Both CPU and GPU. In J. Astola, M. Kameiama, M. Lukac, & M. Stankovic (Eds.), GPU Computing with Applications in Digital Logic. Tampere International Center for Signal Processing.

Strogatz, S. H. (2001). Exploring complex networks. *Nature*, *410*(6825), 268–276. doi:10.1038/35065725 PMID:11258382

Styczynski, M. P., & Stephanopoulos, G. (2005). Overview of computational methods for the inference of gene regulatory networks. *Engineering*, *29*(3), 519–534.

Swain, M., Hunniford, T., Dubitsky, W., Mandel, J., & Palfreyman, N. (2005). Reverse-engineering gene-regulatory networks using evolutionary algorithms and grid computing. *Journal of Clinical Monitoring and Computing*, *19*(4-5), 329–337. doi:10.1007/s10877-005-0678-x PMID:16328947

Theodoridis, S., & Koutroumbas, K. (2008). *Pattern Recognition* (4th ed.). Academic Press.

Thomas, R. (1998). Laws for the dynamics of regulatory networks. *The International Journal of Developmental Biology*, *42*(3), 479–485. PMID:9654035

Thomas, R., Thieffry, D., & Kaufman, M. (1995). Dynamical behaviour of biological regulatory networks-I. Biological role of feedback loops and practical use of the concept of the loop-characteristic states. *Bulletin of Mathematical Biology*, *57*(2), 247–276. doi:10.1007/BF02460618 PMID:7703920

Troyanskaya, O. G. (2005). Putting microarrays in a context: Integrated analysis of diverse biological data. *Briefings in Bioinformatics*, *6*(1), 34–43. doi:10.1093/bib/6.1.34 PMID:15826355

Ulitsky, I., & Shamir, R. (2007). Identification of functional modules using network topology and high-throughput data. *BMC Systems Biology*, *1*(1), 8. doi:10.1186/1752-0509-1-8 PMID:17408515

Velculescu, V. E., Zhang, L., Vogelstein, B., & Kinzler, K. W. (1995). Serial Analysis of Gene Expression. *Science*, *270*(5235), 484–487. doi:10.1126/science.270.5235.484 PMID:7570003

Vicente, F. F. R., & Lopes, F. M. (2014). SFFS-WS: A feature selection algorithm exploring the smallworld properties of GNs. In Pattern Recognition in Bioinformatics, Proceedings, (LNCS), (vol. 8626, pp. 60-71). Berlin: Springer.

Vidal, M. (2005). Interactome modeling. *FEBS Letters*, *579*(8), 1834–1838. doi:10.1016/j.febslet.2005.02.030 PMID:15763560

Voet, D., Voet, J., & Pratt, C. W. (2005). *Fundamentals of Biochemistry: Life at the Molecular Level* (2nd ed.). John Wiley & Sons.

Wang, Z., Gerstein, M., & Snyder, M. (2009). RNA-Seq: A revolutionary tool for transcriptomics. *Nature Reviews. Genetics*, *10*(1), 57–63. doi:10.1038/nrg2484 PMID:19015660

Werhli, A. V., Grzegorczyk, M., & Husmeier, D. (2006). Comparative evaluation of reverse engineering gene regulatory networks with relevance networks, graphical gaussian models and bayesian networks. *Bioinformatics (Oxford, England)*, *22*(20), 2523–2531. doi:10.1093/bioinformatics/btl391 PMID:16844710

Werhli, A. V., & Husmeier, D. (2008). Gene regulatory network reconstruction by bayesian integration of prior knowledge and/or different experimental conditions. *Journal of Bioinformatics and Computational Biology*, *6*(3), 543–572. doi:10.1142/S0219720008003539 PMID:18574862

Xiao, Y. (2009). A tutorial on analysis and simulation of boolean gene regulatory network models. *Current Genomics*, *10*(7), 511–525. doi:10.2174/138920209789208237 PMID:20436877

Yu, H., Paccanaro, A., Trifonov, V., & Gerstein, M. (2006). Predicting interactions in protein networks by completing defective cliques. *Bioinformatics (Oxford, England)*, *22*(7), 823–829. doi:10.1093/bioinformatics/btl014 PMID:16455753

Zhang, Y., Quian, M., Ouyang, Q., Deng, M., Li, F., & Tang, C. (2006, July). Stochastic Model of Yeast Cell-Cycle Network. *Physica D. Nonlinear Phenomena*, *219*(1), 35–39. doi:10.1016/j.physd.2006.05.009

ENDNOTE

1. SFFS-MR is an algorithm which works similarly to the SFFS algorithm, except that the SFFS-MR search starts by considering multiple roots (the best feature and some worst individual features, relying on the intrinsically multivariate prediction phenomenon which may occur), instead of starting only with the best feature as SFFS does.

Chapter 2
Relationships between Models of Genetic Regulatory Networks with Emphasis on Discrete State Stochastic Models

Randip Pal
Texas Tech University, USA

ABSTRACT

Genetic Regulatory Networks (GRNs) represent the interconnections between genomic entities that govern the regulation of gene expression. GRNs have been represented by various types of mathematical models that capture different aspects of the biological system. This chapter discusses the relationships among the most commonly used GRN models that can enable effective integration of diverse types of sub-models. A detailed model in the form of stochastic master equation is described, followed by it coarse-scale and deterministic approximations in the form of Probabilistic Boolean Networks and Ordinary Differential Equation models respectively.

INTRODUCTION

Genetic Regulatory Networks (GRNs) represent the interconnections between genomic entities that govern the regulation of gene expression. Since biological regulatory networks are extremely detailed with numerous interactions, a single mathematical model to represent the whole biological regulatory system is gener-

DOI: 10.4018/978-1-5225-0353-8.ch002

ally not feasible. Depending on the purpose of modeling, the mathematical model representing the GRN brings in a level of abstraction. The focus of the modeling can be capturing interactions between RNA expressions, protein-protein interactions or interactions between metabolites. Usually, only parts of the regulome (i.e. genes, proteins and metabolites involved in gene regulation) such as transcription factors, enhancers, microRNA etc. are made explicit in a mathematical model of a GRN.

To provide further background on genetic regulatory network modeling, we present broad classifications of the models that are commonly used to capture the relationships between genomic entities. GRN models usually belong to either static or dynamical models. Static models are concerned with providing the topology of the network (such as the connectivity of various genomic entities) without attempting to model the dynamics of the interactions. Dynamic models on the other hand not only capture the connectivity of the network entities but also provide the time dependent functional relationships that govern the interactions. Commonly used pathway models as presented in databases such as KEGG and StringDB belong to the category of static models as they primarily capture the connectivity structure without the network dynamics. Examples of dynamic models include ordinary differential equation models to model the average concentration of mRNA expressions (Kampen, 1991) or Dynamic Bayesian Network (Murphy, 2002) models. Based on the model complexity involving inference of connectivity and time dependent relationships, the experimental data requirements for elucidating dynamic models are significantly higher as compared to static models.

To further explore various levels of time dependent behavior of genes or proteins, dynamic network models can be essentially categorized as discrete where the amount of genomic entity is modeled as discrete values or continuous where the genomic entities are models as continuous variables. Since the dynamics of biological entities consist of discrete changes based on the number of molecules created or destroyed, a fine-scale discrete model can potentially faithfully capture the dynamics of the genomic entities. Note that we have categorized the dynamic models as discrete or continuous based on the representation of the genomic entity but a similar categorization can be applied based on discrete or continuous representation of time. Some models such as stochastic master equations (also commonly known as Chemical Master Equations) incorporate time as a continuous quantity whereas models like Boolean Networks update states at discrete time intervals. The next level of characterizations is based on whether the models considered are stochastic or deterministic. A number of studies have shown that the generation of mRNA or protein expression is stochastic in nature (Arkin, Ross, & McAdams, 1998) and thus a stochastic model is suitable to capture the true change in state probability distribution with time. However, available experimental data may limit the inference of the parameters of a stochastic model. For instance, common genomic characterizations using microarrays provide an average observation of the mRNA expressions and are not suitable for inferring the parameters of a detailed stochastic model.

Thus, deterministic models such as differential equation models are often used to represent the average behavior of the biological system (De Jong, 2001) or in other words modeling moments of the distribution rather than the detailed distribution. Dynamic models can be further classified as detailed fine-scale models or coarse-scale models where the detail refers to the level of discretization involved in modeling the gene or protein expression. For instance, our interest may be in modeling the up or down regulation of genes and thus a coarse-scale Boolean Network (Kauffman S. A., 1993; Kauffman S., 1969) with ON and OFF states for each gene will be adequate. However, if our interest is in capturing the individual mRNA molecule generation or degradation, a detailed fine-scale model such as stochastic master equations (Gillespie D. T., 1977; Pal & Bhattacharya, 2010) with an enormous state space will be required. The above categorizations are presented in a graphical form in Figure 1.

Based on the above discussion, we observe that fine-scale stochastic discrete dynamical models such as stochastic master equations can provide a detailed representation of the dynamic interactions but maybe limited by the enormous com-

Figure 1. Categorization of commonly used models of genetic regulatory networks (GRNs)

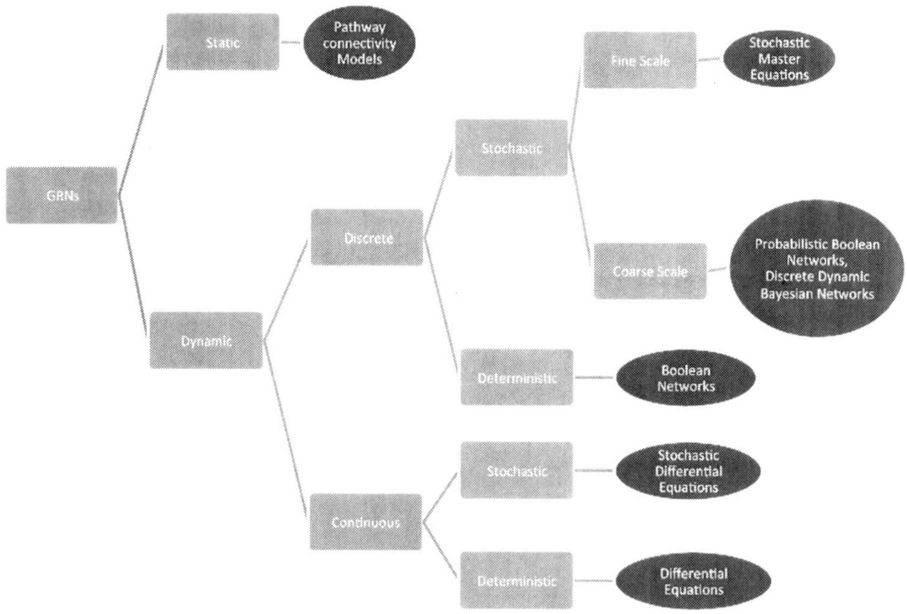

putational complexity involved in its simulation and huge data requirements for estimation of its parameters. This chapter will present a review of stochastic master equation modeling followed by some of its low complexity approximations with focus on reduced state discrete time Probabilistic Boolean Networks. The emphasis will be on the properties maintained by the reduced order models as compared to the original model.

Chemical Master Equation Models

Chemical master equations (CMEs) are considered to be one of the detailed fundamental models for representing dynamic genomic interactions (Pal & Bhattacharya 2010). CMEs model the creation and decay of discrete molecules of involved genomic entities in a stochastic fashion. CME based models have been shown to accurately model processes in bacteriophage lambda that control the switchover from lysogenic to lytic modes of growth (Shea & Ackers, 1985) or model the dynamics of phosphorylation-dephosphorylation cycles with autophosphorylation (Bishop & Qian, 2010) .

To explain the CME models for genetic regulatory networks, let us consider a system with n molecular species and m different reaction channels where the state of the system is defined by $x = [\varphi_1, \varphi_2, \ldots, \varphi_n]$, $x \in N^n$ is a vector of integers representing a specific population of each of the n molecular species. For such a system, given the probability density vector $p(x, t)$ at time t, the probability that the system will be in the state x at time $t + dt$ is given by

$$p(x;t+dt) = p(x;t)(1 - \sum_{\mu=1}^{m} a_\mu(x)dt) + \sum_{\mu=1}^{m} p(x - v_\mu;t)a_\mu(x - v_\mu)dt \tag{1}$$

where $a_\mu(x)dt$ denotes the probability that the μth reaction will happen in a time step of length dt and v_μ is the stoichiometric transition vector. From Equation 1, the differential equation known as *Chemical Master Equation* (CME) (Gillespie D. T., 1977) can be derived:

$$\dot{p}(x;t) = -p(x;t)\sum_{\mu=1}^{m} a_\mu(x) + \sum_{\mu=1}^{m} p(x - v_\mu;t)a_\mu(x - v_\mu) \tag{2}$$

By considering all the reactions beginning or ending at state x, the time derivative of the probability density of state x can be written in vector form as (Munsky & Khammash, 2006):

$$p(x;t)=[p(x;t)......p(x-v_m;t)]\begin{bmatrix} -\sum_{\mu=1}^{m} a_\mu(x) \\ ...... \\ a_m(x-v_m) \\ \end{bmatrix} \tag{3}$$

As n is finite, the number of different possible vectors for x is countable. Thus, we can map all the possible x's $\in N^n$ to a sequence $x_1, x_2, \cdots$ of elements in N. Let us represent that enumeration as X:= $[x_1, x_2, \cdots]$ and rewrite Equation 3 as

$$\dot{P}(X;t) = P(X;t)A \tag{4}$$

where $P(X;\ t)=[p(x_1,\ t),p(x_2,\ t)...]$ is the complete probability density state vector at time t and A is the state reaction matrix. The matrix A has the following properties: it is independent of t, all of its diagonal elements are non-positive, and all of its off-diagonal elements are non-negative and all the rows sum to exactly zero. For the case of finite number of reachable states, the solution can be computed as:

$$P(X;t) = P(X;0)e^{At} \tag{5}$$

An approach to reduce the complexity of simulation of the above solution using Kronecker product analysis and Zassenhaus formula have been considered in (Caglar & Pal, 2013).

CME Model Simulation

The complexity of CME models can be tremendous to allow for direct computation of the analytical solution of the probability density state vector. For instance, n = 5 molecular species with number of molecules ranging from 0 to 99 will have a total of $100^5=10^{10}$ states whose probabilities need to be computed at each time duration. Thus, Monte Carlos based techniques are often utilized to generate the time varying probability density state vectors. The most commonly used technique is known as Gillespie's Stochastic Simulation Algorithm (SSA) (Gillespie D. T., 1977).

The SSA approach is statistically exact in the sense that infinite number of simulations will produce the same distribution as the analytical solution of Equation 2 (Székely Jr. & Burragea, 2014).

The SSA approach simulates multiple trajectories based on the governing CME by selecting the time when the next reaction will happen and what the reaction will be.

SSA for generating a trajectory can be described using the following algorithm (Higham 2008)

1. Evaluate $a_\mu(x)$ at time t for $\mu=1$ to m.
2. Draw random numbers r_1 and r_2 based on uniform (0 1).
3. Reaction j is selected that satisfies $\frac{\sum_{\mu=1}^{j-1} a_\mu(x)}{\sum_{\mu=1}^{m} a_\mu(x)} \le r_1 < \frac{\sum_{\mu=1}^{j} a_\mu(x)}{\sum_{\mu=1}^{m} a_\mu(x)}$
4. Time step for next reaction $\tau = \frac{-\ln(r_2)}{\sum_{\mu=1}^{m} a_\mu(x)}$
5. Update time $t=t+\tau$ and state $x=x+v_j$
6. Repeat steps 1-5 until stopping criteria is met.

The stopping criteria is often an upper limit on time or the number of iterations. Readers are referred to (Gillespie D. T., 1977; Higham, 2008) for further details on SSA algorithm. Software for simulation of SSA are also available such as COPASI (Hoops & et.al, 2006), STOCKS (Kierzek, 2002), BioNets (Adalsteinsson, McMillen, & Elston, 2004) and StochKit2 (Sanft, Wu, Roh, Fu, & Petzold, 2011).

The SSA algorithm is simple to understand and implement but has high computational complexity. Tau Leaping (Gillespie D. T., 2000) is an approximation approach to SSA that considers all reactions occurring in a time interval τ before updating the propensity functions $a_\mu(x)$.

The simulation of CME models can also be facilitated by avoiding inclusion of molecular states that will never be reached or will likely have infinitesimal probabilities. (Cao & Liang, 2008) presents an optimal algorithm for simulating the probability landscape for networks with small copy numbers that avoids attempting infeasible state transitions and provides a concise state transition matrix devoid of redundant information. (Munsky & Khammash, 2006) approximates the solution of CME model by truncating the state space to remove states with minimal probabilities and provides a bound on the difference between the truncated space solution and the true solution. (Cao, Lu, & Liang, 2010) considers a method termed discrete chemical master equation (dCME) that is able to directly compute the full steady state probability distribution of lysogeney maintenance network in phage lambda.

MODEL APPROXIMATIONS

CME to Stochastic Differential Equation

An approximation to the discrete stochastic model in the form of a stochastic DE known as *chemical Langevin Equation (CLE)* (Gillespie D. T., 2000) is often used to make the solution analytically and numerically tractable. The approximation can be represented by the following stochastic DE (Cai & Wang, 2007):

$$x(t+dt) = x(t) + \sum_{\mu=1}^{m} v_{\mu} a_{\mu}(x(t))dt + \sum_{\mu=1}^{m} v_{\mu} \sqrt{a_{\mu}(x(t))} N_{\mu}(t) \tag{5}$$

where $N_{\mu}(t)$ for $\mu = 1, \ldots m$ are independent Gaussian random variables with zero mean and unit variance. The conditions for validity of the approximation require that an infinitesimal time interval *dt* is available under which 1) there is no significant change in the propensity functions and 2) the expected number of occurrences of each reaction in *dt* is significantly greater than one. As shown in (Gillespie D. T., 2000), the first two moments of CME and CLE can be related using the truncation of chemical Kramers-Moyal equations (CKME). CKME can be considered an approximation of CME based on Taylor Series expansion and truncation of CKME will result in chemical Fokker-Plank equation representing the CLE. For details, readers are referred to (Gillespie D. T., 2000).

CME to Differential Equation

The relationship between CME and deterministic Differential Equation (DE) models have been well studied analytically by multiple researchers primarily from the perspective of similarity of the average predictive behavior of the two models (Kurtz, 1972; Kampen, 1991). Since the topic of CME to DE approximations have been well studied, this section will illustrate the relationship from a very basic perspective with simulation examples. For detailed analysis, readers are referred to (Kurtz, 1972; Kampen, 1991).

Example

In this example, we will apply CME to DE approximation technique to a generic switch (similar to a genetic toggle switch model without the detailed gene regulation) where there are two competing proteins U_1 and U_2, each of which inhibits the transcription of the other. The decay reactions $U_1 \rightarrow \varnothing$ and $U_2 \rightarrow \varnothing$ have propensities $c\Psi_1$ and $c\Psi_2$ respectively. The transcription of new copies of the proteins are guided by the reactions $\varnothing \rightarrow U_1$ and $\varnothing \rightarrow U_2$ with propensities b/(b +Ψ_2) and b/(b +Ψ_1) respectively. Our starting Stochastic Master equation model has parameter values $b = 4$ and $c = 0.2$. The corresponding stochastic master equation is

$$p(x;t+dt)=p(x;t)(1-\sum_{\mu=1}^{m} a_\mu(x)dt)+\sum_{\mu=1}^{m} p(x-v_\mu;t)a_\mu(x-v_\mu)dt$$

with $m = 4$ and $v_1 = [-1, 0]$, $v_2 = [1, 0]$, $v_3 = [0, -1]$, $v_4 = [0, 1]$, $a_1([x_1, x_2]) = cx_1$, $a_2([x_1, x_2]) = b/(b+x_2)$, $a_3([x_1, x_2]) = cx_2$, $a_4([x_1, x_2]) = b/(b+x_1)$.

Our corresponding differential equation approximation is:

$$\frac{dy_1(t)}{dt}=\frac{b}{b+y_2(t)}-cy_1(t)$$

$$\frac{dy_2(t)}{dt}=\frac{b}{b+y_1(t)}-cy_2(t)$$

where $y_1(t)$ and $y_2(t)$ approximates $E_{x1}(x_1(t))$ and $E_{x2}(x_2(t))$ respectively. Figure 2 shows

*Figure 2. The average protein 1 expression simulated using stochastic master equation (red *) and Differential Equation approximation (blue dots) for the original genetic toggle switch with parameters b = 4 and c = 0.2*

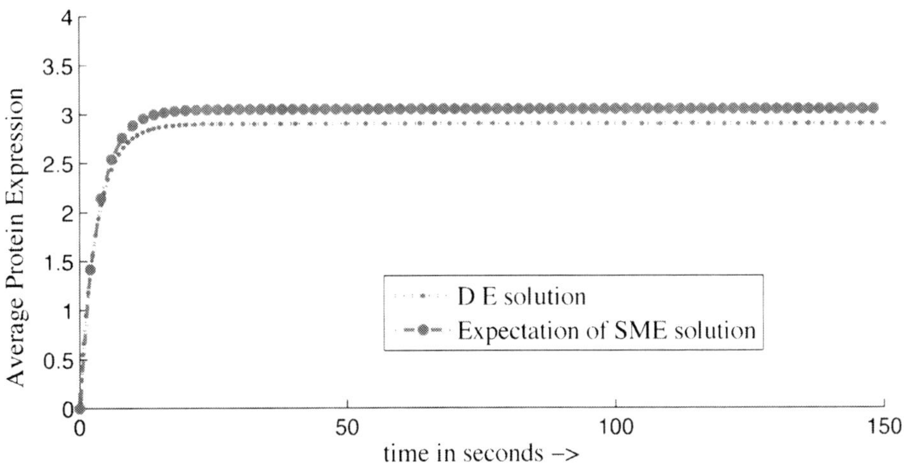

$E_{x1}(x_1(t))$ (red stars) and $y_1(t)$ (blue line) for a starting distribution of $p(X) = [1, 0, 0 \cdots, 0]$ i.e. the number of protein molecules for both protein is zero for the initial state. It can be seen from Figure 4 that the approximation is very close to the expectation of the actual solution. We can solve for the steady state value of y_1 and y_2 analytically based on the equations $dy_1/dt=0$ and $dy_2/dt=0$.

If the steady state values of y_1 and y_2 are represented by y_{1s} and y_{2s}, then

$$\frac{b}{b+y_{2s}} - cy_{1s} = \frac{b}{b+y_{1s}} - cy_{2s} = 0 \tag{10}$$

Solving the equation gives $y_{1s} = \dfrac{-b+\sqrt{b^2+4b/c}}{2}$ which produces $y_{1s} = 2.9$ for the specific parameters of $b = 4$ and $c = 0.2$. The steady state value of 2.9 is also noticeable in Figure 4 for the simulated differential equation. Figure 3 shows the steady state marginal probability distribution of protein 1 simulated using the stochastic master equation. The expectation of this distribution 3.05 is very close to the analytically calculated differential equation steady state 2.9.

To study the effect of the control policy designed on the simpler approximate differential equation model being applied to the original stochastic master equation, we consider a control problem of maintaining the average expression of protein 1 to be around the value *10*. The control input is assumed to decrease the decay of the proteins by decreasing the value of the parameter c to *0.01*. Thus we have two systems, the original with $b = 4$ and $c = 0.2$ and the other corresponding to when control is switched on with parameters $b = 4$ and $c = 0.01$. The response of protein

Figure 3. The steady state probability distribution of Protein 1 for the original genetic toggle switch

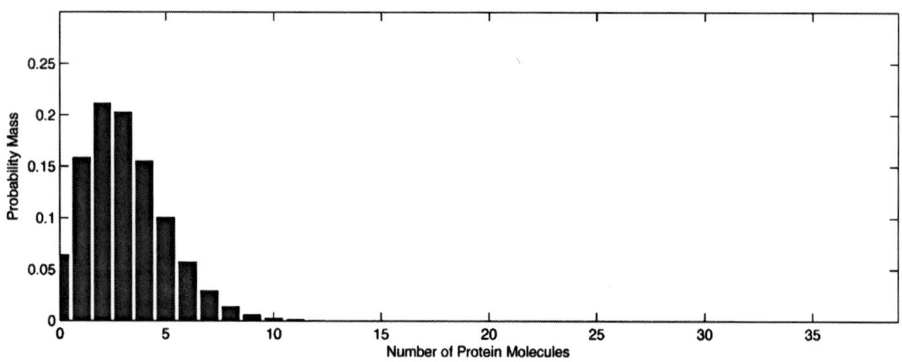

*Figure 4. The average protein expression simulated using stochastic master equation (red *) and Differential Equation approximation (blue dots) for the drug treated genetic toggle switch with parameters b = 4 and c = 0.01*

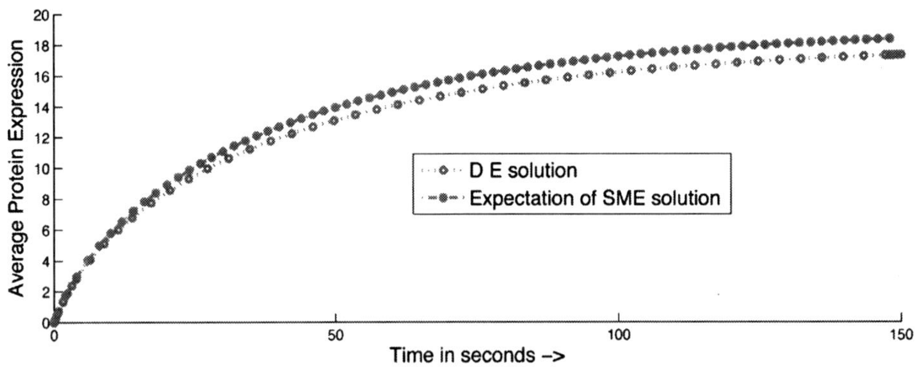

1 of the network corresponding to parameters $b = 4$ and $c = 0.01$ is shown in Figure 4. The red stars denote the expectation of the distribution of protein 1 at every 2 second interval corresponding to the simulation of the stochastic master equation while the blue dots represent the output of the approximate differential equation model. The similarity of both the curves shows the closeness of the approximation. The value of the steady state corresponding to the DE model can be calculated by solving Equation 10 with $b = 4$ and $c = 0.01$ producing a value of $y_{1s} = 18.1$. The expectation of the steady state distribution calculated from the simulated CME model produces a similar value of *19.1*. The steady state probability distribution of protein 1 corresponding to the CME model is plotted in Figure 5.

Figure 5. The steady state probability distribution of Protein 1 for the drug treated genetic toggle switch

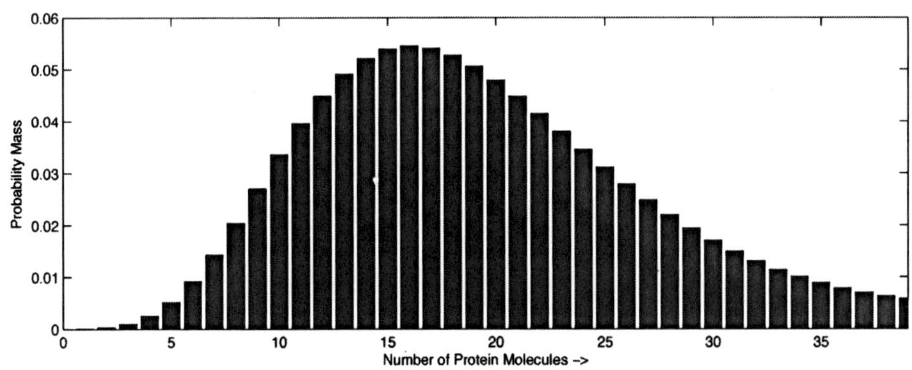

As stated earlier, our control objective is to maintain the average expression of protein 1 to be around the value 10. We design a simple on off controller with deadzone based on the differential equation model which produces the following control policy: use system 1 (original system, b=4, c=0.2) if the average protein expression is above 10.5 and use system 2 (system when control is on, b=4, c=0.01) if the average protein expression is below 9.5. If the protein expression is between 9.5 and 10.5, maintain the earlier control policy. This is done to avoid indeterminate switching around the reference value of 10. The result of this control policy applied to the CME models is shown in Figure 6.

We notice that the control objective has been almost achieved with the average expression staying between 7.5 to 10.5. The variation could have been reduced if we apply control in much finer time interval than the current time interval of 2 seconds. In actual translational medicine, this time will be dependent on a number of factors like technical limitations on the sampling rate and application of control. Another way to approach the control can be by solving for the value of c producing the desired reference output and change the drug amount to achieve that specific value of c. In this case the drug is decreasing the degradation of the protein but not affecting the generation. We may be aware of the relationship between drug amount and parameter c but do not have the analytical solution relating the steady state output to the parameter c. We have to simulate the CME model for various values of c to arrive at the relationship between c and steady state expression. Our differential equation approximation can analytically provide the value of c that produces the desired steady state output. Based on our desired response of 10, we get the required value of $c = \sqrt{\dfrac{b^2}{(20+b)^2 - b^2}}$ to be *0.0286*. Figure 8 shows the average

Figure 6. The average protein expression simulated using stochastic master equation after application of the on-off control policy designed using the differential equation approximation models

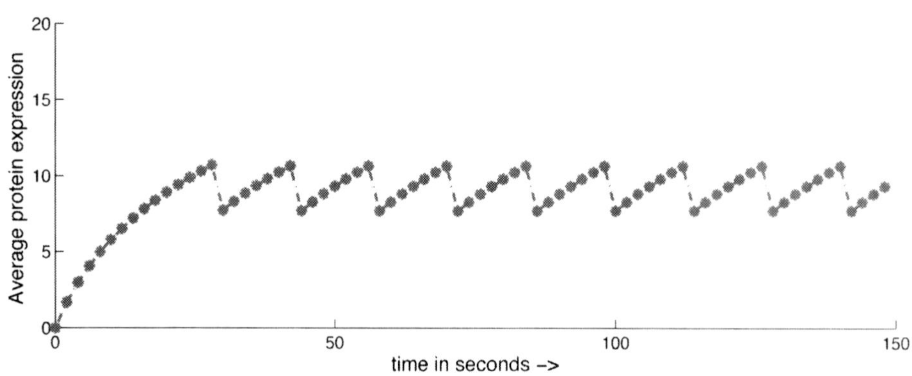

Figure 7. The steady state probability distribution of Protein 1 for the controlled genetic toggle switch

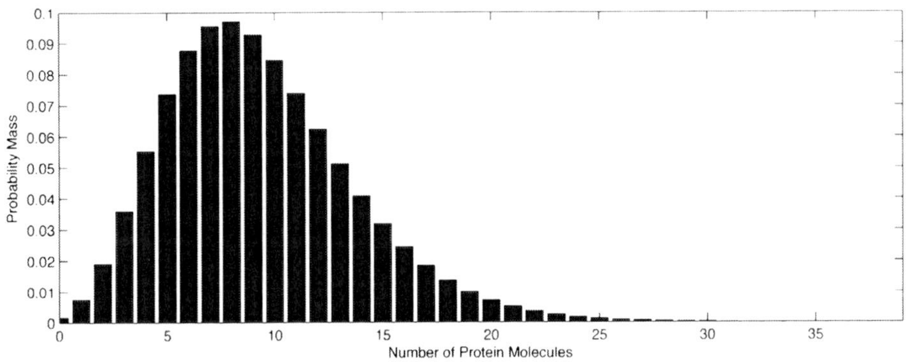

*Figure 8. The average protein expression simulated using stochastic master equation (red *) and Differential Equation approximation (blue x) for the controlled genetic toggle switch with parameters b = 4 and c = 0.0286*

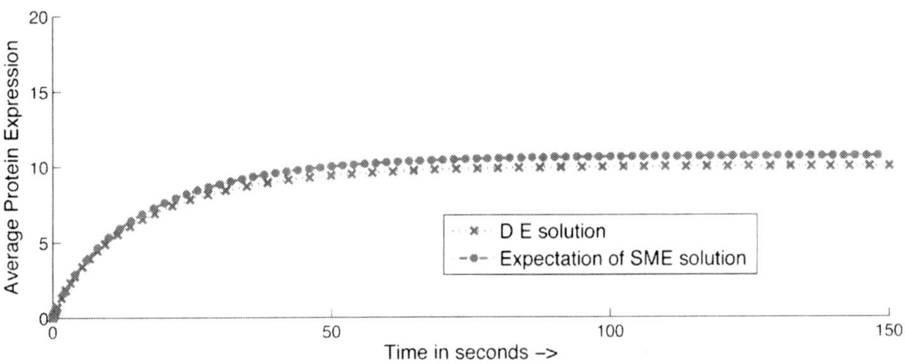

protein expression simulated using stochastic master equation and differential equation model for protein 1 when the parameter c is maintained at *0.286*. The steady state output is very close to the desired output of *10*. In terms of numerical values, the expectation of the CME output for protein 1 at steady state is *10.69*.

CME to Probabilistic Boolean Networks

In an CME model, If we consider the number of possible x_i's to be finite, then the exact solution to the CME can be obtained through the exponential equation

$P(X; t) = P(X, 0)e^{At}$. This kind of situation can arise in biological systems where the number of mRNA/Protein molecules that can be generated are bounded. If for systems, the dimension of X is infinite, we can apply the Finite State Projection approach (Munsky & Khammash, 2006) to arrive at a finite truncation of the state space. Henceforth, we will consider a CME with finite number of states M and our purpose is to transform that model to a smaller scale Probabilistic Boolean Network (PBN) model while maintaining some characteristics of the CME model.

The PBN model (Pal R., 2010) can be considered as a stochastic version of the Boolean Network (BN) (Kauffman S. A., 1993). A BN $B = (V, f)$ on n genes is defined by a set of nodes/genes $V = [x_1,..., x_n]$ and a vector of Boolean functions $\boldsymbol{f} = [f_1,...,f_n]$. The variable $x_i \in [0,1]$ represents the expression level of gene i, with 1 representing high and 0 representing low expression. The vector $\boldsymbol{f}$ represents the regulatory rules between genes. At every time step $t+1$, the value of x_i is predicted by the values of a set W_i of genes at the previous time step t, based on the regulatory function f_i, i.e. $x_i(t+1) = f_i(x_{i_1}(t),......,x_{i_{k_i}}(t))$. The set of genes $W_i = \{x_{i_1},......,x_{i_{k_i}}\}$ is called the predictor set of x_i, and the function f_i is called the predictor function of x_i. The pairs (x_i, W_i), $i = 1, ..., n$ induce a digraph G with edges $x_{i_j} \rightarrow x_i$ representing the structural dependencies among the genes. A state of B is a vector $s = [x_1,..., x_n] \in [0, 1]^n$. All of the possible states of the Boolean network comprise its state space S which combined with the functions in $\boldsymbol{f}$ produces a digraph Γ called the state transition diagram of B. Γ represents the dynamics of the Boolean network and can be identified with a $2^n \times 2^n$ matrix P_n with rows and columns indexed by the states in B and entries $p_{ij} = 1$ or 0 if there is a transition from the state $s_i \rightarrow s_j$ in S.

Since constraints on the experimental data may limit the inference of a unique Boolean Network, the uncertainty in estimation along with the stochastic nature of genomic regulation is captured in the form of a PBN model where each gene can have multiple predictor functions. A *context-sensitive* PBN $A = A^{q,p}(V, \boldsymbol{F}, C)$ is defined by a set of nodes/genes $V = \{x_1,..., x_n\}$, a set of vector-valued Boolean functions $\boldsymbol{F} = \{\boldsymbol{f}_1,...,\boldsymbol{f}_r\}, \boldsymbol{f}_j: \{0,1\}^n \rightarrow \{0,1\}^n, j = 1,..., r$ called *realizations* or *network functions*, a list of selection probabilities $C = \{c_1,..., c_r\}$ for the corresponding realizations, the gene mutation/flipping probability p, and the realization switching probability q (Pal R., 2010). Updating the values of all genes in the network at time t is done synchronously according to the components of the currently used network function, and then the process is repeated. The choice of which network function $\boldsymbol{f}_j$ to apply is governed by a selection procedure. Specifically, at each time point t a random decision is made as to whether to switch the network function for the next transition, with a probability q of a switch being a system parameter. If a decision is made to switch the network function, then a new realization is chosen from among all of the possible realizations $f_j \in F$ of A, according to their individual selection probabilities $c_j \in C$. In other words, each network function f_j represents a deterministic BN B_j and the PBN behaves as a fixed BN until a random decision (with probability of q) is made to change the network function according to the probabilities $\{c_1,\ldots, c_r\}$ from among $\{f_1,\ldots,f_r\}$. In addition to the network switching and selection in the PBN model, there is mechanism which models random gene mutations, i.e. at each time point t there is a probability p of any gene changing its value uniformly randomly. Thus, the PBN model can account for the uncertainties in both data and model selection. The PBN A shares the same state space S with its realizations, and the state transition diagrams Γ_j of the individual B_j's combine naturally into a stochastic state transition diagram Γ representing the dynamics of A. As in the case of deterministic BNs Γ can be identified with a stochastic $2^n\times2^n$ matrix P_n, also known as *transition* matrix, with non-negative entries p_{ij} and having the property $\sum_{j=1}^{n} p_{ij} = 1$, $i = 1,\ldots, n$. Using this matrix, the dynamics of A can be described using the well-developed theory of Markov chains. One should notice that if the probability of gene flipping p is positive then the Markov chain representing the dynamics of the network is ergodic which implies that it possesses a *steady-state probability distribution* π.

Both the CME and PBN models belong to the class of discrete state stochastic dynamic models but differ in the level of discretization (coarse scale for PBN vs. fine-scale for CME) and the discrete (PBNs) or continuous time (CMEs) updating of the states. In this section, our overall goal is to study mappings from CMEs to PBNs that can maintain some desirable properties. As the steady state distribution of the genetic regulatory network is considered representative of the phenotype, our objective will be to maintain the steady state distribution. The finite state continuous time CME model can be approximated by a discrete-time sampled Markov chain by using a suitable time period Δt. The $M \times M$ state transition probabilities of the discrete-time Markov Chain denoted by P_Δ is given by the equation $P_\Delta = e^{A\Delta t}$. We will map these transition probabilities to an $N \times N$ transition probability matrix of a PBN denoted by P_r where N will be 2^n for binary levels and 3^n for the case of ternary levels. We will consider binary PBNs from now onwards but the derivation for more than 2 levels will be similar. Let T_i for $i \in [1, 2, \cdots n]$ denote the thresholds for binarization. For instance, if $x(t) = [\Psi_1, \Psi_2, \ldots, \Psi_n]$, then the corresponding state for the PBN is $y = [\psi_1, \psi_2, \ldots, \psi_n]$ where $\psi_1 = 1$ if $\Psi_i \geq T_i$ or $\psi_i = 0$ if $\Psi_i < T_i$ for all $i \in [1, 2, \cdots n]$. As the states of the PBN will be n-dimensional vectors ranging from 00..0 to 11...1, we can represent them by the decimal equivalent state z_t where

z_t ranges from 1 to 2^n. Let us consider a sequence $0 = a_0, a_1, ..., a_N = M$ such that $a_i - a_{i-1}$ for $i \in \{1, 2, ..., 2^n\}$ denote the number of states in the CME model that map to state i in the PBN model. The states of the CME model will be denoted by w_t ranging from 1 to M. The states will be re-ordered such that the states $a_{i-1} + 1$ to state a_i of the CME will be mapped to state i of the PBN for $i = 1, ..., N$. Note that the subscript Δ refers to the higher dimensional discrete time CME model whereas the subscript r refers to the lower dimensional PBN model.

Let the state of the PBN at time be $z_t = i$ and the state at time $t+1$ be $z_{t+1} = j$. The transition probability of going from state i to state j is given by

$$P_r(i,j) = P(z_{t+1} = j \mid z_t = i) = \sum_{j_1=a_{j-1}+1}^{a_j} P(\omega_{t+1} = j_1, z_t = i)$$

$$= \frac{\sum_{j_1=a_{j-1}+1}^{a_j} P(\omega_{t+1} = j_1, z_t = i)}{P(z_t = i)} = \frac{\sum_{j_1=a_{j-1}+1}^{a_j} \sum_{i_1=a_{i-1}+1}^{a_i} P(\omega_{t+1} = j_1, \omega_t = i_1)}{\sum_{i_2=a_{i-1}+1}^{a_i} P(\omega_t = i_2)}$$

$$= \frac{\sum_{j_1=a_{j-1}+1}^{a_j} \sum_{i_1=a_{i-1}+1}^{a_i} P(\omega_{t+1} = j_1 \mid \omega_t = i_1) P(\omega_t = i_1)}{\sum_{i_2=a_{i-1}+1}^{a_i} P(\omega_t = i_2)} = \frac{\sum_{j_1=a_{j-1}+1}^{a_j} \sum_{i_1=a_{i-1}+1}^{a_i} P_\Delta(i_1, j_1) P(\omega_t = i_1)}{\sum_{i_2=a_{i-1}+1}^{a_i} P(\omega_t = i_2)} \quad (11)$$

We next show that if we use the steady state probabilities of the CME model for $P(\omega_t = i_1)$ in Equation 11, then the steady state probabilities of the resultant PBN, P_r will be the same as the collapsed steady state probabilities of the CME model. Here collapsing refers to the summation of the state probabilities of the detailed model that map to a certain state of the PBN model. For instance, if η represents the M dimensional steady state probability vector for the CME model, then the collapsed steady state vector will be an N dimensional vector represented by

$$\zeta(i) = \sum_{c=a_{i-1}+1}^{a_i} \eta(c) \text{ for } i = 1, ... N.$$

We next prove the equivalence of ζ and the steady state probabilities of the PBN represented by the following state transition probabilities:

$$P_r(i,j)=\frac{\sum_{j_1=a_{j-1}+1}^{a_j}\sum_{i_1=a_{i-1}+1}^{a_i}P_\Delta(i_1,j_1)\eta(i_1)}{\sum_{i_2=a_{i-1}+1}^{a_i}\eta(i_2)} \tag{12}$$

Theorem I: (Pal & Bhattacharya, 2010) *WithEquation 12denoting the transition probabilities of the PBN, the steady state probability distribution vector, π, of the PBN is given by*

$$\pi(i)=\sum_{i_2=a_{i-1}+1}^{a_i}\eta(i_2)=\zeta(i)$$

Proof: Assuming an ergodic Markov Chain, ζ is the steady state distribution of the PBN if it satisfies the equation $\sum_{i=1}^{N}\zeta(i)P_r(i,j)=\zeta(j)$ for each $j\in 1,2,\ldots,N$

Since η is the steady state distribution for the $M\times M$ CME model P_Δ, it satisfies $\eta(j_1)=\sum_{i_0=1}^{M}\eta(i_0)P_\Delta(i_0,j_1)$. Thus

$$\zeta(j)=\sum_{j_1=a_{i-1}+1}^{a_j}\eta(j_1)=\sum_{j_1=a_{i-1}+1}^{a_j}\sum_{i_0=1}^{M}\eta(i_0)P_\Delta(i_0,j_1)=$$

$$\sum_{j_1=a_{j-1}+1}^{a_j}\sum_{i=1}^{N}\sum_{i_1=a_{i-1}+1}^{a_i}\eta(i_1)P_\Delta(i_1,j_1)=\sum_{i=1}^{N}\sum_{i_2=a_{i-1}+1}^{a_i}\eta(i_2)P_r(i,j)=\sum_{i=1}^{N}P_r(i,j)\zeta(i)$$

This shows that ζ satisfies the steady state probability equations of the PBN and is equal to π assuming an ergodic Markov Chain. We have established that if the transition probabilities of the PBN are decided based on Equation 12, then the steady state probabilities of the generated PBN is same as the collapsed steady state probabilities of the original detailed network. Thus, we are able to maintain the collapsed steady-state probabilities, which might be indicative of a phenotype. If we revisit Equation 12, we notice that $P_r(i,j)$ can be calculated from the transitions of the network once it has reached the steady state. If we observe the steady-state transitions of the genetic regulatory network modeled by the CME for a long time and generate the state-transition probabilities of the PBN $P_r(i,j)$ based on the number of transitions occurring from state i to state j divided by the total number of transitions from state i, we will arrive at Equation 12. Thus, the mapping described by Equation 12 exists between a network represented by a CME model and a coarse-scale PBN model when the transition probabilities of the PBN are inferred based on state transition data at steady-state.

The following two examples from (Pal & Bhattacharya, 2010) consider well-studied stochastic master equation models of biological networks for illustrating the maintenance of collapsed steady state behavior when the mapping is based on Equation 12.

Example 1

We apply the CME to PBN approximation to the same genetic toggle switch model presented in the earlier case of CME to DE approximation. We have two competing proteins U_1 and U_2, each of which inhibits the transcription of the other. The decay reactions $U_1 \rightarrow \varnothing$ and $U_2 \rightarrow \varnothing$ have propensities $c\psi_1$ and $c\psi_2$ respectively. The transcription of new copies of the proteins are guided by the reactions $\varnothing \rightarrow U_1$ and $\varnothing \rightarrow U_2$ with propensities $b/(b+\psi_2)$ and $b/(b+\psi_1)$ respectively. Our starting Stochastic Master equation model has parameter values b = 4 and c = 0.05. We truncate the maximum number of protein molecules for U_1 and U_2 to 39 resulting in a total number of 40x40 = 1600 possible states for the CME. Based on the propensity functions and parameter values b=0.4 and c=0.05, the state reaction matrix A_1, is generated and the continuous Markov chain is approximated by a discrete chain using the following equation $P_{\Delta_1} = e^{A_1 \Delta t}$ with a time step of Δt=0.1 seconds. The thresholds for binarization of the proteins U_1 and U_2 are selected to be T_1=T_2=6 molecules. After binarization, the two protein reduced network has 2x2=4 states. We calculated the transition probabilities of the reduced model (PBN) using Equation 12. In our case, the Markov chain is ergodic as all the states are communicating, recurrent and aperiodic. Due to the ergodicity, P_{Δ_1} has a unique steady state probability distribution. The equivalence of the steady state probability distribution π_1 of P_{r_1} and the collapsed steady state probability distribution ζ_1 of P_{Δ_1} is shown in Figure 9 (Pal & Bhattacharya, 2010).

Example 2

In this example, we apply the CME to PBN mapping procedure to a model representing the life cycle of bacteriophage-λ (Hegland, Burden, Santoso, MacNamara, & Booth, 2007) which involves five proteins. The decay reactions $U_i \rightarrow \varnothing$ for

Figure 9. Steady State Probability Distributions $\pi_1$and ζ_1
Pal and Bhattacharya (2010) © 2010 IEEE

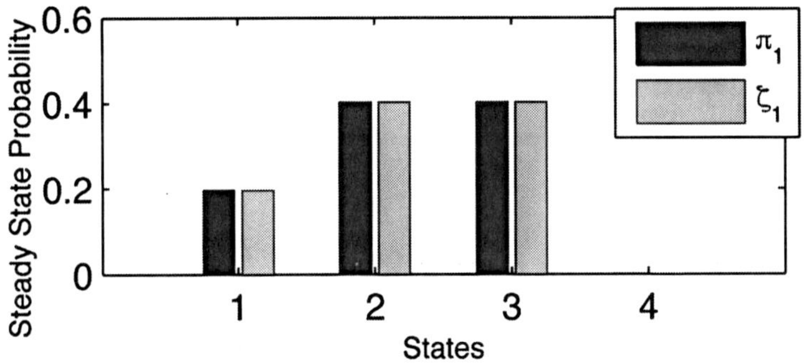

i=1,2,…,5 are with propensities $c_i\Psi_i$ for i=1,2,…,5. The generation reactions with their respective propensities are as follows:

1. $\varnothing \rightarrow U_1$ with propensity $\dfrac{a_1 b_1}{b_1 + \Psi_2}$
2. $\varnothing \rightarrow U_2$ with propensity $\dfrac{(a_2 + \Psi_5)b_1}{b_1 + \Psi_1}$
3. $\varnothing \rightarrow U_3$ with propensity $\dfrac{a_3 b_3 \Psi_2}{b_3 \Psi_2 + 1}$
4. $\varnothing \rightarrow U_4$ with propensity $\dfrac{a_4 b_4 \Psi_3}{b_4 \Psi_3 + 1}$
5. $\varnothing \rightarrow U_5$ with propensity $\dfrac{a_5 b_5 \Psi_3}{b_5 \Psi_3 + 1}$

The generation reactions for the proteins originate from the mechanism of transitions between two phases, lysogenic and lytic, present typically in the life of a bacteriophage-λ. Alterations from one phase to another are regulated by a toggle switch of two competing proteins U_1 and U_2, which inhibit each other's transcription. In the lysogenic phase U_2 is abundant while suppressing the production of U_1 via the reaction 1. The generation of U_2 via reaction 2, on the other hand, depends on the amount of U_5. Under a stressed situation U_5 decreases as the phage transitions into the lytic stage. The parameters a_i, b_i, c_i are taken from (Hegland, Burden, Santoso, MacNamara, & Booth, 2007) and shown in Table 1.

The maximum number of protein molecules for U_1, U_2, U_3, U_4, U_5 are truncated to be at 3, 24, 3, 3, and 2, respectively. Thus, the total number of possible states for

Table 1. Parameters for the example II CME model

	i=1	i=2	i=3	i=4	i=5
a_i	0.5	1	0.15	0.3	0.3
b_i	0.12	0.6	1	1	1
c_i	0.0025	0.0007	0.0231	0.01	0.01

the CME is 4x25x4x4x3=4800. Using the above propensity functions and values for parameters shown in Table I, the state reaction matrix A_1 is computed and the continuous time Markov chain is approximated by a discrete time chain using equation $P_{\Delta_1} = e^{A_1 \Delta t}$ with a time step of Δt=0.1 seconds. The binarization of the proteins is achieved using thresholds 2, 8, 2, 1, and 1, respectively. After binarization, the five protein reduced network has 2^5=32 states. The transition probabilities of the reduced model (PBN) are calculated using Equation 12. Since all the states are communicating, recurrent and aperiodic, the Markov chain P_{Δ_1} is ergodic and hence P_{Δ_1} has a unique steady state probability distribution. The equivalence of the steady state probability distribution π_1 of P_{r1} and the collapsed steady state probability distribution ζ_1 of P_{Δ_1} is shown in Figure 10.

Transient Behavior of Coarse-Scale CME Approximation

The mapping presented in Equation 12 maintains the collapsed steady state behavior without enforcing any constraints on the collapsed transient dynamics. Thus, in this section we will evaluate the deviation of the transient dynamics of the PBN as

Figure 10. Steady State Probability Distributions π_1 and ζ_1 for example II
Pal and Bhattacharya (2010) © 2010 IEEE

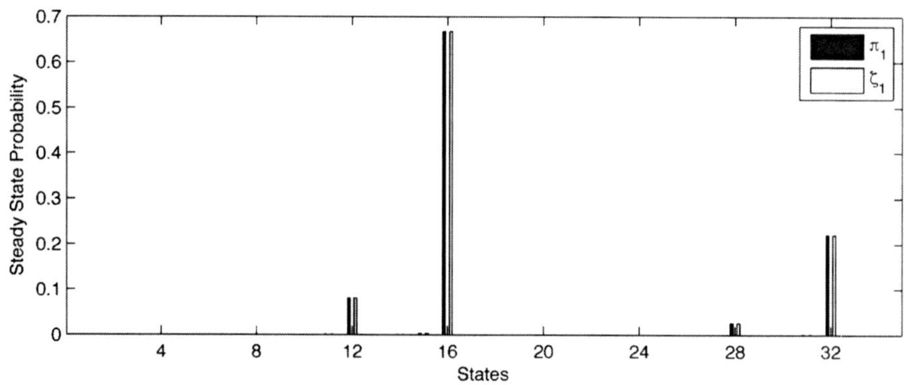

compared to the collapsed CME state probability distribution. Let us consider the difference equation $d_j(t+\Delta t) = |\gamma j_t(t+\Delta t) - \tau j(_t+\Delta t)|$ where $\gamma j(_t)$ denote the collapsed probability of states $aj-_{l+1}$ to $aj_{\ o}$f the CME model at any time t *and* $\tau j(_t)$ denote the probability of state j *at* any time t *for* the PBN model. We can bound the difference equation $dj(_t+\Delta t)$ for $\gamma j(_t)=\tau j(_t)$ in the following way (Pal & Bhattacharya, 2012; Pal, Bhattacharya, & Caglar, 2012):

$$d_j(t+\Delta t) \leq \max_{i_1 \in [1,\dots,N]}(\max_{i_2 \in S_{i_1}} q(i_2, j) - \min_{i_2 \in S_{i_1}} q(i_2, j))$$

where $q(i_2, j) = \sum_{j_1 = a_{j-1}+1}^{a_j} P_\Lambda(i_2, j_1)$.

Similarly, we can extend the difference equation for multiple time steps as follows (Pal & Bhattacharya, 2012):

$$d_j(T\Delta t) \leq \beta(\frac{1-\mu_m^{\ T}}{1-\mu_m})$$

where $\beta = \max_{j \in [1,..N]} \beta_j$ and $\beta_j = \max_{i_1 \in [1,\dots,N]} (\max_{i_2 \in S_{i_1}} q(i_2, j) - \min_{i_2 \in S_{i_1}} q(i_2, j))$ and

$\mu_m = \max_{i_1 \in [1,\dots,N]} (\sum_{i_1=1}^{N} | R(i_1, j) - \rho |)$ where ρ is any constant and

$R(i_1, j) = \min_{i_2 \in [a_{i_1-1}+1,\dots\dots,a_{i_1}]} q(i_2, j)$.

The above-mentioned bounds for single and multiple time steps provide the worst-case bounds that can be difficult to compute as they require the detailed knowledge of the fine-scale transition probability matrix. Thus, a possible approach is to construct stochastic bounds or bounds on the expected deviation based on possible distributions of the entries of the transition probability matrix. Multiple such bounds based on the uniform distribution or distribution centered around different means for the entries of the state transition probability matrix have been considered in (Pal & Bhattacharya, 2012). The next two theorems from (Pal & Bhattacharya, 2012) provide the stochastic bounds for single and multiple steps. The proofs are available in (Pal & Bhattacharya, 2012).

Theorem II (Pal & Bhattacharya, 2012) The expected bound on the deviation β_j for single time step is given by the following equation:

$$E(\beta_j) < 2\sigma(2\log L - \log\log L)^{\frac{1}{2}}$$

where σ denotes the standard deviation for the distribution of $q(i,j)$ and $L = \max_{i=1,\dots,N} | a_i - a_{i-1} |$.

Theorem III (Pal & Bhattacharya, 2012)

$$E(\mu_m) < (N-1)\sqrt{\frac{2}{\pi}}\sigma_1 + (\xi_2 - \xi_1) erf(\frac{\xi_2 - \xi_1}{\sqrt{2}\sigma_2}) + \sqrt{\frac{2}{\pi}}\sigma_2 e^{-(\xi_2 - \xi_1)}$$

where $erf(x) = \frac{2}{\sqrt{\pi}} \int_{t=0}^{x} e^{-t^2} dt$ denotes the error function and the distribution of $R(i,j) = \min_{i_2 \in [a_{i-1}+1,\dots,a_i]} q(i_2, j)$ is approximated by $\aleph(\xi_1, {\sigma_1}^2)$ and $\aleph(\xi_2, {\sigma_2}^2)$ for $i \neq j$ and $i = j$ respectively.

Example 3

In this example, we consider the earlier considered Genetic Toggle Switch in example I with parameters b = 0.8 and c = 0.035 as the initial system. We truncate the maximum number of protein molecules for U_1 and U_2 to 39 resulting in a total number of 40x40 = 1600 possible states for the CME. Based on the propensity functions and parameter values b=0.8 and c=0.035, the state reaction matrix A_1, is generated and the continuous Markov chain is approximated by a discrete chain using the following equation $P_{\Delta_1} = e^{A_1 \Delta t}$ with a time step of Δt=0.5 seconds. The thresholds for binarization of the proteins U_1 and U_2 are selected to be $T_1=T_2=20$ molecules. After binarization, the two protein reduced network has 2x2=4 states. We calculated the transition probabilities of the reduced model (PBN) using Equation 12. The time evolution of the probability of the four states starting from a uniform distribution is shown in Figure 11.

The actual deviation characteristics of the PBN probability and the collapsed CME model are shown in Table II. State denotes the four states (00; 01; 10 and 11) of the reduced model. D_s and D_m denotes the actual maximum deviation in a single time step and multiple time steps, respectively. UB_s and UB_m denotes the stochastic bounds for single time step and multiple time steps, respectively. The ε used to calculate the stochastic bound is $2\frac{(1-d)}{(M-1)}$ with d=0.4 where d is an approximation

Figure 11. Time evolution of probability starting from a uniform distribution for the genetic switch with parameters b = 0.8 and c = 0.035. Red denotes the coarse-scale model evolution and blue + denotes the evolution of the collapsed CME model.
Pal and Bhattacharya (2012) © 2012 IEEE

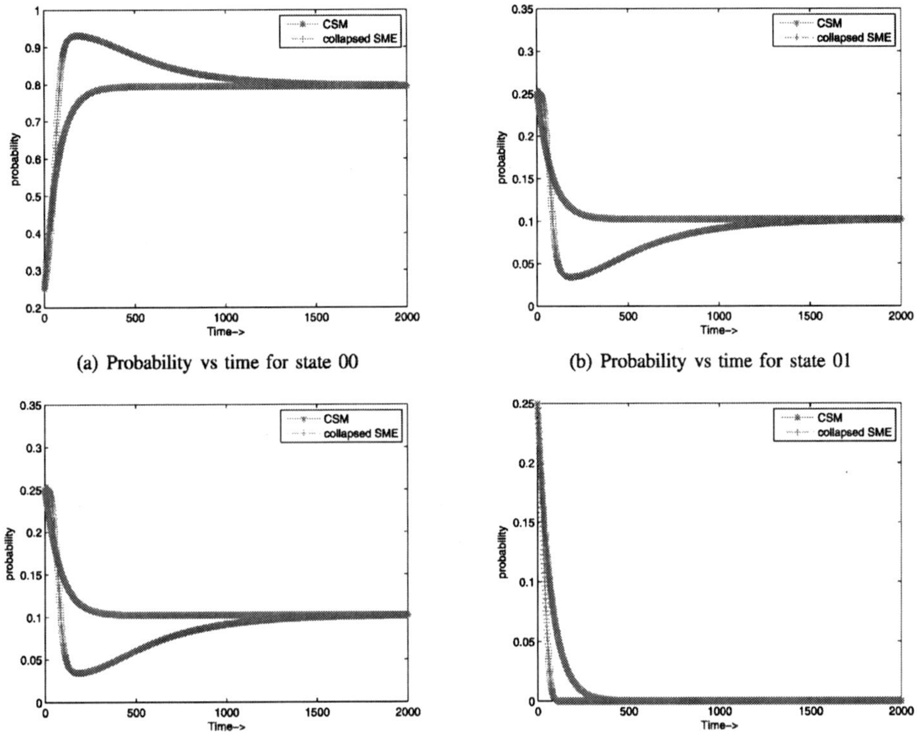

(a) Probability vs time for state 00

(b) Probability vs time for state 01

(c) Probability vs time for state 10

(d) Probability vs time for state 11

of a diagonal entry. We are assuming that each diagonal entry is greater than 0.4 which is reasonable as we have a small δ*t*of 0.25 seconds. The μ_m used to calculate the multiple step bound is estimated as follows: the transition probability matrix is given by $e^{A\delta t}$ where A is the reaction matrix. We approximate the transition probability matrix by I+$A\delta$t *u*sing the first two terms of the series expansion. Based on this approximation, we generated the approximate transition probability matrix and calculated μ s from the approximate matrix. For our case, δt=*0*.25, c=*0*.0035, b=*0*.8, and t1 =$_t$2=$_2$0. To calculate the multistep bound, we used the maximum of μ s 0.8154 as the μ=m. From Table 2, we note that the stochastic bound for multi-steps (0.2596) is very close to the maximum actual deviations for each state (0.2257, 0.0895, 0.0917, 0.0784).

Table 2. Transient Response Bounds for the Genetic Switch Model with Parameters b= 0.8 and c = 0.035 (Pal & Bhattacharya, 2012) © 2012 IEEE

State	Ds	Dm	UBs	UBm
00	0.0054	0.2257	0.0479	0.2596
01	0.0027	0.0895	0.0479	0.2596
10	0.0027	0.0917	0.0479	0.2596
11	0.0016	0.0784	0.0479	0.2596

DISCUSSIONS

In this chapter, we discussed the broad classes of genetic regulatory network models and presented equivalence relationships concerning some of them. We primarily discussed the relationships between CME and PBN type of models that are both discrete state stochastic models with Markovian property but with a hugely different size of the state space. Equivalence relationships among other model types have also been explored in the literature, for instance (Davidich & Bornholdt, 2008) shows that the Boolean Network model is a coarse-grain limit of the differential equation model and demonstrates it on the budding yeast cell cycle network. (Lähdesmäki, Hautaniemi, Shmulevich, & Yli-Harjaa, 2006) illustrates that probabilistic Boolean networks and dynamic Bayesian network models can represent the same multivariate state probability distribution over their common variables. Based on our modeling objectives and available data, different model types can be suitable for different scenarios. For instance, modeling of the basic average inhibition and up regulation of cancer pathways using targeted drugs can often be modeled using Boolean Network based models (Grasso, et al., 2015; Layek, Datta, & Dougherty, 2011; Pal & Berlow, 2012) . However, modeling of detailed stochasticity of gene expression (Kærn, et al. 2005) will require fine scale CME or stochastic differential equation based models. Furthermore, modeling of tumor heterogeneity will either require separate deterministic models for individual sub types or a stochastic model that incorporates the behavior of the diverse subtypes. The most commonly used model of ordinary differential equations is valid when the number of molecules of the modelled entities is sufficiently large. In terms of CME propensity functions, the CME to DE approximation is valid when the propensity functions are linear relationships of state variables (Szallasi, Stelling, & Periwal, 2006).

Furthermore, the mathematical model selected for representing a biological phenomenon will also be dependent on the data type used for inference of the model parameters. For instance, microarray data will provide the average expression of a collection of cells and cannot be used to infer fine-scale stochastic model parameters but are rather suited for DE or coarse scale Boolean Network kind of models. For estimating the parameters of a fine scale stochastic model, we would require individual cell measurements using techniques such as green fluorescent proteins or single cell sequencing. The sampling time period used in data measurements will also constraint the inference of model parameters. For instance, modeling a cyclic biological activity such as circadian rhythm will require measurement time step to be smaller than the period of the biological activity. Some studies measure the gene or protein expression of a cell line at a fixed time and aggregate the responses across multiple samples to reflect the system behavior. This approach can work when the sample behaviors are similar and we have limited variations among the cell lines. However, for inference of dynamic model parameters, time series data is highly desirable as compared to steady state data. The model inference process is often independent of the experimental data generation process and majority of inference analysis is conducted without access to experimental data that is most suitable for high accuracy estimation of model parameters. Model based experiment design can improve the accuracy and decrease uncertainty in inference of model parameters.

One of the benefits of exploring the relationships between various models of GRNs lies in its potential to be used in building whole cell mathematical models where different model types represent the sub-networks. There have been recent attempts at creating whole cell models of smaller organisms such as (Karr, et al., 2012) that creates a whole cell computational model of the life cycle of the human pathogen *Mycoplasma genitalium* by combining 28 dissimilar submodels representing various levels and processes in the cells. (Covert, Xiao, Chen, & Karr, 2008) considers the specific case of integrating flux-balance based central carbon metabolic and transcriptional regulatory model with an ordinary differential equation based detailed model of carbohydrate uptake control for Escherichia coli. (Chandrasekaran & Price, 2010) considers the integration of transcriptional regulatory network with corresponding metabolic network for Escherichia coli and Mycobacterium tuberculosis. Thus, the detailed understanding of relationships between various models of GRNs will enable us to effectively integrate existing sub-models for an organism that have been designed using different model types.

REFERENCES

Adalsteinsson, D., McMillen, D., & Elston, T. C. (2004). Biochemical Network Stochastic Simulator (BioNetS): Software for stochastic modeling of biochemical networks. *BMC Bioinformatics*, *5*(1), 24. doi:10.1186/1471-2105-5-24 PMID:15113411

Arkin, A., Ross, J., & McAdams, H. H. (1998). Stochastic kinetic analysis of developmental pathway bifurcation in phage lambda-infected Escherichia coli cells. *Genetics, 149*(4), 1633–1648. PMID:9691025

Berlow, N., Davis, L., Cantor, E., Seguin, B., Keller, C., & Pal, R. (2013). A new approach for prediction of tumor sensitivity to targeted drugs based on functional data. *BMC Bioinformatics, 14*(1), 239. doi:10.1186/1471-2105-14-239 PMID:23890326

Bishop, L. M., & Qian, H. (2010). Stochastic Bistability and Bifurcation in a Mesoscopic Signaling System with Autocatalytic Kinase. *Biophysical Journal, 98*(1), 1–11. doi:10.1016/j.bpj.2009.09.055 PMID:20074511

Caglar, M. U., & Pal, R. (2013). Stochastic Model Simulation Using Kronecker Product Analysis and Zassenhaus Formula Approximation. *IEEE/ACM Transactions on Computational Biology and Bioinformatics, 10*(5), 1125–1136. doi:10.1109/TCBB.2013.34 PMID:24384703

Cai, X., & Wang, X. (2007). Stochastic Modeling and Simulation of Gene Networks. *IEEE Signal Processing Magazine, 24*(1), 27–36. doi:10.1109/MSP.2007.273051

Cao, Y., & Liang, J. (2008). Optimal enumeration of state space of finitely buffered stochastic molecular networks and exact computation of steady state landscape probability. *BMC Systems Biology, 2*(30). doi:10.1186/1752-0509-2-30 PMID:18373871

Cao, Y., Lu, H. M., & Liang, J. (2010). Probability landscape of heritable and robust epigenetic state of lysogeny in phage lambda. *Proceedings of the National Academy of Sciences of the United States of America, 107*(43), 18445–18450. doi:10.1073/pnas.1001455107 PMID:20937911

Chandrasekaran, S., & Price, N. D. (2010). Probabilistic integrative modeling of genome-scale metabolic and regulatory networks in Escherichia coli and Mycobacterium tuberculosis. *Proceedings of the National Academy of Sciences of the United States of America, 107*(41), 17845–17850. doi:10.1073/pnas.1005139107 PMID:20876091

Covert, M. W., Xiao, N., Chen, T. J., & Karr, J. R. (2008). Integrating metabolic, transcriptional regulatory and signal transduction models in Escherichia coli. *Bioinformatics (Oxford, England), 24*(18), 2044–2050. doi:10.1093/bioinformatics/btn352 PMID:18621757

Davidich, M. I., & Bornholdt, S. (2008). Boolean Network Model Predicts Cell Cycle Sequence of Fission Yeast. *PLoS ONE, 3*(2), e1672.

De Jong, H. (2001). Modeling and Simulation of Genetic Regulatory Systems: A Literature Review. *Journal of Computational Biology*, *9*(1), 67–103. doi:10.1089/10665270252833208 PMID:11911796

Gillespie, D. T. (1977). Exact Stochastic Simulation of Coupled Chemical Reactions. *Journal of Physical Chemistry*, *81*(25), 2340–2361. doi:10.1021/j100540a008

Gillespie, D. T. (2000). The chemical Langevin equation. *The Journal of Chemical Physics*, *113*(1), 297–306. doi:10.1063/1.481811

Gillespie, D. T. (2001). Approximate accelerated stochastic simulation of chemically reacting systems. *The Journal of Chemical Physics*, *115*(4), 1716–1733. doi:10.1063/1.1378322

Grasso, C., Tang, Y., Truffaux, N., Berlow, N. E., Liu, L., Debily, M. A., & Monje, M. et al. (2015). Functionally defined therapeutic targets in diffuse intrinsic pontine glioma. *Nature Medicine*, *21*(6), 555–559. doi:10.1038/nm.3855 PMID:25939062

Hegland, M., Burden, C., Santoso, L., MacNamara, S., & Booth, H. (2007). A solver for the stochastic master equation applied to gene regulatory networks. *Journal of Computational and Applied Mathematics*, *205*(2), 708–724. doi:10.1016/j.cam.2006.02.053

Higham, D. J. (2008). Modeling and Simulating Chemical Reactions. *SIAM Review*, *50*(2), 347–368. doi:10.1137/060666457

Hoops, S., Sahle, S., Gauges, R., Lee, C., Pahle, J., Simus, N., & Kummer, U. et al. (2006). COPASI—a COmplex PAthway SImulator. *Bioinformatics (Oxford, England)*, *22*(24), 3067–3074. doi:10.1093/bioinformatics/btl485 PMID:17032683

Kærn, M., Elston, T. C., Blake, W. J., & Collins, J. J. (2005). Stochasticity in gene expression: From theories to phenotypes. *Nature Reviews. Genetics*, *6*(6), 451–464. doi:10.1038/nrg1615 PMID:15883588

Kampen, N. G. (1991). *Stochastic processes in physics and chemistry*. North Holland.

Karr, J. R., Sanghvi, J. C., Macklin, D. N., Gutschow, M. V., Jacobs, J. M. Jr, Bolival, B. Jr, & Covert, M. W. et al. (2012). A Whole-Cell Computational Model Predicts Phenotype from Genotype. *Cell*, *150*(2), 389–401. doi:10.1016/j.cell.2012.05.044 PMID:22817898

Kauffman, S. (1969). Homeostasis and Differentiation in Random Genetic Control Networks. *Nature*, *224*(5215), 177–178. doi:10.1038/224177a0 PMID:5343519

Kauffman, S. A. (1993). *The Origins of Order*. Oxford, UK: Oxford University Press.

Kierzek, A. M. (2002). STOCKS: STOChastic Kinetic Simulations of biochemical systems with Gillespie algorithm. *Bioinformatics (Oxford, England)*, *18*(3), 470–481. doi:10.1093/bioinformatics/18.3.470 PMID:11934747

Kurtz, T. G. (1972). The Relationship between Stochastic and Deterministic Models for Chemical Reactions. *The Journal of Chemical Physics*, *57*(7), 2976–2978. doi:10.1063/1.1678692

Lähdesmäki, H., Hautaniemi, S., Shmulevich, I., & Yli-Harjaa, O. (2006). Relationships between probabilistic Boolean networks and dynamic Bayesian networks as models of gene regulatory networks. *Signal Processing*, *86*(4), 814–834. doi:10.1016/j.sigpro.2005.06.008 PMID:17415411

Layek, R., Datta, A., & Dougherty, E. (2011). From biological pathways to regulatory networks. *Molecular BioSystems*, *7*(3), 843–851. doi:10.1039/C0MB00263A PMID:21161088

Munsky, B., & Khammash, M. (2006). The finite state projection algorithm for the solution of the chemical master equation. *The Journal of Chemical Physics*, *124*(4), 044104. doi:10.1063/1.2145882 PMID:16460146

Murphy, K. (2002). *Dynamic Bayesian Networks: Representation, Inference and Learning*. Berkeley, CA: University of California.

Pal, R. (2010). Context-sensitive probabilistic Boolean networks: Steady-state properties, reduction, and steady-state approximation. *IEEE Transactions on Signal Processing*, *58*(2), 879–890. doi:10.1109/TSP.2009.2030832

Pal, R., & Berlow, N. (2012). *A kinase inhibition map approach for tumor sensitivity prediction and combination therapy design for targeted drugs. In Pac Symp Biocomput* (pp. 351–362). Kohala Coast: World Scientific.

Pal, R., & Bhattacharya, S. (2010). Characterizing the Effect of Coarse-Scale PBN Modeling on Dynamics and Intervention Performance of Genetic Regulatory Networks Represented by Stochastic Master Equation Models. *IEEE Transactions on Signal Processing*, *58*(6), 3341–3351. doi:10.1109/TSP.2010.2043970

Pal, R., & Bhattacharya, S. (2012). Transient dynamics of reduced-order models of genetic regulatory networks. *IEEE/ACM Transactions on Computational Biology and Bioinformatics*, *9*(4), 1230–1244. doi:10.1109/TCBB.2012.37 PMID:22411891

Pal, R., Bhattacharya, S., & Caglar, M. (2012). Robust approaches for genetic regulatory network modeling and intervention: A review of recent advances. *IEEE Signal Processing Magazine*, *1*(29), 66–76. doi:10.1109/MSP.2011.942818

Sanft, K. R., Wu, S., Roh, M., Fu, J., & Petzold, L. R. (2011). StochKit2: Software for discrete stochastic simulation of biochemical systems with events. *Bioinformatics (Oxford, England)*, *27*(17), 2457–2458. doi:10.1093/bioinformatics/btr401 PMID:21727139

Shea, M., & Ackers, G. (1985). The OR control system of bacteriophage lambda: A physical-chemical model for gene regulation. *Journal of Molecular Biology*, *181*(2), 211–230. doi:10.1016/0022-2836(85)90086-5 PMID:3157005

Szallasi, Z., Stelling, J., & Periwal, V. (2006). *System Modeling in Cellular Biology: From Concepts to Nuts and Bolts*. Cambridge, MA: The MIT Press. doi:10.7551/mitpress/9780262195485.001.0001

Székely, T. Jr, & Burragea, K. (2014). Stochastic simulation in systems biology. *Computational and Structural Biotechnology Journal*, *12*(20-21), 14–25. doi:10.1016/j.csbj.2014.10.003 PMID:25505503

Chapter 3
Modeling Stochastic Gene Regulatory Networks Using Direct Solutions of Chemical Master Equation and Rare Event Sampling

Youfang Cao
Los Alamos National Laboratory, USA

Anna Terebus
University of Illinois – Chicago, USA

Jie Liang
University of Illinois – Chicago, USA

ABSTRACT

Stochasticity plays important roles in many biological networks. A fundamental framework for studying the full stochasticity is the Discrete Chemical Master Equation (dCME). Under this framework, the combination of copy numbers of molecular species defines the microstate of the molecular interactions in the network. The probability distribution over these microstates provide a full description of the properties of a stochastic molecular network. However, it is challenging to solve a dCME. In this chapter, we will first discuss how to derive approximation methods including Fokker-Planck equation and the chemical Langevin equation from the dCME. We

DOI: 10.4018/978-1-5225-0353-8.ch003

also discuss the widely used stochastic simulation method. After that, we focus on the direct solutions to the dCME. We first discuss the Finite State Projection (FSP) method, and then introduce the recently developed finite buffer method (fb-dCME) for directly solving both steady state and time-evolving probability landscape of dCME. We show the advantages of the fb-dCME method using two realistic gene regulatory networks.

1. INTRODUCTION

Molecular components in biological systems, such as proteins, DNA, RNAs, and substrates, are often interacting with each other in a complex reaction network to perform certain biological functions. These networks of interacting biomolecules are at the heart of the regulations of many critical cellular processes, from the regulation of gene expression (Arkin, Ross, & McAdams, 1998; Hasty, Pradines, Dolnik, & Collins, 2000; Levin, 2003; McAdams & Arkin, 1997; Ozbudak, Thattai, Kurtser, Grossman, & van Oudenaarden, 2002), signal transduction (Samoilov, Plyasunov, & Arkin, 2005), to the differentiation of stem cells (Ogawa, 1989). The biological networks are intrinsically stochastic due to thermal fluctuations (McCullagh et al., 2009). The intrinsic stochasticity in these cellular processes originates from reactions involving small copy numbers of molecules. It frequently occur in a cell when molecular concentrations are in the range of 0.1 μM to 1 nM (typically from about 10 to 100 copies in a cell) (Arkin et al., 1998). For example, the regulation of transcriptions depends on the binding of often a few proteins to a promoter site; the synthesis of protein peptides on ribosome involves a small copy number of molecules; and patterns of cell differentiation depend on initial small copy number events. In these biological processes, fluctuations due to the stochastic behavior intrinsic in small copy number events play important roles.

The importance of stochasticity in cellular functions has been well recognized (Mettetal, Muzzey, Pedraza, Ozbudak, & van Oudenaarden, 2006; Ozbudak et al., 2002; Paulsson & Ehrenberg, 2000; Volfson et al., 2006; Zhou, Chen, & Aihara, 2005). Studies of network models show that stochasticity is important for magnifying signal, sharpening discrimination, and inducing multistability (Paulsson & Ehrenberg, 2000). Understanding the stochastic nature and its consequences for cellular processes involving molecular species of small copy numbers in a network is an important problem. A fundamental framework for studying the full stochasticity is the discrete chemical master equation (dCME). Under this framework, the combination of copy numbers of molecular species defines the microscopic state of the molecular interactions in the network. By explicitly treating microscopic states of reactants, reactions can all be effectively modeled as transitions between microstates, with transition rates determined by the physical properties of the molecules and the cell environment. The reaction trajectories can be modeled in the framework of the dCME, which describes continuous time Markov chains (CTMC) equivalent to the Kolmogorov equation. The probability distribution or probability landscape over these microstates and its time-evolving behavior provide a full description of the properties of a stochastic molecular network.

However, it is challenging to solve a dCME that involves a nontrivial number of species. Analytical solutions of the chemical master equation exist only for very simple cases (Vellela & Qian, 2009). A widely used method is to carry out Monte Carlo simulations of the chemical master equation using the Stochastic Simulation Algorithm (SSA), also known as Gillespie's algorithm (Gillespie, 1977). Although the SSA approach has found wide applications, it is ineffective in simulating rare events, as most of computing time is spent on following high-probability paths (Daigle, Roh, Gillespie, & Petzold, 2011; Roh, Daigle, Gillespie, & Petzold, 2011; Roh, Gillespie, & Petzold, 2010). A recently developed importance path sampling technique, the Adaptively Biased Sequential Importance Sampling (ABSIS) has been shown to significantly improve rare event sampling efficiency (Cao & Liang, 2013). Alternatively, the chemical master equation can also be approximated using the Fokker-Planck equation (FPE) and the chemical Langevin equation (CLE) (Gillespie, 2000).

Direct solutions to the dCMEs can provide exact results, however complete identification and characterization of the state space is prerequisite, which can rapidly become intractable due to the explosion of the size of the discrete state space. Currently, the state space of a network cannot be fully characterized in general. The conventional hypercube approach predefines the maximum copy number of each reactant, and bound the state space by the product of the maximum numbers. However, the size of state spaces generated with the hypercube approach will quickly inflate to enormity, which makes it intrinsically inefficient as many unreachable states from an initial condition are included. An alternative approach is to generate states by carrying out stochastic simulations. One can simply follow explicitly simulated reaction events to whatever microstates the system reaches, such as the *N-reachability* approach used in Ref (Munsky & Khammash, 2006). However, this approach cannot guarantee that all reachable states are included, especially when the system contains rare transitions. Therefore, it cannot guarantee full characterization of rare events.

Munsky and Khammash developed the finite state projection (FSP) method to directly solve the time evolution of the dCME by using an absorbing state to truncate the state space generated by the *N-reachability* approach (Munsky & Khammash, 2006; Munsky & Khammash, 2007). However, the absorbing state in the FSP method makes it unable to solve the steady state probability landscape of dCME, which is of great importance in studying the behavior of reaction networks.

In this chapter, we first discuss the basic theoretical framework of the probability landscape of a stochastic network and the underlying discrete chemical master equation (dCME). We then discuss the connections between dCME and other modeling methods often used in studying stochastic biological networks. We further discuss the formulation of the continuous chemical master equation (cCME), which approximates the dCME, as well as its further simplifications in the form of Fokker-Planck and Langevin models. This is followed by a discussion of the approach of Monte Carlo simulations to study stochastic network, with the Gillespie's stochastic simulation algorithm discussed in some details. We also discuss a few recent developments on rare event probability estimation using biased path sampling. After discussing the advantage and challenges to directly solve the dCME, we describe a recently developed approach to significantly reduce the size of the state space, and therefore efficiently solve the steady state and time evolution probability landscapes of the dCME, namely the *finite buffer* method for direct solution of dCME (Cao & Liang, 2008; Cao, Lu, & Liang, 2010). We then describe two realistic gene regulatory networks to show the effectiveness of the fb-dCME method.

2. DISCRETE CHEMICAL MASTER EQUATION FRAMEWORK FOR MODELING STOCHASTIC NETWORKS

2.1 Discrete Chemical Master Equation (dCME)

Here we first describe the chemical master equation on a discrete state space. The dCME describes the gain and loss of probability associated with each microstate due to chemical reactions. The chemical reactions can be thought as jump processes that bring the system from one combination of copy number of molecular species (micro state) to a different combination of copy number of molecular species once a reaction occurs. The dCME describes the change of probability of different microstates connected by such jump processes due to reactions.

Molecular species and reactions: We assume a system with m molecular species $\{X_1, X_2, \ldots, X_m\}$, where X_i is the i-th molecular species. There are n chemical reactions $\boldsymbol{R} = (R_1, R_2, \ldots, R_n)$ that can happen in the system, and each has a rate constant $\boldsymbol{r} = (r_1, r_2, \cdots, r_n)$. We denote the copy number of the -th molecular species as . The combination of the copy numbers at time t is a vector of nonnegative integers and is denoted as $\boldsymbol{x}(t) = (x_1(t), x_2(t), \ldots, x_m(t)) \in \mathbb{Z}_0^+$. We call

$\boldsymbol{x}(t)$ the *microstate* of the system at time t. The probability for the system to be in state $\boldsymbol{x}(t)$ is $p(\boldsymbol{x},t)$. The collection of all possible microstates consists of the *state space* of the system: $\boldsymbol{S} = \{\boldsymbol{x}(t)|t \to \infty\}$. Its size is denoted as $|\boldsymbol{S}|$. The probability distribution over the state space $\boldsymbol{S}$ at time is the *probability landscape* $p(\boldsymbol{S},t)$. The time-evolving and steady state probability landscape provides a full description of the properties of a stochastic molecular network (Ao, Kown, & Qian, 2007; Cao & Liang, 2008; Cao et al., 2010; Kim & Wang, 2007; Schultz, Onuchic, & Wolynes, 2007).

Stoichiometry: A chemical reaction R_k takes the general form:

$$c_{1,k}X_1 + c_{2,k}X_2 + \ldots + c_{m,k}X_m \xrightarrow{r_k} c'_{1,k}X_1 + c'_{2,k}X_2 + \ldots + c'_{m,k}X_m, \tag{1}$$

where r_k is the rate constant associated with reaction R_k. The reaction brings the system from the microstate $\boldsymbol{x}_j$ to $\boldsymbol{x}_i$. The difference between $\boldsymbol{x}_j$ and $\boldsymbol{x}_i$ is the stoichiometry vector $\boldsymbol{s}_k$ of reaction R_k:

$$\boldsymbol{s}_k = \boldsymbol{x}_i - \boldsymbol{x}_j = \left(c'_{1,k} - c_{1,k}, c'_{2,k} - c_{2,k}, \ldots, c'_{m,k} - c_{m,k}\right). \tag{2}$$

Here $\boldsymbol{s}_k$ can admit 0 entries if a molecular species does not participate in the reaction, so $\boldsymbol{s}_k$ has the same dimension as that of the microstate.

As an example, the reaction $A + 2B \to C$ reduces the number of *A* and *B* by 1 and 2, respectively, and increase the number of *C* by 1. Its stoichiometry vector has $c_A = -1,\ c_B = -2, c_C = +1$, and $s = (-1,-2,+1)$. If there are other molecular species, their coefficients are all 0 for this reaction. By treating microscopic states of reactants explicitly, linear and nonlinear reactions, such as synthesis, degradation, dimeric binding, and multimerization, can all be modeled as transitions between microstates.

Reaction rate: The rate of the k-th reaction that transition the system from state $\boldsymbol{x}_j$ to state $\boldsymbol{x}_i$ is determined by the intrinsic reaction rate constant r_k, and the available copy numbers of each reactant in the current state $\boldsymbol{x}_j$:

$$A_k\left(\boldsymbol{x}_i, \boldsymbol{x}_j\right) = A_k\left(\boldsymbol{x}_j\right) = r_k \prod_{l=1}^{m} \binom{x_l}{c_{l,k}}, \tag{3}$$

assuming the convention $\begin{pmatrix} 0 \\ 0 \end{pmatrix} = 1$. If the k-th reaction can lead the system from state $\boldsymbol{x}_j$ to state $\boldsymbol{x}_i$, we have $A_k(\boldsymbol{x}_i, \boldsymbol{x}_j) > 0$, otherwise $A_k(\boldsymbol{x}_i, \boldsymbol{x}_j) = 0$. In most cases, only one reaction connects two microstates. However, since in principle more than one reaction may connect two states, we have the overall reaction rate that brings the system from $\boldsymbol{x}_j$ to $\boldsymbol{x}_i$ as:

$$A(\boldsymbol{x}_i, \boldsymbol{x}_j) = \sum_{R_k \in \boldsymbol{R}} A_k(\boldsymbol{x}_i, \boldsymbol{x}_j),$$

where $A(\boldsymbol{x}_i, \boldsymbol{x}_j)$ represents the transition probability function per unit time from $\boldsymbol{x}_j$ to $\boldsymbol{x}_i$. Overall, we have the transition rate matrix: $\boldsymbol{A} = \{A(\boldsymbol{x}_i, \boldsymbol{x}_j)\}$, where the diagonal elements are defined as: $A(\boldsymbol{x}_i, \boldsymbol{x}_i) = -\sum_{i \neq j} A(\boldsymbol{x}_i, \boldsymbol{x}_j)$.

Discrete Chemical Maser Equation. The discrete chemical master equation can then be written as:

$$\frac{dp(\boldsymbol{x},t)}{dt} = \sum_{\boldsymbol{x}' \neq \boldsymbol{x}} \left[A(\boldsymbol{x},\boldsymbol{x}')p(\boldsymbol{x}',t) - A(\boldsymbol{x}',\boldsymbol{x})p(\boldsymbol{x},t) \right]. \tag{4}$$

Note here we regard the probability $p(\boldsymbol{x},t)$ of a microstate $\boldsymbol{x}$ is continuous in time, while the states are all discrete. We call this the discrete chemical master equation (dCME). The dCME in this form fully account for the probabilities of jumps between states, regardless whether the copy number components of $\boldsymbol{x}$ and $\boldsymbol{x}'$ are small or large. It gives a full account for the stochasticity due to small copy number events.

In matrix form, the dCME can be written as:

$$\frac{d\boldsymbol{p}(\boldsymbol{S},t)}{dt} = \boldsymbol{A}\boldsymbol{p}(\boldsymbol{S},t), \tag{5}$$

where $\boldsymbol{A} \in \mathbb{R}^{|\boldsymbol{S}| \times |\boldsymbol{S}|}$ is the rate matrix formed by the collection of all $A(\boldsymbol{x}_i, \boldsymbol{x}_j)$: $\boldsymbol{A} = \{A(\boldsymbol{x}_i, \boldsymbol{x}_j)\}, \boldsymbol{x}_i, \boldsymbol{x}_j \in \boldsymbol{S}$, and $\boldsymbol{p}(\boldsymbol{S},t)$ is the probability distribution vector of the dCME over the state space $\boldsymbol{S}$ at time t. The dCME describes the gain and loss in probability associated with each microstate due to chemical reactions. These chemical reactions can be regarded as jump processes upon firings of reactions, which bring the system from one combination of copy number of molecular species to a different combination of copy number of molecular species. The dCME fully accounts for the stochastic jumps between states, regardless whether the copy numbers $\boldsymbol{x}_i$ and $\boldsymbol{x}_j$ are small or large. The overall stochasticity due to small copy number events is therefore fully described. It provides a fundamental framework to study stochastic molecular networks (Gillespie, 1977; Van Kampen, 2007).

2.1.1 Continuous Chemical Master Equation

If we treat the state space as continuous, that is, we assume the amount of a molecular species x_i is measured by a real value (such as concentration) instead of a copy number, the micro-state $\boldsymbol{x}(t)$ becomes a real-valued vector $\boldsymbol{x}(t) \in \mathbb{R}^m$. We have the chemical master equation equivalent to Eqn. (4) on continuous state space as:

$$\frac{dp(\boldsymbol{x},t)}{dt} = \int_0^{\infty} \left[A(\boldsymbol{x},\boldsymbol{x}') p(\boldsymbol{x}',t) - A(\boldsymbol{x}',\boldsymbol{x}) p(\boldsymbol{x},t) \right] d\boldsymbol{x}', \tag{6}$$

where the kernel $A(\boldsymbol{x},\boldsymbol{x}')$ represents the transition probability function per unit time from $\boldsymbol{x}'$ to $\boldsymbol{x}$. The CME in this form is equivalent to the Chapman-Kolmogorov equation frequently used to describe continuous Markov processes.

Remark. The continuous state space version of the CME requires a strong assumption. It is only appropriate if one can assume that the difference in the amount of molecules in neighboring states is infinitesimally small, which is valid only if the copy number of the molecular species in the system are much larger than 1, and larger than the changes in the numbers of molecules when a reaction occurs. The continuous CME therefore cannot be used when the total amount of molecules involved is very small, for example, in systems of single or a handful of particles. In these cases, the discrete CME should be used, as it does not contain any intrinsic singularity difficulties.

2.1.2 Relationship to Law of Mass Action Equations

Based on the above definition of biological networks, we can now derive the deterministic mass action equations from the discrete chemical master equation, which describe the time dynamics of the mean value of concentration of each molecular species. By multiplying the copy number of each molecular species x_i to both sides of the Eqn. (4), and taking summation over all elements from zero to infinity, we can obtain the average concentration x_i of species X_i:

$$\frac{d\langle x_i\rangle}{dt}=\frac{d\left(\sum_{x\in S}x_i p(\boldsymbol{x},t)\right)}{dt}=$$

$$\sum_{k=1}^{m}\left(\sum_{x_1=0}^{\infty}\sum_{x_2=0}^{\infty}\cdots\sum_{x_n=0}^{\infty}x_i A_k\left(\boldsymbol{x}-\boldsymbol{s}_k\right)p\left(\boldsymbol{x}-\boldsymbol{s}_k,t\right)\right)-\sum_{k=1}^{m}\left(\sum_{x_1=0}^{\infty}\sum_{x_2=0}^{\infty}\cdots\sum_{x_n=0}^{\infty}x_i A_k\left(\boldsymbol{x}\right)p\left(\boldsymbol{x},t\right)\right) \tag{7}$$

After adjusting the indexes of the terms at the right hand side, we can obtain:

$$\frac{d\langle x_i\rangle}{dt}=\sum_{k=1}^{m}\left(\sum_{x_1=0}^{\infty}\sum_{x_2=0}^{\infty}\cdots\sum_{x_n=0}^{\infty}\boldsymbol{s}_{ik}A_k\left(\boldsymbol{x}\right)p\left(\boldsymbol{x},t\right)\right). \tag{8}$$

As $\langle A_k(\boldsymbol{x})\rangle=A_k(\boldsymbol{x})p(\boldsymbol{x},t)$, we therefore obtain the deterministic equation of the mean concentration of molecular species X_i as:

$$\frac{d\langle x_i\rangle}{dt}=\sum_{k=1}^{m}\boldsymbol{s}_{ik}\langle A_k(\boldsymbol{x})\rangle. \tag{9}$$

The same equation can be derived from all molecular species $X_i, i=1,\cdots,n$. If we can assume the equality $\langle A_k(\boldsymbol{x})\rangle=A_k(\langle\boldsymbol{x}\rangle)$, then the above equation can be used to accurately describe the average behavior of each molecular species $X_i, i=1,\cdots,n$. However, equality $\langle A_k(\boldsymbol{x})\rangle=A_k(\langle\boldsymbol{x}\rangle)$ only holds when $A_k(\boldsymbol{x})$ is a linear function of $\boldsymbol{x}$. Whenever a reaction has two or more reactants, its reaction rate $A_k(\boldsymbol{x})$ becomes a nonlinear function of $\boldsymbol{x}$, therefore $\langle A_k(\boldsymbol{x})\rangle\neq A_k(\langle\boldsymbol{x}\rangle)$. According to van Kampen, a Taylor expansion of $A_k(\boldsymbol{x})$ can be used to approximate the $\langle A_k(\boldsymbol{x})\rangle$ (Van Kampen, 2007):

$$\langle A_k(\boldsymbol{x})\rangle=A_k(\langle\boldsymbol{x}\rangle)+\frac{1}{2}\left\langle(\boldsymbol{x}-\boldsymbol{x})^2\right\rangle A_k''(\langle\boldsymbol{x}\rangle)+\cdots. \tag{10}$$

By truncating the second order and higher terms, we can approximate the mean reaction rates as:

$$\langle A_k(\boldsymbol{x})\rangle \approx A_k(\langle \boldsymbol{x}\rangle). \tag{11}$$

Therefore, we obtain the deterministic mass action equation as:

$$\frac{dx_i}{dt} = \sum_{k=1}^{m} s_{ik} A_k(\boldsymbol{x}). \tag{12}$$

We can rewrite the equation in matrix form as:

$$\boldsymbol{x}' = \boldsymbol{s}\boldsymbol{A}(\boldsymbol{x}), \tag{13}$$

where is the stoichiometry matrix of the reaction network, and $\boldsymbol{A}(\boldsymbol{x}) = (A_1(\boldsymbol{x}), \cdots, A_m(\boldsymbol{x}))^T$ is a vector of rate functions of all reactions in the network. The above mass action kinetic equations show the relationship between discrete chemical master equation and its corresponding deterministic equations. This is consistent with Kurtz and Keizer's theoretical studies about the relationship between the stochastic and deterministic models of chemical reactions (Keizer, 1977; Kurtz, 1972).

2.2 Approximations

2.2.1 Fokker-Planck Equation

Solving the CME requires knowledge of all details of the transition kernel $A(\boldsymbol{x}, \boldsymbol{x}')$, including both the state space and the transition rates between them. Currently, there exist no general analytical solution to the CME, except for the simplest problems (Schultz et al., 2007; Vellela & Qian, 2007). Solving the CME numerically is also challenging. Nevertheless, for systems of small and moderate size, there exists algorithm that can enumerate optimally the state space, with $A(\boldsymbol{x}, \boldsymbol{x}')$ fully characterized, and the discrete CME can be solved numerically (Cao & Liang, 2008).

One can approximate CME with the Fokker-Planck Equation (FPE). Similar to the continuous CME, the FPE also describe the evolution of probabilities of system states over time, but with the transition kernel in the CME replaced by a differential operator of second order. Unlike the discrete CME, the states of the system are continuous macroscopic states, and FPE does not describe the discrete jump processes (Van Kampen, 2007). We follow the disposition of van Kampen (Van Kampen,

2007) and briefly describe how FPE is related to CME, focusing on the additional assumptions and approximations involved beyond those made for continuous CME.

2.2.1.1 Assumptions of Fokker-Planck Equation

The first assumption is that the jumps between states must be small, namely, the "before" and the "after" states are in a close neighborhood: $\left|x_j - x_i\right| < \epsilon$, where $\epsilon \in \mathbb{R}$ is infinitesimal. Second, the transition probability varies slowly: $A(x,x') \approx A(x+\epsilon, x'+\epsilon)$. Third, the probability $p(x,t)$ must also vary slowly: $p(x,t) \approx p(x+\epsilon,t)$. A consequence of these assumptions is that the transition kernel $A(x,x')$ is differentiable to a higher order.

With these assumptions, the first term in Eqn. (6) can be approximated, where the full detail of the transition kernel $A(x,x')$ from x' to x is needed. The goal is to replace $A(x,x')$ with its Taylor expansion centered around x. For this, we first reparameterize $A(x,x')$ as $A(x;s)$, where $s = x - x'$. Eqn. (6) can be re-written as:

$$\frac{dp(x,t)}{dt} = \int A(x-s;s)\,p(x-s,t)\,ds - p(x,t)\int A(x;-s)\,ds. \tag{14}$$

2.2.1.2 Kramers-Moyal Expansion

The first term of Eqn. (6) is then expanded around using Taylor expansion as:

$$\begin{aligned}\int A(x-s;s)\,p(x-s,t)\,ds = {} & \int A(x;s)\,p(x,t)\,ds - \sum_i \frac{\partial}{\partial x_i}\left[\int s\cdot A(x;s)\,ds\cdot p(x,t)\right] \\ & + \frac{1}{2}\sum_{i,j}\frac{\partial^2}{\partial x_i \partial x_j}\left[\int s^2\cdot A(x;s)\,ds\cdot p(x,t)\right] \\ & - \frac{1}{3!}\sum_{i,j,k}\frac{\partial^3}{\partial x_i \partial x_j \partial x_k}\left[\int s^3\cdot A(x,s)\,ds\cdot p(x,t)\right] + \cdots\end{aligned} \tag{15}$$

Putting it back to Eqn. (6), and approximating by dropping higher than second order terms, we obtain the Fokker-Planck equation:

$$\frac{\partial p(x,t)}{\partial t} = -\sum_i \frac{\partial}{\partial x_i}\left[\int s\cdot A(x;s)\,ds\cdot p(x,t)\right] + \frac{1}{2}\sum_{i,j}\frac{\partial^2}{\partial x_i \partial x_j}\left[\int s^2\cdot A(x;s)\,ds\cdot p(x,t)\right] \tag{16}$$

Using a simpler notation, we have:

$$\frac{dp(\boldsymbol{x},t)}{dt} = -\sum_i \frac{\partial}{\partial x_i}\left[F(\boldsymbol{x})p(\boldsymbol{x},t)\right] + \frac{1}{2}\sum_{i,j}\frac{\partial^2}{\partial x_i \partial x_j}\left[G(\boldsymbol{x})p(\boldsymbol{x},t)\right]. \tag{17}$$

Where $F(\boldsymbol{x}) = \int \boldsymbol{s}\cdot A(\boldsymbol{x};\boldsymbol{s})d\boldsymbol{s}$ and $G(\boldsymbol{x}) = \int \boldsymbol{s}^2 \cdot A(\boldsymbol{x};\boldsymbol{s})d\boldsymbol{s}$.

2.2.2 Chemical Langevin Equation

When the macroscopic behavior of the system can be determined, a general approach to study stochastic dynamic network is to follow the Langevin equation by combining a deterministic term describing the macroscopic behavior of the network and a diffusion term describing the stochastic fluctuation.

From the dCME, we can derive the average behavior of the network using the Mayo expansion as:

$$\frac{d\overline{\boldsymbol{x}}(t)}{dt} = \sum_x \frac{d\left[x\cdot p(\boldsymbol{x},t)\right]}{dt} = \frac{d\sum_x\left[x\cdot p(\boldsymbol{x},t)\right]}{dt}, \tag{18}$$

where $\overline{\boldsymbol{x}}(t) = \sum_x x\cdot p(\boldsymbol{x},t)$ is the vector of expected mean value of the amount of molecular species. Van Kampen and others have shown that the following formula could be derived from the dCME for the mean amount of molecular species (Gillespie, 2007; Melykuti, Burrage, & Zygalakis, 2010; Van Kampen, 2007):

$$\frac{d\overline{\boldsymbol{x}}(t)}{dt} = \sum_{k\in \boldsymbol{R}} s_k \overline{A}_k(\boldsymbol{x}), \tag{19}$$

where s_k and $A_k(\boldsymbol{x})$ are defined in Eqn. (2) and Eqn. (3). This gives the same expression as that of corresponding ordinary differential equations (ODE) of mass action kinetics of the same chemical reactions. The detailed derivation is shown in following sections.

Random fluctuations in the copy numbers of molecules occur because of the random jumps due to the spontaneous firing of reactions. Such reactions will introduce changes to the copy number of molecular species, *e.g.*, by the amount of $\boldsymbol{s}_k$ of the k-th reaction. Assuming that the jump is small, namely,

$\boldsymbol{x}(t+\Delta t)=\boldsymbol{x}(t)+\boldsymbol{s}_k \approx \boldsymbol{x}(t)$ if reaction $\boldsymbol{s}_k$ occurs during an infinitesimally small time interval Δt. This assumption would result in unchanged reaction rate:

$$A_k\left(\boldsymbol{x}(t+\Delta t)\right) \approx A_k\left(\boldsymbol{x}(t)\right) = r_k \prod_{l=1}^{n} \binom{x_l}{c_{lk}}. \tag{20}$$

With this assumption, the vector of the amount of molecular species $\boldsymbol{x}(t+\Delta t)$ at time $t+\Delta t$ can be written as:

$$\boldsymbol{x}(t+\Delta t) = \boldsymbol{x}(t) + \sum_{R_k \in \boldsymbol{R}} n_k\left(\boldsymbol{x}, \Delta t\right) \cdot \boldsymbol{s}_k. \tag{21}$$

where $n_k(\boldsymbol{x}, \Delta t)$ is the number of reaction R_k occurs during the period Δt. Based on the assumption $\boldsymbol{x}(t+\Delta t)=\boldsymbol{x}(t)$, the copy numbers of molecular species in Eqn. (21) do not change during Δt, and the reaction rates also do not change during Δt. Therefore, all reactions occurring during Δt could be considered independent of each other.

This assumption is valid only if the copy numbers in $\boldsymbol{x}(t)$ are all large, so the stoichiometry coefficients c_i forming the jump vector $\boldsymbol{s}_k$ are all comparatively small. This assumption clearly breaks down when the copy number of molecular species is not significantly larger than the stoichiometry coefficients, and therefore cannot be used to describe systems of a handful of particles.

A reasonable model for the random variable $n_k(\boldsymbol{x}, \Delta t)$ is that of the Poisson distribution: $n_k \sim P(\lambda_k)$ with $\lambda_k = A_k(\boldsymbol{x}(t)) \cdot \Delta t$. With the additional assumption that Δt is sufficiently long such that a large number (>>1) of reactions occur during Δt, the Poisson distribution for the number of spontaneous reactions can be approximated by a Gaussian random variable (Gillespie, 2000). The approximation will be very accurate when λ_k is sufficiently large, *e.g.*, $\lambda_k > 1000$. Therefore, we now have $n_k \sim N(\mu, \sigma^2)$, with $\mu = \sigma^2 = \lambda_k$, or alternatively, $n_k \sim \lambda_k + \lambda_k^{\frac{1}{2}} N(0,1) = A_k(\boldsymbol{x}(t)) \cdot \Delta t + \left[A_k(\boldsymbol{x}(t)) \cdot \Delta t\right]^{\frac{1}{2}} N(0,1)$. With these assumptions, the fluctuations of the amount of molecules follow m independent Gaussian processes, one for each reaction:

$$x(t+\Delta t) = x(t) + \sum_{R_k \in \boldsymbol{R}} A_k(x(t)) \cdot \Delta t \cdot s_k + \sum_{R_k \in \boldsymbol{R}} \left[A_k(x(t)) \cdot \Delta t\right]^{\frac{1}{2}} \cdot s_k \cdot N(0,1). \tag{22}$$

This leads to the following equation:

$$\frac{\partial x(t)}{dt} = \sum_{R_k \in \boldsymbol{R}} A_k(x(t)) \cdot s_k + \sum_{R_k \in \boldsymbol{R}} \left[A_k(x(t))\right]^{\frac{1}{2}} \cdot s_k \cdot \left[\frac{1}{\partial t^{\frac{1}{2}}} \cdot N(0,1)\right]. \tag{23}$$

We denote $G(t) = \frac{1}{\partial t^{\frac{1}{2}}} \cdot N(0,1) = N\left(0, \frac{1}{\partial t}\right)$, we have the Langevin equation:

$$\frac{\partial x(t)}{\partial t} = \sum_{R_k \in \boldsymbol{R}} A_k(x(t)) \cdot s_k + \sum_{R_k \in \boldsymbol{R}} \left[A_k(x(t))\right]^{\frac{1}{2}} \cdot s_k \cdot G(t). \tag{24}$$

This is same as the chemical Langevin equation described by Gillespie (Gillespie, 2000).

2.2.3 Accuracy of Chemical Fokker-Planck and Langevin Equation

As shown above, due to the second-order truncation in the Kramers-Moyal expansion during the derivation of the Fokker-Planck and Langevin equation from the chemical master equation, the accuracy and error of these equations need to be clarified. Using the system-size expansion, Grima *et al.* (Grima, Thomas, & Straube, 2011) have shown that the estimates for the mean and variance of concentrations using the chemical Fokker-Planck equation are accurate to order $\Omega^{-\frac{3}{2}}$ for reaction networks which do not obey detailed balance, and at least accurate to order Ω^{-2} for systems obeying detailed balance, where Ω is the characteristic size of the system.

2.3 Monte-Carlo Simulation of the dCME

A widely used method to study stochastic networks is to carry out Monte Carlo simulations. By following the trajectories of reactions, one can gather statistics of reaction events at different time to gain understanding of the network behavior (Gillespie, 1977). We discuss the underlying algorithm, called the *stochastic simulation algorithm* (SSA), which is also known as *Gillespie's* algorithm.

Reaction probability: We denote the probability that after the last reaction at t, the next reaction, which happens to be the k-th reaction, occurs during an infinitesimally small time interval dt after a time interval Δt of no reaction occurring to be:

$$p\left[\boldsymbol{x}\left(t + \Delta t + dt\right), \Delta t, R_k | \boldsymbol{x}\left(t\right)\right] dt. \tag{25}$$

If we divide the time interval Δt into H equal subintervals, and assume that the occurrence of reactions following a Poisson process, the probability that none of the n reactions have occurred during the time prior to the end of a small time interval $\epsilon = \Delta t / H$ is:

$$\prod_{k=1}^{n}\left[1 - A_k\left(\boldsymbol{x}(t)\right)\epsilon\right] \approx \sum_{k=1}^{n}\left[1 - A_k\left(\boldsymbol{x}(t)\right)\epsilon\right]. \tag{26}$$

As the probability of no reactions for each of the H intervals is the same, no reactions have occurred during "t is:

$$\lim_{H\to\infty}\sum_{k=1}^{n}\left[1 - A_k\left(\boldsymbol{x}(t)\right)\epsilon\right]^H = e^{-A(\boldsymbol{x}(t))" t}. \tag{27}$$

As no reaction has occurred during Δt, the microstate remains the same. Therefore, the instantaneous state transition probability for reaction R_k occurring during the infinitesimal interval $\left[t + \Delta t, t + \Delta t + dt\right]$ is $A_k\left[\boldsymbol{x}\left(t + \Delta t\right)\right] dt = A_k\left(\boldsymbol{x}\left(t\right)\right) dt.$ We have:

$$p\left[\boldsymbol{x}\left(t + \Delta t + dt\right), \Delta t, R_k | \boldsymbol{x}\left(t\right)\right] = A_k\left[\boldsymbol{x}\left(t\right)\right] e^{-A(\boldsymbol{x}(t))\Delta t} dt. \tag{28}$$

Reaction trajectory. Let the microstate of the system at time t to be $\boldsymbol{x}\left(t\right)$. After time interval Δt, reaction R_k occurs at an infinitesimally small time interval dt at $t + \Delta t$, and the system is brought to the state $\boldsymbol{x}\left(t + \Delta t + dt\right)$. We assume the reactions are occurring instantaneously, *i.e.* $dt \to 0$, so we use $\boldsymbol{x}\left(t + \Delta t\right)$ to denote the new microstate after the reaction occurred in dt. We can observe the trajectory of a sequence of such reactions. Starting from state $\boldsymbol{x}\left(0\right)$ at time $t_0 = 0$, after a series of time intervals $\left(\Delta t_0, \Delta t_1, \cdots, \Delta t_{T-1}\right)$, the system reaches the state $\boldsymbol{x}\left(t_T\right)$, after traversing the sequence of states $\left(\boldsymbol{x}(t_0),\boldsymbol{x}(t_1),\cdots,\boldsymbol{x}(t_{T-1})\right)$, with a reaction sequence $\left(R_{k_0},R_{k_1},\cdots,R_{k_{T-1}}\right)$ occurring along the way. Let $\left(t_1,t_2,\cdots,t_T\right)$ be the sequence of time points when a reaction occurs. The trajectory of reactions can be denoted as:

$$\left(\boldsymbol{x}(t_0);\boldsymbol{x}(t_1),R_{k_0};\cdots;\boldsymbol{x}(t_{T-1}),R_{k_{T-2}};\boldsymbol{x}(t_T),R_{k_{T-1}}\right).$$

Alternatively, we can denote the time intervals by its increments $\left(\Delta t_0,\Delta t_1,\cdots,\Delta t_{T-1}\right)$.

Probability of a reaction trajectory. Assuming a Markovian process, namely, future reactions depends only on the current state but not on any past state, the probability associated with a time trajectory is:

$$\begin{aligned}&\pi\left[\boldsymbol{x}(t_0);\boldsymbol{x}(t_1),R_{k_0};\cdots;\boldsymbol{x}(t_{T-1}),R_{k_{T-2}};\boldsymbol{x}(t_T),R_{k_{T-1}}\right]\\&=\pi\left[\boldsymbol{x}(t_1),\Delta t_0,R_{k_0}|\boldsymbol{x}(t_0)\right]\cdot\pi\left[\boldsymbol{x}(t_2),\Delta t_1,R_{k_1}|\boldsymbol{x}(t_1)\right]\cdots\pi\left[\boldsymbol{x}(t_T),\Delta t_{T-1},R_{k_{T-1}}|\boldsymbol{x}(t_{T-1})\right].\end{aligned} \tag{29}$$

In principle, the probability of starting from state $\boldsymbol{x}(t_0)$ and reaching state $\boldsymbol{x}(t_T)$ can then be obtained by integrating over all possible paths:

$$\pi\left[\boldsymbol{x}(t_T)|\boldsymbol{x}(x_0)\right]=\sum_{(\Delta t_0,\cdots,\Delta t_{T-1}),(R_{k_0},\cdots,R_{k_{T-1}})}\pi\left[\boldsymbol{x}(t_0);\boldsymbol{x}(t_1),R_{k_0};\cdots;\boldsymbol{x}(t_{T-1}),R_{k_{T-2}};\boldsymbol{x}(t_T),R_{k_{T-1}}\right] \tag{30}$$

2.3.1 Stochastic Simulation Algorithm

If we can generate many independent samples of reaction trajectories that follow a proper probabilistic model starting from the same initial state, we can study the behavior of the stochastic network by analyzing these sample trajectories. The stochastic simulation algorithm (SSA) or Gillespie's algorithm was designed to perform such simulations (Gillespie, 1977). It is summarized in the Algorithm in Figure 1.

Generating random variables for selecting the time interval $\boldsymbol{\Delta t}$ and reaction $\boldsymbol{R_k}$. A key component in Gillespie's algorithm is to generate a pair of random variables $\left(\Delta t,k\right)$. The Δt will be the time interval until next reaction occurs, and the number *k* will specify the reaction R_k that will actually occur next. We have: $\pi\left(\Delta t,k\right)=\pi_1\left(\Delta t\right)\cdot\pi_2\left(R_k|\Delta t\right)$, where $\pi_1\left(\Delta t\right)$ is the probability that the next

Figure 1. Stochastic Simulation Algorithm

Algorithm 1 Stochastic Simulation

Set initial microstate: $x(0) \leftarrow \{x_1(0), \cdots, x_m(0)\}$
Set total simulation time: t_T.
Set $t = 0$.
while $t < t_T$ **do**
  Compute rates $A_k(x(t))$ for all reactions R_k, $k = 1, \cdots, n$, in current state $x(t)$.
  Let $A(x(t)) = \sum_k A_k(x(t))$.
  Generate two uniformly distributed random numbers $(r_1, r_2) \sim U[0, 1]$.
  Compute $\Delta t = \frac{1}{A(x(t))} \ln \frac{1}{r_1}$.
  Find an integer $1 \le k \le m$ such that $\sum_{i=1}^{k-1} A_i(x(t)) < r_2 A(x(t)) \le \sum_{i=1}^{k} A_i(x(t))$.
  Update the microstate of the system $x(t + \Delta t) = x(t) + s_k$, where s_k is the stoichiometry vector of the occurred reaction R_k in Δt.
  Update $t \leftarrow t + \Delta t$.
end while

reaction, regardless which specific ones, will occur in time interval $\left[t + \Delta t, t + \Delta t + dt\right]$, and $\pi_2(R_k \mid \Delta t)$ is the probability that the next reaction will be the k-th reaction. As $\pi_1\left(\Delta t\right) = \sum_i \pi\left(i, \Delta t\right)$, where $\pi\left(i, \Delta t\right)$ is the probability that reaction R_i occurs at time $t + \Delta t + dt$, we have $\pi_2\left(R_k|\Delta t\right) = \dfrac{\pi\left(k, \Delta t\right)}{\sum_i \pi\left(i, \Delta t\right)}$.

As we assume a model of Poisson process, we have:

$$\pi_1\left(\Delta t\right) = A\left(\boldsymbol{x}\left(t\right)\right) e^{-A(\boldsymbol{x}(t))\Delta t}, \text{ and } \pi_2\left(R_k|\Delta t\right) = \frac{A_k\left(\boldsymbol{x}\left(t\right)\right)}{A\left(\boldsymbol{x}\left(t\right)\right)}. \tag{31}$$

That is, if we can generate a random variable Δt following $\pi_1\left(\Delta t\right)$, and another random integer k according to $\pi_2\left(R_k|\Delta t\right)$, the resulting pair $\left(\Delta t, k\right)$ will follow the desired distribution $\pi\left(\Delta t, k\right)$.

Assume we can generate a random number r following the uniform distribution $r \sim U\left[0,1\right]$. A general approach to obtain a random variable ρ that follows a distribution F is to calculate the transformation of r through the inverse function $F^{-}: \rho = F^{-}\left(r\right)$. Since $\pi_1\left(\Delta t\right) = e^{-A(x(t))\Delta t}$, we can have:

$$\Delta t = \frac{1}{A\left(\boldsymbol{x}\left(t\right)\right)} \ln \frac{1}{r_1}, \text{ where } r_1 \sim U\left[0,1\right]. \tag{32}$$

To sample the next reaction R_k, we can again first generate a uniformly distributed random number $r_2 \sim U[0,1]$. We can then take the k-th reaction such that:

$$\sum_{i=1}^{k-1} A_i(x(t)) < r_2 A(x(t)) \leq \sum_{i=1}^{k} A_i(x(t)), \text{ where } A(x(t)) = \sum_{i=1}^{m} A_i(x(t)). \tag{33}$$

Another approach to generate a pair of random number $(\Delta t, k)$ is to first calculate the probability at time $t + \Delta t$ for a reaction R_i to occur during an infinitesimal time interval $[t + \Delta t, t + \Delta t + dt]$, assuming that there were no changes between $x(t)$ and $x(t + \Delta t)$, namely, there is no reaction occurred. We can generate a tentative reaction time Δt_l for reaction R_l as:

$$\Delta t_l = \frac{1}{A_l(x(t))} \ln \frac{1}{r_l}, \text{ where } r_l \sim U[0,1]. \tag{34}$$

From this set of random number pairs $(\Delta t_l, l)$ for all $l = 1, \cdots m$, we select the pair of random numbers with the shortest Δt, at which the next reaction R_k would occur:

$$\Delta t = \min(\Delta t), \text{ and } k = \arg_l \min(\Delta t_l). \tag{35}$$

Remark. There are a number of issues in carrying out studies using stochastic simulation, as adequate sampling is challenging when the network becomes complex. There is no general guarantee that simulation can provide a full account of the network stochasticity, as it is difficult to determine whether simulations are extensive enough for accurate statistics. It is also difficult to determine whether adequate sampling has been achieved for individual trajectory. In addition, it is often difficult to characterize rare events that may be biologically important, as simulations follow high probability paths. Much recent work has been focused on improving SSA, for example, by introducing data structure so the generation of the two random variables of Δt and reaction *k* is more efficient. In addition, an approach to speed up SSA is to find the best time step Δt such that the copy numbers of the molecular species, hence the reaction rates, do not change much, so the simulation can leap forward with large time step (Cao, Gillespie, & Petzold, 2006).

2.3.2 Biased Sampling for Rare Events

Recent interests in introducing bias in selection of the next reaction, and in altering the reaction rate showed promise in improved sampling of rare events (Daigle et al., 2011; Kuwahara & Mura, 2008; Roh et al., 2011). The techniques of importance sampling and reweighting can improve sampling efficiency significantly. They have been widely used in equilibrium sampling where the condition of detailed balance holds (Torrie & Valleau, 1977). However, stochastic processes in reaction networks are generally not time reversible and the condition of detailed balance is not valid. Kuwahara and Mura developed the weighted SSA (wSSA) algorithm by applying the importance sampling technique to study stochastic reaction networks, in which each reaction rate is biased by a pre-determined constant, with the overall summation of reaction rates unchanged (Kuwahara & Mura, 2008). As the probability for reaction selection can be biased such that rare events are sampled more frequently while the time scale of the underlying reactions is maintained, significantly improved sampling efficiency for rare events was reported (Daigle et al., 2011; Kuwahara & Mura, 2008; Roh et al., 2011). However, the choice of bias constants strongly affects the effectiveness of wSSA. When there are many reactions and the network is complex, the heuristic approach of determining bias constants by examining the reactions does not work (Kuwahara & Mura, 2008). As there is no general guidance in how bias constants should be chosen, poor choices may lead to estimations that are less accurate than the original SSA (Daigle et al., 2011).

Daigle *et al.* developed the doubly-weighted SSA (dwSSA) algorithm, in which a multilevel cross-entropy (CE) method is used iteratively to provide estimates of bias constants (Daigle et al., 2011). This is achieved by running long trial simulations until a fraction of the sampled trajectories reaches the target states (Daigle et al., 2011). With this automated estimation, both reaction selection and the underlying time scale of reactions can be biased (Daigle et al., 2011).

A drawback of methods using constant biases such as wSSA and dwSSA is that the bias coefficients are global and state-independent, and are not influenced by the concentrations of molecules which evolve with time. As the apparent rate of a reaction can vary dramatically depending on the copy number of molecules, the degree of bias for a reaction therefore need to be adjusted according to the available copy numbers of reactants. With globally fixed bias constants, a network with reactions of a wide range of rates will have over- and under-biased reactions, depending on the states of the system. As a result, estimated properties of a network will have large variance, making these methods unsuitable for complex networks (Roh et al., 2010).

Roh *et al.* developed a state-dependent biasing wSSA method (swSSA) (Roh et al., 2010). By empirically classifying reactions into groups of favored, disfavored, and neutral reactions, biases in selection probability for reactions in the first two groups are calculated in a state-dependent fashion. The swSSA method can have better estimation accuracy and efficiency than the wSSA method (Roh et al., 2010), at the expense of about twice as many biasing parameters as that of the wSSA (Roh et al., 2010). Roh *et al.* further developed the state-dependent doubly weighted SSA method (sdwSSA), where reactions are further grouped into bins according to their selection probabilities, and are assigned different bias constants, which are automatically estimated using the cross-entropy method (Roh et al., 2011). However, the number of parameters to be estimated using sdwSSA is much larger than that of wSSA, dwSSA, and swSSA. For example, about 20 bias constants need to be estimated for a simple reversible isomerization system with only two reactions (Roh et al., 2011). Estimating a large number of bias constants needed for complex networks becomes difficult.

2.3.3 Adaptively Biased Sequential Importance Sampling for Rare Events (ABSIS)

To more efficiently sample rare events in biological networks, a novel biased sampling method named *adaptively biased sequential importance sampling* (ABSIS) has been developed by adaptively adjusted bias of reactions based on look-ahead strategy. The method has shown that barrier-crossing can be engineered for efficient and accurate sampling of rare events (Cao & Liang, 2013). Based on the principle of sequential importance sampling, the ABSIS approach adopts the look-ahead strategy, a technique well-established in polymer and protein studies (Liang, Zhang, & Chen, 2002; Lin, Chen, & Liu, 2009; Meirovitch, 1982, 1988), to gather future information for design of bias parameters to enable effective barrier crossings (Liang et al., 2002; Lin et al., 2009; Liu, Chen, & Logvinenko, 2001; Meirovitch, 1982, 1988). By enumerating short paths from the current state, bias coefficients are can be dynamically generated according to the need for barrier-crossing to effectively steer the sample paths towards the rare event destination. Unlike the dwSSA and sdwSSA methods, in which biases are fixed constants after parameter estimation, the biases in ABSIS for each reaction is dynamically determined based on exact calculation of the total probability of short κ-step forward- and backward-moving reaction paths, without the need of binning reaction rates. Reactions with higher probability of forward-moving are then encouraged, and reactions with higher probability of backward-moving are discouraged. Regardless of the number of reactions in the networks, the ABSIS only need to assign two bias parameters for the whole network: the degree to encourage forward-moving reactions and the degree to discourage backward-moving reactions, which both can be estimated through an efficient parameter estimation algorithm.

3. DIRECT SOLUTIONS TO THE dCME

The discrete chemical master equation provides a fundamental framework for studying the full stochasticity of biological networks (Gillespie, 1977). Under the definition of biological networks, the combination of copy numbers of molecular species defines the microstate of the molecular interactions in the network. By treating microscopic states of molecular species explicitly, linear and nonlinear reactions (such as synthesis, degradation, dimeric binding, and multimerization) can all be effectively modeled as transitions between microstates, with transition rates determined by the physical properties of the molecules and the cell environment. The probability distribution or potential landscape (Ao et al., 2007; Cao & Liang, 2008; Cao et al., 2010) over these microstates and its time-evolving and steady state behavior provide a full description of the properties of a stochastic molecular network.

3.1 Finite State Projection (FSP) Method for Direct Solution of Time Evolution

As biological networks often involve synthesis reactions, such as protein synthesis, the size of the state space can be potentially infinite. The finite state projection (FSP) method is based on a truncated projection of the state space and uses numerical techniques to compute direct solution to the dCME [29, 33]. Munsky and Khammash made two insightful observations. Denote two sets of indices of the microstates being chosen as J_1 and J_2, and assume $J_1 \subseteq J_2$. The truncated state spaces obtained by selecting states from J_1 and J_2 are $\boldsymbol{S}_{J_1}$ and $\boldsymbol{S}_{J_2}$, and the corresponding rate matrices are $\boldsymbol{A}_{J_1}$ and $\boldsymbol{A}_{J_2}$, respectively. The first observation is:

$$\left[e^{A_{J_2}}\right]_{J_1} \geq e^{A_{J_1}} \geq 0. \tag{36}$$

This assures that the time evolution probability distribution solved on the larger state space $\boldsymbol{S}_{J_2}$ is not smaller than that on $\boldsymbol{S}_{J_1}$:

$$\left[e^{A_{J_2}t}\right]_{J_1} \boldsymbol{p}\left(\boldsymbol{S}_{J_1},0\right) \geq e^{A_{J_1}t} \boldsymbol{p}\left(\boldsymbol{S}_{J_1},0\right). \tag{37}$$

This inequality implies that by increasing the size of the selected subset of states, the approximation improves monotonically. The second observation is, if one obtains a truncated state space by selecting states contained in the index set J, and if $1^T e^{tA_J} p\left(\boldsymbol{S}_J,0\right) \geq 1-\epsilon$ for $\epsilon > 0$, then:

$$e^{tA_J} p(S_J,0) \le p(S_J,t) \le e^{tA_J} p(S_J,0) + \epsilon 1. \tag{38}$$

That is, starting with the initial probability vector $p(S_J,0)$ on the truncated state space, we can compute the probability vector in the truncated space as $e^{tA_J} p(S_J,0)$ at time t using the truncated rate matrix A_J. If the inner-product of this vector with 1 is no less than $1-\epsilon$, the difference of this vector from the projected true vector $p(S_J,t)$ of the true probability $p(S,t)$ is also no more than $\epsilon 1$. This inequality guarantees that the approximation obtained with truncated state space will never exceed the actual solution, and its error is bounded by ϵ (Munsky & Khammash, 2006).

These key observations led to the Finite State Projection algorithm, which iteratively adds more states to an initial truncated state space, until the approximation error is within a prescribed bound (Munsky & Khammash, 2006). The original Finite State Projection method was further extended (Munsky & Khammash, 2007), and it was recommended that the initial non-sparse probability vector $p(S,0)$ should be determined by running a few steps of stochastic simulation discussed in a previous section.

Although the error due to state space truncation can be estimated (Munsky & Khammash, 2006), there are no systematic guidance and general strategy as to which states and how many of them should be incorporated to a finite projection to most effectively improve the approximation accuracy (Munsky & Khammash, 2006). Furthermore, the introduction of an absorption state where all truncated states are projected to will lead to accumulation of errors as time proceeds. As the absorbing state would eventually absorb all probability mass, the FSP method therefore is not appropriate for computing the steady state probabilistic landscape, as the approximation of the absorbing state will lead to errors that increase to 1 in the steady state.

3.2 Challenges to Direct Solutions: Enormous State Space

A complete identification and characterization of the space of the microstates is prerequisite to directly solve the dCME and obtain the full probability landscapes of a network. However, it is challenging to enumerate an optimal state space of a network.

A straight-forward method is to enumerate all microstates in the whole hyper-cube space limited by the maximum copy numbers of reactants. The size of the state space enumerated using such a hyper-cube method is bounded by the product of the maximum copy numbers, which can be quickly inflated to enormity. For example, if there are 10 species, and each with a maximum of 30 molecules in the network, without taking into consideration of the details of the network, the hyper-cube approach will generate a state space with $30^{10} = 5.9 \times 10^{14}$ states, which would easily exceed the memory capacity of most computers. However, the state space can be significantly reduced if detailed reaction schemes are considered during state space enumeration. There are many states that the system can never reach from a certain initial state. There are many states that no reaction can happen, therefore not needed. Alternatively, one can follow explicitly simulated reaction events by carrying out stochastic simulations. However, this approach cannot guarantee that all reachable states will be explored, therefore cannot guarantee full characterization of rare events. Moreover, it can be highly inefficient in enumerating microstate, because most of computing time will be spent on revisiting those high-probability states.

To address this challenging issue, Cao and Liang have developed a novel method, namely the *finite buffer* method (Cao & Liang, 2008), to significantly reduce the size of the state space of arbitrary biological network. The state space enumeration algorithm has been shown to be optimal in terms of time and space complexity. The state space enumerated using the *finite buffer* method has been dramatically reduced comparing with the conventional hyper-cube approach (*e.g.* a reduction of factor 10^9 can be achieved in the MAPK network). Furthermore, the finite buffer method uses a reflective boundary, rather than the absorbing boundary state in the FSP method, therefore both time-evolving and steady state probability landscapes can be solved using the state space generated with the *finite buffer* method. Furthermore, the reflective boundary adopted in the finite buffer method allows the quantification for the truncation error in the steady state probability landscape (to be published).

3.3 State Space Enumeration and Transition Rate Matrix Construction

The technique of optimally enumerating microstates for a given initial condition now allows certain realistic systems to be studied using dCME, under the condition of finite buffer (Cao & Liang, 2008). Below we describe how microstates can be optimally enumerated. For a network with n molecular species and m reactions, we generate all microstates that the network can reach starting from a given initial condition, under the *finite buffer* constraint. We use a buffer queue of finite capacity to represent a reservoir of molecular tokens. Each time when a synthesis reaction occurs, a buffer token is spent to generate the new molecule, and each time when a degradation reaction happens, a buffer token is released and deposited back to the buffer queue. Synthesis reaction is allowed to occur only if the buffer queue is not exhausted. This is necessary due to the limitation of computing resources. As the microstate of a specific combination of copy numbers is $\boldsymbol{x} = (x_1, \cdots, x_m)$, we add an additional virtual buffer species B and its copy number b to denote the number of currently available tokens in the buffer queue, namely, the number of net new molecules that can still be synthesized at current microstate. Therefore, the microstate vector of the network can be augmented as $\boldsymbol{x} = (x_1, \cdots, x_m, b)$ by adding the copy number of buffer token B. A synthesis reaction can only occur when $b > 0$ in the *finite buffer* state enumeration algorithm.

Under these conditions, the set of all possible microstates that can be reached from an initial condition constitute the state space of the system. The set of allowed transitions is $\boldsymbol{T} = (T_{i,j})$, in which $t_{i,j}$ maps the microstate $\boldsymbol{x}_j$ to the microstate $\boldsymbol{x}_i$ by one reaction. The initial state of the reaction system is given as: $\boldsymbol{x}(0) = (x_1(0), x_2(0), \cdots, x_m(0), b(0))$, where $x_i(0)$ is the initial copy number of the i-th molecular species at time $t = 0$, and $b(0)$ is the predefined buffer capacity.

The *finite buffer* algorithm for enumerating the state space is summarized in the Algorithm in Figure 2. The state space and the reactions constitute a huge network with microstates as nodes and reactions as edges. The state space enumeration can then be projected to a network traversal problem. After initialization, the *finite buffer* algorithm starts with the given initial microstate $\boldsymbol{x}(0)$ to generate all possible microstates with no consideration of reaction rates. Each reaction is examined in turn to determine if this reaction can occur from the current microstate. If so, and if the buffer is not used up, the state that this reaction leads to is generated. If the newly generated state has never been visited before, we declare it as a new state and add it to our collection of states for the state space. We repeat this process for all new states, with the aid of a stack data structure (Cormen, Leiserson, & Rivest, 1990). This process terminates when all new states are exhausted (Cao & Liang, 2008). Note that a queue data structure can also be used in place of the stack. Both data structure generate the exact same state space, but with different sequences, as using the stack corresponds to a depth-first search (DFS) of the network, while the queue corresponds to a breadth-first search (BFS) (Cormen et al., 1990).

Under the finite buffer constraint, the time complexity of this algorithm is optimal. Since only unseen state will be pushed onto the stack, every state is pushed and popped at most once, and each state will be generated/visited at most twice before it is popped from the stack. As access to each state and to push/pop operations take $O(1)$ time, the total time required for the stack operations is $O(|\Omega|)$, where Ω is the state space. As the *finite buffer* algorithm examines all reactions for

Figure 2. The finite buffer state space enumeration algorithm

```
Algorithm 2 Finite Buffer State Space Enumeration(M, R, B)
Network model: N ← {M, R};
Initial condition: x(0) ← {x_1(0), x_2(0), ..., x_m(0)};
Set the value of buffer capacity: b ← B;
Initialize a state space: S ← ∅;
Stack ST ← ∅;   Push(ST, x(0));   StateGenerated ← FALSE
while ST ≠ ∅ do
  x ← Pop (ST);
  if x ∉ S then
    S ← S ∪ x;
    for k = 1 to n do
      if reaction R_k can occur in state x then
        if reaction R_k is a synthesis reaction and generates u new molecules then
          b ← b − u
          if b ≥ 0 then
            Generate state y = x + s_k, where s_k is the stoichiometry of R_k;
            StateGenerated ← TRUE
          end if
        else
          if reaction R_k is a degradation reaction and breaks down r molecules then
            b ← b + r
          end if
          Generate state y = x + s_k;
          StateGenerated ← TRUE
        end if
        if (StateGenerated = TRUE) then
          Push(ST, y);
          StateGenerated ← FALSE;
        end if
      end if
    end for
  end if
end while
Output S.
```

each reached state, the complexity of total time required is $O\left(m\left|\Omega\right|\right)$, where m is usually a modest constant (e.g. < 50). Based on the same argument, it is also easy to see that the algorithm is optimal in storage, as only valid states and valid transitions are recorded. Using this algorithm, all states reachable from an initial condition within the finite buffer constraint will be accounted for, and no irrelevant states will be included. Furthermore, all possible transitions will be recorded, and no infeasible transitions will be attempted (Cao & Liang, 2008).

In the *finite buffer* algorithm, the reaction rates $\left\{A_{i,j}\right\}$ between any two microstates $\boldsymbol{x}_j$ and $\boldsymbol{x}_i$ in the transition rate matrix $\boldsymbol{A}$ for the dCME are computed following the same approach outlined in references (Gillespie, 1977; Munsky & Khammash, 2006; Schultz et al., 2007). We give further details in later sections on how this is done for two realistic networks. The transition rate matrix $\boldsymbol{A}$ is highly sparse. We will also discuss numerical methods for finding the solutions to the dCME.

With this optimal method for dramatically reducing the state space and enumerating the microstates of a finite system, numerical methods for solving large linear systems can be applied to efficiently solve the dCME. Realistic biological systems can now be directly studied, such as the decision network of phage lambda (Cao & Liang, 2008; Cao et al., 2010).

3.3.1 Open and Close Network, Buffer Capacity

The finite buffer method can be applied to efficiently enumerate state spaces for both open and closed networks. In closed networks, buffer queues are not necessary for state space enumeration, only the initial state is required. In open networks, the capacity of the buffer queue introduced in the finite buffer method to limit the synthesis reactions actually put a restriction on the total mass of the network. That is, the state space enumerated using the finite buffer method consists of microstates that can be reached from the initial state and with total mass smaller than the limit of buffer capacity. Note that, by introducing a buffer queue to control the synthesis reaction, an open network is effectively transformed into a closed network, if considering the buffer queue as a virtual molecule.

3.3.2 State Space Enumeration as a Graph Traverse Problem

The state space and the transitions under a given initial condition can be considered as a directed graph $\boldsymbol{G}=\left(\boldsymbol{S},\boldsymbol{T}\right)$, in which microstates are vertices, and allowed transitions between any two states are edges, *i.e.*, reactions connecting two states. Two vertices $\boldsymbol{x}_j \in \boldsymbol{S}$ and $\boldsymbol{x}_i \in \boldsymbol{S}$ are connected by a directed edge $t_{i,j} \in \boldsymbol{T}$ if and only if $\boldsymbol{x}_j$ can be transformed to $\boldsymbol{x}_i$ through a reaction. As only states reachable from the specified initial state through one or more steps of reactions are concerned, the directed graph G is a connected graph.

It can be shown that the finite buffer algorithm must generate a finite set of states when starting from an initial state with a specified buffer capacity. Our algorithm implicitly generates the graph $\boldsymbol{G}$. Due to the finite set of reactions R in the network,

G will have a finite number of total child branches at any finite steps away from the initial state. Suppose the algorithm will not terminate in finite steps. Since each state is only visited no more than twice in the finite buffer algorithm, G must have an unlimited depth. That is, there must exist an infinite path with all different microstate in the graph G that starts from the initial state and extends to infinity. This is impossible for any initial condition, as each molecular species has a limited initial copy number, and the buffer capacity limits the number of new molecules that can be synthesized in open systems. Therefore, such infinite paths do not exist. The finite buffer algorithm therefore must terminate in finite steps to generate a finite state space.

3.4 Steady State Probability Landscape

The steady state probability landscape over the state space can be computed by directly solving the linear equation $A\pi = 0$, where A is the transition rate matrix constructed using the finite buffer method, and π is the steady state probability distribution over the enumerated state space. The steady state probability distribution π is a high-dimensional vector, it can be further projected to a lower dimensional space of biological interest. A number of iterative numerical techniques can be applied to solve the equation, such as the Gauss-Seidel method (Golub & Van Loan, 1996) and the bi-conjugate gradient stabilized method (BiCGSTAB) (Saad, 2003).

Alternatively, the transition rate matrix A can be transformed into a probability transition matrix M following a uniformization step: $M = I + A\Delta t$, where I is the identify matrix, and $\Delta t \leq 1 / \max\left(-A_{ii}\right)$ is the discrete time unit (Kachalo, Lu, & Liang, 2006; Stewart, 1994). Therefore, the steady state probability distribution can be equivalently obtained by solving the equation $\pi = M\pi$, which corresponds to the eigenvector of M with respect to the eigenvalue . The Arnoldi method implemented in the software ARPACK can be used to solve the eigenvalue-eigenvector problem and compute the steady state distribution π (Lehoucq, Sorensen, & Yang, 1998).

3.5 Time Evolving Probability Landscape

The time evolving probability landscape derived from the dCME can be expressed in the form of a matrix exponential: $p(t) = e^{At} p(0)$, where $p(0)$ is the initial probability landscape and A is the transition rate matrix over the enumerated state space. Once $p(0)$ is specified, $p(t)$ can be calculated using numerical methods such as the Krylov subspace projection method, *e.g.*, as implemented in the EXPOKIT package by Sidje (Sidje, 1998). Other state-of-the-art numerical techniques can also be applied (Kazeev, Khammash, Nip, & Schwab, 2014).

3.6 Software Availability and Online Simulation Tools

A C++ language implementation for the finite buffer method software package has been made publicly available. One can download the source codes and executable binary programs from the URL http://tanto.bioe.uic.edu/dcme/. The software package includes four separate programs to perform the functions of (1) finite buffer state space enumeration, (2) transition matrix construction, (3) steady state probability landscape solution using the Successive Over Relaxation method, and (4) time evolution probability landscape solution using the EXPOKIT method (Sidje, 1998), respectively. The website also includes simple examples to demonstrate the formats of input files.

For easier use of the tool, an online GUI simulation tool has been built recently for the finite buffer method at the nanoHUB computing grid. The online tool is also publicly available. One can open the URL https://nanohub.org/tools/fbsdcme in any Java supported browser, and click the "Lunch Tool" button to start the finite buffer software. To use the finite buffer software, one needs to input a SBML format reaction network file and an initial condition file. The format for the initial condition file has been demonstrated in the finite buffer method website (http://tanto.bioe.uic.edu/dcme/). In the SBML file of reaction network, the buffer species must be explicitly included in the species list, and also incorporated into the reactant list of all synthesis reactions and the product list of all degradation reactions.

4. EXAMPLES OF REALISTIC GENE REGULATORY NETWORKS

We now use two biological examples, namely the genetic toggle switch and the phage lambda lysogeny and lysis circuit, to show how stochastic gene regulatory networks can be modeled by directly solving the underlying discrete chemical master equation. We show how state spaces are enumerated and transition rate matrices are constructed using the finite buffer algorithm. We also show steady state and time evolving probability landscapes computed using the fb-dCME method, and we discuss interesting biological relevance. We show that the finite buffer dCME method can be used to answer important biological questions in stochastic gene regulatory networks.

4.1 Genetic Toggle Switch

Toggle switch is one of the smallest genetic networks that can present bistability (Blake, KAErn, Cantor, & Collins, 2003; Schultz et al., 2007). It is a small network consisting of two genes, say, A and B. Single copies of gene A and gene B in the chromosome each encoding a protein product. The protein product of each gene represses the other gene: When two protein monomers associate, they bind to the appropriate operator site and repress the transcription of the other gene.

The molecular species and the network topology of a toggle switch model are shown in Figure 3. The reactions include: the synthesis and degradation of proteins A and B, with reaction constants denoted as s and d, respectively; the binding and unbinding of the operator site of one gene by the protein products of the other gene at rate b and u, respectively (Eqn. (39)). Here the reactions rates are chosen as: $d = 1s^{-1}$, $s = 100s^{-1}$, $u = 0.1s^{-1}$, $b = 0.00001s^{-1}$. The binding states of the two operator sites are "on-on/unbound-unbound", "on-off/unbound-bound", "off-on/bound-unbound", and "off-off/bound-bound". The synthesis rates of both proteins A and B depend on the binding state of the operator sites (Cao & Liang, 2008; Schultz et al., 2007).

$$GeneA \xrightarrow{s} GeneA + ProteinA,$$

$$GeneB \xrightarrow{s} GeneB + ProteinB,$$

$$ProteinA \xrightarrow{d} \varnothing,\ ProteinB \xrightarrow{d} \varnothing,$$

Figure 3. Genetic toggle switch

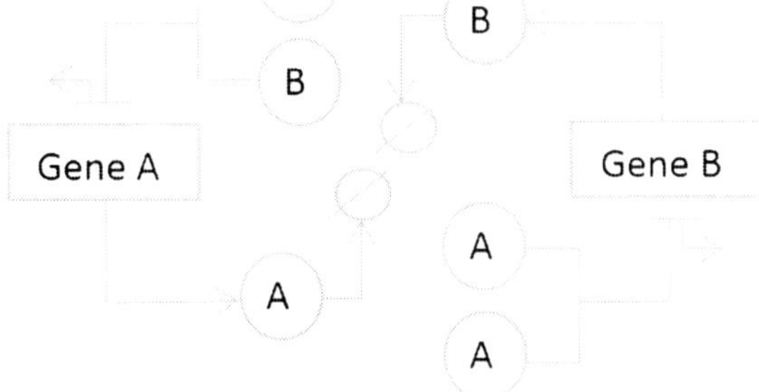

$$2ProteinA + GeneB \xrightarrow{b} BoundGeneB,$$

$$2ProteinB + GeneA \xrightarrow{b} BoundGeneA,$$

$$BoundGeneB \xrightarrow{u} 2ProteinA + GeneB,$$

$$BoundGeneA \xrightarrow{u} 2ProteinB + GeneA. \quad (39)$$

4.1.1 State Space Enumeration

We first calculate the state spaces under the initial condition of 1 copy of unbound gene A, 1 copy of unbound gene B, 0 copies of bound gene A and bound gene B, and 1 copies of their protein products. We set the buffer size to different copies of total protein A and protein B combined that can be synthesized. When the buffer capacity is 20, the size of the state space is 764. At buffer capacity of 200, 400, and 800 copies of proteins, the size increases to 79,604 states, 319,204 states, and 1,278,404 states, respectively. A comparison of state space size between the *finite buffer* method and the conventional hypercube method is shown in Table 1. The finite buffer method has shown dramatic reductions in state space sizes. Moreover, the factor of reduction is increasing with the buffer capacity. In the following computing results, we use the buffer capacity 300, and the finite buffer method generates

Table 1. State space sizes of toggle switch network using different methods

Buffer Capacity	Finite Buffer Method	Hypercube Method	Reduction Factor
50	4,904	1.5625×10^{10}	3.1862×10^{6}
100	19,804	1.0×10^{12}	5.0495×10^{7}
200	79,604	6.4×10^{13}	8.0398×10^{8}
300	179,404	7.29×10^{14}	4.0635×10^{9}
400	319,204	4.096×10^{15}	1.2832×10^{10}

a state space with 179,404 states. Therefore, the transition rate matrix A is a 179,404 dimensional sparse matrix.

4.1.2 Steady State Probability Landscape with Four Distinct Peaks

The exact probably landscape of the toggle switch model at steady state can be computed numerically. We choose the initial condition as: 1 copy of unbound gene A, 1 copy of unbound gene B, 0 copies of bound gene A and bound gene B, 0 copies of their protein products, and the buffer size for the total protein A and protein B combined to be $b = 300$, meaning the total copy number of proteins A and B combined in the system cannot exceed 300 at any time. We can increase or decrease the buffer capacity to include or exclude states. We then enumerate the state space of the toggle switch using the finite buffer algorithm. The steady state probability landscape of the network can then be computed (Figure 4). The Figure 4 shows the steady state probability landscape projected on the 2D space of protein A and B copy numbers. With these chosen parameters, the toggle switch exhibits a clear multi-stability in the steady state probability landscape. The toggle switch shows four distinct macroscopic states, corresponding to the "ON/ON", "ON/OFF", "OFF/

Figure 4. Steady state probability landscape of toggle switch

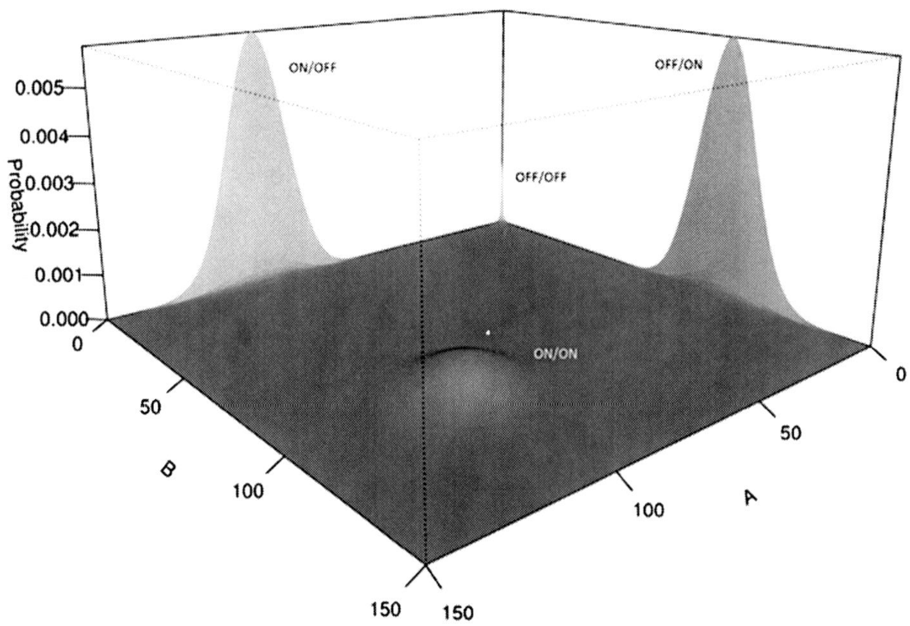

ON" and "OFF/OFF" configurations of the genetic circuits. Among them, the on/off and off/on states are more prominent, and the on/on state has a low probability, and the off/off state is severely unstable showing a very narrow and spiky peak (Cao & Liang, 2008). This result has important biological implications in cancer biology (Shiraishi, Matsuyama, & Kitano, 2010) and stem cell biology (Bu et al., 2013).

4.1.3 Time Evolving Probability Landscapes

Using the same state space and transition rate matrix, we can also directly compute the exact time-evolving probably landscape of the toggle switch at different time points. Starting from the uniform distribution, the time-evolving probability landscape is computed using the EXPOKIT package with 0.01 second time interval. Figure 5 shows the snapshots of the time-evolving probability landscape at 6 different time points: $t = 0$, 0.5, 1, 2, 4, and 25 seconds, respectively. The time-evolving probability landscapes have shown complex behavior of probability mass flows on the projected A-B 2D plane. At the very beginning, the probability of all states are equal ($t = 0$), and then a multi-layered peak arises in the center of the plane ($t = 0.5$), which is quickly separated into four thick peaks ($t = 1$). Gradually, the four thick peaks form the four different macroscopic states ($t = 2$, 4, and 25), and eventually converge to the steady state probability landscape ($t = 25$). Interestingly, the relative heights of the four peaks are changing during the time evolution. For example, four peaks have nearly same probabilities at about $t = 2$,

Figure 5. Time evolving probability landscapes of toggle switch starting from uniform distribution

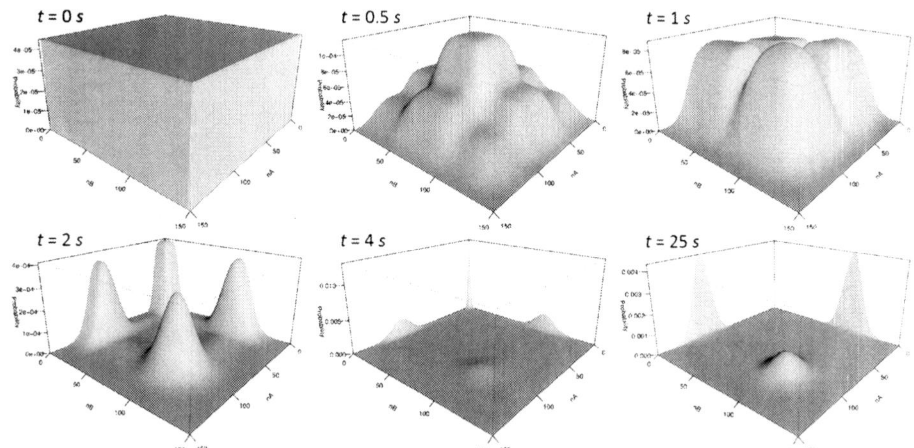

but the probability of the off/off state becomes a few folds higher than the other three peaks at $t = 4s$, and eventually the probabilities of the on/off and off/on states take most of the probability mass. The dynamic evolution of the landscape over time may be important in explaining biological behaviors of genetic circuits.

4.2 Phage Lambda Lysogeny-Lysis Epigenetic Switch

Bacteriophage lambda is a virus that infects *E. coli* cells. Of central importance is the molecular circuitry that controls phage lambda to choose between two productive modes of development, namely, the lysogenic phase and the lytic phase. In the lysogenic phase, phage lambda represses its developmental function, integrates its DNA into the chromosome of the host *E. coli* bacterium, and is replicated in cell cycles for potentially many generations. When threatening DNA damage occurs, for example, when UV irradiation increases, phage lambda switches from the epigenetic state of lysogeny to the lytic phase and undergoes massive replications in a single cell cycle, releases $50 \sim 100$ progeny phages upon lysis of the *E. coli* cell. This switching process is called prophage induction (Ptashne, 2004).

The molecular network that controls the choice between these two different physiological states has been studied extensively (Anderson & Yang, 2008; Arkin et al., 1998; Aurell, Brown, Johanson, & Sneppen, 2002; Jacob & Monod, 1961; Johnson et al., 1981; Little, Shepley, & Wert, 1999; Ptashne, 2004; Ptashne et al., 1976; Shea & Ackers, 1985). All of the major molecular components of the network have been identified, binding constants and reaction rates characterized, and there is a good experimental understanding of the general mechanism of the molecular switch (Ptashne, 2004). Theoretical studies have also contributed to the illumination of the central role of stochasticity (Arkin et al., 1998) and the stability of lysogen against spontaneous switching (Aurell et al., 2002; Zhu, Yin, Hood, & Ao, 2004).

To study how lysogeny is maintained and how it transitions to the lytic state, one can use a simplified stochastic model for the molecular regulatory network that controls the epigenetic switch in phage lambda (Figure 6) (Cao et al., 2010). Using a total of 54 biochemical reactions involving 13 molecular species, this model explicitly includes key components, essential reactions, and cooperativities of the phage lambda decision circuitry. The effects of UV irradiation can be modeled by increasing the CI degradation rates k_d due to the response of the SOS system. This epigenetic network model can reach around 1.7 million microstates. The steady state probability associated with each of these microstates can be computed from dCME after the microstates are enumerated using the finite buffer algorithm (Cao et al., 2010).

Figure 6. Phage lambda lysogeny-lysis genetic circuit

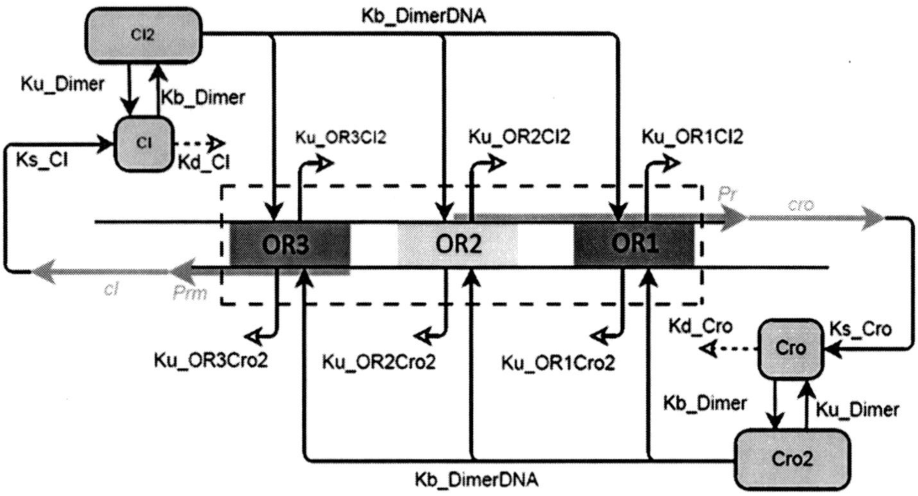

4.2.1 Steady State Probability Landscapes Can Help to Reveal Control Mechanisms

Figure 7 (row 1) shows the probability landscape of the phage lambda at five different UV irradiation conditions, each modeled with a different CI degradation rate k_d. Although there are 13 molecular species, the landscapes can be projected to the 2-dimensional subspace and record the total copy numbers of CI2 dimer and Cro2 dimer molecules. With a high copy number of CI2 repressor, the lysogenic phase of the phage lambda is maintained, whereas a high copy number of Cro2 protein signifies the lytic phase (Johnson et al., 1981). A clear picture of the landscape in lysogeny, at the start of transition, during mid-transition, at the end of transition, and in lysis can be seen (Figure 7).

The stochastic network models can also be used to aid in understanding of the mechanism of how the decision network works. It is well known that cooperativity among proteins play important roles. After removing all cooperativities between neighboring proteins in the model, phage lambda cannot enter lysogeny regardless the dosage of the UV irradiation. However, when the cooperativity ΔG_{12} between two CI dimers when binding to operator sites OR1 and OR2 are restored, the lysogeny is largely restored (Figure 7, row 2). In contrast, if all other cooperativities are restored except ΔG_{12}, phage lambda still lacks the ability to enter the lysogeny phase (Figure 7, row 3). These calculations suggest that the cooperativity ΔG_{12} plays key roles in maintaining the stability of the circuit.

Figure 7. The probability landscape of the epigenetic circuits of lysogeny maintenance in phage lambda

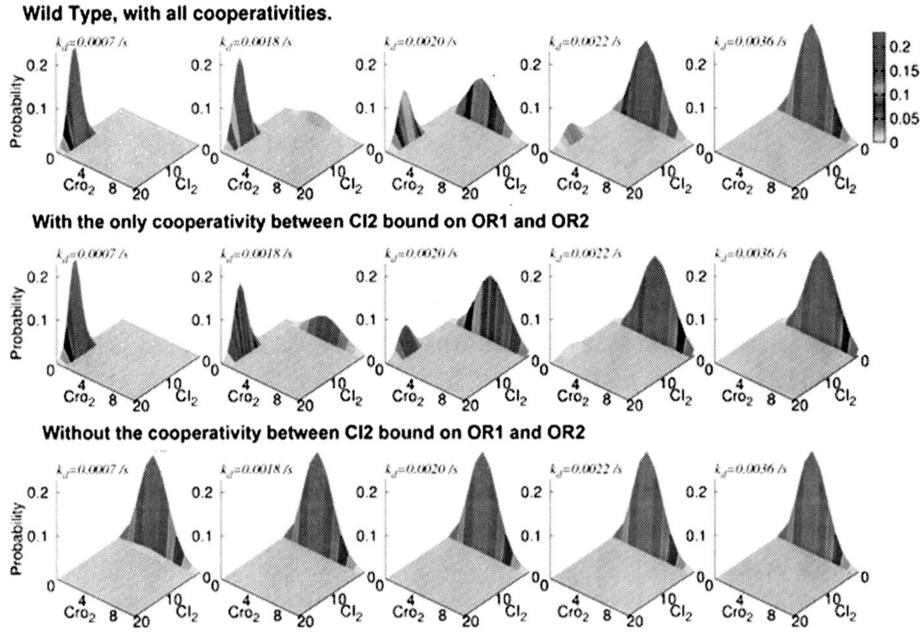

An important property of biological stochastic network is its robustness against changes in the molecular components of the epigenetic network. Experimental studies showed that when the ordering of operator sites are changed, mutants of phage lambda have markedly different tolerance to UV irradiation. Calculations from solving the dCME model showed that the wild-type lysogeny has a high threshold towards lysis, and is overall insensitive to small fluctuation of UV dosage, if it is below certain threshold (Figure 8a). That is, the switching network of phage lambda is very stable and is strongly buffered with a high threshold against fluctuations in CI degradation rate due to environmental changes in UV irradiation. This high threshold against environmental fluctuations is important for the self-perpetuating nature of the epigenetic state of *E. coli* cells, allowing lysogeny to be passed on to its offspring. Once the degradation rate of CI reaches a threshold, phage lambda switches very efficiently to the lytic phase, and this efficiency is not built at the expense of stability against random fluctuation. Wild type phage lambda therefore can integrate signaling in the form of different CI degradation rates and can distinguish a true signal above the high threshold from random noise fluctuating below this threshold.

Figure 8. Instability, shallow threshold, and switching inefficiency of the network against fluctuation in UV irradiation in mutant phage lambda

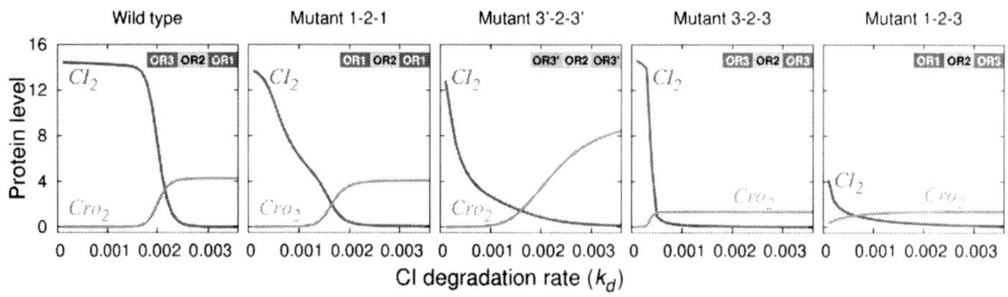

In contrast, all mutant variants exhibit the behavior of a hair trigger, and require much less UV irradiation for the onset of lysis induction (Figure 8b-e). In addition, they are "leaky", and respond in a graded fashion towards increase UV irradiation, instead of the well-behaved threshold behavior observed in wild type phage lambda. In the case of mutant 1-2-3, the mutant phage lambda cannot enter lysogenic state. These results are in full agreement with experimental findings (Cao et al., 2010; Little et al., 1999).

4.2.2 Time Evolving Probability Landscapes

The time-evolving probability landscape is computed using the EXPOKIT package starting from the uniform distribution. Figure 9 shows the snapshots of the time-evolving probability landscape at 4 different time points at: $t = 10$, 200, 1000, and $200,000$ second, respectively. The CI degradation rate is fixed as $k_d = 0.0025s^{-1}$. The time-evolving probability landscapes have shown the behavior of probability mass flows on the projected CI-Cro plane. The probability mass first flows towards the state CI=0, Cro=0, and forms a peak there at about $t = 1000s$. However, two stable peaks are eventually formed separately, and the landscape converges to the steady state probability landscape ($t > 200,000s$).

5. DISCUSSIONS

In this chapter, we have discussed the significance of the discrete chemical master equation (dCME) as a theoretical framework for modeling nonlinear, biochemical reaction networks (Gillespie, 1977). This formulation provides a foundation to study stochastic phenomena in biological networks. Its role is analogous to that of the Schrödinger equation in quantum mechanics (Beard & Qian, 2008). Develop-

Figure 9. Time evolving probability landscapes of phage lambda genetic circuit at different time points

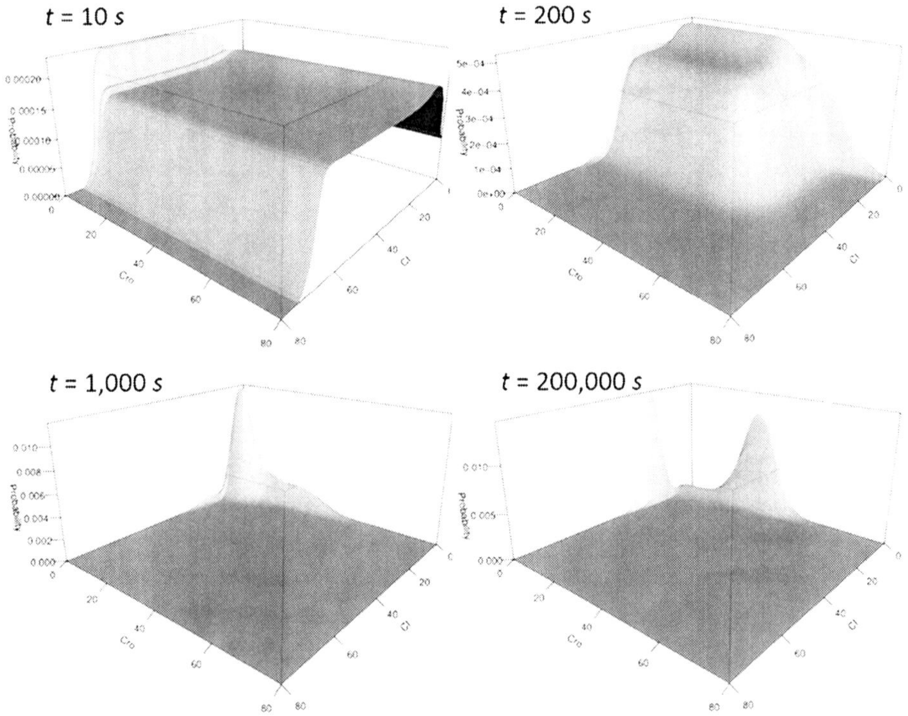

ing computational solutions to the dCME has important implications, just as the development of computational techniques for solving the Schrödinger equation for systems with many atoms is (Kohanoff, 2006; Kohn, Sham, & others, 1965). By computing the time-evolving probability landscape of cellular stochastic networks, we may gain understanding of the possible mechanisms of cellular states, as well as the inheritable phenotypes with a distributive epigenetic code, in which the network architecture and its landscape dictate the physiological meta-states of the cell under different conditions (Ptashne, 2007; Zhu et al., 2004).

Solving a given dCME, however, is a computationally challenging task at the present time. We have outlined several key difficulties, as well as some of the progresses that have been made so far. The finite buffer algorithm allows direct numerical solution to the dCME for both steady state and time-evolving probability landscapes. It can be applied to study stochasticity of systems with a handful of particles, as well as larger networks arising from very realistic biological problem, such as that of the lysogeny-lysis control circuit of the phage lambda (Cao et al., 2010). As an exact method, the finite buffer method can also be used to study model systems of finite size to gain insight into stochastic behavior of networks. The stochastic simulation algorithm offers the approach of studying the stochastic network through simulations (Gillespie, 1977). The formulation of stochastic differential equation such as the Langevin equation allows exploration of more complex stochastic systems, at the expense of less rigorous assumptions and perhaps more errors.

An important task is to integrate different stochastic methods for efficient computational solution of complex stochastic networks at large scale, with controlled accuracy. For example, one may apply the finite buffer algorithm to solve dCME directly for certain critical parts of the network, where rare events need to be assessed very accurately. Other parts of the network where general stochastic behavior needs to be determined can be studied using the Langevin equation. In addition, one may also wish to apply the stochastic simulation algorithm to certain parts of the network to probe their behavior. Furthermore, one may wish to apply ordinary differential equation (ODE) models to study parts of the system where copy numbers of molecules are large and there are little stochastic effects.

A great challenge is to develop a general strategy so the best methods can be applied to specific parts of the network and the results integrated to provide an overall picture of the stochastic dynamic behavior of the network. It would be desirable that the resulting errors due to approximations of varying degree are bounded within a tolerance level, while maintaining necessary computing speed and resource requirement. A further challenge is to develop such hybrid methods to compute the overall spatio-temporal stochastic dynamic properties of different cellular compartments, multi-cellular tissue, with consideration of different spatial distribution or gradient of molecules such as oxygen, nutrient, morphogens, and other signaling factors, all with stochasticity appropriately considered.

REFERENCES

Anderson, L. M., & Yang, H. (2008). DNA looping can enhance lysogenic CI transcription in phage lambda. *Proceedings of the National Academy of Sciences of the United States of America, 105*(15), 5827–5832. doi:10.1073/pnas.0705570105 PMID:18391225

Ao, P., Kown, C., & Qian, H. (2007). On the existence of potential landscape in the evolution of complex systems. *Complexity, 12*(4), 19–27. doi:10.1002/cplx.20171

Arkin, A., Ross, J., & McAdams, H. H. (1998). Stochastic kinetic analysis of developmental pathway bifurcation in phage lambda-infected Escherichia coli cells. *Genetics, 149*(4), 1633–1648. PMID:9691025

Aurell, E., Brown, S., Johanson, J., & Sneppen, K. (2002). Stability puzzles in phage lambda. *Physical Review E: Statistical, Nonlinear, and Soft Matter Physics, 65*(5), 051914. doi:10.1103/PhysRevE.65.051914 PMID:12059600

Beard, D. A., & Qian, H. (2008). *Chemical biophysics: quantitative analysis of cellular systems*. Cambridge University Press. doi:10.1017/CBO9780511803345

Blake, W. J., KÆrn, M., Cantor, C. R., & Collins, J. J. (2003). Noise in eukaryotic gene expression. *Nature, 422*(6932), 633–637. doi:10.1038/nature01546 PMID:12687005

Bu, P., Chen, K. Y., Chen, J. H., Wang, L., Walters, J., Shin, Y. J., & Shen, X. et al. (2013). A microRNA miR-34a-regulated bimodal switch targets Notch in colon cancer stem cells. *Cell Stem Cell, 12*(5), 602–615. doi:10.1016/j.stem.2013.03.002 PMID:23642368

Cao, Y., Gillespie, D. T., & Petzold, L. R. (2006). Efficient step size selection for the tau-leaping simulation method. *The Journal of Chemical Physics, 124*(4), 044109. doi:10.1063/1.2159468 PMID:16460151

Cao, Y., & Liang, J. (2008). Optimal enumeration of state space of finitely buffered stochastic molecular networks and exact computation of steady state landscape probability. *BMC Systems Biology, 2*(1), 30. doi:10.1186/1752-0509-2-30 PMID:18373871

Cao, Y., & Liang, J. (2013). Adaptively biased sequential importance sampling for rare events in reaction networks with comparison to exact solutions from finite buffer dCME method. *The Journal of Chemical Physics, 139*(2), 025101. doi:10.1063/1.4811286 PMID:23862966

Cao, Y., Lu, H. M., & Liang, J. (2010). Probability landscape of heritable and robust epigenetic state of lysogeny in phage lambda. *Proceedings of the National Academy of Sciences of the United States of America, 107*(43), 18445–18450. doi:10.1073/pnas.1001455107 PMID:20937911

Cormen, T. H., Leiserson, C. E., & Rivest, R. L. (1990). *Introduction to algorithms*. Cambridge, MA: The MIT Press.

Daigle, B. J. Jr, Roh, M. K., Gillespie, D. T., & Petzold, L. R. (2011). Automated estimation of rare event probabilities in biochemical systems. *The Journal of Chemical Physics, 134*(4), 044110. doi:10.1063/1.3522769 PMID:21280690

Gillespie, D. T. (1977). Exact stochastic simulation of coupled chemical reactions. *Journal of Physical Chemistry, 81*(25), 2340–2361. doi:10.1021/j100540a008

Gillespie, D. T. (2000). The chemical Langevin equation. *The Journal of Chemical Physics, 113*(1), 297–306. doi:10.1063/1.481811

Gillespie, D. T. (2007). Stochastic simulation of chemical kinetics. *Annual Review of Physical Chemistry, 58*(1), 35–55. doi:10.1146/annurev.physchem.58.032806.104637 PMID:17037977

Golub, G. H., & Van Loan, C. F. (1996). *Matrix computations* (3rd ed.). Baltimore, MD: Johns Hopkins University Press.

Grima, R., Thomas, P., & Straube, A. V. (2011). How accurate are the nonlinear chemical Fokker-Planck and chemical Langevin equations? *The Journal of Chemical Physics, 135*(8), 084103. doi:10.1063/1.3625958 PMID:21895155

Hasty, J., Pradines, J., Dolnik, M., & Collins, J. J. (2000). Noise-based switches and amplifiers for gene expression. *Proceedings of the National Academy of Sciences of the United States of America, 97*(5), 2075–2080. doi:10.1073/pnas.040411297 PMID:10681449

Jacob, F., & Monod, J. (1961). Genetic regulatory mechanisms in the synthesis of proteins+. *Journal of Molecular Biology, 3*(3), 318–356. doi:10.1016/S0022-2836(61)80072-7 PMID:13718526

Johnson, A. D., Poteete, A. R., Lauer, G., Sauer, R. T., Ackers, G. K., & Ptashne, M. (1981). lambda Repressor and cro--components of an efficient molecular switch. *Nature, 294*(5838), 217–223. doi:10.1038/294217a0 PMID:6457992

Kachalo, S., Lu, H., & Liang, J. (2006). Protein folding dynamics via quantification of kinematic energy landscape. *Physical Review Letters, 96*(5), 058106. doi:10.1103/PhysRevLett.96.058106 PMID:16487000

Kazeev, V., Khammash, M., Nip, M., & Schwab, C. (2014). Direct solution of the Chemical Master Equation using quantized tensor trains. *PLoS Computational Biology, 10*(3), e1003359. doi:10.1371/journal.pcbi.1003359 PMID:24626049

Keizer, J. (1977). On the macroscopic equivalence of descriptions of fluctuations for chemical reactions. *Journal of Mathematical Physics, 18*(7), 1316–1321. doi:10.1063/1.523422

Kim, K. Y., & Wang, J. (2007). Potential energy landscape and robustness of a gene regulatory network: Toggle switch. *PLoS Computational Biology, 3*(3), e60. doi:10.1371/journal.pcbi.0030060 PMID:17397255

Kohanoff, J. (2006). *Electronic structure calculations for solids and molecules: theory and computational methods*. Cambridge Univ Pr. doi:10.1017/CBO9780511755613

Kohn, W., & Sham, L. et al. (1965). Self-consistent equations including exchange and correlation effects. *Physical Review*, *140*(4A), A1133–A1138. doi:10.1103/PhysRev.140.A1133

Kurtz, T. G. (1972). The relationship between stochastic and deterministic models for chemical reactions. *The Journal of Chemical Physics*, *57*(7), 2976–2978. doi:10.1063/1.1678692

Kuwahara, H., & Mura, I. (2008). An efficient and exact stochastic simulation method to analyze rare events in biochemical systems. *The Journal of Chemical Physics*, *129*(16), 165101. doi:10.1063/1.2987701 PMID:19045316

Lehoucq, R., Sorensen, D., & Yang, C. (1998). *Arpack users' guide: Solution of large scale eigenvalue problems with implicitly restarted Arnoldi methods*. Philadelphia: SIAM. doi:10.1137/1.9780898719628

Levin, M. D. (2003). Noise in gene expression as the source of non-genetic individuality in the chemotactic response of Escherichia coli. *FEBS Letters*, *550*(1-3), 135–138. doi:10.1016/S0014-5793(03)00857-3 PMID:12935899

Liang, J., Zhang, J., & Chen, R. (2002). Statistical geometry of packing defects of lattice chain polymer from enumeration and sequential Monte Carlo method. *The Journal of Chemical Physics*, *117*(7), 3511–3521. doi:10.1063/1.1493772

Lin, M., Chen, R., & Liu, J. S. (2009). *Lookahead Strategies for Sequential Monte Carlo*. Retrieved from Little, J. W., Shepley, D. P., & Wert, D. W. (1999). Robustness of a gene regulatory circuit. *The EMBO Journal*, *18*(15), 4299–4307.

Liu, J. S., Chen, R., & Logvinenko, T. (2001). A theoretical framework for sequential importance sampling and resampling. *Sequential Monte Carlo Methods in Practice*, 225-246.

McAdams, H. H., & Arkin, A. (1997). Stochastic mechanisms in gene expression. *Proceedings of the National Academy of Sciences of the United States of America*, *94*(3), 814–819. doi:10.1073/pnas.94.3.814 PMID:9023339

McCullagh, E., Farlow, J., Fuller, C., Girard, J., Lipinski-Kruszka, J., Lu, D., & El-Samad, H. et al. (2009). Not all quiet on the noise front. *Nature Chemical Biology*, *5*(10), 699–704. doi:10.1038/nchembio.222 PMID:19763097

Meirovitch, H. (1982). A new method of simulation of real chains: Scanning future steps. *Journal of Physics. A, Mathematical and General*, 15.

Meirovitch, H. (1988). Statistical properties of the scanning simulation method for polymer chains. *J. Chemical Phisics, 89*.

Melykuti, B., Burrage, K., & Zygalakis, K. C. (2010). Fast stochastic simulation of biochemical reaction systems by alternative formulations of the chemical Langevin equation. *The Journal of Chemical Physics*, *132*(16), 164109. doi:10.1063/1.3380661 PMID:20441260

Mettetal, J. T., Muzzey, D., Pedraza, J. M., Ozbudak, E. M., & van Oudenaarden, A. (2006). Predicting stochastic gene expression dynamics in single cells. *Proceedings of the National Academy of Sciences of the United States of America*, *103*(19), 7304–7309. doi:10.1073/pnas.0509874103 PMID:16648266

Munsky, B., & Khammash, M. (2006). The finite state projection algorithm for the solution of the chemical master equation. *The Journal of Chemical Physics*, *124*(4), 044104. doi:10.1063/1.2145882 PMID:16460146

Munsky, B., & Khammash, M. (2007). A multiple time interval finite state projection algorithm for the solution to the chemical master equation. *Journal of Computational Physics*, *226*(1), 818–835. doi:10.1016/j.jcp.2007.05.016

Ogawa, M. (1989). Hemopoietic stem cells: Stochastic differentiation and humoral control of proliferation. *Environmental Health Perspectives*, *80*, 199–207. doi:10.1289/ehp.8980199 PMID:2647480

Ozbudak, E. M., Thattai, M., Kurtser, I., Grossman, A. D., & van Oudenaarden, A. (2002). Regulation of noise in the expression of a single gene. *Nature Genetics*, *31*(1), 69–73. doi:10.1038/ng869 PMID:11967532

Paulsson, J., & Ehrenberg, M. (2000). Random signal fluctuations can reduce random fluctuations in regulated components of chemical regulatory networks. *Physical Review Letters*, *84*(23), 5447–5450. doi:10.1103/PhysRevLett.84.5447 PMID:10990965

Ptashne, M. (2004). A Genetic Switch: Phage Lambda Revisited (3rd ed.). Cold Spring Harbor Laboratory Press.

Ptashne, M. (2007). On the use of the word 'epigenetic'. *Current Biology*, *17*(7), R233–R236. doi:10.1016/j.cub.2007.02.030 PMID:17407749

Ptashne, M., Backman, K., Humayun, M., Jeffrey, A., Maurer, R., Meyer, B., & Sauer, R. (1976). Autoregulation and function of a repressor in bacteriophage lambda. *Science*, *194*(4261), 156–161. doi:10.1126/science.959843 PMID:959843

Roh, M. K., Daigle, B. J. Jr, Gillespie, D. T., & Petzold, L. R. (2011). State-dependent doubly weighted stochastic simulation algorithm for automatic characterization of stochastic biochemical rare events. *The Journal of Chemical Physics*, *135*(23), 234108. doi:10.1063/1.3668100 PMID:22191865

Roh, M. K., Gillespie, D. T., & Petzold, L. R. (2010). State-dependent biasing method for importance sampling in the weighted stochastic simulation algorithm. *The Journal of Chemical Physics*, *133*(17), 174106. doi:10.1063/1.3493460 PMID:21054005

Saad, Y. (2003). *Iterative methods for sparse linear systems*. Siam. doi:10.1137/1.9780898718003

Samoilov, M., Plyasunov, S., & Arkin, A. (2005). Stochastic amplification and signaling in enzymatic futile cycles through noise-induced bistability with oscillations. *Proceedings of the National Academy of Sciences of the United States of America*, *102*(7), 2310–2315. doi:10.1073/pnas.0406841102 PMID:15701703

Schultz, D., Onuchic, J. N., & Wolynes, P. G. (2007). Understanding stochastic simulations of the smallest genetic networks. *The Journal of Chemical Physics*, *126*(24), 245102. doi:10.1063/1.2741544 PMID:17614590

Shea, M. A., & Ackers, G. K. (1985). The OR control system of bacteriophage lambda. A physical-chemical model for gene regulation. *Journal of Molecular Biology*, *181*(2), 211–230. doi:10.1016/0022-2836(85)90086-5 PMID:3157005

Shiraishi, T., Matsuyama, S., & Kitano, H. (2010). Large-scale analysis of network bistability for human cancers. *PLoS Computational Biology*, *6*(7), e1000851. doi:10.1371/journal.pcbi.1000851 PMID:20628618

Sidje, R. B. (1998). Expokit: A software package for computing matrix exponentials. *ACM Transactions on Mathematical Software*, *24*(1), 130–156. doi:10.1145/285861.285868

Stewart, W. J. (1994). *Introduction to the numerical solution of Markov chains* (Vol. 41). Princeton University Press Princeton.

Torrie, G. M., & Valleau, J. P. (1977). Nonphysical sampling distributions in Monte Carlo free-energy estimation: Umbrella sampling. *Journal of Computational Physics*, *23*(2), 187–199. doi:10.1016/0021-9991(77)90121-8

Van Kampen, N. G. (2007). Stochastic processes in physics and chemistry (3rd ed.). Elsevier Science and Technology.

Vellela, M., & Qian, H. (2007). A Quasistationary Analysis of a Stochastic Chemical Reaction: Keizer's Paradox. *Bulletin of Mathematical Biology*, *69*(5), 1727–1746. doi:10.1007/s11538-006-9188-3 PMID:17318672

Vellela, M., & Qian, H. (2009). Stochastic dynamics and non-equilibrium thermodynamics of a bistable chemical system: The Schlogl model revisited. *Journal of the Royal Society, Interface*, *6*(39), 925–940. doi:10.1098/rsif.2008.0476 PMID:19095615

Volfson, D., Marciniak, J., Blake, W. J., Ostroff, N., Tsimring, L. S., & Hasty, J. (2006). Origins of extrinsic variability in eukaryotic gene expression. *Nature*, *439*(7078), 861–864. doi:10.1038/nature04281 PMID:16372021

Zhou, T., Chen, L., & Aihara, K. (2005). Molecular communication through stochastic synchronization induced by extracellular fluctuations. *Physical Review Letters*, *95*(17), 178103. doi:10.1103/PhysRevLett.95.178103 PMID:16383875

Zhu, X. M., Yin, L., Hood, L., & Ao, P. (2004). Robustness, stability and efficiency of phage lambda genetic switch: Dynamical structure analysis. *Journal of Bioinformatics and Computational Biology*, *2*(4), 785–817. doi:10.1142/S0219720004000946 PMID:15617166

Chapter 4
Structural Intervention and External Control for Markovian Regulatory Network Models

Xiaoning Qian
Texas A&M, USA

Ranadip Pal
Texas Tech University, USA

ABSTRACT

In order to derive system-based methods to control dynamic behavior of biological systems of interest for future gene-based intervention therapeutics, two basic categories of intervention strategies have been studied based on the Markov chain theory and Markov decision processes: Structural intervention by function perturbation and external control based on state perturbation. The chapter reviews the existing network analysis and control methods in these two categories and discusses their extensions for more robust and clinically relevant intervention strategies considering collateral damages from intervention.

1. INTRODUCTION

Mathematical modeling and dynamic analysis of gene regulatory networks enable mathematical avenues to study potential systems therapeutics to intervene regulatory networks so that the modeled system dynamics can be altered for desired systems

DOI: 10.4018/978-1-5225-0353-8.ch004

behavior, which is one of the utmost important goals of systems biology from mathematical modeling of biological networks. Network intervention till today has been mostly investigated in the context of probabilistic Boolean networks (PBNs) (Shmulevich & Dougherty, 2007; Shmulevich & Dougherty, 2010; Dougherty, Pal, Qian, Bittner, & Datta, 2010). Two basic categories of intervention strategies have been studied based on the Markov chain theory and Markov decision processes: Structural intervention and external control based on state perturbation.

Structural intervention or function perturbation involves changing regulatory relationships or wiring among genes in networks (Shmulevich, Dougherty, & Zhang, 2002a; Xiao & Dougherty, 2007; Qian & Dougherty, 2008). The wiring changes cause the changes in the rule-based structure in PBNs and hence state transition of the underlying Markov chains, which further lead to the alteration in long-run network behavior characterized by their steady-state distributions. Given a class of potential structural changes ($\mathbf{u}_s$) in mathematical network models that simulate practical therapeutic strategies by the use of drugs to act on gene products for pathway blockage, for example using small interfering RNAs (siRNAs), the mathematical problem is to derive the structural intervention resulting in an optimal alteration of the steady-state distribution. In order to characterize the dynamic changes caused by structural perturbations and derive effective intervention strategies, we need to mathematical address two critical issues: (1) Efficient characterization of the effects caused by perturbation and (2) search for the optimal perturbation(s) that provides the most desirable steady-state distribution after intervention. We have recently adopted the classical Markov chain perturbation theory to address these problems.

External control by state perturbation is generally based on flipping the state of a control gene, hence the concentration of either its corresponding mRNA expression or protein product will be knocked down or over expressed, for beneficial network dynamical change. Corresponding gene expression changes are forced by external perturbations to modulate the dynamics. This type of network intervention models feasible medical treatments for potential therapeutics such as gene knockdown or chemotherapy. Typically, optimal external control is derived in the framework of Markov decision process (Shmulevich, Dougherty, & Zhang, 2002b; Datta, Choudhary, Bittner, & Dougherty, 2003; Bertsekas, 2005; Pal, Datta, & Dougherty, 2006; Datta, Pal, Choudhary, & Dougherty, 2007) for both finite-horizon and infinite-horizon control policies, in which the control action is based on an objective function that rewards beneficial gene activity. Mathematically, we are interested in deriving a control policy u_g for a given control gene g or a set of control genes when necessary. At a time point t if the network state is at x, the expression state for g is flipped if $u_g(x) = 1$, denoting the control is on. Otherwise when $u_g(x) = 0$, the state of the control gene g remains unchanged. To achieve beneficial network behavior, we minimize the average cost function reflecting the undesirability of the gene activities by deriving the optimal control policy u_g. For the infinite-horizon problem, stationary control policies can be solved by dynamic programming with networks of reasonable size (Pal *et al.*, 2006; Datta *et al.*, 2007). But for large networks, the optimal solution is often computationally prohibitive. Hence, greedy algorithms are necessary to alter the steady-state network behavior to lessen the likelihood of being in undesirable states, for instance, metastatic states in cancer.

In Figure 1, we schematically illustrate these two types of intervention strategies. We will start with a brief review on PBNs for modeling gene regulatory networks in Section 2. Both methods for structural intervention (Section 3) and external control (Section 4) will be discussed, followed by further extensions considering both potential side effects (Section 5) and robustness of intervention strategies (Section 6) for practical usage.

2. BACKGROUNDS

One way to model dynamic behavior of gene regulatory networks is to consider the time evolution of gene expression profiles as dependent stochastic processes regulated by each other. The dynamics of each individual gene, for example, the *i*th gene, involved in a given biological process, can be manifested by the temporal changes of its gene expression level: $\boldsymbol{x}_i(t)$. Due to both external stimuli and internal perturbations to other genes involved in the process, the gene expression level can be modeled as a stochastic process. For simplicity, we may discretize the gene expression into binary values: $\boldsymbol{x}_i = 1$, which denotes that the gene is turned "ON"; or $\boldsymbol{x}_i = 0$, representing the gene is "OFF". Assuming that there are $\boldsymbol{n}$ genes involved in a biological process of interest, a gene regulatory network model studies the dynamics of $\boldsymbol{n}$ gene expression profiles: $\mathbf{X}(t) = \{\boldsymbol{x}_1(t), \boldsymbol{x}_2(t), \ldots, \boldsymbol{x}_i(t), \ldots, \boldsymbol{x}_n(t)\}$.

Figure 1. Schematic illustration of two network intervention strategies: (a) Structural intervention or function perturbation; (b) External control by state perturbation

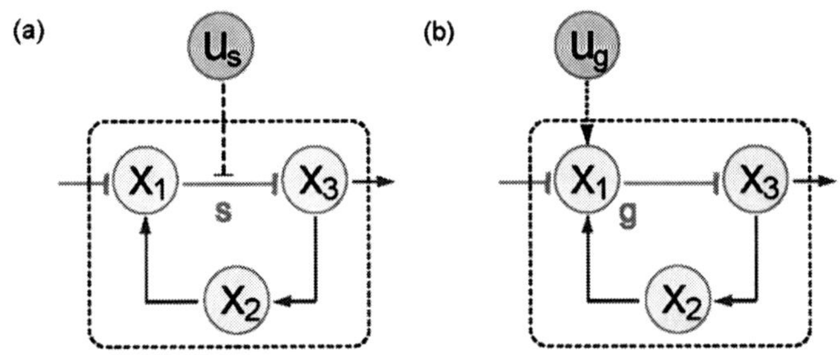

A finite Markov chain (Kemény & Snell, 1983) with the state space $S=\{0, 1\}^n$ can model such dependent stochastic processes with observations obtained at different stages of the biological process that can be indexed by time. The time evolution of involved gene expressions levels is determined by the regulatory functional relationships among these genes in the model. A basic type of regulatory functional relationships can be modeled through a logic-rule-based structure that governs the state transition X(t) → X(t+1). *Boolean Networks (BN)* (Kauffman, 1993) enables such a rule-based structure through a graph representation G(V, f), in which V = $\{x_1, x_2,\ldots, x_i,\ldots, x_n\}$ denotes the set of involved genes and $\mathbf{f} = \{f_1, f_2, \ldots, f_i,\ldots, f_n\}$ captures the regulatory functional relationships among n genes. Given such a graph representation, each individual gene x_i is regulated by a corresponding regulatory set R_i in V: $x_i(t + 1) = f_i(X(t)) = f_i(R_i(t))$. As schematically illustrated in Figure 1, the expression change of x_3 is determined by its regulatory set $R_i = \{x_1\}$: $x_3(t + 1) = f_3(x_1(t))$, in which f_3 will be dependent on different types of logic function rules, including "strong inhibition rule" and "majority vote rule" that have been often implemented in literature (Li, Long, Lu, Ouyang, & Tang, 2004; Lau, Ganguli, & Tang, 2007). For example, if we take the "strong inhibition rule", we can write $x_3(t+1) = -x_1(t)\wedge x_3(t)$, in which "–" and "∧" denote the "NOT" and "AND" logic operators. In this chapter, we focus on the basic model structure that assumes that the finite Markov chains possess the first-order time-homogeneous dependence structure. Hence, for arbitrary time points t and t', *one-step transition probability* is sufficient to characterize the systems dynamics:

$$P(\mathbf{X}(t+1) = s_j|\mathbf{X}(t) = s_i) = P(\mathbf{X}(t'+1) = s_j|\mathbf{X}(t') = s_i) = p_{ij},$$

In which $s_i, s_j \in S$ denote the gene expression profiles for the whole network and transition probability is dependent and can be derived based on the logic regulatory functions **f**. With these transition probabilities, we can define a state transition probability matrix to fully specify the finite Markov chain: $P = (p_{ij})$. $i,j\in\{0,1,\ldots,2^n-1\}$. BNs are deterministic and the entries in the transition probability matrix are either 0 or 1. The dynamic behavior of a BN is modeled by the deterministic state transitions and the system will eventually enter and cycle endlessly through one state or a set of states called an *attractor cycle*. The set of all states that reach a particular attractor cycle forms its corresponding *basin of attraction (BOA)*. These attractors can either by interpreted as different cell types (Kauffman, 1969; Kauffman 1993) or different states of cell growth, differentiation, and apoptosis (Huang, 1999).

To further model stochasticity in biological processes, a *Boolean network with perturbation (BNp)* assumes that there is an (identically and independently distributed) i.i.d. random perturbation vector $\eta \in\{0, 1\}^n$ such that the ith gene flips its state if the corresponding entry η_i of the perturbation vector is equal to 1. A

BNp inherits the attractor structure from the original BN without perturbation, the difference being that random state-flips can cause a BNp to jump out of an attractor cycle, perhaps then transitioning to a different attractor cycle prior to another perturbation. With the perturbation probability for possible random flips $p = \Pr(\eta_i = 1)$ identically for all genes, we have state transition $\mathbf{X}(t+1) = \mathbf{f}(\mathbf{X}(t))$ determined by regulatory functions with a probability of $(1-p)^n$ while $\mathbf{X}(t+1) = \mathbf{X}(t) \oplus \eta$ with random perturbation probability $1-(1-p)^n$, where $\oplus$ is component-wise addition modulo 2. Its state transition probability matrix, P, can be derived accordingly (Shmulevich, Dougherty, Kim, & Zhang, 2002; Shmulevich & Dougherty, 2007; Shmulevich & Dougherty, 2010; Dougherty, *et al.*, 2010). Random state-flipping perturbation makes the underlying finite Markov chain of a BNp irreducible and ergodic. Hence, the network possesses a steady-state distribution $\boldsymbol{\pi}$ that satisfies $\boldsymbol{\pi}^T = \boldsymbol{\pi}^T P$ with its ith entry $\pi(i)$ denoting the steady-state probability that the system stays at the state s_i in the long run.

In biological systems, molecular interactions as well as transcriptional regulations often depend on both cellular internal states and external environmental conditions. In order to further capture possible changes under different *contexts*, we can construct one BNp for each possible context and a family of m BNps together constitute a *Probabilistic Boolean Network (PBN)*, with corresponding probabilities governing the selection of each BNp at different time points (Shmulevich, *et al.*, 2002; Shmulevich & Dougherty, 2007; Shmulevich & Dougherty, 2010; Dougherty, *et al.*, 2010). Each constituent BNp or context is characterized by its corresponding logic-rule-based structure, $\mathbf{f}^l$, $1 \leq l \leq m$. At a given time point, there is a positive probability q of switching from the current governing constituent BNp to another. A PBN inherits the attractor cycles of its constituent BNps. If $q = 1$, the PBN is said to be *instantaneously random*; if $q < 1$, the PBN is said to be *context-sensitive*. The underlying Markov chains of PBNs are still irreducible and ergodic. Based on assumed regulatory functions and probability structures, we can derive the state transition probability matrices, P, as well as the corresponding steady-state distributions, $\boldsymbol{\pi}$, for fully specified PBNs (Shmulevich & Dougherty, 2007; Shmulevich & Dougherty, 2010; Dougherty, *et al.*, 2010).

3. STRUCTURAL INTERVENTION

As explained, the underlying network dynamics in PBNs can be modeled as finite Markov chains determined by corresponding regulatory Boolean functions and the related probabilistic parameters. Hence, network intervention can be investigated using the classical Markov chain perturbation theory (Kemény & Snell, 1983; Schweitzer, 1968; Hunter, 1986; Hunter, 2005; Qian & Dougherty, 2008). We study structural intervention by assuming that we can block the regulation between two genes in networks. Changing the rule-based structure in a PBN has a fundamental impact by changing the network dynamics permanently so that it alters the long-term abnormal network behavior. In order to derive optimal structural intervention, we first need to characterize the long-run effect of perturbations to network dynamics caused by the wiring changes.

Under the framework of Markov chain perturbation, following the wiring perturbation, the transition matrix P and the corresponding steady-state distribution $\boldsymbol{\pi}$ for the original network are changed to P_a and $\boldsymbol{\pi}_a$ after intervention, respectively. The key issue is to apply computationally efficient matrix computation for the altered steady-state distribution, $\boldsymbol{\pi}_a$, in terms of $\boldsymbol{\pi}$ and P so that an optimal perturbation can be determined directly from $\boldsymbol{\pi}$ and P (Qian & Dougherty, 2008).

Potential rule perturbations can be mathematically represented by a function perturbation matrix E with $P_a = P + \mathrm{E}$.

As the underlying Markov chain of a PBN is always irreducible and ergodic by the model assumptions, both the original and perturbed PBN have their steady-state distributions satisfying: $\boldsymbol{\pi}^{\mathrm{T}} = \boldsymbol{\pi}^{\mathrm{T}}P$, and $\boldsymbol{\pi}_a^{\mathrm{T}} = \boldsymbol{\pi}_a^{\mathrm{T}}P_a$. We have the following Theorem 1:

Theorem 1: For an irreducible and ergodic finite Markov chain with its corresponding state transition matrix P, steady-state distribution $\boldsymbol{\pi}$, and fundament matrix $Z = (I - P + e\boldsymbol{\pi}^{\mathrm{T}})^{-1}$, we can compute the perturbed steady-state distribution $\boldsymbol{\pi}_a$ from the perturbation to the state transition by the perturbation matrix **E** by the following equation:

$$\pi_a^T = \pi^T(\mathrm{I - EZ})^{-1} \tag{1}$$

Proof: As $\boldsymbol{\pi}^{\mathrm{T}} = \boldsymbol{\pi}^{\mathrm{T}}P$, and $\boldsymbol{\pi}_a^{\mathrm{T}} = \boldsymbol{\pi}_a^{\mathrm{T}}P_a$, we have

$$\pi_a^T - \pi^T = \pi_a^T P_a - \pi^T P = \pi_a^T(P+E) - \pi^T P = (\pi_a - \pi)^T P + \pi_a^T E,$$

Hence, $(\pi_a - \pi)^T(I - P) = \pi_a^T E$, where I is the identity matrix. We further notice that

$$(\pi_a - \pi)^T e\pi^T = \pi_a^T e\pi^T - \pi^T e\pi^T = 0,$$

since any steady-state distribution satisfies the normalization condition $\boldsymbol{\pi}^{\mathrm{T}}e = 1$, where e is the all one vector. By adding the left-hand term in the above equation, we have

$$(\pi_a - \pi)^T(I - P + e\pi^T) = \pi_a^T E,$$

Multiplying both sides by the fundamental matrix $Z = (I - P + e\pi^T)^{-1}$, we finally have

$$(\pi_a - \pi)^T(I - P + e\pi^T)Z = (\pi_a - \pi)^T(I - P + e\pi^T)(I - P + e\pi^T)^{-1} = (\pi_a - \pi)^T = \pi_a^T EZ.$$

To represent the perturbed steady-state distribution based on the quantities in the original finite Markov chain, we rewrite the above equation:

$$\pi_a^T - \pi_a^T EZ = \pi_a^T(I - EZ) = \pi^T \Rightarrow \pi_a^T = \pi^T(I - EZ)^{-1}.$$

Similar results have been given in (Schweitzer, 1968; Hunter, 2005; Qian & Dougherty, 2008). For a given function perturbation characterized by the perturbation matrix E, the efficient computation of the perturbed steady-state distribution π_a lies at efficient computation of the inverse matrix $(I–EZ)^{-1}$. Clearly, for an arbitrary perturbation matrix E, the computational complexity of deriving π_a lies will be cubic to the state space size of the Markov chain as it involves this inverse matrix, even if we have all the exact quantities in the original Markov chain. However, in (Schweitzer, 1968; Hunter, 2005; Qian & Dougherty, 2008; Qian, Yoon, & Dougherty, 2012), we have seen that there exist much more efficient solutions when E is a general rank-*k* matrix with $k << 2^n$, which is the size of the state space of the underlying Markov chain in a PBN with *n* variables corresponding to genes to model.

3.1 Perturbed Steady-State Distribution by Sherman-Morrison-Woodbury Formula

We now derive analytic solutions to efficiently compute the perturbed steady-state distribution π_a for general rank-*k* function perturbations with $E = UV^T$, in which E is a rank-*k* matrix in the decomposed version UV^T: U and V are both $2^n \times k$ matrices satisfying the constraints to guarantee that the perturbed state transition matrix P_a remains a stochastic matrix for legitimate structural intervention in PBNs: $P_a e=e$. Since $P_e=e$ as P is stochastic as the original transition matrix and $P_a = P + E$, we need to have the constraints $UV^Te=0$. Based on Theorem 1, we have

$$\pi_a^T = \pi^T(\mathrm{I - EZ})^{-1} = \pi^T(\mathrm{I - UV^T Z})^{-1}.$$

This form immediately calls for the efficient computation of the inverse matrix $(I–UV^TZ)^{-1}$ based on the Sherman-Morrison-Woodbury (SWM) formula (Sherman & Morrison, 1950), especially when $k \ll 2^n$. We now first introduce the SMW formula for computing the inverse matrix with the general form $(A+BC^T)^{-1}$ with general matrices A, B, and C:

Proposition 1: Given matrices A, B, and C,

$$(A + BC^T)^{-1} = A^{-1} - A^{-1}B(I + C^T A^{-1}B)^{-1}C^T A^{-1}.$$

With this, we set A = I, B = – U, and C = Z^TV. Substituting these into the SMW formula, we have

$$(A + BC^T)^{-1} = (I - UV^T Z)^{-1} = I + U(I - V^T ZU)^{-1}V^T Z. \tag{2}$$

Note that the main computation of computing the inverse matrix $(I–UV^TZ)^{-1}$ now is transferred to compute the new inverse matrix $(I–UV^TZU)^{-1}$. The dimension of the original matrix $I–EZ=I–UV^TZ$ is $2^n \times 2^n$. After transformation using the SMW formula, the inverse of this large matrix has been changed to computing the inverse of $I–V^TZU$, which is instead a $k \times k$ matrix since both U and V are $2^n \times k$ matrices. Hence, for an arbitrary rank-k structural intervention, the man factor for the computational complexity to compute $\pi_a^T = \pi^T (I - EZ)^{-1}$---the inverse of I–EZ---has been greatly reduced from $O(2^{3n})$ to $O(k^3)$ when $k \ll 2^n$. By substituting (2) into the Equation (1), we have

$$\pi_a^T = \pi^T + \pi^T U(I - V^T ZU)^{-1}V^T Z. \tag{3}$$

Hence, the steady-state distribution change is $\Delta\pi^T = \pi^T U(I - V^T ZU)^{-1}V^T Z$, which is the final analytic form by the given quantities in the original network together with the rank-k decomposition of the perturbation matrix. Our derived analytic solution based on the SMW formula is much more efficient compared to the direct computation based on (1).

Special Case: Rank-One Perturbations

We now consider a special case with rank-one perturbations, in which the perturbation matrix can be represented by a rank-one matrix $E = uv^T$, where u and v are two arbitrary 2^n dimensional vectors. Here, to make sure that the perturbed state transition matrix $P_a = P + E$ is a stochastic matrix, we need to have $v^Te=0$ as the constraint. Based on the derived Equation (3), we now have

Theorem 3: For an irreducible and ergodic finite Markov chain with its corresponding state transition matrix P, steady-state distribution $\boldsymbol{\pi}$, and fundament matrix $Z = (I - P + e\boldsymbol{\pi}^T)^{-1}$, the perturbed steady-state distribution $\boldsymbol{\pi}_a$ from any rank-one perturbation $E = uv^T$ with $v^Te=0$ has a closed-form analytic solution:

$$\pi_a^T = \pi^T + \frac{\pi^T uv^T Z}{1 - v^T Zu}. \tag{4}$$

The result has appeared in Hunter's publications for general Markov chains (Hunter, 1986; Hunter, 2005) and in Qian and Dougherty's work (Qian & Dougherty, 2008) for structural intervention in PBNs. To the best of our knowledge, most of the previous work has focused on this special case of rank-one perturbations and both of the derivations were based on that the fundamental matrix is a g-inverse of $I - P$. We show that, using the SMW formula, we can derive a general analytic solution for arbitrary rank-k perturbations.

An important special case occurs when the transition mechanisms in a Markov chain before and after perturbation differ only in one state, say the kth state. In this case, the perturbation matrix $E = uv^T$ has non-zero values only in its kth row. Hence, $u=e_k$, an elementary vector with its kth entry equal to 1 and 0 elsewhere. Substituting this to (4) yields

$$\pi_a^T = \pi^T + \frac{\pi^T uv^T Z}{1 - v^T Zu} = \pi^T + \frac{\pi^T e_k v^T Z}{1 - v^T Ze_k}.$$

Denoting $\beta^T = v^T Z$, we can rewrite the above equation to get:

$$\pi_a^T = \pi^T + \frac{\pi^T e_k v^T Z}{1 - v^T Ze_k} = \pi^T + \frac{\pi_k}{1 - \beta_k}\beta^T, \tag{5}$$

in which β_k is the kth element in the vector β. This special case has appeared in Schweitzer (1968) with the different derivation.

Based on (5), we can compute the steady-state probability $\pi_a^{\,i}$ after perturbation for the ith state in the Markov chain:

$$\pi_a^{\ i} = \pi^i + \frac{\pi^k}{1-\beta_k}\beta^i,$$

where $\beta^i = \sum_j v^j z_{ji}$ and v is in fact the difference vector for the kth rows between the perturbed state transition matrix and the original transition matrix. Using the relationship between the mean first passage time (MFPT) and the fundamental matrix (Kemény & Snell, 1983; Schweitzer, 1968; Hunter, 2005; Qian & Dougherty, 2008), $z_{ij}=z_{jj}-\pi^j m_{ij}$, where m_{ij} is the MFPT from the state to the state, we further have

$$\beta^i = \sum_j v^j (z_{ii} - \pi^i m_{ji}) = z_{ii}\sum_j v^j - \pi^i \sum_j v^j m_{ji}.$$

Since $v^T e = \sum_j v^j = 0$, we have $\beta^i = -\pi^i \sum_j v^j m_{ji}$. and

$$\pi_a^{\ i} = \pi^i + \frac{\pi^k}{1-\beta^k}\beta^i = \pi^i - \frac{\pi^k \pi^i \sum_j v^j m_{ji}}{1+\pi^k \sum_j v^j m_{jk}} = \pi^i \frac{1-\pi^k \sum_j v^j m_{ji}}{1+\pi^k \sum_j v^j m_{jk}}.$$

Clearly, for a given Markov chain and perturbation, the perturbed steady-state distribution can be fully characterized by either the fundamental matrix Z or the MFPT matrix $M = (m_{ij})$.

For the most special case, where v has only two non-zero entries for the rth and sth states, we have $P_{kr}^a = P_{kr} - \varepsilon$ and $P_{ks}^a = P_{ks} + \varepsilon$, where ε is the only perturbation to the original transition matrix P with $v^r=-\varepsilon$ and $v^s=\varepsilon$. In this case, we have

$$\pi_a^{\ i} = \pi^i + \frac{\pi^k \varepsilon(z_{si} - z_{ri})}{1-\varepsilon(z_{sk} - z_{rk})}; or \quad \pi_a^{\ i} = \pi^i \frac{1-\pi^k \varepsilon(m_{si} - m_{ri})}{1+\pi^k \varepsilon(m_{sk} - m_{rk})}. \tag{6}$$

3.2 Perturbations in PBNs

With these derived results in general finite Markov chains, we now can derive the perturbation results in PBNs. Due to potential notation abuse, we note that we use the bold font to denote the gene states while we write them in the regular font to represent the relative positions in vectors and matrices when we use them as the subscripts for their entries.

We first adapt the preceding solution to characterize the long-run effects from perturbations to the simple class of PBNs: BNps, in which we only have one context network. A single perturbation on a BNp is a flip for one target gene **g** at the response side for the *k*th state, originally determined by the Boolean function $f_g(\boldsymbol{k})$. Based on the probability structure of BNps as described in **Section 2**, the transition probability from the *k*th state to the *i*th state is given by:

$$p_{ki} = 1_{\mathbf{f(k)=i}}(1-p)^n + 1_{\mathbf{f(k)\neq i}}\, p^{\eta(\mathbf{k,i})}(1-p)^{n-\eta(\mathbf{k,i})},$$

in which *p* is the perturbation probability in the BNp model; *n* is the number of genes in the BNp; η(**k,i**) is the Hamming distance between the binary presentations of the *k*th state and *i*th state; the first term takes care of the transition determined by logic functions and the second term represents the potential transitions from random gene perturbation. If we ignore random gene perturbations, then each row in the original transition matrix *P* corresponds to a single input state (***k***) and the *k*th row consists of all zeros except for the entry in the column of the output state **f**(***k***). In other words, we can in fact decompose $P=F(1-p)^n+H$, where *H* is the random perturbation part determined by pairwise Hamming distances and *F* is a binary matrix determined by logic functions in the BNp. We note that the structural or function perturbations to BNps will not affect the second term of random gene perturbations. Function perturbations only change the binary matrix *F*. A one-bit permutation for input state ***k*** means that there is a network function transformation from original function *F* to F^a with $\boldsymbol{s} = \mathbf{f}^a(\boldsymbol{k}) \neq \mathbf{f}(\boldsymbol{k}) = \boldsymbol{r}$ but for all $\boldsymbol{i} \neq \boldsymbol{k}$, $\mathbf{f}^a(\boldsymbol{i}) = \mathbf{f}(\boldsymbol{i})$. Hence, we have $F_{ks}^a = 1$ while $F_{kr}^a = 0$, whereas originally $F_{ks}=0$ while $F_{kr}=1$. The rest of the matrix keeps unchanged after one-bit perturbations. Taking this back to the transition matrix *P*, we have only two terms changed: $p_{kr}^a = p_{kr} - \varepsilon$ and $p_{ks}^a = p_{ks} + \varepsilon$ and ε is decided by the change to *F* and the perturbation probability *p*: $\varepsilon=(1-p)^n$. With that, we can now compute the perturbed steady-state distribution by one-bit function perturbation based on (6):

$$\pi_a^{\ i} = \pi^i + \frac{\pi^k \varepsilon(z_{si} - z_{ri})}{1-\varepsilon(z_{sk} - z_{rk})} = \pi^i + \frac{\pi^k (z_{si} - z_{ri})(1-p)^n}{1-(z_{sk} - z_{rk})(1-p)^n}.$$

and

$$\pi_a^{\ i} = \pi^i \frac{1-\pi^k \varepsilon(m_{si} - m_{ri})}{1+\pi^k \varepsilon(m_{sk} - m_{rk})} = \pi^i \frac{1-\pi^k (m_{si} - m_{ri})(1-p)^n}{1+\pi^k (m_{sk} - m_{rk})(1-p)^n}.$$

From the above equation, we can compute

$$\pi_a^i - \pi^i = \pi^i \pi^k (1-p)^n \frac{(m_{ri} - m_{si}) + (m_{rk} - m_{sk})}{1 + \pi^k (m_{sk} - m_{rk})(1-p)^n}.$$

The above equation shows that the perturbation depends on whether the states ***k*** and ***i*** are absorbing or transient. In addition, if ***r*** and ***s*** are either transient states or attractor states belonging to the same attractor cycle, then the perturbation to long-run dynamics should be small. More generally, properties of the perturbation are revealed by the mean first passage times (which are related to state status in the Markov chain).

This framework can be extended to more sophisticated function perturbations for structural intervention in PBNs by applying the results from general rank-k perturbations. For example, if we now consider function perturbation to a single regulatory function f_g for gene **g**. Assume gene **g** has K_g regulators in the network. Although we only perturb f_g to change only the value for one target gene, the output states could change between multiple states and there will be changes to at most 2 2^{n-K_g} rows of the original transition matrix *P*. In this case, we can implement rank-k perturbation results in (3) with $k = 2^{n-K_g}$. But when the output states only change from ***r*** to ***s***, then we still have a rank-one perturbation with the perturbation matrix $E = uv^T$, where

$u = e_{K_g}$ with its corresponding 2^{n-K_g} entries equal to 1 and the others being 0 and

$$v^T = \{0,\ldots,0,\underbrace{-\varepsilon}_{rth},0,\ldots,0,\underbrace{\varepsilon}_{sth},0,\ldots,0\}.$$

Substituting into (4), the perturbed steady-state distribution can be written as

$$\pi_a^T = \pi^T + \frac{\pi^T e_{K_g} v^T Z}{1 - v^T Z e_{K_g}}.$$

We now extend the results to instantaneously random PBNs. We can again decompose the original transition matrix by $P=F(1-p)^n+H$, where F is now determined by both the corresponding logic functions in constituent context networks and the selection probabilities. Each row in F corresponds to a single input state ***i*** and that row contains a number of positive values summing to one with the remaining values being zero. The value in the corresponding column j is the sum of all context-selection probabilities c_l such that $\mathbf{f}^l(\mathbf{i})=\mathbf{j}$, where $\mathbf{f}^l(\bullet)$ is the network function for the lth constituent network. A one-bit perturbation for input state $\boldsymbol{k}$ corresponds to a change in one network function $\mathbf{f}^l(\mathbf{k})$, so that exactly two columns are changed. This means that there is a network function transformation $\mathbf{f}^l(\bullet)$ to $\mathbf{f}^{l''}(\cdot)$ with $\mathbf{s} = \mathbf{f}^{l''}(\mathbf{k}) \neq \mathbf{f}^{l}(\mathbf{k}) = \mathbf{r}$ while all the rest of network functions keep unchanged. Therefore, there is an increase of the corresponding network selection probability c_l in the column corresponding to the output state $\boldsymbol{s}$ and a decrease of c_l in the column of the original output state $\boldsymbol{r}$, and all other entries in the state transition matrix are unchanged. Hence, we can proceed in the same manner as with a one-bit perturbation for a BNp as described above.

If we are not limited to changing the logic functions corresponding to the input state $\boldsymbol{k}$ in one single constituent network, the situation becomes more complicated because there could be several changes in the transition matrix on the corresponding row. For example, we change the functions in two constituent BNs, l and w. There are two transformations from the perturbation: $\mathbf{f}^l(\bullet)$ to $\mathbf{f}^{l''}(\cdot)$ and $\mathbf{f}^w(\bullet)$ to $\mathbf{f}^{w''}(\cdot)$ with $\mathbf{f}^{l''}(\mathbf{k}) \neq \mathbf{f}^{l}(\mathbf{k})$ and $\mathbf{f}^{w''}(\mathbf{k}) \neq \mathbf{f}^{w}(\mathbf{k})$. Consequently, there could be up to four changes for input state $\boldsymbol{k}$, which in general leads to more than two non-zero entries in v^T of the rank-one representation of the perturbation matrix E. Nevertheless, we can still directly compute the perturbed steady-state distribution based on (4).

For a context-sensitive PBN, each state of the underlying Markov chain is of the form (l, $\boldsymbol{i}$), where l denotes the lth constituent BN and $\boldsymbol{i}$ the gene state vector. If there are m contexts and n genes, then the size of the transition matrix is $m2^n \times m2^n$. A one-bit perturbation for an input state (w, k) corresponds to a change in network function $\mathbf{f}^w(\bullet)$, so that exactly two columns are changed, both being columns of the matrix corresponding to the wth constituent BN. Hence, we can proceed in the same manner as with a rank-one perturbation for a BNp using (3). To change d constituent BN network functions for the input gene state $\boldsymbol{k}$ leads to one-bit perturbations to d states, (w_1, $\boldsymbol{k}$), (w_2, $\boldsymbol{k}$), …, (w_d, $\boldsymbol{k}$) in the underlying Markov chain. Relative to the matrix this means a rank-d perturbation. This can be solved based on (3).

3.3 Structural Intervention Based on Perturbed Steady-State Distributions

The objective of network intervention can be directing cellular dynamics away from certain states (for example, metastatic cancerous phenotypes). The goal of structural intervention is to alter the long-run dynamics of the network by altering the network wiring or the involved logic functions in PBNs. Given a set ***U*** of undesired states, we want to find the optimal one-bit perturbation to minimize:

$$\arg\min_{v,\mathbf{k}} \left\{ \sum_{i\in U} \pi_a^i = \sum_{i\in U} \left[\pi^i + \frac{\pi^{\mathbf{k}} \sum_j v^j z_{ji}}{1 - \sum_j v^j z_{j\mathbf{k}}} \right] \right\},$$

which is equivalent to the following problem:

$$\arg\min_{v,\mathbf{k}} \sum_{i\in U} \frac{\pi^{\mathbf{k}} \sum_j v^j z_{ji}}{1 - \sum_j v^j z_{j\mathbf{k}}},$$

Analogously, we can also maximize the probability mass of a set ***D*** of desired states:

$$\arg\max_{v,\mathbf{k}} \sum_{i\in D} \frac{\pi^{\mathbf{k}} \sum_j v^j z_{ji}}{1 - \sum_j v^j z_{j\mathbf{k}}},$$

We note here that *U* and *D* need not form a partition of the state space because there may be “don’t care” or “ambiguous” states. Recently, we have proposed new bi-objective intervention strategies to minimize the undesirable steady-state mass while avoid introducing too high steady-state mass on “ambiguous” states as they may correspond to phenotypes with collateral damage to cells (Qian & Dougherty, 2012; Qian, 2012), we will discuss this in more detail in Section 5.

Clearly, the search for the optimal one-bit structural intervention is computational expensive as the search space is exponential to the size of the network. The extension to multiple changes by either rank-*k* perturbation or applying the iterative expression in (Qian & Dougherty, 2008) will be more computationally expensive. Hence, network reduction is necessary to make the optimization practical for intervening in real-world networks.

Melanoma Example

To illustrate structural intervention, we consider a gene regulatory network developed from data collected in a study of metastatic melanoma (Bittner *et al.*, 2000; Weeraratna *et al.*, 2002), in which the upregulation of gene WNT5A was found to be highly discriminating between cells with properties typically associated with high metastatic competence versus those with low metastatic competence (Dissanayake *et al.*, 2008). This suggests intervention to down regulate WNT5A to reduce the risk of a melanoma metastasizing, an undesirable outcome. We consider a 7-gene network containing the genes WNT5A, pirin, S100P, RET1, MART1, HADHB and STC2 (see Qian & Dougherty (2008) for network details) whose steady-state distribution is shown in Figure 2(a), which shows the steady-state masses for the undesirable (WNT5A up-regulated) and desirable (WNT5A down-regulated) states. The states are ordered as WNT5A, pirin, S100P, RET1, MART1, HADHB and STC2, with WNT5A as the most significant bit (MSB) and STC2 as the least significant bit (LSB). Note that the states from 1–64 have WNT5A 0 and hence are desirable states, as compared to states 65–128 that have WNT5A 1 and hence are undesirable. Our goal is to reduce the probability of undesirable states while at the same time not introducing probability mass for states possessing negligible probability in the original network because such states represent unknown phenotypes that may correspond to collateral damage. Using what we have derived above, an optimal structural change to the regulatory network based on these criteria is deduced which alters the regulation of RET1. The resulting steady-state distribution is given in Figure 2(b).

Figure 2. Steady-state distributions of a seven-gene melanoma network (Qian & Dougherty, 2008) modeled as a BNp: (a) Before structural intervention; (b) After intervention.

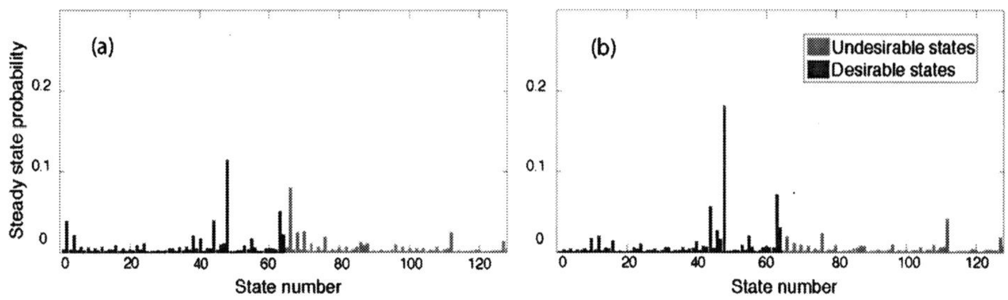

4. EXTERNAL CONTROL OF PBNs

In addition to structural intervention focusing on steady-state probabilities of entering either desirable (D) or undesirable states (U), PBNs also can model the transient dynamics of gene regulatory networks. Once a probability distribution vector has been specified for the initial states, the probability distribution vector w(t) evolves according to $w(t+1)^T = w(t)^T P$. It ultimately converges to the steady-state distribution π for PBNs as explained earlier. From this perspective, PBNs are *descriptive* in nature. External control through external or auxiliary variables can be designed to control the evolution of the probability distribution vector. For treatment or intervention purposes, we are interested in working with PBNs in such a *prescriptive* fashion, where the transition probabilities of the associated Markov chain depend on certain auxiliary variables, whose values can be chosen to make the probability distribution vector evolve in some desirable manner.

The use of such auxiliary variables makes sense from a biological perspective. For instance, in the case of diseases like cancer, auxiliary treatment inputs such as radiation and chemotherapy may be employed to move the state probability distribution vector away from one associated with uncontrolled cell proliferation or markedly reduced apoptosis. The auxiliary variables could also include genes that serve as external master-regulators for all the genes in the network. The values of the individual control inputs can be changed from one time step to another in an effort to make the network behave in a desirable fashion.

Suppose we have *bexternal control inputs* as auxiliary variables, $u_1, u_2, \ldots, u_b$. To be consistent with the binary nature of the expression status of individual genes in a PBN, we will assume that each of these auxiliary variables can take on only the binary values 0 or 1 in this chapter. One can represent the control input status at time t using a decimal integer $v(t)$ ranging from 1 to 2^b, so that $B = \{1, \ldots, 2^b\}$ is the set of possible control actions. As shown in Datta, Choudhary, Bittner, & Dougherty (2004), the one-step evolution of the probability distribution vector in the case of a PBN containing 2^n states with control inputs takes place according to the equation

$$w(t+1)=w(t)P(v(t)) \tag{7}$$

where $w(t)$ is the 2^n dimensional state probability distribution vector and $P(v(t))$ is the $2^n \times 2^n$*control-dependent transition probability matrix*. The system in (7) can be equivalently represented as a stationary discrete-time dynamic system

$$x(t+1) = f(z(t), v(t), d(t)),\ t=0,1,\ldots \tag{8}$$

where for all t, the state $z(t)$ is an element of $\boldsymbol{S}$, the control input $v(t)$ is an element of B, the disturbance $d(t)$ is an element of a space D and f: $S \times B \times D \mapsto S$. The disturbance $d(t)$ is manifested in terms of the change of network context based on the switching probability q or the random flip of states due to the perturbation probability p. At a given time t, $d(t)$ is independent of prior disturbances $d(0)$, $d(1)$, ..., $d(t-1)$. We will interchangeably use either the representation (7) or (8) depending on their suitability for a particular context or a particular derivation.

Since the transition probability matrix is a function of the control input $v(t)$, the evolution of the probability distribution vector of the PBN with control now depends not only on the initial distribution vector but also on the values of the control input at different time points. Intuitively, it appears possible to make the states of the network evolve in a desirable fashion by appropriately choosing the control input at each time point.

4.1 Finite Horizon Control

We start with the finite horizon control formulation. Suppose that we have a priori determined period of fixed length M over which the control input is to be applied. We are interested in controlling the dynamic behavior of the PBN over the interval $t = 0, 1, 2, ..., M-1$. Suppose at time point t, the state of the PBN is given by $z(t)$ and the corresponding control input is $v(t)$. (We emphasize that $z(t)$ is equivalent to the actual state $\mathbf{X}(t)$ when no external control is applied.) Then we can define a cost $C_t(z(t), v(t))$ as being the cost of applying the control input $v(t)$ when the state is $z(t)$. The other component of the finite-horizon cost is the cost associated with the terminal state $z(M)$. Owing to the probabilistic nature of the evolution, the terminal state $z(M)$ is a random variable that can possibly take on any of the states in $\boldsymbol{S}$. We assign higher terminal costs to the undesirable states $\boldsymbol{U}$. For instance, a state associated with rapid cell proliferation leading to cancer should be associated with a high terminal penalty while a state associated with normal behavior should be assigned a low terminal penalty. For the purposes of this section, we will assume that the assignment of terminal penalties has been carried out and we have at our disposal a terminal penalty $C_M(z(M))$ that is a function of the terminal state.

With these definitions, the finite-horizon cost to be minimized is given by

$$E\left[\sum_{t=0}^{M-1} C_t\left(z(t),v(t)\right)+C_M\left(z(M)\right)\middle|z(0)\right] \tag{9}$$

To proceed further, let us assume that at time t the control input $v(t)$ is a function of the current state $z(t)$, namely,

$$v(t)=\mu_t(z(t)) \tag{10}$$

where $\mu_t: S \mapsto B$. The optimal control problem can now be stated. Given an initial state $z(0)$, find a control law $\pi = \{\mu_0, \mu_1, \dots, \mu_{M-1}\}$ that minimizes the cost functional

$$J_0(z(0)) = E\left[\sum_{t=0}^{M-1} C_t(z(t),v(t)) + C_M(z(M))\right] \tag{11}$$

subject to the constraint

$$\Pr\left(z(t+1)=j \mid z(t)=i, v(t)=v\right) = p_{ij}(v) \tag{12}$$

where $p_{ij}(v)$ is the ith row, jth column entry of the matrix $P(v)$.

Solution Using Dynamic Programming

Optimal control problems described by (11) and (12) can be solved by dynamic programming, pioneered by Bellman in the 1960s (Bellman, 1952; Datta *et al.*, 2004; Bertsekas, 2005). When an optimal strategy exists, the principle of optimality asserts: if one searches for an optimal strategy over a subset of the original number of steps, then this new optimal strategy will be given by the overall optimal strategy, restricted to the steps being considered. It can be used to arrive at the dynamic programming solution to (11) and (12):

$$J_M(z(M)) = C_M(z(M)) \tag{13}$$

$$J_t(z(t)) = \min_{v(t)\in B}\left\{C_t(z(t),v(t)) + \sum_{j\in S} p_{z(t),j}(v(t))\, J_{t+1}(j)\right\} \tag{14}$$

for $t = 0, 1, \dots, M-1$.

Melanoma Example

Illustrative examples to show optimal control design for instantaneously random PBNs are available in Datta *et al.* (2004). We here provide an example of applying dynamic programming over finite time steps for a context-sensitive PBN derived from actual gene expression data (Pal, Datta, Bittner, & Dougherty, 2005; Datta *et al.*, 2007) for the same melanoma example in the previous section. From the gene expression profiling study of metastatic melanoma (Bittner *et al.*, 2000; Weeraratna *et al.*, 2002), it is biologically known that WNT5A ceasing to be down regulated is strongly predictive of the onset of metastasis. Hence, our objective to intervene this network is to externally down-regulate the WNT5A gene. Considering the seven-gene network with WNT5A, pirin, S100P, RET1, MART1, HADHB and STC2, we have used the Bayesian connectivity-based approach of Zhou *et al.* (2004) to construct four highly probable BNs that are used as the constituent BNs in the PBN, with their selection probabilities based on their Bayesian scores. The control strategy of this section has been applied to the designed PBN with pirin chosen as the control gene **g** and $p = q = 0.01$. Figure 3 shows the expected cost for a finite horizon problem of length 5 originating from each of the 128 states (Note that the choice of $M = 5$ is arbitrary.) The cost of control is assumed to be 0.5 and the states are assigned a

Figure 3. Cost function comparison with and without finite horizon external control on the melanoma network with the priori determined control period M=5

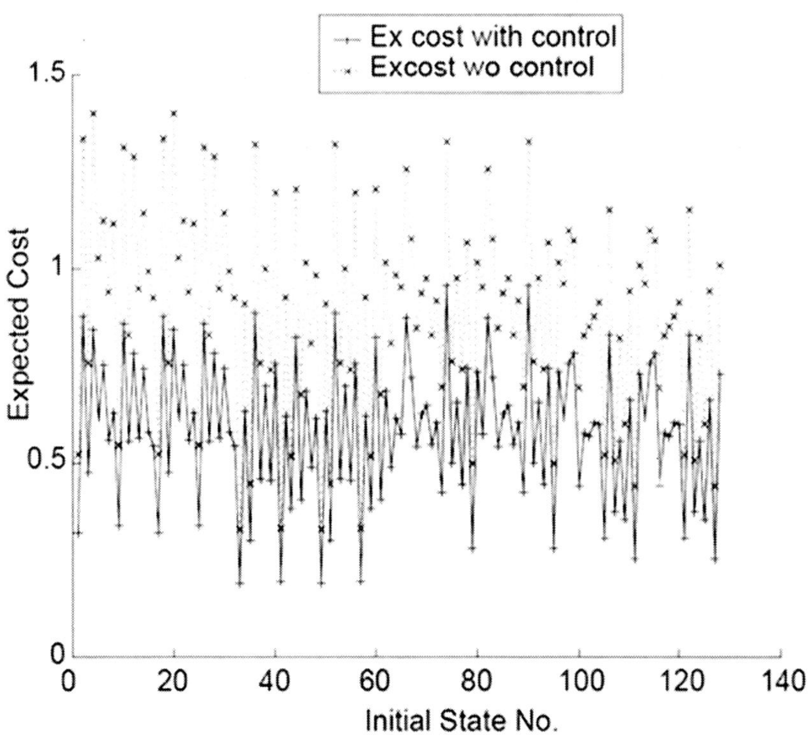

terminal penalty of 5 if WNT5A is 1 and 0 if WNT5A is 0. Here the choice of the numbers 0 and 5 is arbitrary but it does reflect our attempt to capture the intuitive notion that states where WNT5A equals one are less desirable than those where WNT5A equals 0. The control objective is to down-regulate the WNT5A gene. From Figure 3, it is clear that the expected cost with control is much lower than that without control, which agrees with our objective.

4.2 Infinite Horizon Control

The finite horizon external control approach has focused on manipulating external control variables that affect the transition probabilities of a PBN to desirably affect its transient dynamic evolution over a finite time horizon. These short-term policies are not always effective in changing the steady-state behavior of the PBN, even though they can change the dynamical performance of the network for a small number of stages. Motivated by this, in Pal *et al.* (2006), intervention via external control variables in PBNs over an infinite length of time has been proposed. We focus on the stationary control policy that does not change from one time step to the next because implementations of such stationary policies are often simple and stationary policies can lead to steady-state distribution shift from undesirable states to desirable ones. We next present a tutorial discussion of the results obtained.

The problem formulation and results summarized earlier for the finite horizon case serve to motivate the infinite horizon developments here. Consider the finite horizon cost function being minimized in (11) and suppose that the control horizon characterized by M is made larger and larger and in the limit we would like for it to tend to infinity. In trying to do so, we immediately encounter a number of potential obstacles that did not arise in the finite horizon case. First, in the infinite horizon problem, the control horizon is infinite and therefore there is no terminal state or its associated terminal penalty. Consequently, for the infinite horizon case, the cost per stage should incorporate the penalty of the state along with the cost of control. $C_t(z(t),v(t))$ of the finite horizon problem should now be replaced by $\tilde{C}(z(t),v(t),d(t))$ so that the per stage cost takes into account the origin, the control, and the destination state. Second, in the finite horizon problem, the summation in (11) is a finite one and so the quantity being minimized is finite. If we let the control horizon go to infinity, there is a possibility that the summation of the one-stage costs may go to infinity (for all controls) leading to an ill-posed optimization problem. To make the optimization problem well posed, the cost considered in (11) has to be modified before letting the length M of the control horizon tend to infinity. There are a couple of approaches for doing this. We will use the approach referred to in the literature as the problem of total cost with discounted and bounded cost per stage.

In the rest of this section, we will be denoting $w(t)$, $z(t)$, $v(t)$, $d(t)$ by w_t, z_t, v_t, d_t respectively, mainly for the purpose of simplifying the notation.

We assume that the cost per stage $\tilde{C}(z_t, v_t, d_t)$ is bounded $\forall\, z_t \in \boldsymbol{S}$, $d_t \in \mathrm{D}$ and $v_t \in \mathrm{B}$ and adiscounting factor $\alpha \in (0, 1)$ is introduced in the cost to make sure that the limit of the cost sum converges as the horizon length goes to infinity. More specifically, our objective is to find a policy $\mu = \{u_0, u_1, \ldots\ldots\}$, where μ_t: $S \mapsto B$, t = 0, 1,……, that minimizes the cost function

$$J_{\mu}\left(z_0\right)=\lim_{M \to \infty} E\left[\sum_{t=0}^{M-1} \alpha^t \tilde{C}\left(z_t, \mu_t\left(z_t\right), d_t\right)\right] \tag{15}$$

where the cost per stage $\tilde{C}: S \times B \times D \mapsto \mathbb{R}$ is given. (Note that a Markov chain can be modeled by $z_{t+1} = d_t$ (Bertsekas, 2005). Hence, the destination state is the same as the disturbance d_t.) In the general formulation, the inclusion of α in the cost captures the fact that costs incurring at a later time are less significant. In the case of cancer treatment, $\alpha < 1$ signifies that the condition of the patient in the initial stages of treatment is more important than the condition at a later stage, or in other words, the reward for improving the condition of the patient in the present is more significant than the reward obtained from similar improvement at a later stage. This approach is reasonable if we keep in mind the expected life span of the patient. Another approach for ensuring that the infinite horizon cost is well defined is referred to as the average cost per stage formulation and has been considered in Pal *et al.* (2006).

We next present the solution to the discounted and bounded cost per stage problem. Let us denote by Π the set of all admissible policies μ, i.e., the set of all sequences of functions $\mu = \{u_0, u_1, \ldots\ldots\}$. The optimal cost function is defined by

$$J^*\left(z\right)=\min_{\mu \in \Pi} J_{\mu}(z), \qquad z \in S \tag{16}$$

A stationary policy is an admissible policy of the form μ taking the same control policy function with respect to the current state z_t: $\mu = \{u, u, \ldots\ldots\}$, which is not changing with time. Its corresponding cost function is denoted by J_μ. We say that the stationary policy $\mu = \{u, u, \ldots\ldots\}$ is optimal if $J_\mu(z) = J^*(z)$ for all states z.

Optimal Control Solution

We solve the problem of minimizing the cost in (15) under the assumption that the cost per stage $\tilde{C}(z,v,d)$ is bounded, i.e., $\exists\, \tau > 0$ such that $\tilde{C}$ satisfies $\left|\tilde{C}(z,v,d)\right| \leq \tau$, for all $(z,v,d) \in S \times B \times D$. In the case of context-sensitive PBNs, $\tilde{C}(z,v,d)$ is bounded since the control and disturbance spaces are finite.

Observe that if we set $C_M(z_M) = 0, \forall z_M \in S$ and $C_t(z_t, v_t) = \alpha^t C(z_t, v_t)$ (where $C(z_t, v_t)$ is the expectation of $\tilde{C}(z,v,d)$ over d_t) in the finite horizon problem of (11) and let $M \to \infty$, then we obtain the infinite horizon cost function considered in (15). Thus it seems reasonable that the finite horizon solution described by (13) and (14) could provide a basis for arriving at the solution of the optimization problem in (16) where J_μ is given by (15). A formal derivation of this connection is given in (Bertsekas, 2005; Pal *et al.*, 2006). Here we simply state the result and present an intuitive justification for it.

Towards this end, note that (14) in the dynamic programming algorithm basically describes how the optimal cost J_{t+1} propagates backwards in time to the optimal cost J_t in the finite horizon problem of (11). For the cost function considered in (15), it is clear that the cost J_{t+1} must be discounted by the factor α while being propagated to the previous stage. Consequently, for the new optimal infinite horizon control problem, (14) will have to be replaced by

$$J_t(i) = \min_{v \in B} \left[C(i,v) + \alpha \sum_{j \in S} p_{ij}(v) J_{t+1}(j) \right] \tag{17}$$

The above equation motivates the introduction of the following mapping. For any cost function $J : S \mapsto \mathbb{R}$, define the mapping $TJ : S \mapsto \mathbb{R}$ by

$$TJ(i) = \min_{v \in B} \left[C(i,v) + \alpha \sum_{j \in S} p_{ij}(v) J(j) \right], \quad i \in S. \tag{18}$$

Note that TJ is the optimal cost function for the one-stage (finite horizon) problem that has stage cost C and terminal cost αJ.

It can be shown that the optimal cost function J^* is the unique fixed point of the map and the iteration $J_{t+1} = TJ_t$ converges to J^* as $t \to \infty$. This provides us with a computational algorithm for determining the optimal cost function by running the recursion

$$J_{t+1} = TJ_t, \; t = 0, 1, 2, \ldots \tag{19}$$

for any bounded initial cost function $J_0 : S \mapsto \mathbb{R}$. The iteration described in (19) above is referred to as the value iteration procedure since at every stage we are iterating on the values of the cost function and the optimal policy simply falls out as a byproduct when the iteration converges to the optimal value of the cost function. The details of the optimality and convergence of the value iteration procedure, along with that of another computational algorithm referred to as policy iteration, are available in Bertsekas (2005) and Pal *et al.* (2006).

Melanoma Example

In this section, we apply the infinite horizon control policy to a context-sensitive PBN derived from the same gene expression data as before. The network contains the seven genes WNT5A, pirin, S100P, RET1, MART1, HADHB and STC2. To obtain the PBN, we have used the algorithms described in Pal, Ivanov, Datta, Bittner, & Dougherty (2005) to construct four highly probable BNs to use as the constituent BNs in the PBN. The control strategy of this section is applied to the designed PBN with pirin chosen as the control gene ($v = 2$ signifying the state of pirin is reversed and $v = 1$ signifying no intervention) and $p = q = 0.01$ (see the p and q definitions). Since our objective is to down regulate the WNT5A gene, a higher penalty of five is assigned for destination states having WNT5a up-regulated. Also for a given WNT5A status for the destination state, a higher penalty is assigned when the control is active (control cost assumed to be one) versus when it is not. Further details on the penalty assignment are available in Pal *et al.* (2006).

Figure 4 (a) shows the steady-state distribution of the PBN using the obtained stationary policy, and Figure 4(b) shows the original PBN steady-state distribution for comparison purposes. The steady-state distribution shift demonstrated in Figures 4(a) and (b) shows that the stationary external control policy has enabled

Figure 4. (a) Steady state distribution using discounted cost stationary policy and (b) original steady state distribution without control
Source: Pal et al. (2006)

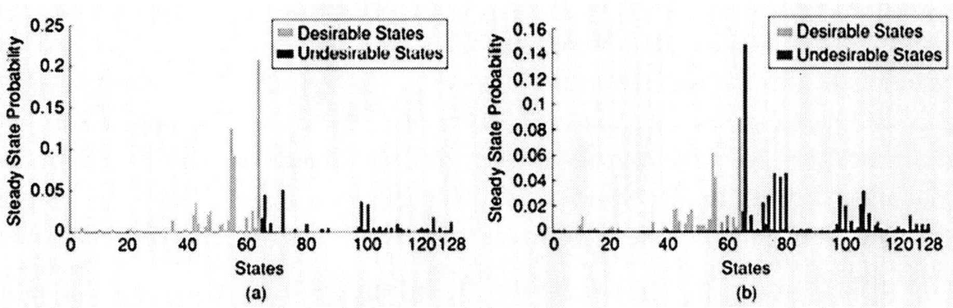

us to move the probability mass from the bad states to states with lower metastatic competence. For example, state 66 (WNT5A is 1) has a high probability mass (0.15) in the original steady state but stationary control has reduced its steady-state mass to 0.04. Similarly, the probability mass of state 64 (desirable state) is high when using the stationary policy.

4.3 Greedy Control

Finally, for both structural and external control strategies we have covered so far, we can explicitly derive corresponding perturbed state transition probability matrix $P(\mathbf{u}_s)$ or $P(\mathbf{u}_g)$. By Theorems 1 and 3, we can obtain an explicit expression of the corresponding perturbed steady-state distribution $\boldsymbol{\pi}_a$. Based on these results, we also have derived greedy stationary control policies in Qian, Ivanov, Ghaffari, & Dougherty (2009), which can help efficiently check and find the best gene to control in genetic therapeutic strategies.

For derived greedy stationary control policies, the task is to have a perturbed steady-state distribution $\boldsymbol{\pi}_a$ after either structural or external control that possesses the minimum undesirable mass (the total steady-state probability mass of undesirable states U): $\pi_a^U = \sum_{z \in U} \pi_a(z)$, where $\boldsymbol{\pi}_a(z)$ denotes the steady-state probability at a given state $z \in S$. Different greedy control policies have been proposed based on this long-run objective to directly reduce the undesirable mass (Qian *et al.*, 2009). In addition to directly investigating steady-state distributions, different other criteria of network behavior can be used as the surrogates to shift network dynamics towards beneficial directions. For example, due to the close relationships to steady-state distributions, mean first passage time (MFPT) of the underlying Markov chain of PBNs can be considered as the intervention goal to increase or decrease time to reach undesirable or desirable states respectively (Shmulevich *et al.*, 2002b; Vahedi, Faryabi, Chamberland, Datta, & Dougherty, 2008). Steady-state distributions are also tightly dependent on inherited attractor cycles and topological structure of corresponding basins of attraction (BOA) (Brun, Dougherty, & Shmulevich, 2005), greedy control policies considering such BOA structures also have been derived to beneficially alter network dynamics (Qian *et al.*, 2009, Qian, 2012). Compared to previous optimal control policies, the derivation of these greedy control policies is straightforward and more computationally efficient. Based on empirical results in (Shmulevich *et al.*, 2002b; Vahedi *et al.*, 2008; Qian *et al.*, 2009, Qian, 2012), these greedy control policies have been shown to also be effective in reducing the undesirable mass.

5. PHENOTYPICALLY CONSTRAINED INTERVENTION

As we reviewed in previous sections, both optimal intervention strategies and greedy control policies based on Markovian regulatory network models can help derive effective intervention strategies to alter the systematic behavior of biological systems. However, as we recently pointed out in Qian & Dougherty (2012), most of the existing strategies focus on reducing the risk of entering some pre-defined undesirable states U, typically based on the expression states of certain genetic markers as we have witnessed in the melanoma example. All the other states in the state space of network dynamics, the complement of U, are simply considered as desirable states D as they have not been discovered to be associated with known pathology. It is clear that the problem has been over-simplified and in reality not all the states in D represent healthy phenotypes. With the ultimate objective of identifying potential drug targets and developing corresponding therapeutics for complex diseases from network modeling and analysis (Shmulevich *et al.*, 2002; Pal *et al.*, 2006; Akutsu, Hayashida, Ching, & Ng, 2007; Datta *et al.*, 2007; Qian & Dougherty, 2008; Ching *et al.*, 2009; Qian *et al.*, 2009; Cong, Ching, Tsing, & Leung, 2010; Shmulevich & Dougherty, 2010), the derived intervention simply by avoiding undesirable states may not be practical and the resulting intervention may direct significant network dynamics to ambiguous states D_a, a subset of D, which may lead to collateral damage by intervention.

In order to control such potential collateral damage, we have derived new stationary control policies that can shift steady-state probability from undesirable states U, but with additional phenotypic constraints to limit the dynamic shift to these ambiguous states D_a with unknown phenotypic effects (Qian & Dougherty, 2012). Specifically, with a given control gene $\mathbf{g}$, stationary control policies $u_g(z)$, denoting whether an intervention should be applied to flip the expression state of $\mathbf{g}$ at any network state $z \in S$, aim to alter the underlying state transitions to have a new controlled steady-state distribution $\boldsymbol{\pi}_a$, with a minimum undesirable steady-state mass and the ambiguous mass at an acceptable level. Motivated by this, we can formulate the following constrained optimization problem for phenotypically constrained intervention:

$$\begin{aligned} &\min_{u_g} \left\{\pi_a^U = \sum_{z \in U} \pi_a(z)\right\}, \\ &s.t. \quad \pi_a(z) \le \gamma, \quad \forall z \in D_a \end{aligned} \qquad (20),$$

where γ is the allowed maximum level for the steady-state probability for any ambiguous state. With this formulation, we extended original greedy control policies based on steady-state distributions to only allow *admissible control actions*, for example, gene state changes for external control, that will guarantee that the ambiguous states have their steady-state probabilities upper bounded by γ, so that the maximum amount of risk to introducing steady-state probability mass at an ambiguous state is constrained to be low enough without causing significant collateral damage from intervention.

Due to the requirement of the explicit construction of a controlled state probability transition matrix of the underlying Markov chain to compute π_a, these steady-state distribution based phenotypically constrained control policies has a computational complexity $O(2^{2n})$ for networks with n genes (Qian, Ghaffari, Ivanov, & Dougherty, 2010; Qian & Dougherty, 2012). Although different reduction mappings can be implemented to derive sub-optimal intervention strategies (Qian *et al.*, 2010), many of these intervention strategies do not scale well with the size of networks n. One way to address this challenge is to directly analyze the structural properties of the basins of attraction (BOAs) without explicitly computing steady-state distributions (Qian, 2012). As we mentioned earlier in this chapter, PBNs preserve structures of the attractors from context BNs, which have been considered to represent biologically meaningful phenotypes (Kauffman, 1969) and the steady-state probabilities of attractors are dependent on the size and structure of BOAs (Brun *et al.*, 2005). We have derived such BOA-based greedy control policies that directly explore the structural properties of BOAs and avoid creating any new attractor cycles so that this phenotypically constrained BOA-based intervention strategy only shifts significant steady-state probabilities among the states within the existing attractor cycles of the original network models. As transient (non-attractor) states typically have negligible steady-state probabilities, by constraining from creating new attractor cycles from intervention, this new BOA-based control policy will not shift steady-state probability mass to ambiguous states D_a as they will remain as transient states after intervention. Hence, steady-state probability mass will be directed to known desirable states and naturally controls potential side effects by intervention. The detailed derivation and description of this phenotypically constrained BOA-based policy can be found in Qian (2012).

Simulation Experiments

We have simulated several random ensembles of BNps with different number of genes with $n = 6$, 8, and 10. These randomly simulated BNps have similar logic and probabilistic structures, in which we set the perturbation probability $p = 0.001$ and the maximum input degree or the maximum cardinality of regulatory sets for each gene is 3. We randomly generate 1,000 BNps in each ensemble and compare phenotypically constrained greedy control policies and the corresponding original greedy control policies without constraints on steady-state mass of ambiguous states (Qian *et al.*, 2009; Qian & Dougherty, 2012; Qian, 2012). We denote the corresponding BOA-based control policies by BOA (without phenotypic constraints) and conBOA (with constraints); steady-state distribution based control policies by SSD (without constraints) and conSSD (with constraints). Without loss of generality, the control gene g is chosen to be the last gene and the expression state of the first gene x_1 determines that the network state x is either desirable (x ∈ D) with x_1 = 1 or undesirable (x ∈ U) with x_1 = 0. The transient states of the original BNs are all considered as ambiguous states in D_a. For conSSD, the allowed acceptance level of ambiguous mass γ is set to 0.001.

Table 1 provides both the original and perturbed undesirable mass as well as ambiguous mass that are averaged within different simulated ensembles before and after applying different control policies. Due to the additional imposed phenotypic constraints to avoid ambiguous mass, both conBOA and conSSD have the loss of beneficial shifts compared to their corresponding unconstrained greedy control policies; however, constrained control policies can still achieve about 50% reduction of undesirable mass. More importantly, when we look at the change of ambiguous mass, both original BOA and SSD control policies without phenotypic constraints shift a large percentage of undesirable steady-state mass to ambiguous states, whereas the constrained control policies limit steady-state mass for the ambiguous states in $\boldsymbol{D}_a$. Comparing the effectiveness of the constrained control policies shows that the new constrained BOA policy (conBOA) performs competitively to conSSD especially considering the reduced computational time as the BOA policy does not require explicitly computing the perturbed steady-state distributions.

Table 1. Average undesirable and ambiguous masses before and after applying different control policies on random ensembles of BNps.

	Average Undesirable Mass			Average Ambiguous Mass		
Control	**n=6**	**n=8**	**n=10**	**n=6**	**n=8**	**n=10**
Original	0.5006	0.5097	0.5127	0.0144	0.0259	0.0389
BOA	0.2960	0.3125	0.3272	0.2008	0.2552	0.2735
SSD	0.2719	0.2928	0.3114	0.3104	0.3595	0.3879
conBOA	0.3973	0.4248	0.4299	0.0148	0.0269	0.0411
conSSD	0.4295	0.4363	0.4360	0.0264	0.0529	0.0765

6. ROBUST AND ADAPTIVE CONTROL OF PBNs

To explain the issue of robust design of control policies for Markovian models, let us revisit the nominal control problem. When the transition probabilities are exactly known, the states make transitions according to $(P^u)_{u\in\Pi}$. In this case, once a control input is chosen, the resulting controlled transition probability matrix is uniquely determined. Let $\mu = \{u_1, u_2, ...\}$ represent a generic control policy and Π represent the set of all possible μ's, i.e., the set of all possible control policies. Let $J_{\mu,\omega}$ denote the expected total cost for the discounted cost infinite-horizon problem under control policy μ and transitions ω

$$J_{\mu,\omega}(z_0) = \lim_{M\to\infty} E_{z_{t+1},t=0,1,...}\left[\sum_{t=0}^{M-1} \alpha^t \tilde{C}(z_t, \mu_t(z_t), d_t)\right]$$

In case of uncertainties, we can parameterize the class of transitions as $\Omega := (P_\alpha^u)_{u\in\Pi,\alpha\in F_\alpha}$, where F_α is the noise parameter distribution. One of the ways to approach robust intervention in presence of uncertainties is to consider the worst-case scenario such that therapeutic intervention is not detrimental even for the worst patient (Iyengar, 2005). A minimax (worst case) intervention policy is defined as a policy whose worst performance over the uncertainty class Ω is best among all admissible policies. The minimax robust policy (Pal, Datta, & Dougherty, 2008), denoted by μ_{mm}, is the one that satisfies

$$\Phi(\Pi, \Omega, z_0) = \min_{\mu\in\Pi} \max_{\omega\in\Omega} J_{\mu,\omega}(z_0).$$

There is no guarantee that a minimax robust solution will exist for all distributions. For some uncertainty classes, analytical solutions are feasible. In Pal *et al.* (2008), the minimax robust solutions for three different uncertainty classes are worked out. The uncertainties are characterized by 1) transition probabilities in an interval, 2) finite number of transition probability matrices, and 3) upper bound on the 1-norm of the difference in the actual and estimated transition probabilities. However, the condition of rectangularity (Iyengar, 2005) required for some of the analytical solutions are hardly valid in practice. The worst-case robust policy design approach is typically conservative because it gives too much importance to the scenarios that hardly occur in practice. When the objective is to avoid extremely undesirable results, a minimax design is suitable, but when the objective is to improve the expected chances of success, a Bayesian approach will be preferable. A Bayesian robust policy minimizes the average cost over the uncertainty class Ω. Let

μ_b denote the policy designed to be optimal at point b of the parameter distribution i.e. μ_b minimizes $J_{\mu,\omega_b}(z_0)$. Here, ω_b refers to the probability transition matrix at point b. Let $\gamma_a(\mu_b) = E_{z_0}\left[J_{\mu_b,\omega_a}(z_0)\right]$ denote the expected cost per state at point a of the parameter distribution for the intervention policy μ_b, where the expectation over the states z_0 of the network is taken to arrive at a single value for representing the cost of a policy. Consequently, the expected cost of $\gamma_a(\mu_b)$ over the distribution F_a is given by

$$E_a[\gamma_a(\mu_b)] = E_a\left[E_{z_0}\left[J_{\mu_b,\omega_a}(z_0)\right]\right],$$

where E_a denotes the expectation relative to the parameter distribution. A Bayesian control policy, denoted by μ_{b*}, is one that minimizes $E_a[\gamma_a(\mu_b)]$. The parameter value $b*$ that achieves the minimum is called a maximally robust parameter value and the corresponding policy is called the Bayesian robust policy (Pal, Datta, & Dougherty, 2009). If the actual parameter for the distribution is b and one optimizes at b, then there is no gain, and perhaps a loss, in using the Bayesian robust policy because $\gamma_a(\mu_b) \leq \gamma_a(\mu_{b^\cdot})$; however, in the face of uncertainty regarding the actual parameter, the gain in applying the Bayesian robust policy is $E_a[\gamma_a(\mu_b)] - E_a\left[\gamma_a(\mu_{b^\cdot})\right]$, which is always nonnegative and can be substantial for parameter values far from the maximally robust one. The conservativeness of the minimax approach as compared to the Bayesian approach is shown in Pal *et al.* (2009) through simulations using PBNs generated from gene expression data collected in a study of metastatic melanoma. Figure 5 (a) shows the cost for policies designed at a particular noise parameter value σ_b, when applied to all possible values of the noise parameter. Figure 5(b) shows that the expected cost in using the Bayesian policy (μ_{b*}) is much less than the expected cost in using the minimax policy (μ_{mm}).

We now consider an adaptive intervention strategy that can be used in the presence of model uncertainty. We assume that the underlying network is modeled by a member of a known finite family of PBNs and we have no *a priori* knowledge about which member of that family models the actual network. In such a situation, a natural approach is to estimate the model identity on-line and then use policy iteration to determine the corresponding controller. This is the principle of adaptive control and considerable theoretical research has been aimed at showing that *certainty equivalence* schemes can provide the required performance (Kumar & Varaiya, 1986; Ioannou & Sun, 2996).

Figure 5. (a) Cost for policies designed at σ_b over the set of all possible σ_b's. (b) Expected cost for various policies
Source: Pal et al. (2009)

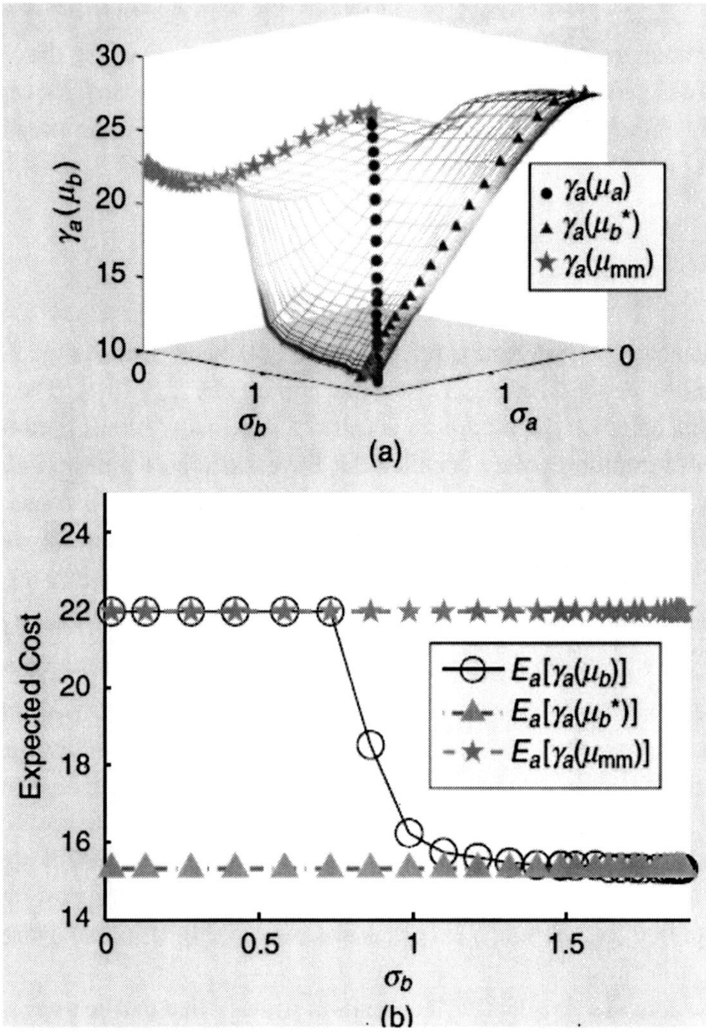

There are a number of ways in which one can possess a list of PBNs and thereby be presented with the problem of adaptively determining a model number. Several inference procedures produce PBNs by way of first producing Boolean networks satisfying some desired relation to the data. In Pal *et al.* (2005), Boolean networks are constructed whose attractor structures coincide with data points assumed to be in attractors in the true biological network, along with the networks satisfying certain constraints, such as the number of predictors. Then one or more PBNs are constructed from these Boolean networks by comparing the steady-state distributions of potentially inferred PBNs with the full set of experimental data. In Zhou *et al.* (2004), Boolean networks are inferred by first using a Bayesian approach to generate regulatory graphs (topologies) most compatible with the data and then inferring the predictors via a nonlinear perceptron model, using a reversible jump Markov chain Monte Carlo (MCMC) method. Then one or more PBNs are constructed from the Boolean networks by using Bayesian scores.

Thus, different filtering techniques can lead to different PBNs. In each of the preceding cases, rather than settling on a single PBN model when applying control, one can take the view that there is a list of potential PBNs and that new data are to be used to adaptively determine the control policy. Moreover, in the cases of Pal *et al.* (2005) and Zhou *et al.* (2004), one might not even form a PBN and simply treat the problem in the framework of a collection of Boolean networks, in which the adaptation is aimed at selecting a control policy for the governing Boolean network, a view compatible with our proposed algorithms. This latter view, that one has a collection of Boolean networks, absent a PBN structure, was taken in Choudhary, Datta, Bittner, & Dougherty (2006), where a finite-horizon control policy was determined that performed optimally relative to the family of networks. Here we next present an adaptive approach (Layek, Datta, Pal, & Dougherty, 2009).

Suppose the family of controlled PBNs is parameterized by the parameter $\alpha \in A$ where, for any $\alpha \in A, \sum_{j \in S} p(i,j,u,\alpha) = 1$ for any $(i,u) \in S \times B$. Here, $p(i,j,u,\alpha)$ denotes $p_{ij}(u)$ when the model α has been selected. The only constraint on A is that every element of A results in a set of bona fide transition probabilities. The cardinality, $|A|$, of A determines the total number of possible PBNs. For each $\alpha \in A$, let $J^*(\alpha)$ be the optimal long-term average cost obtained for model α using the above stationary infinite horizon control method and let $\phi(\bullet,u)$: $S \mapsto B$ be the corresponding control law attaining it. Let $f: \mathbb{R} \mapsto \mathbb{R}$, $o: \mathbb{Z} \mapsto \mathbb{R}$ and constant l be defined as follows: f is a strictly monotonically increasing continuous function such that $f(\inf_{\alpha \in A} o(t)t^{-\theta}) > 0$; o is any function such that $\lim_{t \to \infty} o(t)t^{-\theta}$ is a positive finite number for some $\theta \in (0,1)$ and l is any integer such that $l > |S|+1$. As an example, f can be the logarithm function and $o(t) = 2\sqrt{t}$ with $\theta=0.5$ The adaptive controller consists of two separate operations, estimation and control:

- **Estimator:** At each time step 0, l, $2l$, $3l$, .. kl, $(k+1)l$,...., estimate α by

$$\hat{\alpha}_t := \arg\max_{\alpha \in A} \bar{D}_t(\alpha)$$

where $\overline{D}_t(\alpha) := K\prod_{(i,j,u)\in F^c} p(i,j,u,\alpha)^{n_t(i,j,u)}$. For each model α, we assume that we have the prior knowledge about the set of forbidden transitions F, for which the corresponding transition probabilities $p(i,j,u,\alpha)=0$. F^c denotes the complement of F. In addition,

$$K = \frac{1}{f\left(J^*(\alpha)\right)^{o(t)}};$$

and

$$n_t(i,j,u) = 1 + \sum_{s=0}^{t-1} I(z_s = i, z_{s+1} = j, u_s = u),$$

which can be interpreted as measuring the number of times that a transition occurs from i to j under control u. Here $I(.)$ denotes the indicator function.

- **Controller:** At time kl, knowing the parameter estimate $\hat{\alpha}_{kl}$, we can find the optimal cost function $J^*\left(\hat{\alpha}_{kl}\right)$ and the optimal control law $\phi\left(z_t, \hat{\alpha}_{kl}\right)$ which will be used for the next l time steps. The parameter estimate is kept constant at $\hat{\alpha}_{kl}$ between time steps kl and $(k+1)l-1$.

The presented algorithm is based on the adaptive control algorithm developed in Kumar & Woei (1982).

7. DISCUSSION

As biological functions arise as coordinated interactions among genes and gene products, the current fundamental challenges in computational network biology are to study molecular interactions and their relations with biological functions, understand the underlying cellular mechanisms, simulate and predict long-term systematic behavior under perturbation, and translate into future therapeutic outcomes for complex diseases, including cancer. To achieve these goals, it is our belief that appropriate mathematical models and corresponding computational tools for large-scale systems analysis of biological networks will play critical roles in interpreting the collected high-throughput biological data for a better understanding of cellular dynamic properties and generating new hypotheses for therapeutic strategy design by applying control theory to these biological systems. Despite the current successes in mathematical modeling and deriving corresponding network intervention strategies of gene regulatory networks, there are still significant challenges that we face to translate these computational methods to clinical use for disease treatment.

First, due to the inherent double exponential model complexity increasing with the number of genes in gene regulatory networks, control algorithms for genetic therapeutics have to be scalable with the size of the network model. One possible direction is to develop efficient approximate control policies with effective model reduction. Methods of network reduction and their control issues have been extensively studied in the literature (Dougherty & Shmulevich, 2003; Ivanov & Dougherty, 2004; Ivanov *et al.*, 2007; Ivanov *et al.*, 2010; Ghaffari *et al.*, 2010; Qian *et al.*, 2010).

In addition to the pressing need to develop highly scalable control algorithms for gene regulatory networks of reasonable size, the new control algorithms have to be developed with the caution that we only have incomplete and often inaccurate information on actual regulatory systems in cells. Appropriate uncertainty quantification of gene regulatory networks (Yoon, Qian, & Dougherty, 2013) to capture the critical properties and dynamics that are pertinent to our aim for network intervention as well as robust intervention strategies (Pal *et al.*, 2008; Layek *et al.*, 2009; Pal *et al.*, 2009; Yousefi & Ivanov, 2013) in the presence of substantial uncertainty will play critical roles in future research to achieve our ultimate operational objective of deriving systems biology methods for future intervention therapeutics.

REFERENCES

Akutsu, T., Hayashida, M., Ching, W.-K., & Ng, M. K. (2007). Control of Boolean networks: Hardness results and algorithms for tree structured networks. *Journal of Theoretical Biology*, *244*(4), 670–679. doi:10.1016/j.jtbi.2006.09.023 PMID:17069859

Bellman, R. (1952). On the theory of dynamic programming. *Proceedings of the National Academy of Sciences of the United States of America*, *38*(8), 716–719. doi:10.1073/pnas.38.8.716 PMID:16589166

Bertsekas, D. P. (2005). *Dynamic Programming and Optimal Control*. Belmont, MA: Athena Scientific.

Bittner, M., Meltzer, P., Chen, Y., Jiang, Y., Seftor, E., Hendrix, M., & Trent, J. et al. (2000). Molecular classification of cutaneous malignant melanoma by gene expression profiling. *Nature*, *406*(6795), 536–540. PMID:10952317

Brun, M., Dougherty, E. R., & Shmulevich, I. (2005). Steady-state probabilities for attractors in probabilistic Boolean networks. *Signal Processing*, *85*(10), 1993–2013. doi:10.1016/j.sigpro.2005.02.016

Ching, W.-K., Zhang, S., Jiao, Y., Akutsu, T., Tsing, N., & Wong, A. (2009). Optimal control policy for probabilistic Boolean networks with hard constraints. *IET Systems Biology*, *3*(2), 90–99. doi:10.1049/iet-syb.2008.0120 PMID:19292563

Choudhary, A., Datta, A., Bittner, M. L., & Dougherty, E. R. (2006). Intervention in a family of Boolean networks. *Bioinformatics (Oxford, England)*, *22*(2), 226–232. doi:10.1093/bioinformatics/bti765 PMID:16286362

Cong, Y., Ching, W.-K., Tsing, N., & Leung, H. (2010). On finite-horizon control of genetic regulatory networks with multiple hard-constraints. *BMC Systems Biology*, *4*(Suppl 2), S14. doi:10.1186/1752-0509-4-S2-S14 PMID:20840728

Datta, A., Choudhary, A., Bittner, M. L., & Dougherty, E. R. (2003). External Control in Markovian Genetic Regulatory Networks. *Machine Learning*, *52*(1), 169–191. doi:10.1023/A:1023909812213

Datta, A., Choudhary, A., Bittner, M. L., & Dougherty, E. R. (2004). External control in Markovian genetic regulatory networks: The imperfect information case. *Bioinformatics (Oxford, England)*, *20*(6), 924–930. doi:10.1093/bioinformatics/bth008 PMID:14751971

Datta, A., Pal, R., Choudhary, A., & Dougherty, E. R. (2007). Control approaches for probabilistic gene regulatory networks. *IEEE Signal Processing Magazine*, *24*(1), 54–63. doi:10.1109/MSP.2007.273057

Dissanayake, S. K., Olkhanud, P. B., O'Connell, M. P., Carter, A., French, A. D., Camilli, T. C., & Weeraratna, A. T. et al. (2008). Wnt5A regulates expression of tumor-associated antigens in melanoma via changes in signal transducers and activators of transcription 3 phosphorylation. *Cancer Research*, *68*(24), 10205–10214. doi:10.1158/0008-5472.CAN-08-2149 PMID:19074888

Dougherty, E. R., Pal, R., Qian, X., Bittner, M. L., & Datta, A. (2010). Stationary and structural control in gene regulatory networks: Basic concepts. *International Journal of Systems Science*, *41*(1), 5–16. doi:10.1080/00207720903144560

Dougherty, E. R., & Shmulevich, I. (2003). Mappings between probabilistic Boolean networks. *Signal Processing*, *83*(4), 799–809. doi:10.1016/S0165-1684(02)00480-2

Ghaffari, N., Ivanov, I., Qian, X., & Dougherty, E. R. (2010). A CoD-based reduction algorithm for designing stationary control policies on Boolean networks. *Bioinformatics (Oxford, England)*, *26*(12), 1556–1563. doi:10.1093/bioinformatics/btq225 PMID:20421196

Huang, S. (1999). Gene expression profiling, genetic networks and cellular states: An integrating concept for tumorigenesis and drug discovery. *Journal of Molecular Medicine*, *77*(6), 469–480. doi:10.1007/s001099900023 PMID:10475062

Hunter, J. J. (1986). Stationary distributions of perturbed Markov-chains. *Linear Algebra and Its Applications*, *82*, 201–214. doi:10.1016/0024-3795(86)90152-7

Hunter, J. J. (2005). Stationary distributions and mean first passage times of perturbed Markov chains. *Linear Algebra and Its Applications*, *410*, 217–243. doi:10.1016/j.laa.2005.08.005

Ioannou, P. A., & Sun, J. (1996). *Robust Adaptive Control*. Upper Saddle River, NJ: PTR Prentice-Hall.

Ivanov, I., & Dougherty, E. R. (2004). Reduction mappings between probabilistic Boolean networks. *EURASIP J. Acoust. Signal Process.*, *1*(1), 125–131. doi:10.1155/S1110865704309182

Ivanov, I., Pal, R., & Dougherty, E. R. (2007). Dynamics preserving size reduction mappings for probabilistic Boolean networks. *IEEE Transactions on Signal Processing*, *55*(5), 2310–2322. doi:10.1109/TSP.2006.890929

Ivanov, I., Simeonov, P., Ghaffari, N., Qian, X., & Dougherty, E. R. (2010). Selection policy-induced reduction mappings for Boolean networks. *IEEE Transactions on Signal Processing*, *58*(9), 4871–4882. doi:10.1109/TSP.2010.2050314

Iyengar, G. N. (2005). Robust dynamic programming. *Mathematics of Operations Research*, *30*(2), 257–280. doi:10.1287/moor.1040.0129

Kauffman, S. (1993). *The Origins of Order: Self-organization and Selection in Evolution*. Oxford University Press.

Kauffman, S. A. (1969). Metabolic stability and epigenesis in randomly constructed genetic nets. *Journal of Theoretical Biology*, *22*(3), 437–467. doi:10.1016/0022-5193(69)90015-0 PMID:5803332

Keméy, J. G., & Snell, J. L. (1983). *Finite Markov Chains*. New York: Springer-Verlag.

Kumar, P. R., & Varaiya, P. (1986). *Stochastic Systems: Estimation, Identification and Adaptive Control*. Prentice-Hall.

Kumar, P. R., & Woei, L. (1982). Optimal adaptive controllers for unknown Markov chains. *IEEE Transactions on Automatic Control*, *27*(4), 765–774. doi:10.1109/TAC.1982.1103017

Lau, K., Ganguli, S., & Tang, C. (2007). Function constrains network architecture and dynamics: A case study on the yeast cell cycle Boolean network. *Physical Review E: Statistical, Nonlinear, and Soft Matter Physics*, *75*(5), 051907. doi:10.1103/PhysRevE.75.051907 PMID:17677098

Layek, R., Datta, A., Pal, R., & Dougherty, E. R. (2009). Adaptive intervention in probabilistic Boolean networks. *Bioinformatics (Oxford, England)*, *25*(16), 2042–2048. doi:10.1093/bioinformatics/btp349 PMID:19505946

Li, F., Long, T., Lu, Y., Ouyang, Q., & Tang, C. (2004). The yeast cell-cycle network is robustly designed. *Proceedings of the National Academy of Sciences of the United States of America*, *101*(14), 4781–4786. doi:10.1073/pnas.0305937101 PMID:15037758

Pal, R., Datta, A., Bittner, M. L., & Dougherty, E. R. (2005). Intervention in context-sensitive probabilistic Boolean networks. *Bioinformatics (Oxford, England)*, *21*(7), 1211–1218. doi:10.1093/bioinformatics/bti131 PMID:15531600

Pal, R., Datta, A., & Dougherty, E. R. (2006). Optimal infinite-horizon control for probabilistic Boolean networks. *IEEE Transactions on Signal Processing*, *54*(6), 2375–2387. doi:10.1109/TSP.2006.873740

Pal, R., Datta, A., & Dougherty, E. R. (2008). Robust intervention in probabilistic Boolean networks. *IEEE Transactions on Signal Processing*, *56*(3), 1280–1294. doi:10.1109/TSP.2007.908964

Pal, R., Datta, A., & Dougherty, E. R. (2009). Bayesian Robustness in the Control of Gene Regulatory Networks. *IEEE Transactions on Signal Processing*, *57*(9), 3667–3678. doi:10.1109/TSP.2009.2022872

Pal, R., Ivanov, I., Datta, A., Bittner, M. L., & Dougherty, E. R. (2005). Generating Boolean networks with a prescribed attractor structure. *Bioinformatics (Oxford, England)*, *21*(21), 4021–4025. doi:10.1093/bioinformatics/bti664 PMID:16150807

Qian, X. (2012). A new phenotypically constrained control policy in Boolean networks based on basins of attraction. In *Proceedings of the IEEE International Workshop on Genomic Signal Processing and Statistics (GENSIPS)*. Washington, DC: IEEE. doi:10.1109/GENSIPS.2012.6507725

Qian, X., & Dougherty, E. R. (2008). Effect of Function Perturbation on the Steady-State Distribution of Genetic Regulatory Networks: Optimal Structural Intervention. *IEEE Transactions on Signal Processing*, *56*(10), 4966–4976. doi:10.1109/TSP.2008.928089

Qian, X., & Dougherty, E. R. (2012). Intervention in Gene Regulatory Networks via Phenotypically Constrained Control Policies Based on Long-Run Behavior. *IEEE/ ACM Transactions on Computational Biology and Bioinformatics*, *9*(1), 123–136. doi:10.1109/TCBB.2011.107 PMID:21788674

Qian, X., Ghaffari, N., Ivanov, I., & Dougherty, E. R. (2010). State reduction for network intervention in probabilistic Boolean networks. *Bioinformatics (Oxford, England)*, *26*(24), 3098–3104. doi:10.1093/bioinformatics/btq575 PMID:20956246

Qian, X., Ivanov, I., Ghaffari, N., & Dougherty, E. R. (2009). Intervention in gene regulatory networks via greedy control policies based on long-run behavior. *BMC Systems Biology*, *3*(1), 61. doi:10.1186/1752-0509-3-61 PMID:19527511

Qian, X., Yoon, B.-J., & Dougherty, E. R. (2012). Structural intervention of gene regulatory networks by general rank-k matrix perturbation. In *Proceedings of the 37th IEEE International Conference on Acoustics, Speech, and Signal Processing (ICASSP)*. Kyoto, Japan: IEEE. doi:10.1109/ICASSP.2012.6287987

Schweitzer, P. J. (1968). Perturbation theory and finite Markov chains. *Journal of Applied Probability*, *5*(2), 401–413. doi:10.2307/3212261

Sherman, J., & Morrison, W. J. (1950). Adjustment of an inverse matrix corresponding to a change in one element of a given matrix. *Annals of Mathematical Statistics*, *21*(1), 124–127. doi:10.1214/aoms/1177729893

Shmulevich, I., & Dougherty, E. R. (2007). *Genomic Signal Processing*. Princeton University Press. doi:10.1515/9781400865260

Shmulevich, I., & Dougherty, E. R. (2010). *Probabilistic Boolean Networks: The Modeling and Control of Gene Regulatory Networks*. New York: SIAM Press. doi:10.1137/1.9780898717631

Shmulevich, I., Dougherty, E. R., Kim, S., & Zhang, W. (2002). Probabilistic Boolean networks: A rule-based uncertainty model for gene regulatory networks. *Bioinformatics (Oxford, England)*, *18*(2), 261–274. doi:10.1093/bioinformatics/18.2.261 PMID:11847074

Shmulevich, I., Dougherty, E. R., & Zhang, W. (2002a). Control of stationary behavior in probabilistic Boolean networks by means of structural intervention. *Journal of Biological System*, *10*(04), 431–445. doi:10.1142/S0218339002000706

Shmulevich, I., Dougherty, E. R., & Zhang, W. (2002b). Gene perturbation and intervention in probabilistic Boolean networks. *Bioinformatics (Oxford, England)*, *18*(10), 1319–1331. doi:10.1093/bioinformatics/18.10.1319 PMID:12376376

Vahedi, G., Faryabi, B., Chamberland, J.-F., Datta, A., & Dougherty, E. R. (2008). Intervention in gene regulatory networks via a stationary mean-first-passage-time control policy. *IEEE Transactions on Bio-Medical Engineering*, *55*(10), 2319–2331. doi:10.1109/TBME.2008.925677 PMID:18838357

Weeraratna, A. T., Jiang, Y., Hostetter, G., Rosenblatt, K., Duray, P., Bittner, M., & Trent, J. M. (2002). Wnt5a signaling directly affects cell motility and invasion of metastatic melanoma. *Cancer Cell*, *1*(3), 279–288. doi:10.1016/S1535-6108(02)00045-4 PMID:12086864

Xiao, Y., & Dougherty, E. R. (2007). The impact of function perturbations in Boolean networks. *Bioinformatics (Oxford, England)*, *23*(10), 1265–1273. doi:10.1093/bioinformatics/btm093 PMID:17379691

Yoon, B.-J., Qian, X., & Dougherty, E. R. (2013). Quantifying the objective cost of uncertainty in complex dynamical systems. *IEEE Transactions on Signal Processing*, *61*(9), 2256–2266. doi:10.1109/TSP.2013.2251336

Yousefi, M. R., & Ivanov, I. (2013). Optimal control of gene regulatory networks with uncertain intervention effects. In *Global Conference on Signal and Information Processing (GlobalSIP)* (pp. 105–108). IEEE. doi:10.1109/GlobalSIP.2013.6736826

Zhou, X., Wang, X., Pal, R., Ivanov, I., Bittner, M. L., & Dougherty, E. R. (2004). A Bayesian connectivity-based approach to constructing probabilistic gene regulatory networks. *Bioinformatics (Oxford, England)*, *20*(17), 2918–2927. doi:10.1093/bioinformatics/bth318 PMID:15145802

Chapter 5
Optimal Intervention Methods for Markovian Gene Regulatory Networks

Mohammadmahdi Rezaei Yousefi
The Ohio State University, USA

ABSTRACT

A central problem in translational medicine is to provide a framework for deriving and studying effective intervention methods to elicit desired steady-state behavior for a gene regulatory network of interest with Markovian dynamics. Heretofore, two rather different external control approaches have been taken. The first optimizes a subjectively defined cost function while modeling treatment constraints; therefore, desirable shift of the steady-state mass is a by-product. The second approach, on the other hand, focuses solely on the steady-state behavior of the network and provides the maximal shift achievable. Although both approaches are optimal with respect to their objectives, the choice of which to use depends on the treatment goals.

INTRODUCTION

In order for a cancer treatment to be successful one must account for the patient's condition, the nature of the disease, its dynamics, the effect of antitumor drugs or radiation, the duration of effect and side effects. Each of these is a major field in the medical community, and a vast amount effort has been devoted to their char-

DOI: 10.4018/978-1-5225-0353-8.ch005

acterization. Typically, the main objective of many treatment schemes is to find the right dose schedules while maximizing the benefit to toxicity ratio for a large portion of the patient population (Simon & Norton, 2006). To this end, treatments usually involve some kind of protocol and several drug candidates acting on various gene products with the aim of mitigating undesirable gene functions in cancer-related genetic pathways. Cancer therapies, for example chemotherapy, are usually administered in cycles, where each cycle begins by delivering a drug that kills cells which divide rapidly. The drug is effective for some period, on both normal and tumor cells, after which there is usually a rest period allowing the body to recover from the adverse effects of the drug (Brunton, Chabner, & Knollmann, 2010). To maximize the chance of eradicating a tumor, one should deliver the most effective drug at a sufficient dosage over as short a time as possible (Simon & Norton, 2006). However, since the toxicity of a drug could be intolerable to the patient, we desire a therapeutic method that acts very rapidly and with high efficiency so that a large percentage of tumor cells die or shift into a mode where they can no longer proliferate. The method should, at the same time, prolong patients' lives with reasonable quality during and after treatment.

The duration of effective response for a given set of drugs depends on many factors that may vary among a patient population or even in a population of cells within an individual. Achieving an acceptable success rate is tied to three key factors: (1) building accurate models for the underlying dynamical system governing the cancer and the drug response; (2) forcing cancerous cells to either initiate cell death or terminate differentiation as early as possible; and (3) choosing a suitable intervention strategy according to the states of the tumor and patient at the beginning of each therapeutic cycle (Yousefi, Datta, & Dougherty, 2013).

A gene regulatory network (GRN) refers to a class of models describing the multivariate functional relationships among a cohort of genes or their products. From a graphical perspective, genes are nodes in this network and edges describe regulatory relationships between genes. A GRN can be alternatively viewed as an input-output system, where the inputs are signals from signaling pathways or transcription factors, and the outputs are gene expression levels at a given time (Huang, Tienda-Luna, & Wang, 2009). A regulatory network should be able to predict the behavior of the collection of genes of interest under different conditions and perturbations (Dougherty & Shmulevich, 2012; Pe'er & Hacohen, 2011). These networks aim to model cellular control and how abnormal cell functions result from one or several breakdowns in the regulatory mechanisms, making them an important part of translational medicine (Dougherty, Pal, Qian, Bittner, & Datta, 2010).

Building a model for the underlying GRN is the first step toward designing optimal intervention. There has been a considerable amount of work on developing models and inference methods that provide accuracy, robustness, adaptivity and a system-level view, including Bayesian networks (Friedman, 2004), signed directed graphs (Kyoda, Baba, Onami, & Kitano, 2004), Petri nets (Chaouiya, Remy, & Thieffry, 2008; Matsuno, Doi, Nagasaki, & Miyano, 2000), differential equations (De Jong, 2002), stochastic differential equations (Chen, Wang, Tseng, Huang, & Kao, 2005; El Samad, Khammash, Petzold, & Gillespie, 2005), Boolean networks (Kauffman, 1969) and probabilistic Boolean networks (PBNs) (Shmulevich, Dougherty, Kim, & Zhang, 2002). Each modeling paradigm has its own strengths and weaknesses. For example, Bayesian networks are descriptive and require a large number of samples for inference. Boolean models are predictive, but they are very simplistic and come short when a more refined characterization of a regulatory process is needed. Although ordinary differential equations can capture quantitative molecular or physical details of the system, predictions are usually accurate only when the number of molecular species involved is large. When regulatory molecules are produced in small concentrations, fluctuations in the number of molecules have a significant impact on the process dynamics, and therefore a stochastic differential equation model is more appropriate (Cao & Samuels, 2009; Shmulevich & Aitchison, 2009).

Modeling GRNs via Markovian dynamical networks, such as PBNs, has received much attention recently because they are simple and can capture uncertainty, intrinsic or extrinsic, due to measurement noise in the data, latent variables or stochastic interactions among genes or gene products at different levels. Another feature of PBNs, which makes them attractive for studying intervention methods that affect the dynamics of a GRN, is that they can be represented as finite-state Markov chains. Each state in the resulting Markov chain represents a *gene activity profile* (GAP) and state transitions are made according to a *transition probability matrix* (TPM). To be self-contained, this chapter will give a review of PBNs.

The set of all GAPs constitutes the state space of the network, which can be then partitioned into two disjoint sets of *desirable* and *undesirable* cellular states according to the expression levels of a given set of genes. An undesirable state can be associated with a phenotype representative of a cancerous state; for example, cell growth in the absence of growth factors is undesirable (Faryabi, Vahedi, Chamberland, Datta, & Dougherty, 2008). Once a model for the underlying GRN is assumed and desirable and undesirable states are identified, the goal in optimal intervention design is to find therapeutic strategies (control policies) associated with some predefined objective functions such that the evolution of the network is optimally steered from undesirable states to desirable ones (Datta & Dougherty, 2006; Dougherty et al., 2010; Dougherty & Datta, 2005). This problem has been extensively studied within two basic intervention approaches in the context of PBNs, namely structural intervention and external control.

Structural intervention, in its simplest form, involves a one-time change in the regulatory pathway of interest. Given a cancer-related network and a collection of potential structural interventions, the objective is to single out the one resulting in a maximally desirable alteration in the steady-state of the network (Qian & Dougherty, 2008; Shmulevich, Dougherty, & Zhang, 2002; Xiao & Dougherty, 2007).

External control, on the other hand, is generally based on changing the expression status of a set of specific genes or gene products, called the *control* genes, at each decision epoch. The first proposed intervention approach in the context of PBNs was to determine an optimal single-gene perturbation to the network based on first-passage times in Markov chains (Shmulevich, Dougherty, & Zhang, 2002). Attention, later, turned to dynamic-programming-based finite-horizon (Datta, Choudhary, Bittner, & Dougherty, 2003) and infinite-horizon external control methods (Pal, Datta, & Dougherty, 2006). This chapter will be solely focused on external control methods, and will utilize the rich theory of stochastic optimal control.

Decisions regarding possible interventions are usually assumed to be made at every state transition of the network (Datta et al., 2003; Pal et al., 2006). To model dosing constraint, a cyclic approach to intervention may be implemented in which there is a sequence of treatment windows and treatment is allowed only at the beginning of each window (Vahedi, Faryabi, Chamberland, Datta, & Dougherty, 2009). These protocols ignore an important practical aspect of therapy: Due to biological and micro-environmental variability within tumor cells, the action period of an administered drug can vary among a population of patients, so that treatments may have an unknown duration of action. Accounting for this practical modeling assumption introduces an acyclic structure in the intervention protocol. Furthermore, in practice, a physician might monitor a patient at every stage of treatment and decide whether to intervene or not, and if the decision is to intervene, then the patient's body will be under the influence of the drug for some duration, usually followed by a recovery phase. Based on a recently proposed model for tumor growth inhibition, we will first characterize the variability of the length of drug action. Then, we will present a unified approach to both cyclic and acyclic intervention models for PBNs. We accomplish this by placing the problem in the general framework Markov decision processes with partially controlled decision intervals (White, 1978), also known as *semi-Markov* models (Bertsekas, 2005), with an infinite horizon discounting cost objective function. We present a methodology to devise optimal solutions for this class of sequential-decision-making problems and locate optimal intervention policies for synthetically generated gene regulatory networks, and a mutated mammalian cell-cycle network.

Usually, optimal intervention methods are defined relative to a cost value that is assigned through a subjective process. Thus the long-run effect of such methods on a PBN becomes a by-product of the optimization procedure. Since a patient's phenotype is associated with the steady-state behavior of their underlying gene regulatory network, from a practical perspective, it is advantageous to define optimality directly in terms of long-run behavior, which in turn would relieve the physician or biologist from the burden of having to define a subjective cost function (Dougherty et al., 2010; Yousefi & Dougherty, 2013). To address this, we will formulate optimization on the steady-state distribution of PBNs to maximally alter undesirable phenotypes. In addition to desirable and undesirable states, there may exist some "ambiguous" states that are either unrelated to the condition of interest or not fully understood, and thus bear some perceived risk. The framework is flexible enough to account for such cases, where we put constraints on the amount of mass we are willing to accept at different types of states. We will demonstrate the performance of optimal policies using synthetically generated networks as well as a real mammalian cell cycle network and a metastatic melanoma network.

There are other approaches for the control of biological systems. For example, an integrated approach to modeling and control of biological systems has been presented in (Julius et al., 2008), where the authors propose a stochastic hybrid model of the lactose regulation system of *E. coli* bacteria. The dynamics of this model can be further abstracted into a much simpler two-state continuous-time Markov chain, which then can be used to design a feedback controller that drives a certain amount of bacteria to an induced state by manipulating some chemical concentration in the environment (Julius et al., 2008). A recent approach to controlling GRN dynamics has been proposed in (Zañudo & Albert, 2015), where an asynchronous updating scheme for a Boolean network is considered as the model. By integrating the structural and functional information available in the network, the authors propose a procedure to find control targets that directly stabilize the expression or activity of a few select components so that the cell dynamics are driven towards a desired fate or away from an undesired fate. The utility of this approach is demonstrated on a leukemia signaling network and a network controlling the differentiation of T helper cells (Zañudo & Albert, 2015). We refer the interested reader to (Iglesias & Ingalls, 2010; Queinnec, Garcia, Niculescu, & Tarbouriech, 2007) for more examples on the application of control theory in biology.

MARKOVIAN GENE REGULATORY NETWORKS

A "good" model for a GRN should be able to predict the behavior of the main processes, pertinent to the phenotypes of interest, that occur inside a cell or within a population of cells under different conditions and perturbations (Dougherty & Shmulevich, 2012; Pe'er & Hacohen, 2011). It should also identify cellular control mechanisms that lead to normal and abnormal cell functions. Although one can construct complex regulatory models that describe genetic interactions at a fine grain level, this would require a precise understanding of the underlying biological processes and extensive amounts of experimental data to infer the model. Due to these limitations, we may need to move toward coarser models that incorporate uncertainty, rather than complexity, at different levels.

PBNs, as a class of Markovian networks, model uncertainty in gene activity in a GRN (Shmulevich, Dougherty, Kim, et al., 2002). They are essentially a probabilistic extension of a class of rule-based models for the GRNs known as the Boolean networks (Kauffman, 1969). We will first give a brief review of Boolean networks and then describe a class of PBNs we will be using throughout this chapter. We will also discuss how PBNs can incorporate external intervention inputs, and address two types of controlled PBNs that are central in our study of network intervention.

Boolean Networks

A Boolean network assumes a regulatory graph representing multivariate interactions among a set of genes or gene products. An interaction is specified by a directed edge starting from a predictor gene and ending at an influenced gene. Every gene that is directly connected to another gene may serve as its predictor. A Boolean regulatory rule defines how the predictors of a gene determine its expression level. We restrict ourselves to binary levels, where it is assumed that the expression of each gene is quantized to two levels: zero or one. Zero corresponds to an unexpressed (OFF) gene and one corresponds to an expressed (ON) gene. The mathematical theory extends directly to networks with discrete-valued nodes. We now formally define a Boolean network.

Definition 1: A Boolean network is fully characterized by a set of n nodes, $V=\{v^1,v^2,\ldots,v^n\}$, (representing genes $\{g^1,g^2,\ldots,g^n\}$ or their products) and a collection of Boolean functions $F=\{f_1,f_2,\ldots,f_n\}$ describing the functional relationships between the nodes. Each Boolean function $f_i:\{0,1\}^{j_i}\rightarrow\{0,1\}$ determines the value of node i at time $t+1$ given the value of its predictors at time t by

$$v^i_{t+1}=f_i(v^{i1}_t,v^{i2}_t,\ldots,v^{ij_i}_t), \tag{1}$$

where $\{v^{i1}_t,v^{i2}_t,\ldots,v^{ij_i}_t\}$ corresponds to values of j_i predictors of gene g^i, $\{g^{i1},g^{i2},\ldots,g^{ij_i}\}$, at time t. In a Boolean network, all genes are assumed to update synchronously, so that the network evolves according to a genome-wide GAP, $\mathbf{v}_t=(v^1_t,v^2_t,\ldots,v^n_t)$ at each time t, forming a *state space* of size 2^n. One may alternatively convert each $\mathbf{v}_t$

to a decimal value by $x_t = \sum_{i=1}^{n} 2^{n-i} v_t^i$, where $x \in \mathcal{S} = \{0, 1, \ldots, 2^n - 1\}$ is a *state*. Given an initial state, a Boolean network will eventually enter and stay in a set of states called an *attractor cycle*. The set of states leading to an attractor cycle is known as a *basin of attraction (BOA)* (Dougherty et al., 2010). Once we have a state space, we can identify phenotypes as the attractors of a Boolean network. We label states as being *desirable* or *undesirable* depending on the type and severity of the phenotype they are associated with. The set of desirable and undesirable states are denoted by $\mathcal{D}$ and $\mathcal{U}$, respectively.

Probabilistic Boolean Networks

Due to latent variables, unknown interactions among genes and measurements noise, one may introduce uncertainty into a Boolean network by allowing a collection of several contextual regulatory functions, instead of a single F, as well as random perturbation of the nodes at every time epoch (Shmulevich, Dougherty, Kim, et al., 2002). The resulting model is a PBN formally defined below.

Definition 2: A probabilistic Boolean network is characterized by a set of n nodes, $V=\{v^1,v^2,\ldots,v^n\}$, and a set of m constituent Boolean networks, $\mathbf{F}=\{F^1,F^2,\ldots, F^m\}$, called *contexts*, a selection probability vector $S=(s^1,s^2,\ldots,s^m)$ over $\mathbf{F}$, a network switching probability q and a random gene perturbation probability p. Random switching and gene perturbation are assumed to be mutually exclusive.

At any time point, with probability q, the network evolution may switch from a constituent Boolean network to another according to the selection probability vector. It is assumed that the probability of switching to a new constituent network is independent of the current network, and in particular the current network may switch to itself. In addition, with probability p, the current state of each gene in the network can be randomly flipped. The PBN is said to be *context-sensitive* if $q<1$, the interpretation being that there are latent variables outside the network whose changes cause the model network to behave stochastically (Brun, Dougherty, & Shmulevich, 2005). If $q=1$, the PBN is said to be *instantaneously-random*, the interpretation being that the uncertainty in the PBN arises from uncertainty in model inference (Shmulevich, Dougherty, Kim, et al., 2002). With sufficiently small perturbation probability p, the long-run behavior of a PBN will reflect the attractor structures within the context Boolean networks (Brun et al., 2005). Throughout this chapter, we assume instantaneously random PBNs.

Transition rules of any instantaneously random PBN can be modeled by a homogeneous Markov chain, whose states are the GAPs of the underlying regulatory network (Shmulevich, Dougherty, Kim, et al., 2002). Let $\{Z_t \in \mathcal{S},\ t = 0,1,...\}$ denote the stochastic process of the state of a PBN. For all t=0,1,..., originating from state $x \in \mathcal{S}$, the successor state $y \in \mathcal{S}$ is selected randomly according to the TPM $\mathcal{P}$, whose (x,y) element can be calculated readily from the following equation:

$$[\mathcal{P}]_{xy} = p_{xy} = P(Z_{t+1} = y \mid Z_t = x) = (1-p)^n \sum_{l=1}^{m} s^l 1_{F^l(x)=y} + 1_{D(x,y)\neq 0}(1-p)^{n-D(x,y)} p^{D(x,y)}, \tag{2}$$

where $D(x,y)$ denotes the Hamming distance of x and y in their binary representations. Due to random gene perturbation, the equivalent Markov chain is ergodic and has a unique invariant probability vector, π, equal to its steady-state distribution.

Controlled PBNs

We assume that PBNs admit external control inputs that may take their value from a set of actions specifying the type of intervention on a set of genes (or their products), which directly influences the dynamics of the GRN of interest. We refer to these genes as the *control genes*, and discuss two types of control inputs that have been used in the literature. One can easily extend the theory to other arbitrary control inputs as long as they influence the network dynamics. We assume there is a finite set of control actions, $\mathcal{A}$, where each element $a \in \mathcal{A}$ specifies a control input that effects the expression level of a control gene, g^c, for some $c \in \{1,2,\ldots,n\}$. Since a decision maker might resort to random mechanisms for taking actions, we assume that the actions are permitted to be random. Let $\{A_k \in \mathcal{A}, k = 0,1,...\}$ denote the stochastic process of actions taken.

The first, and simpler, class of control inputs are called *binary* inputs, where one allows for choosing decisions from a set $\mathcal{A}=\{0,1\}$ at every transition (or decision epoch) of the process. The control action a=0 results in no-intervention, while a=1 toggles the current state of the control gene. The TPM for the controlled network under a binary input is denoted by $\mathcal{P}(a)$ and defined as

$$[\mathcal{P}(a)]_{xy} = p_{xy}(a) = P(Z_{t+1} = y \mid Z_t = x, A_t = a), \tag{3}$$

which is the probability of going to y at time t+1 starting from state x and taking action a at time t. Let $\tilde{x}$ have the same binary representation as x except with bit ν^c flipped. For a=1, we have

$$\left[\mathcal{P}(1)\right]_{xy} = p_{xy}(1) = P(Z_{t+1} = y \mid Z_t = x, A_t = 1) = P(Z_{t+1} = y \mid Z_t = \tilde{x}) = p_{\tilde{x}y}, \tag{4}$$

where $p_{\tilde{x}y}$ is defined in (2). Thus, the control action a=1 replaces the row corresponding to the state x by the row corresponding to the state $\tilde{x}$ in the original TPM of the underlying Markov chain. For a=0, we simply have $\mathcal{P}(0) = \mathcal{P}$ from (2).

The second class of control inputs are referred to as *persisting effect* inputs, which find application in the intervention models, where the decision maker is not allowed to make decisions at every transition of the network, but only at certain time epochs. Furthermore, once a decision is made it may last for several time epochs until the next decision epoch. To model this effect, one can assume that the external control forcibly sets the expression of the control gene to a certain level. A simple and common choice for $\mathcal{A}$ then would be $\{0,-1,1\}$, where 0 indicates no-intervention, and -1 or 1 indicate that the control gene is externally set to 0 or 1, respectively. The network evolves without alteration when a=0, i.e., the transition rules are found from (2). However, for $a \in \{-1,1\}$ the network evolves according to its underlying regulatory functions, but at the same time the control gene is forcibly kept at a fixed state. Hence, depending on the choice of control, some of the states become *accessible* while the others become *inaccessible*, where only transitions between accessible states are permitted. Hence, in the presence of intervention, we will have a PBN with restricted transition rules.

We define $\mathcal{S}_a$ as the set of accessible states when an action $a \in \mathcal{A}$ is in effect. Consequently, the set of inaccessible states for an action $a \in \mathcal{A}$ is $\mathcal{S} \setminus \mathcal{S}_a$. For example, assuming that g^1 is the control gene, which corresponds to the most significant bit in the binary representation of the states, we get

$$\begin{cases} \mathcal{S}_0 = \mathcal{S}, \\ \mathcal{S}_{-1} = \left\{0, 1, 2, \ldots, 2^{n-1} - 1\right\}, \\ \mathcal{S}_1 = \left\{2^{n-1}, 2^{n-1} + 1, \ldots, 2^n - 1\right\}. \end{cases} \tag{5}$$

During a *control phase*, when the network is completely under the influence of a control input $a \in \{-1,1\}$, only transitions between states in $\mathcal{S}_a$ are allowed. Therefore, all the probability mass of possible transitions from an inaccessible state to an accessible state would be absorbed by its corresponding accessible state. We denote the TPM of the controlled network during this phase by $Q(a)$, given by

$$\left[\mathcal{Q}(a)\right]_{xy} = q_{xy}(a) - \begin{cases} p_{xy} + p_{x\tilde{y}}, & \text{if } x, y \in \mathcal{S}_a \\ 0, & \text{if } x, y \notin \mathcal{S}_a \end{cases} \tag{6}$$

where $a \in \{-1, 1\}$ and p_{xy} is defined in (2). For $a=0$, we have $\mathcal{Q}(0) = \mathcal{P}$.

Next we address two types of transient phases, *uncontrolled-to-controlled* and *controlled-to-uncontrolled*, and find the TPMs during these phases. These transient effects last for only one time epoch until the network enters or leaves the controlled phase. When uncontrolled, the network can start from any original state in $\mathcal{S}$; however, under the control action a it must arrive at a state in $\mathcal{S}_a$ following the transition rules of the controlled network. Using a similar argument as in the control phase, we can calculate the TPM of the network in the uncontrolled to controlled transient phase, denoted by $\mathcal{R}(a)$, as

$$\left[\mathcal{R}(a)\right]_{xy} = r_{xy}(a) = r_{\tilde{x}y}(a) = \begin{cases} p_{xy} + p_{x\tilde{y}}, & \text{if } x, y \in \mathcal{S}_a \\ 0, & \text{if } y \notin \mathcal{S}_a \end{cases} \tag{7}$$

where $a \in \{-1, 1\}$ and p_{xy} is defined in (2). For $a=0$, we have $\mathcal{R}(0) = \mathcal{P}$. During the controlled to uncontrolled transient phase, the control action transitions from being effective to ineffective; therefore the state must originate in $\mathcal{S}_a$, depending on the control action taken earlier, but then may transition to any state in the full state space, $\mathcal{S}$. Hence, the network transition probabilities follow that of the original TPM, $\mathcal{P}$.

Cost-Dependent Control

Optimal cost-dependent finite- and infinite-horizon intervention methods assume that there is a bounded cost function associated with every state and control decision pair. Although the choice of cost function is subjective and depending on the treatment goals, typically one assigns non-negative costs to an intervention decision and/or if the next state is an undesirable state.

It has been argued that short term control methods are not always effective as they do not change the steady state behavior of the PBN (Pal et al., 2006); therefore, infinite-horizon methods are preferred, unless one specifically is interested in temporarily changing network dynamics. Optimal infinite-horizon control policies can be formulated relative to two different objective functions: the expected total discounted cost and the expected average cost. Whereas the latter is only concerned with the long-run behavior of the network, the former puts more emphasis on short-term effects of the intervention decisions, although both will change the steady-state behavior of the PBN.

One of the earliest and simplest control methods permits making decisions, at every epoch, to intervene or let the PBN follow its uncontrolled transition rules (Datta et al., 2003; Pal et al., 2006). The optimal control methods for both types of objective functions are stationary, meaning that they are independent of time and only dependent on the state of the network. We refer to this class of control models as *one-transition* models. Figure 1 depicts an example.

Generalizing the class of one-transition methods in order to account for dosing constraint in chemotherapy, (Vahedi et al., 2009) defines cyclic strategies in which treatment windows are assumed to be every W time epochs. A one-time decision is allowed only at the beginning of each treatment window; thereafter, the network evolves for $W-1$ epochs without any intervention effects. Similar cost functions as to those in (Pal et al., 2006) are considered and the goal is to minimize the expected total discounted cost. Optimal control methods for this problem are stationary. Figure 2 shows a schematic of this *W-transition* intervention model.

Figure 1. Intervention is allowed at each time epoch. Blunt arrows stand for no-intervention; normal arrows stand for up-regulation; reverse arrows stand for down-regulation.

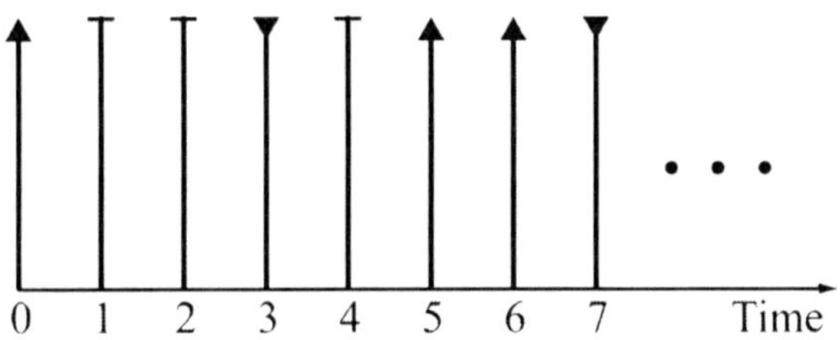

Figure 2. Intervention is only permitted at the beginning of each window of W=3 epochs. Thereafter, the network evolves freely with no intervention for W–1 time epochs. Blunt arrows stand for no-intervention; normal arrows stand for up-regulation; reverse arrows stand for down-regulation.

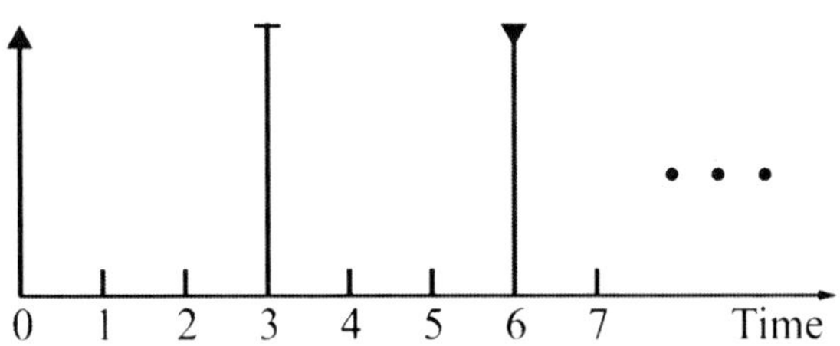

The effect of external intervention, specifically those of antitumor drugs, on tumor cells have been extensively analyzed in the pharmacology literature (Cojocaru & Agur, 1992; de Vladar & González, 2004; Himmelstein & Bischoff, 1973). The pharmacokinetics and pharmacodynamics (PK/PD) of a given drug are separately modeled and then linked together to arrive at a model for the time-course *concentration-effect*. In short, PK uses mathematical models to explain and predict drug concentration in the body (for example, in the blood stream or the tumor site) over time. PD, on the other hand, links the effects of a drug and its mechanism of action to the drug concentration level or the total integrated effective concentration during the course of treatment (Simon & Norton, 2006). It is common to use differential equations (e.g. multi-compartment models) for PK, which in many cases are fairly accurate. However, modeling PD is more challenging as the time-course effects of a drug are not easy to measure. PK/PD linking, with the objective of determining how much variability in drug effectiveness is attributable to measurable drug concentrations, can be done via several functions, including linear models, sigmoid-E_{max} models, etc. (Dayneka, Garg, & Jusko, 1993).

The *W*-transition policy ignores the drug's duration of action, and the fact that persistent effects exists and must be accounted for in the intervention model. To alleviate this limitation, we build upon recent work for tumor growth inhibition in response to a given drug, where a measure for the duration of effective response as a function of parameters is provided (Magni et al., 2008; Magni, Simeoni, Poggesi, Rocchetti, & De Nicolao, 2006). In the new intervention model, the duration of drug action, following each application, is assumed to be random. The randomness is attributed to biological variability among tumor cells, which results in different action periods for a drug. Figure 3 shows an example of this intervention model,

Figure 3. The drug, once administered, has an action period with a random length, followed by a recovery phase. For example, at epoch 0, the decision is to up-regulate. The drug has a persistent effect of length three epochs (shown by the shaded bars). Similarly, at epoch 4 the decision is to down-regulate, but now the drug effect lasts two epochs. The recovery phase is one epoch. Blunt arrows stand for no-intervention; normal arrows stand for up-regulation; reverse arrows stand for down-regulation.

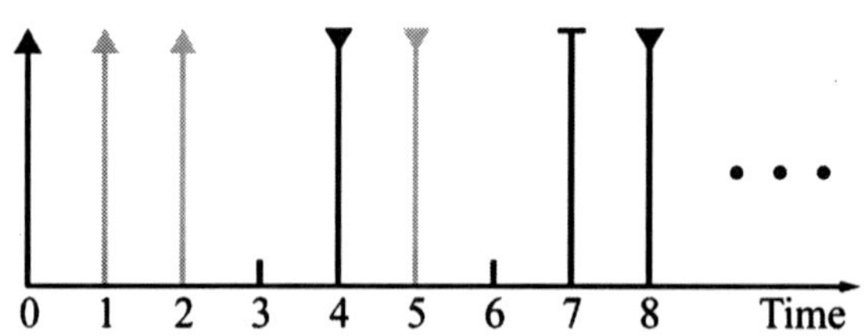

referred to as the *random-length* intervention model, where the effect is persistent over a random number of consecutive epochs.

This framework includes several special cases of intervention models:

1. **One-Transition:** Decisions are allowed at every epoch, and drug effect vanishes immediately.
2. ***W*-Transition:** Decisions are allowed at every W epochs, and drug effect vanishes immediately.
3. **Cyclic Control with Fixed-Length Duration of Action:** Decisions are allowed at every W epochs, and drug effect vanishes after $L \leq W$ epochs.
4. **Acyclic Control with Fixed-Length Duration of Action:** Similar to the cyclic model, but when the control decision is *no intervention*, the next decision is made immediately at the next epoch.

We will present the optimal intervention policy for this new intervention model. Our formulation uses the theory of MDPs with partially controlled decision intervals, where given the current state and the action taken, the time interval between successive decisions can be random with a known probability distribution (White, 1978). Similar to the cost function defined for the one-transition model, a predefined cost at each stage is applied when a control action is taken and/or the next state is in the set of undesirable states. The objective is to devise a control strategy that minimizes the expected total discounted cost.

Standard Discounting Cost MDP

We will first review basic MDPs corresponding to the one-transition model. For t=0,1,... denoting the time, let $Z_t \in \mathcal{S}$ and $A_t \in \mathcal{A}$ be a controlled Markov chain, where $\mathcal{S}$ is a finite set of all possible states the Markov chain can occupy, and $\mathcal{A}$ is the set of all possible actions available to the decision maker. The TPM of this chain is defined by

$$p_{xy}(a) = P(Z_{t+1} = y \mid Z_t = x, A_t = a). \tag{8}$$

A *control policy*, $\mu=\{\mu_t\}$ for t=0,1,... is sequence of instructions for taking actions that steers the dynamics of the process. Let $\{z_t\}$ and $\{a_t\}$ be the sequence of observed states and actions, respectively. We allow for μ_t to be a function of the entire history of the system up to time t, denoted by $h_t = (z_0, a_0, z_1, a_1, \ldots, z_t, a_t)$. Once

h_{t-1} is available then some action $a \in \mathcal{A}$ is chosen with probability $\mu_t(a \mid h_{t-1}, z_t)$, where $\mu_t(a \mid h_{t-1}, z_t) \in [0,1]$, and

$$\sum_a \mu_t(a \mid h_{t-1}, z_t) = 1. \tag{9}$$

Let $\mathcal{M}$ be set of all control policies μ. The transition probabilities, $p_{xy}(a)$, the initial state $Z_0=x$ of the Markov chain and any given policy μ determine a unique probability distribution, P_x^μ, over the space of all trajectories of states and actions [derman1970finite]. We denote by E_x^μ the corresponding expectation relative to P_x^μ. Let $g : \mathcal{S} \times \mathcal{A} \times \mathcal{S} \to \Re$ be the immediate cost function, which is assumed non-negative and bounded for all *x*, *a*, and *y*. Given a discount factor $\lambda \in (0,1)$ and an initial state *x*, the goal is to find a control policy minimizing the expected total discounted cost defined by

$$J(x,\mu) = \lim_{N \to \infty} E_x^\mu \left\{ \sum_{t=0}^{N} \lambda^t g(Z_t, A_t, Z_{t-1}) \right\}. \tag{10}$$

The discount factor guarantees convergence of (10) (Derman, 1970). The optimal cost function, J^*, is defined as

$$J^*(x) = \min_{\mu \in \mathcal{M}} J(x,\mu), \tag{11}$$

for every $x \in \mathcal{S}$ and a policy μ* yielding the minimum is called an *optimal policy*. It turns out that one need not consider the entire set $\mathcal{M}$ when searching for the optimal policy; instead it suffices to restrict the search to the set of stationary and deterministic policies, denoted by $\mathcal{M}_D$. Policies, μ_t, in $\mathcal{M}_D$ are the same for all *t*=0,1,..., depend only on the current state, and at each state are equal to one for a single element in $\mathcal{A}$ (Derman, 1970). These three conditions imply that μ* is a single-valued mapping from $\mathcal{S}$ to $\mathcal{A}$. On the basis of the principles mentioned above, one can formulate requirements to minimize the expected cost over an infinite horizon in terms of a functional equation for the finite horizon case. The resulting optimal policy tells us which action to take at each decision epoch, given the state. The functional equation takes the form,

$$J_{k-1}(x) = \min_a \left\{ \tilde{g}(x,a) + \lambda \sum_y p_{xy}(a) J_k(y) \right\}, \tag{12}$$

for k=0,1,... and $x \in \mathcal{S}$, where $\tilde{g}(x,a) = \sum_{y} p_{xy}(a) g(x,a,y)$. Note that, in the above equation, k does not index forward time; rather, it indicates the number of remaining time epochs. In this new reformulation, $J_{k+1}(x)$ is the minimal expected discounted cost over k+1 remaining epochs when the initial state is x and the cost at the final epoch is $J_0(x)$, which is assumed to be bounded for all $x \in \mathcal{S}$.

Starting from state x and by choosing action a, we will move to state y at the beginning of the next epoch with probability $p_{xy}(a)$. Now that the chain is at state y, all decisions for the remaining k epochs are assumed to be optimal. Thus, by definition, the optimal expected discounted cost for these k epochs will be $J_k(y)$ relative to the beginning of the kth epoch. Since this is looking one step ahead, it will be discounted and, therefore, we get a cost of $\lambda J_k(y)$. But at epoch k+1 we do not yet know y, so we must take the expected value of $\lambda J_k(y)$ relative to y. The immediate expected cost at the k+1st epoch is $\tilde{g}(x,a)$ and hence, given x and a, the term on the R.H.S of (12) is the optimal expected discounted cost at epoch k+1.

Moving to the infinite-horizon domain, and inspired by (12), one can apply the dynamic programming mapping, $\mathfrak{T}$, transforming any function $J : \mathcal{S} \rightarrow \Re$ into a new function $\mathfrak{T}J : \mathcal{S} \rightarrow \Re$, defined by

$$(\mathfrak{T}J)(x) = \min_{a} \left\{ \tilde{g}(x,a) + \lambda \sum_{y} p_{xy}(a) J(y) \right\}, \tag{13}$$

for $x \in \mathcal{S}$. We can determine the optimal policy with the help of the following theorems, resulting in an iterative algorithm called *value iteration*. Proofs of these statements can be found in (Bertsekas, 2005).

Theorem 1.1 (Convergence): For any bounded function $J : \mathcal{S} \rightarrow \Re$ and all $x \in \mathcal{S}$, the optimal cost function satisfies

$$J^{*}(x) = \lim_{N \rightarrow \infty} (\mathfrak{T}^{N} J)(x), \tag{14}$$

where $\mathfrak{T}^{N}$ is the composition of the mapping $\mathfrak{T}$ with itself N times.

Theorem 1.2 (Bellman's Optimality Equation): For all $x \in \mathcal{S}$, the optimal cost function J^* satisfies

$$J^{*}(x) = \min_{a} \left\{ \tilde{g}(x,a) + \lambda \sum_{y} p_{xy}(a) J^{*}(y) \right\}. \tag{15}$$

Furthermore, J^* is the unique solution of this equation in the set of all bounded functions.

Theorem 1.3 (Necessary and Sufficient Condition for Optimality): A stationary and deterministic policy μ is optimal if and only if μ(x) attains the minimum in Bellman's optimality equation for every $x \in \mathcal{S}$.

The first part of Theorem 1.2 says that $J^* = \mathbf{T}J^*$. In other words, J^* is the fixed point of $\mathbf{T}$. Theorem 1.1 states that J^* can be iteratively found by successively computing $\mathfrak{T}J, \mathfrak{T}^2J, \ldots$ for any bounded initial cost function J. The convergence of this algorithm is guaranteed due to discounting, where the limit is J^*. Hence, the value iteration algorithm can be continued until some stopping criterion is reached. The last theorem assures that the resulting optimal policy is stationary and deterministic.

Discounting Cost MDP with Partially Controlled Decision Intervals

In a standard MDP, the time index and decision epochs are the same. However, for a MDP with partially controlled decision intervals decision epochs might include multiple time units (White, 1978). Define $\mathcal{L} = \{1, 2, \ldots, K\}$, for a fixed $K \geq 1$, to be the largest time interval between two consecutive decisions epochs. Let t=0,1,... index the decision epochs. We denote by $L_t \in \mathcal{L}$ and $\{l_t\}$ the sequences of random variables representing the length of the time intervals between consecutive decision epochs and the corresponding observed values, respectively. We assume $L_0 \equiv 0$. The transition probabilities for the Markov chain with partially controlled decision intervals are characterized by

$$p^l_{xy}(a) = P(L_{t+1} = l, Z_{t+1} = y \mid Z_t = x, A_t = a), \tag{16}$$

for any t=0,1,..., where $Z_t \in \mathcal{S}$ is the sequence of random states, $A_t \in \mathcal{A}$ the sequence of actions, and $\{z_t\}$ and $\{a_t\}$ are the corresponding sequences of observed states and actions, respectively. $p^l_{xy}(a)$ is the probability that if we start at a state $x \in \mathcal{S}$ and the decision maker takes an action $a \in \mathcal{A}$, the next call for a decision will take place $l \in \mathcal{L}$ time epochs later and we arrive at state $y \in \mathcal{S}$.

We define a *control policy* μ to be a sequence $\{\mu_t\}$ of instructions for taking actions at each decision epoch t. Each μ_t may be a function of the entire history of the system up to decision epoch t, which we define by

$$h_t = (l_0, z_0, a_0, l_1, z_1, a_1, \ldots, l_t, z_t, a_t). \tag{17}$$

Once we have h_{t-1}, we will choose an action $a \in \mathcal{A}$ with probability $\mu_t(a \mid h_{t-1}, l_t, z_t) \in [0,1]$, such that

$$\sum_a \mu_t(a \mid h_{t-1}, l_t, z_t) = 1. \tag{18}$$

Denote the set of all control policies μ by $\mathcal{M}$. Similar to the standard MDP, the transition probabilities, $p_{xy}^l(a)$, the initial state, Z_0=*x*, and the policy, μ, determine P_x^μ over the space of all trajectories of the states, actions and decision epochs. Let E_x^μ denote the corresponding expectation operator. We define $g(x,a,l,y) : \mathcal{S} \times \mathcal{A} \times \mathcal{L} \times \mathcal{S} \to \Re$ to be the immediate cost function over a decision epoch of length *l* when a transition from state *x* under control *a* to state *y* is made in *l* subsequent periods. We assume *g* is non-negative and bounded for all tuples (*x,a,l,y*). The objective function is the expected total discounted cost, defined by

$$J(x,\mu) = \lim_{N\to\infty} E_x^\mu \left\{ \sum_{t=0}^{N} \lambda^{\left(\sum_{m=0}^{t} L_m\right)} g(Z_t, A_t, L_{t+1}, Z_{t+1}) \right\}, \tag{19}$$

for every state $x \in \mathcal{S}$ and $\mu \in \mathcal{M}$, and the goal is to minimize it over the space of policies $\mathcal{M}$. (White, 1978) formulated the following equivalent functional equation for $k \geq K-1$ and all $x \in \mathcal{S}$:

$$J_{k+1}(x) = \min_a \left\{ \tilde{g}(x,a) + \sum_y \sum_l \lambda^l p_{xy}^l(a) J_{k+1-l}(y) \right\}, \tag{20}$$

$$\tilde{g}(x,a) = \sum_y \sum_l p_{xy}^l(a) g(x,a,l,y), \tag{21}$$

and λ∈(0,1). The functions $J_k(x)$ are assumed given and bounded for all $1 \leq k \leq K-2$ and $x \in \mathcal{S}$. In, $J_{k+1}(x)$ is the minimal expected discounted cost beginning in state *x* with *k*+1 remaining epochs and $\tilde{g}(x,a)$ is the immediate expected cost at state *x* and for control action *a*. If the next decision epoch happens *l* time epochs later and leads to arrival at state *y* the minimal expected discounted cost (relative to that point in time) is $J_{k+1-l}(y)$, which when discounted to the present time by a factor λ^t, and when averaged over the joint distribution of y and l, gives the required result by minimizing over all actions $a \in \mathcal{A}$ (White, 1978).

We define the dynamic programming mapping, $\mathfrak{T}$, by

$$(\mathfrak{T}J)(x) = \min_{a} \left\{ \tilde{g}(x,a) + \sum_{y} \sum_{l} \lambda^{l} p_{xy}^{l}(a) J(y) \right\}, \tag{22}$$

for all $x \in \mathcal{S}$. Finding the fixed point of the mapping $\mathfrak{T}$, and the minimizing control action at each state x, gives the control policy minimizing the objective function (19). The following three theorems provide the solution to this problem.

Theorem 1.4 (Convergence): For any bounded function $J : \mathcal{S} \to \mathbb{R}$ and all $x \in \mathcal{S}$, the optimal cost function satisfies

$$J^{*}(x) = \lim_{N \to \infty} (\mathfrak{T}^{N} J)(x). \tag{23}$$

Theorem 1.5 (Bellman's Optimality Equation): For all $x \in \mathcal{S}$, the optimal cost function J^* satisfies

$$J^{*}(x) = \min_{a} \left\{ \tilde{g}(x,a) + \sum_{y} \sum_{l} \lambda^{l} p_{xy}^{l}(a) J^{*}(y) \right\}. \tag{24}$$

Equivalently, $J^{*} = \mathfrak{T}J^{*}$, and J^* is the unique solution of this equation in the set of all bounded functions.

Theorem 1.6 (Necessary and Sufficient Condition for Optimality): A stationary and deterministic policy μ is optimal if and only if $\mu(x)$ attains the minimum in Bellman's optimality equation for every $x \in \mathcal{S}$.

Proofs of these statements can be found in (Yousefi, Datta, & Dougherty, 2012). Similar to the standard MDP, the value iteration algorithm can be used to find the optimal policy.

Random-Length Intervention Model

Successful treatment, both at the individual level or the population level, requires modeling uncertainties in the underlying dynamical system corresponding to the tumor, its interaction with the drug, and the costs associated with the treatment

(Heppner, 1984). For each treatment cycle, samples must be taken from patients to determine the state of the tumor. Depending on the state, a decision must be made as to what type of intervention is necessary. An intervention at the beginning of a treatment cycle usually implies administering an antitumor drug that initiates the process of tumor growth inhibition.

As a special case of the finite-dimensional growth model, (Magni et al., 2008, 2006) propose a semi-empirical tumor growth inhibition model that exhibits good prediction for tumor growth for *in vivo* experiments. Once the effect of the drug on the dynamics of an unperturbed tumor is modeled, under some simplifying assumptions, they show that, in a single treatment window, tumor growth is delayed. The extent of this effect approximately depends on the proportionality constant of a *damage* function, which models the desirable effect of the drug on the tumor cells, the tumor growth rate at early stages of development, and the integral of the drug concentration-time curve. We denote by T_e the length of the drug effect on tumor cells over a single treatment window. Taking into account the variability of response to the antitumor drug within a population, we assume a four-parameter beta distribution for T_e,

$$T_e \sim \mathrm{Beta}(\alpha, \beta, \underline{T}_e, \overline{T}_e). \tag{25}$$

defined on the interval $(\underline{T}_e, \overline{T}_e)$, where $\alpha,\beta>0$ are two real numbers characterizing the shape of this distribution. The parameters of this distribution may be set after consulting with pharmacologists and drug developers.

Assuming PBNs as our reference model for the overall dynamical behavior of tumor cells in the population and external control inputs with persisting effect with $\mathcal{A}=\{0,1,-1\}$, we denote by $T_e(a)$ the length of the window during which the corresponding control gene, directly affected by the drug, will be invariant depending on the type of intervention action $a \in \mathcal{A}$. We also assume that there is a fixed-length rest window of length $T_r(a)$, following $T_e(a)$, to let the side effects of the drug gradually subside. In addition, in the case of no-intervention, i.e. $a=0$, the next decision will be called after $T_r(0)$ units of time have elapsed since $T_e(0)$ is zero. Note that transitions in a PBN occur in discrete time, while $T_e(a)$ and $T_r(a)$ are continuous-time quantities. A simple resolution would be to construct a probability mass function over a discrete set of levels for these variables. We denote the discretized variables by $\overline{T_e(a)}$ and $\overline{T_r(a)}$.

We define a cost function over the set of control actions and the set of states a PBN may occupy. The total cost for the entire course of treatment is obtained by accumulating the discounted cost incurred at each transition of the network. The discounted cost emphasizes that obtaining a response at earlier stages is favored over later stages. The cost functions directly reflect our objectives; hence, they should be assigned by physicians in conjunction with pharmacologists according to their understanding of the treatment objectives.

To solve the optimal intervention problem for this particular model, we need only calculate $p^l_{xy}(a)$ and $\tilde{g}(x,a)$ for every *x,y,a,* and *l,* and use the mathematical machinery for the MDPs with partially controlled decision intervals. To this end, we first define $\mathcal{L}_w(a) = \{0,1,2,\ldots,W(a)\}$, where *W*(*a*) denotes the maximum length of response to the intervention action *a*, and we define

$$\mathcal{L}(a) = \left\{l : l = \overline{T_e(a)} + \overline{T_r(a)} \text{and} \overline{T_e(a)} \in \mathcal{L}_w(a)\right\}. \tag{26}$$

The probability mass function of $\overline{T_e(a)}$ is defined as $P_{l_w}(a) = P\left(\overline{T_e(a)} = l_w\right)$ for all $l_w \in \mathcal{L}_w(a)$ and $a \in \mathcal{A}$. Note $P_{l_w}(a) \geq 0$ and $\sum_{l_w} P_{l_w}(a) = 1$. Thus, the total length of each treatment cycle (drug action and rest together) is a discrete random variable whose probability mass function is $P_l(a) = P_{l_w}(a)$, for every $l \in \mathcal{L}(a)$ and $a \in \mathcal{A}$.

One structural constraint of the problem is that if no intervention occurs, i.e. $\overline{T_e(0)} = 0$, then the next decision epoch will be called deterministically $\overline{T_r(0)}$ time epochs later on. Thus, we have $P_l(0)=1$ for $l = \overline{T_r(0)}$. Another structural constraint is that if any intervention occurs we have $\overline{T_e(a)} \geq 1$, meaning that $P_l(a)=0$ for $l = \overline{T_r(a)}$ when $a \in \mathcal{A} \setminus \{0\}$. Therefore,

$$p^l_{ij}(0) = \begin{cases} p_{ij}^{\left(\overline{T_r(0)}\right)}; & \text{if } l = \overline{T_r(0)} \\ 0, & \text{otherwise} \end{cases} \tag{27}$$

$$p^l_{xy}(a) = P_l(a) \sum_{x'} r_{xx'}(a) \sum_{x''} q_{x'x''}^{\left(l-\overline{T_r(a)}-1\right)}(a) p_{x''y}^{\left(\overline{T_r(a)}\right)}, \tag{28}$$

for all $a \in \mathcal{A} \setminus \{0\}$ and all $l \in \mathcal{L}(a)$. The terms p_{xy}, q_{xy}, and r_{xy} are found from (2), (6) and (7), respectively, and a superscript (*n*), as in $p^{(n)}_{xy}$, denotes the transition probability of going from state *x* to *y* in *n* epochs.

In our setting, the immediate cost function $\tilde{g}(x,a,l,y)$ is also an expected discounted cost accumulated over the entire window. Let $g_{xy}(a)$ denote the one-transition cost function, which depends on the current state *x*, next state *y* and control *a*. We assume that $g_{xy}(a)$ is non-negative and bounded for all $x, y \in \mathcal{S}$ and $a \in \mathcal{A}$, and postulate the following structure for it:

$$g_{xy}(a) = \begin{cases} C + c(a), & \text{if } y \in \mathcal{U} \\ c(a), & \text{if } y \in \mathcal{D} \end{cases} \tag{29}$$

where *c*(*a*)=*c* for all $a \in \mathcal{A} \setminus \{0\}$ and zero otherwise. It is straightforward to show that

$$\tilde{g}(x,a) = \begin{cases} \sum_{m=0}^{\overline{T_r(0)}-1} \lambda^m \sum_y p_{xy}^{(m+1)} g_{xy}(a) & \text{for } a = 0 \\ \sum_{l \in \mathcal{L}(a)} P_l(a) \tilde{g}(x,a,l) & \text{for } a \in \mathcal{A} \setminus \{0\} \end{cases} \tag{30}$$

$$\begin{aligned} \tilde{g}(x,a,l) = {} & \sum_y r_{xy}(a) g_{xy}(a) + \sum_{m=1}^{l-\overline{T_r(a)}-1} \lambda^m \sum_{x'} r_{xx'}(a) \sum_y q_{x'y}^{(m)}(a) g_{x'y}(a) \\ & + \sum_{m=l-\overline{T_r(a)}}^{l-1} \lambda^m \sum_{x'} r_{xx'}(a) \sum_{x''} q_{x'x''}^{(l-\overline{T_r(a)}-1)}(a) \sum_y p_{x''y}^{(m-l+\overline{T_r(a)}+1)} g_{x''y}(0), \end{aligned} \tag{31}$$

for all $l \in \mathcal{L}(a)$. Substituting $p_{xy}^{l}(a)$ and $\tilde{g}(x,a)$ from, and, respectively, into and using the value iteration algorithm one can find the optimal control policy.

Simulation on Networks

To illustrate the effectiveness of our approach, we run several simulations on both synthetically generated PBNs and a PBN corresponding to a mammalian cell cycle network. Throughout, we choose the following values for the parameters:

$$\text{Cost function parameters: } \begin{cases} \lambda = 0.95 \\ C = 2 \\ c = 0.1 \end{cases} \quad \cdots \quad \text{Drug effect parameters: } \begin{cases} W(-1) = W(1) = W \\ \overline{W = 10} \\ \overline{T_r(a)} \equiv \overline{T_r} \in \{1,2,5,10,20\} \\ T_e(-1) = T_e(1) = T_e \\ T_e \sim \text{Beta}(10,30,0,10) \end{cases}$$

We also discretize the beta distributions into ten levels. The resulting discrete distributions are shown in Figure 4.

We evaluate six intervention strategies, each based on different assumptions and constraints:

M1 (One-Transition): Decisions are made at each time epoch and the drug duration of effect is assumed to be 1;

M2 (Cyclic $(W + \overline{T}_r)$-Transition without Drug Effect): Decisions are made at every $(W + \overline{T}_r)$ time epochs and the drug duration of effect is assumed to be 1;

M3 (Acyclic $(W + \overline{T}_r)$-Transition without Drug Effect): Decisions are made at every $(W + \overline{T}_r)$ time epochs if $a \in \{-1,1\}$ and $\overline{T}_r$ time epochs otherwise. The drug duration of effect is assumed to be 1;

M4 (Cyclic $(W + \overline{T}_r)$-Transition with Drug Effect): Similar to M2 but the drug response is modeled with fixed length of *W*;

M5 (Acyclic $(W + \overline{T}_r)$-Transition with Drug Effect): Similar to M3 but the drug response is modeled with fixed length of *W*;

M6 (Random-Length Drug Effect): The drug is correctly modeled with a random-length effect, and we allow for a rest period of length $\overline{T}_r$.

We refer to the optimal policies derived for each model by the name of that model. Once control policies are found, we apply them with the random-length drug

Figure 4. Quantized beta distributions for $\overline{T}_e$ with T_e ~ Beta(10, 30, 0, 10)

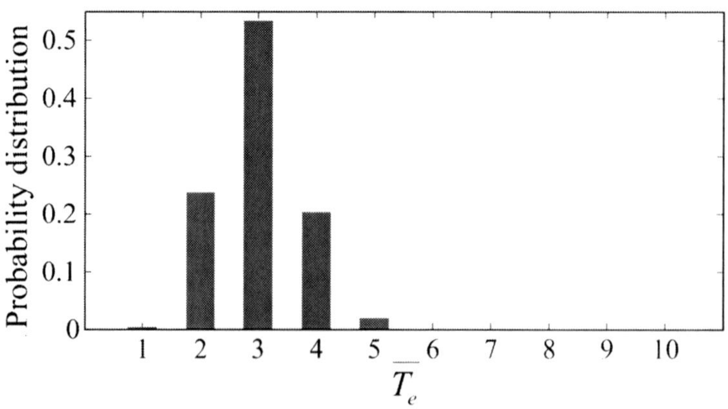

effect (M6) serving as the true model. Note that M1-M5 policies are suboptimal for this model. Hence, our simulation will show how much, relative to the suboptimal policies, one can improve upon when using the optimal policy.

Synthetic Networks

We consider instantaneously random PBNs with seven genes. We also assume each PBN consists of four equally likely constituent Boolean networks with the maximum number of predictors for each Boolean function set to three. When generating Boolean updating functions for each gene, we set the output of each Boolean function, for each input, to one with a small probability, which itself is drawn randomly from a beta distribution with mean 0.1 and standard deviation 0.01. The gene perturbation probability is set to p=0.01. Once BNs are generated, we compute the TPM of each PBN. We define undesirable states, $\mathcal{U}$, as those for which the least significant bit in their binary representation is one, and assume that the control gene is the most significant bit of states.

For each set of simulation parameters, we generate 2900 random PBNs. For each randomly generated PBN in a set of simulations, we find the optimal policies for M1-M6, implement them on the M6 model by simulating the intervention processes over a time-course of length 1500 time epochs, and accumulate the discounted cost incurred at each state-control pair observed. We repeat this process 50000 times, each with a random (uniformly generated) initial state, and compute the expected total discounted cost.

Figure 5 shows the effect of different values for $\overline{T_r}$ on the expected total discounted cost, averaged over 2900 random networks. In this case, $\overline{T_r}$ will most likely take on values from {2,3,4} as shown in Figure 4 which indicates a relatively small variability in the length of drug response. We observe that M6 is optimal on average with a small margin compared to the suboptimal policies.

Although suboptimal policies appear to perform similarly to the optimal one on average, there are networks for which the optimal policy clearly outperforms the suboptimal ones. We quantify the amount of reduction in the cost, for each random network, achieved by the optimal intervention policy relative to the suboptimal ones via the following formula:

$$\Delta(\mu) = 100 \times \frac{J(\mu) - J(\mu^{M6})}{J(\mu)}, \tag{32}$$

Figure 5. Average cost of different policies across 2900 random PBNs for $\overline{T}_r \in \{1, 2, 5, 10, 20\}$

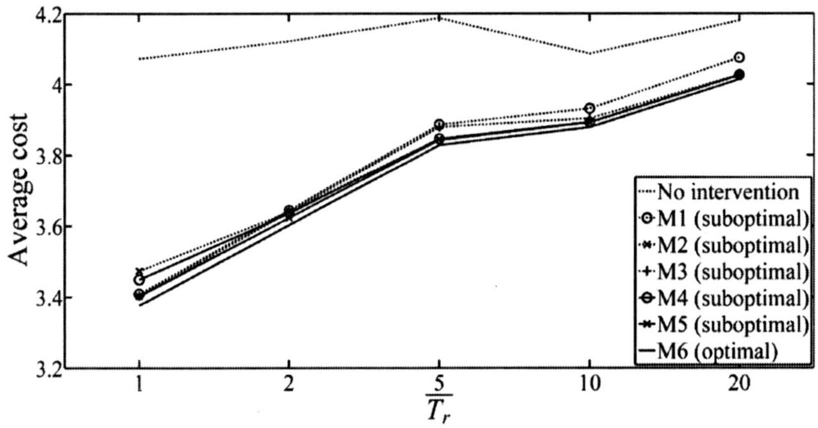

where $J(\mu)$ denotes the expected total discounted cost when a policy μ is used and $J(\mu^{M6})$ denotes the expected total discounted cost when under the optimal policy, M6. Because the networks are randomly generated Δ will also be random. We graph the empirical complementary cumulative distribution function (CDF) of Δ for different suboptimal polices when $\overline{T}_r \in \{1, 5, 20\}$.

The complementary CDF graph in Figure 6 reveals a high probability that the optimal policy noticeably outperforms others. Interestingly, we observe very little improvement over the M2 and M3 policies for larger $\overline{T}_r$. In this case, $\overline{T}_e$ is likely to be much smaller than $\overline{T}_r$, so that the policies resulting from M2 and M3 may induce similar cost compared to M6.

Real Network

This network models normal biological processes that occur during the mammalian cell cycle, whose disruption can lead to uncontrolled cell proliferation. It was originally constructed in (Faure, Naldi, Chaouiya, & Thieffry, 2006) to provide a logical framework for comparing different strategies to explore the dynamical properties of the core mechanisms controlling the mammalian cell cycle. The network has ten genes (or proteins), CycD, Rb, p27, E2F, CycE, CycA, Cdc20, Cdh1, UbcH10 and CycB, and the regulatory relationships are assumed to be Boolean. Table 1 summarizes these regulatory functions. Three key components in this network are Cyclin D (CycD), retinoblastoma (Rb) and p27. Extracellular signals, under normal

Figure 6. Empirical complementary CDF of the improvement over different intervention policies across 2900 random PBNs: (Top-left) $\overline{T}_r = 1$ *; (Top-right)* $\overline{T}_r = 5$ *; (Bottom)* $\overline{T}_r = 20$

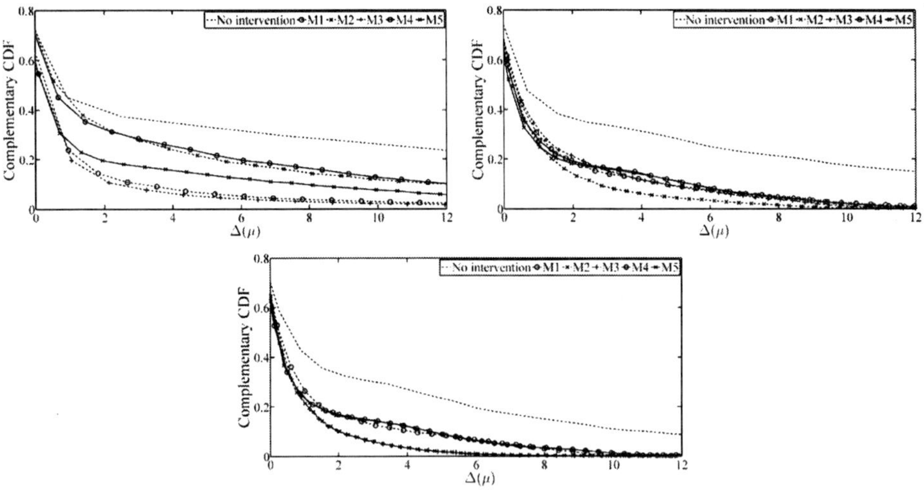

conditions, coordinate cell division with overall growth by controlling the activation of CycD. The tumor suppressor protein, Rb, plays a central role in the negative control of the cell cycle and in tumor progression (Giacinti & Giordano, 2006). Rb is active in the absence of certain cyclin-dependent kinases. When present, however, these cyclins inhibit Rb through phosphorylation. p27 is also active in the absence of the cyclins. An active p27 blocks the action of CycE or CycA and, hence, Rb can also be active even if the cyclins are present, resulting in a mechanism that stops uncontrolled cell division.

This Boolean network can also be used to study the consequences of various mutations. For instance, a possible mutation of p27 (always OFF) may introduce detrimental effects on a cell's normal regulation (Faure et al., 2006). A situation can arise, where both CycD and Rb might be inactive and cell division occurs repeatedly in the absence of any growth factor. This suggests labeling the states in which both Rb and CycD are down-regulated, under a mutated p27, as undesirable. Table 2 presents the Boolean functions of this mutated cell cycle network. The network consists of nine genes, CycD, Rb, E2F, CycE, CycA, Cdc20, Cdh1, UbcH10 and CycB, ordered from the most significant bit to the least significant bit in the binary representation. Depending on the state of the extracellular signal, which determines CycD being ON or OFF, we obtain two contexts. We first build an instantaneously random PBN where these two contexts described above are equally

Table 1. Boolean network of a wild-type mammalian cell cycle

Node Name	Gene Name	Predictor Function
v^1	CycD	Extracellular signal
v^2	Rb	$\left(\overline{\text{CycD}}\ \textbf{AND}\ \overline{\text{CycE}}\ \textbf{AND}\ \overline{\text{CycA}}\ \textbf{AND}\ \overline{\text{CycB}}\right)$ $\textbf{OR}\ \left(\text{p27}\ \textbf{AND}\ \overline{\text{CycD}}\ \textbf{AND}\ \overline{\text{CycB}}\right)$
v^3	E2F	$\left(\overline{\text{Rb}}\ \textbf{AND}\ \overline{\text{CycA}}\ \textbf{AND}\ \overline{\text{CycB}}\right)\ \textbf{OR}\ \left(\text{p27}\ \textbf{AND}\ \overline{\text{Rb}}\ \textbf{AND}\ \overline{\text{CycB}}\right)$
v^4	CycE	$\text{E2F}\ \textbf{AND}\ \overline{\text{Rb}}$
v^5	CycA	$\left(\text{E2F}\ \textbf{AND}\ \overline{\text{Rb}}\ \textbf{AND}\ \overline{\text{Cdc20}}\ \textbf{AND}\ \left(\overline{\text{Cdh1}\ \textbf{AND}\ \text{UbcH10}}\right)\right)$ $\textbf{OR}\ \left(\text{CycA}\ \textbf{AND}\ \overline{\text{Rb}}\ \textbf{AND}\ \overline{\text{Cdc20}}\ \textbf{AND}\ \left(\overline{\text{Cdh1}\ \textbf{AND}\ \text{UbcH10}}\right)\right)$
v^6	p27	$\left(\overline{\text{CycD}}\ \textbf{AND}\ \overline{\text{CycE}}\ \textbf{AND}\ \overline{\text{CycA}}\ \textbf{AND}\ \overline{\text{CycB}}\right)$ $\textbf{OR}\ \left(\text{p27}\ \textbf{AND}\ \overline{\text{CycE}\ \textbf{AND}\ \text{CycA}}\ \textbf{AND}\ \overline{\text{CycB}}\ \textbf{AND}\ \overline{\text{CycD}}\right)$
v^7	Cdc20	CycB
v^8	Cdh1	$\left(\overline{\text{CycA}}\ \textbf{AND}\ \overline{\text{CycB}}\right)\ \textbf{OR}\ \text{Cdc20}$
v^9	UbcH10	$\left(\text{Cdh1}\ \textbf{AND}\ \text{UbcH10}\ \textbf{AND}\ \left(\text{Cdc20}\ \textbf{OR}\ \text{CycA}\ \textbf{OR}\ \text{CycB}\right)\right)\textbf{OR}\ \overline{\text{Cdh1}}$
v^{10}	CycB	$\overline{\text{Cdc20}}\ \textbf{AND}\ \overline{\text{Cdh1}}$

likely. We assume that the gene perturbation probability is 0.01, and define U to be those states for which CycD and Rb are simultaneously down-regulated. This reflects the undesirability of cell growth in the absence of growth factors. We also choose E2F as the control gene.

Similar to the synthetic networks, we find the M1-M6 policies and apply them to the M6 intervention model. We implement a similar procedure for estimating the expected total discounted cost, but with a small difference where we iterate the time course simulation 250000 times, each with a random initial state, and set the length of the time course simulation to 2500 time epochs. Figure 7 shows performance

Table 2. Boolean network of a mutated mammalian cell cycle network

Node Name	Gene Name	Predictor Function
v^1	CycD	Extracellular signal
v^2	Rb	$\left(\overline{\text{CycD}}\ \textbf{AND}\ \overline{\text{CycE}}\ \textbf{AND}\ \overline{\text{CycA}}\ \textbf{AND}\ \overline{\text{CycB}}\right)$
v^3	E2F	$\left(\overline{\text{Rb}}\ \textbf{AND}\ \overline{\text{CycA}}\ \textbf{AND}\ \overline{\text{CycB}}\right)$
v^4	CycE	$\text{E2F}\ \textbf{AND}\ \overline{\text{Rb}}$
v^5	CycA	$\left(\text{E2F}\ \textbf{AND}\ \overline{\text{Rb}}\ \textbf{AND}\ \overline{\text{Cdc20}}\ \textbf{AND}\ \left(\overline{\text{Cdh1}\ \textbf{AND}\ \text{UbcH10}}\right)\right)$ $\textbf{OR}\ \left(\text{CycA}\ \textbf{AND}\ \overline{\text{Rb}}\ \textbf{AND}\ \overline{\text{Cdc20}}\ \textbf{AND}\ \left(\overline{\text{Cdh1}\ \textbf{AND}\ \text{UbcH10}}\right)\right)$
v^6	Cdc20	CycB
v^7	Cdh1	$\left(\overline{\text{CycA}}\ \textbf{AND}\ \overline{\text{CycB}}\right)\ \textbf{OR}\ \text{Cdc20}$
v^8	UbcH10	$\left(\text{Cdh1}\ \textbf{AND}\ \text{UbcH10}\ \textbf{AND}\ \left(\text{Cdc20}\ \textbf{OR}\ \text{CycA}\ \textbf{OR}\ \text{CycB}\right)\right)\ \textbf{OR}\ \overline{\text{Cdh1}}$
v^9	CycB	$\overline{\text{Cdc20}}\ \textbf{AND}\ \overline{\text{Cdh1}}$

Figure 7. Average total discounted cost of control policies derived from different intervention models for the cell cycle network and $\overline{T}_r \in \{1, 2, 5, 10, 20\}$

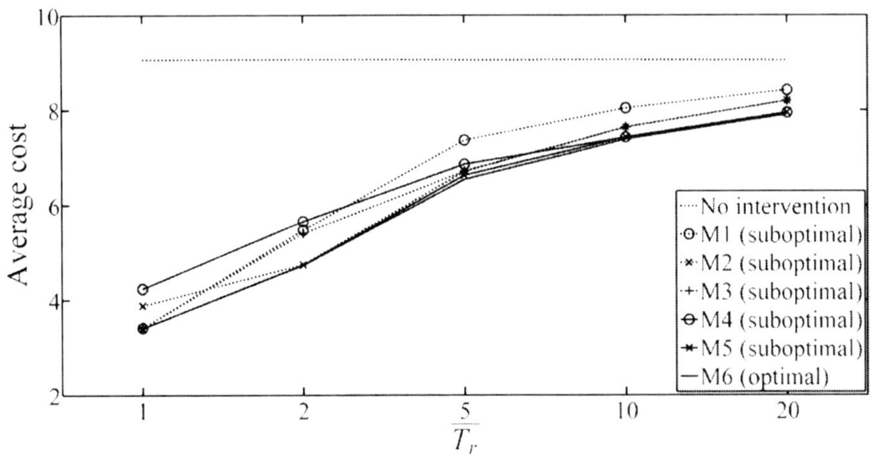

comparison results for each intervention policy based on the expected total discounted cost for different sets of parameters. The performance results are consistent with the results obtained from simulations on the synthetic networks.

Cost-Independent Control

In the previous section, we assigned a large cost to undesirable states and aimed to minimize the long-run discounted cost of being in these states. In systems biology, it is common to model phenotypes as the steady-state behavior of the underlying regulatory network. Thus, from a practical perspective, in many cases our primary interest may be to optimize the long-run occupation of undesirable states, and not to minimize a subjective cost function. Direct optimization has been addressed to some extent in (Qian & Dougherty, 2012; Qian, Ivanov, Ghaffari, & Dougherty, 2009), where the authors propose several classes of greedy stationary policies to reduce the mass corresponding to undesirable states and increase the mass corresponding to desirable states in the steady state distribution, subject to some constraints.

In this section, we will review these suboptimal methods and present an optimal solution to the maximal phenotype alteration problem. Our solution provides an optimal policy with a computational complexity equivalent to solving a linear program (LP) optimization problem. We also address a generalization of this optimization procedure, where the amount of mass moved to some ambiguous states may be constrained. This constrained optimization is especially important in therapeutic methods since one would like to avoid introducing new probability mass to states associated with unknown phenotypes as much as possible (Qian & Dougherty, 2012).

Suboptimal Phenotype Alteration

Consider an instantaneously random PBN with a single control gene g^c and a binary control input. One of the earliest suboptimal methods for maximally altering the long-run behavior of a PBN is a greedy algorithm based on *mean-first-passage-time (MFPT)* (Vahedi, Faryabi, Chamberland, Datta, & Dougherty, 2008), which locates a policy in $\mathcal{M}_D$ based on the heuristic that reaching desirable states and leaving undesirable states as soon as possible is preferable. One of the advantages of this algorithm is that it is model-free and can be directly designed from time-course data without having to infer the TPM of the network.

Suppose that up-regulation of the gene g^1 is undesirable. The TPM of the uncontrolled PBN can be partitioned into four sub-matrices as follows:

$$\mathcal{P}(0) = \begin{pmatrix} \mathcal{P}_{\mathcal{DD}} \mathcal{P}_{\mathcal{DU}} \\ \mathcal{P}_{\mathcal{UD}} \mathcal{P}_{\mathcal{UU}} \end{pmatrix}. \tag{33}$$

One can compute the MFPS vectors $MFPT_{\mathcal{D}}$ (the MFPTs from each state in $\mathcal{U}$ to any state in $\mathcal{D}$) and $MFPT_{\mathcal{U}}$ (the MFPTs from each state in $\mathcal{D}$ to any state in $\mathcal{U}$) by solving the following system of linear equations (Kemény & Snell, 1960):

$$\begin{cases} MFPT_{\mathcal{D}} = e + \mathcal{P}_{\mathcal{DD}} MFPT_{\mathcal{D}} \\ MFPT_{\mathcal{U}} = e + \mathcal{P}_{\mathcal{UU}} MFPT_{\mathcal{U}} \end{cases} \tag{34}$$

where *e* is a column vector of 1's with the appropriate length (with *n* genes, *e* has 2^{n-1} elements). We denote by $\mu_{\text{MFPT}} : \mathcal{S} \to \mathcal{A}$ the control policy derived from the following procedure: choose a number γ. For every $x \in \mathcal{U}$, if $MFPT_{\mathcal{D}}(x) - MFPT_{\mathcal{D}}(\tilde{x}) > \gamma$, then $\mu_{\text{MFPT}}(x)=1$; otherwise, $\mu_{\text{MFPT}}(x)=0$. Using this rule, one reduces the expected time to reach the desirable states. Analogously, for every $x \in \mathcal{D}$, if $MFPT_{\mathcal{U}}(\tilde{x}) - MFPT_{\mathcal{U}}(x) > \gamma$, then $\mu_{\text{MFPT}}(x)=1$; otherwise, $\mu_{\text{MFPT}}(x)=0$. Thus, the expected time until a passage to undesirable states is increased. The tuning parameter γ is set to a large when the ratio of the cost of control to the cost of the undesirable states is higher so that the control is applied less frequently; otherwise, under no constraints on the control frequency, we set $\gamma=0$ (Vahedi et al., 2008).

Another greedy method for solving this problem is called a *basin of attraction (BOA)* control policy, which belongs to the set of stationary and deterministic policies (Qian et al., 2009). For any state *x*, let $\mathcal{L}^j(x)$ be the set of attractors for the basin containing a state *x* in the *j*th constituent Boolean network, keeping in mind that a state *x* belongs to exactly one basin of attraction in each constituent Boolean network. Let

$$\mathcal{B}(x) = \bigcup_{j=1}^{m} \mathcal{L}^j(x) \tag{35}$$

be the union of attractors for the basins of *x* over all of the constituent Boolean networks. For each constituent Boolean network, we calculate the minimal distance of each state in $\mathcal{S}$ to states in $\mathcal{D}$ or in $\mathcal{U}$, and denote their weighted average, with respect to the Boolean networks selection probabilities, by $d_{\mathcal{D}}$ and $d_{\mathcal{U}}$, respectively (Qian et al., 2009).

A BOA control policy, $\mu_{\mathrm{BOA}} : \mathcal{S} \to \mathcal{A}$, is derived via the following algorithm for every $x \in \mathcal{S}$. Suppose $x \in \mathcal{U}$. If $\mathcal{B}(x)$ does not contains any desirable attractors and at the same time $\mathcal{B}(\tilde{x})$ contains desirable attractors, then $\mu_{\mathrm{BOA}}(x)=1$. By this, the network moves to $\tilde{x}$ in which the likelihood of arriving into desirable attractors is increased. If both or neither of $\mathcal{B}(x)$ and $\mathcal{B}(\tilde{x})$ contain desirable attractors, then $\mu_{\mathrm{BOA}}(x)=1$ if $d_{\mathcal{D}}(x) > d_{\mathcal{D}}(\tilde{x})$, $\mu_{\mathrm{BOA}}(x)=0$ if $d_{\mathcal{D}}(x) < d_{\mathcal{D}}(\tilde{x})$, or otherwise, $\mu_{\mathrm{BOA}}(x)=0$. Now, suppose $x \in \mathcal{D}$. If $\mathcal{B}(x)$ contains undesirable attractors and at the same time $\mathcal{B}(\tilde{x})$ does not contain any undesirable attractors, $\mu_{\mathrm{BOA}}(x)=1$, so that the chance of entering undesirable attractors is reduced. If both or neither of $\mathcal{B}(x)$ and $\mathcal{B}(\tilde{x})$ contain undesirable attractors, then $\mu_{\mathrm{BOA}}(x)=1$ if $d_{\mathcal{U}}(\tilde{x}) > d_{\mathcal{U}}(x)$, or $\mu_{\mathrm{BOA}}(x)=0$ if $d_{\mathcal{U}}(\tilde{x}) < d_{\mathcal{U}}(x)$. Otherwise, $\mu_{\mathrm{BOA}}(x)=0$ (Qian et al., 2009).

A third algorithm, the *steady-state distribution (SSD)* control policy, is based on the theory of rank-one perturbations of regulatory functions, where one can readily calculate the exact shifted stationary mass (Qian & Dougherty, 2008; Qian et al., 2009). Let π and $\mathcal{Z} = [I - \mathcal{P}(0) + e\pi]^{-1}$ be the steady-state row vector and the fundamental matrix of the unperturbed network, respectively, where I is the identity matrix e is a column vector of 1's with the appropriate dimensions. The perturbed steady-state distribution can be computed by:

$$\hat{\pi}^{x} = \pi - \frac{\pi_{x}(\mathcal{P}_{x}(0) - \mathcal{P}_{\tilde{x}}(0))\mathcal{Z}}{1 + (\mathcal{P}_{x}(0) - \mathcal{P}_{\tilde{x}}(0))\mathcal{Z}^{x}}, \tag{36}$$

where $\mathcal{P}_{x}$ is a row in $\mathcal{P}$ corresponding to the states x, $\mathcal{Z}^{x}$ is a column in $\mathcal{Z}$ corresponding to state x, π_{x} is the stationary mass at state x, and $\hat{\pi}^{x}$ is the steady-state vector after we flip the control gene at state x.

A SSD control policy, .., is found via the following procedure. For every pair of states x and $\tilde{x}$, compute $\pi_{\mathcal{U}} = \sum_{y \in \mathcal{U}} \pi_{y}$, $\hat{\pi}^{x}_{\mathcal{U}} = \sum_{y \in \mathcal{U}} \hat{\pi}^{x}_{y}$ and $\hat{\pi}^{\tilde{x}}_{\mathcal{U}} = \sum_{y \in \mathcal{U}} \hat{\pi}^{x}_{y}$. If $\pi_{\mathcal{U}} \le \min(\hat{\pi}^{x}_{\mathcal{U}}, \hat{\pi}^{\tilde{x}}_{\mathcal{U}})$ then $\mu_{\mathrm{SSD}}(x) = \mu_{\mathrm{SSD}}(\tilde{x}) = 0$; otherwise, if $\hat{\pi}^{x}_{\mathcal{U}} \le \hat{\pi}^{\tilde{x}}_{\mathcal{U}}$, then $\mu_{\mathrm{SSD}}(x)=1$ and $\mu_{\mathrm{SSD}}(\tilde{x}) = 0$, and if $\hat{\pi}^{x}_{\mathcal{U}} \ge \hat{\pi}^{\tilde{x}}_{\mathcal{U}}$, then $\mu_{\mathrm{SSD}}(x)=0$ and $\mu_{\mathrm{SSD}}(\tilde{x}) = 1$.

When deriving the SSD control policy, all beneficial one-row perturbations are found independently. However, one can also iteratively find a state and corresponding control action having a one-row perturbation resulting in the largest reduction of undesirable mass. This is the *conservative steady-state distribution* (CSSD) control policy. We begin by assigning no control to all states. At each iteration, one selects a state x and corresponding control action at state x with a maximal reduction in undesirable mass. The algorithm runs iteratively until intervention at any additional state actually increases the stationary mass of undesirable states. One can theoretically guarantee that the located control policy will always induce undesirable stationary mass less than or equal to that of the uncontrolled network (Qian et al., 2009). By construction, μ_{CSSD} is a stationary and deterministic policy. Table 3 presents the algorithm, where π and $\mathcal{Z}$ denote the steady-state distribution and the fundamental matrix of the unperturbed network, respectively.

None of the control policies discussed thus far check if there is an introduction of mass at states representing complications or harmful side effects to healthy cells, even though these states are labeled as desirable. It is therefore prudent to restrict the shift of mass to only those states associated with known healthy phenotypes. To meet this new criterion, we partition $\mathcal{D}$ further into two categories of .., representing known healthy states, and $\mathcal{D}_{\Lambda}$, representing *ambiguous* states where we have little or no prior knowledge and thus may lead to complications or damages (Qian & Dougherty, 2012).

The goal is now to find a control policy that shifts the undesirable probability mass as much as possible and at the same time keeps the ambiguous probability mass of each state below certain level. The *constrained SSD (conSSD)* and *constrained CSSD (conCSSD)* control policies aim to find such (suboptimal) policies (Qian & Dougherty, 2012).

We denote by $\mu_{conSSD} : \mathcal{S} \rightarrow \mathcal{A}$ and $\mu_{conCSSD} : \mathcal{S} \rightarrow \mathcal{A}$ the conSSD and conCSSD control policies, respectively. The algorithms presented in Table 4 and Table 5 are similar to those of SSD and CSSD, respectively, except that they take into account an upper-bound, $\delta \geq 0$, on the long-run probability mass of ambiguous states after control. This will represent the maximum amount of risk one is willing to accept

Table 3. CSSD control policy algorithm (Qian et al., 2009)

```
Set π̲=π and Z̲ = Z ; Assign the set of states with no control to the set of all states: O = S ;
Repeat
    Δ=0;
    For all states x ∈ O do
        Compute Δπ_u = π̲_u − π̲̂'_u ;
        If Δπ_u > Δ̲ then
            Δ̲ = Δπ_u ; Assign the best state to control: x̲=x; π̲ = π̂' ;
    End
    If Δ̲>0 then
        μ_CSSD(x̲)=1 and μ_CSSD(x̲̃) = 0 ; O = O \ {x̲, x̲̃} ; Update Z̲ for the rank-one
        perturbed network accordingly (refer to (Qian et al., 2009) for the updating equation);
Until Δ̲=0;
```

Table 4. conSSD control policy algorithm (Qian & Dougherty, 2012)

For all pairs of states $x, \tilde{x} \in \mathcal{O}$
 Compute $\hat{\pi}^{x}$ and $\hat{\pi}^{\tilde{x}}$;
 If $\max_{y \in \mathcal{D}_A} \hat{\pi}^{x}_{y} > \delta$ then
  $\hat{\pi}^{x}_{\mathcal{U}} = 1.0$; // Make the control action not admissible
 Else
  $\hat{\pi}^{x}_{\mathcal{U}} = \sum_{y \in \mathcal{U}} \hat{\pi}^{x}_{y}$;
 If $\max_{y \in \mathcal{D}_A} \hat{\pi}^{\tilde{x}}_{y} > \delta$ then
  $\hat{\pi}^{\tilde{x}}_{\mathcal{U}} = 1.0$; // Make the control action not admissible
 Else
  $\hat{\pi}^{\tilde{x}}_{\mathcal{U}} = \sum_{y \in \mathcal{U}} \hat{\pi}^{\tilde{x}}_{y}$;
 If $\pi_{\mathcal{U}} \leq \min(\hat{\pi}^{x}_{\mathcal{U}}, \hat{\pi}^{\tilde{x}}_{\mathcal{U}})$
  $\mu_{\text{conSSD}}(x) = \mu_{\text{conSSD}}(\tilde{x}) = 0$;
 Else
  If $\hat{\pi}^{x}_{\mathcal{U}} \leq \hat{\pi}^{\tilde{x}}_{\mathcal{U}}$
   $\mu_{\text{conSSD}}(x)=1$ and $\mu_{\text{conSSD}}(\tilde{x}) = 0$;
  Else
   $\mu_{\text{conSSD}}(x)=0$ and $\mu_{\text{conSSD}}(\tilde{x}) = 1$;
End

Table 5. conCSSD control policy algorithm (Qian & Dougherty, 2012)

Set $\underline{\pi}=\pi$ and $\underline{\mathcal{Z}} = \mathcal{Z}$; Assign the set of states with no control to the set of all states: $\mathcal{O} = \mathcal{S}$;
Repeat
 $\underline{\Delta}=-0.1$;
 For all states $x \in \mathcal{O}$ do
  Compute $\Delta\pi_{\mathcal{U}} = \underline{\pi}_{\mathcal{U}} - \hat{\underline{\pi}}^{x}_{\mathcal{U}}$;
  If $\max_{y \in \mathcal{D}_A} \hat{\underline{\pi}}^{x}_{y} > \delta$ then
   $\Delta\pi_{\mathcal{U}} = -0.1$; // Make the control action not admissible
  If $\Delta\pi_{\mathcal{U}} > \underline{\Delta}$ then
   $\underline{\Delta} = \Delta\pi_{\mathcal{U}}$; Assign the best state to control: $\underline{x}=x$; $\underline{\pi} = \hat{\pi}^{x}$;
 End
 If $\underline{\Delta}>0$ then
  $\mu_{\text{conCSSD}}(\underline{x})=1$ and $\mu_{\text{conCSSD}}(\tilde{\underline{x}}) = 0$; $\mathcal{O} = \mathcal{O} \setminus \{\underline{x}, \tilde{\underline{x}}\}$; Update $\underline{\mathcal{Z}}$ for the rank-one perturbed network accordingly (refer to (Qian et al., 2009) for the updating equation);
Until $\underline{\Delta} \leq 0$;

with regard to introducing probability mass at an ambiguous state while implementing the intervention policy. In practice, expert medical knowledge would be employed to identify the set $\mathcal{D}_A$ and upper bound δ (Qian & Dougherty, 2012). As a consequence of adding this stringent criterion, it is expected that these constrained control policies become less effective in terms of shifting the undesirable mass.

Optimal Phenotype Alteration

We now find optimal intervention policies with respect to the objectives and constraints. As we will see, the problem of maximal phenotype alteration is a special case of constrained MDPs (Altman, 1999). Therefore, we begin with a brief review of MPDs, where we focus on the average cost objective function. In general, the decision maker minimizes one objective function subject to constraints on possibly one or more additional objective functions. The basic components of a constrained MDP with average cost are the same as those of discounted cost, but with additional cost functions. Therefore, we adopt the same notations, and only define a new immediate cost function as $r : \mathcal{S} \times \mathcal{A} \times \mathcal{S} \to \Re^D$, which is a *D*-dimensional nonnegative and bounded vector corresponding to *D* constraints. We also denote the TPM of the controlled chain by $\mathcal{P}(a)$, and assume that the chain is ergodic under any deterministic policy. For an initial state $x \in \mathcal{S}$ and a control policy $\mu = \{\mu_1, \mu_2, \ldots\} \in \mathcal{M}$, the infinite-horizon expected average costs are defined by

$$G(x, \mu) = \limsup_{N \to \infty} E_x^{\mu} \frac{1}{N+1} \sum_{t=0}^{N} g(Z_t, A_t, Z_{t+1}), \tag{37}$$

$$R_d(x, \mu) = \limsup_{N \to \infty} E_x^{\mu} \frac{1}{N+1} \sum_{t=0}^{N} r_d(Z_t, A_t, Z_{t+1}), \text{ for } d = 1, 2, \ldots, D. \tag{38}$$

Given $C_1, C_2, \ldots, C_D \geq 0$, we can formulate the optimization problem in its most general form for any initial state $x \in \mathcal{S}$,

$$\begin{aligned} &\min_{\mu \in \mathcal{M}} G(x, \mu) \\ &\text{subject to } R_d(x, \mu) \leq C_d \text{ for all } d = 1, 2, \ldots, D. \end{aligned} \tag{39}$$

All policies that satisfy the constraints are called feasible. Let $G^*(x)$ denote a solution of (39). A feasible policy μ* is optimal if $G^*(x) = G(x, \mu^*)$, for all $x \in \mathcal{S}$.

Although, the search space $\mathcal{M}$ is infinite (possibly uncountable), we consider a set $\mathcal{M}_S \subset \mathcal{M}$, of stationary policies without loss of optimality. $\mathcal{M}_S$ includes those policies for which μ_t is time invariant (not dependent on t) and only a function of z_t for any t. In other words, the probabilities that a control policy μ assigns to different actions only depends on the current state observed and not the time. Note that the set of stationary and deterministic policies, $\mathcal{M}_D$, is a subset of $\mathcal{M}_S$.

It can be shown that may be transformed into a LP with its dual being (Altman, 1999):

$$\min_{\{\nu_{xa}\}} \sum_x \sum_a \nu_{xa} g(x,a)$$
$$\text{subject to } \begin{cases} \sum_x \sum_a \nu_{xa} r_d(x,a) \le C_d, \text{ for all } d = 1,2,\ldots,D \\ \sum_a \nu_{xa} = \sum_y \sum_a \nu_{ya} p_{yx}(a), \text{ for all } x \in \mathcal{S} \\ \sum_x \sum_a \nu_{xa} = 1 \\ \nu_{xa} \ge 0, \text{ for all } x \in \mathcal{S},\ a \in \mathcal{A} \end{cases} \tag{40}$$

where $g(x,a) = \sum_y p_{xy}(a) g(x,a,y)$ and $r_d(x,a) = \sum_y p_{xy}(a) r_d(x,a,y)$ for each d=1,2,…,D. Letting $\left\{\nu_{xa}^*\right\}$ be a minimizing argument of (40), one can recover an optimal stationary policy $\mu^* \in \mathcal{M}_S$ by

$$\mu^*(a \mid x) = \frac{\nu_{xa}^*}{\sum_a \nu_{xa}^*}. \tag{41}$$

If D>0 and is feasible, then μ* requires randomization (choosing between the actions in $\mathcal{A}$ through a randomized mechanism) in at most D states (Altman, 1999; Ross, 1989). However, if there are no constraints on the expected average cost criteria, i.e., D=0, then there is no randomization and $\mu^* \in \mathcal{M}_D$.

In our maximal phenotype alteration problem, we desire an intervention policy that optimally shifts the long-run probability mass of undesirable states towards desirable states, without introducing new mass at ambiguous states. This objective essentially concerns the long-run expected frequencies of observing undesirable states which is encoded in the variables $\{\nu_{xa}\}$ marginalized over the actions, i.e., $\left\{\sum_a \nu_{xa}\right\}$.

Let $\bar{\mu}(x) \equiv 0$ denote the stationary and deterministic control policy of *no-intervention*. For any policy $\mu \in \mathcal{M}_S$, the amount of *shift* in the aggregated probability of undesirable states is defined as

$$\Delta\pi_{\mathcal{U}}(\mu) = \pi_{\mathcal{U}}(\bar{\mu}) - \pi_{\mathcal{U}}(\mu), \tag{42}$$

where $\pi_{\mathcal{U}}(\mu) = \sum_{x \in \mathcal{U}} \pi_x(\mu)$ and $\pi(\mu)$ is the steady-state row vector of the controlled Markov chains governed under policy μ. In general, $-1 \leq \Delta\pi_{\mathcal{U}}(\mu) \leq 1$ and our goal is to maximize it subject to some constraints on the elements of $\pi(\mu)$, which are usually in the form of $\pi_x(\mu) \leq \gamma_x$ for every $x \in \mathcal{D}_A$ and constants $\gamma_x \geq 0$.

First, note that maximizing $\Delta\pi_{\mathcal{U}}(\mu)$ is equivalent to minimizing $\pi_{\mathcal{U}}(\mu)$. Now, let the immediate cost functions g and r_d be defined as

$$g(x, a, y) = \begin{cases} 1, & \text{if } x \in \mathcal{U} \\ 0, & \text{otherwise} \end{cases} \tag{43}$$

$$r_d(x, a, y) = \begin{cases} 1, & \text{if } x \in \mathcal{D}_A \\ 0, & \text{otherwise} \end{cases} \tag{44}$$

for any action $a \in \mathcal{A}$, any state $y \in \mathcal{S}$ and all d. Using these cost functions, we will count the number of visits to states in $\mathcal{U}$ and $\mathcal{D}_A$. Thus, for a policy $\mu \in \mathcal{M}_S$ and an initial state x, equations (37) and (38) provide in the steady-state mass of undesirable and ambiguous states, respectively. The constrained maximal phenotype alteration problem can thus be formulated as

$$\min_{\{\nu_{xa}\}} \sum_x \sum_a \nu_{xa}$$
$$\text{subject to} \begin{cases} \sum_a \nu_{xa} \leq \gamma_x, & \text{for each } x \in \mathcal{D}_A \\ \sum_a \nu_{xa} = \sum_y \sum_a \nu_{ya} p_{yx}(a), & \text{for all } x \in \mathcal{S} \\ \sum_x \sum_a \nu_{xa} = 1 & \\ \nu_{xa} \geq 0, & \text{for all } x \in \mathcal{S},\ a \in \mathcal{A} \end{cases} \tag{45}$$

Once the solution $\left\{\nu^{*}_{xa}\right\}$ is found, one can recover the optimal stationary policy using (41). If there are no ambiguous states, then the optimal policy is stationary and deterministic. We refer to this policy as the *optimal unconstrained (UC)* policy and denote it by μ^{*}_{UC}. If there are $|\mathcal{D}_{\mathrm{A}}|$ ambiguous states then the optimal policy may not be deterministic; however, it belongs to $\mathcal{M}_{\mathrm{S}}$ and requires randomization in at most $|\mathcal{D}_{\mathrm{A}}|$ states. In this case, we call this policy *optimal phenotypically constrained (PC)* and denote it by μ^{*}_{PC}.

Simulation on Networks

We test our method on synthetically generated networks and a real network modeling metastatic melanoma.

Synthetic Networks

For phenotypically constrained problems and for any PBN, we define $\mathcal{D}_{\mathrm{A}} = \{x \in \mathcal{D} : \pi_x \leq 1/2^n\}$, where π is the steady-state vector of the uncontrolled network. We also set $\gamma_x = \max_{x \in \mathcal{D}_{\mathrm{A}}} \pi_x$ for all $x \in \mathcal{D}_{\mathrm{A}}$.

We synthetically generate 2500 instantaneously random PBNs with seven genes and report both the average performance across all networks as well as the empirical complementary CDF of

$$\Delta(\mu) = \Delta\pi_{\mathcal{U}}(\mu^{*}) - \Delta\pi_{\mathcal{U}}(\mu) = \pi_{\mathcal{U}}(\mu) - \pi_{\mathcal{U}}(\mu^{*}). \tag{46}$$

Here, $\Delta(\mu)$ quantifies the gain made by using an optimal policy, μ^*, over a suboptimal one, μ. Clearly, $0 \leq \Delta(\mu) \leq 1$. For each network, we calculate $\Delta(\mu)$ for all six suboptimal policies discussed earlier in this section and both optimal policies, for both the unconstrained and constrained problems, separately.

We assume that each PBN is generated from four equally likely constituent Boolean networks with the maximum number of predictors for each Boolean function set to three. We assume that the probability that each Boolean regulatory function takes on the value 1 is a random variable whose values are taken from a beta distribution with mean $\theta \in \{0.25, 0.5, 0.75\}$ and standard deviation 0.01. This quantity controls the dynamics of a randomly generated Boolean network. We also assume a random gene perturbation probability of 0.001. We define undesirable states, $\mathcal{U}$, to be those for which the most significant bit in their binary representation is zero, and assume that the control gene is the most significant bit of states.

Figure 8 shows the shift in the probability mass of undesirable states induced by the optimal and suboptimal policies, defined in (42), averaged across all PBNs for each values of θ. These graphs clearly illustrate the optimality of μ^*_{UC} and μ^*_{PC} compared to their suboptimal counterparts. Among the unconstrained suboptimal policies, μ_{CSSD} yields the best results. However, μ_{conCSSD} does not induce as much shift as μ^*_{PC}, which is due to the fact that the search space of the conCSSD policy belongs to the class $\mathcal{M}_{\mathrm{D}}$ of policies, whereas μ^*_{PC} belongs to the larger class $\mathcal{M}_{\mathrm{S}}$. It is worth noting that the amount of average shift induced by μ^*_{PC} is significantly less than μ^*_{UC}, which can be attributed to the additional constraints imposed on the optimization problem.

Figure 9 presents the empirical complementary CDF of Δ(μ) for different suboptimal interventions. The graphs indicate that μ^*_{UC} and μ_{CSSD} induce similar amounts of shift with a high probability. It is also evident that μ^*_{PC} outperforms μ_{conCSSD} with a (realistically) significant probability; for instance, with probability 0.1 using μ^*_{PC} may yield an improvement of more than 5% over μ_{conCSSD}.

Real Network

It has been demonstrated that by manipulating the concentration level of the Wnt5a protein secreted by a melanoma cell line, one can directly affect the metastatic status of the cell as measured by standard *in vitro* assays for metastasis (Bittner et al., 2000; Dissanayake et al., 2008; Weeraratna et al., 2002), which suggests that reducing the WNT5A gene's action could reduce the risk of a melanoma metastasizing. Therefore, it is desirable to down-regulate WNT5A as much as possible through appropriate

Figure 8. Average shift for different policies across 2500 random PBNs with seven genes: (Left) unconstrained policies; (Right) Phenotypically constrained policies

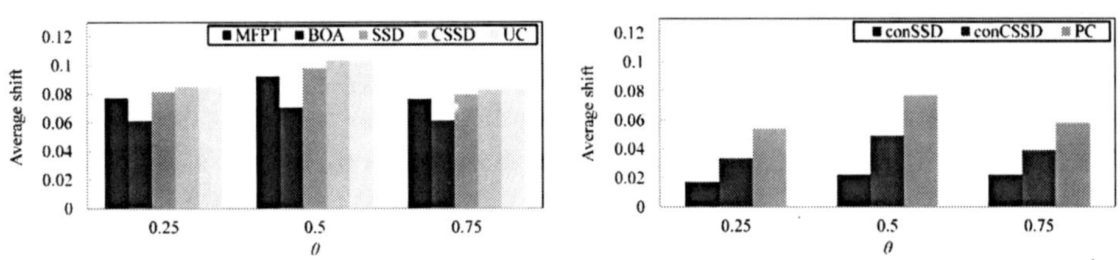

intervention mechanisms (Datta et al., 2003; Qian & Dougherty, 2008).

Figure 9. Empirical complementary CDF of $\Delta(\mu)$ for different intervention policies: (Top-left) θ=0.25 and μ^*_{UC}; (Top-right) θ=0.25 and μ^*_{PC}; (Middle-left) θ=0.5 and μ^*_{UC}; (Middle-right) θ=0.5 and μ^*_{PC}; (Bottom-left) θ=0.75 and μ^*_{UC}; (Bottom-right) θ=0.75 and μ^*_{PC}

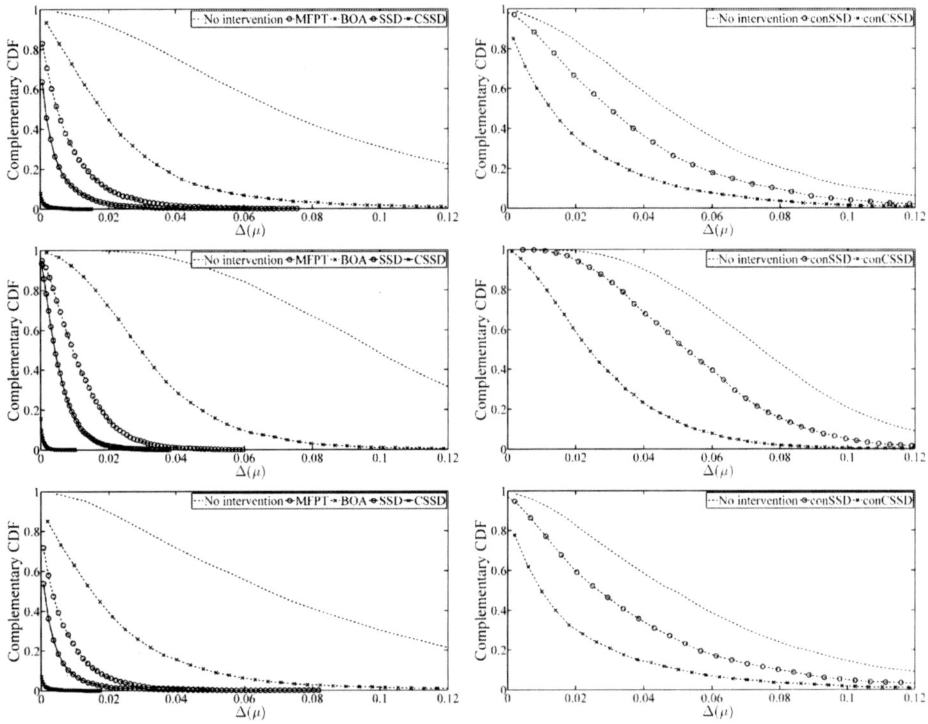

We use a Boolean network composed of ten nodes extracted from a set of 587 genes (Kim et al., 2002). The Boolean regulatory relationships between these genes/ proteins are presented in Table 6. Note there is a single context when we construct a PBN from this network. We label states with an up-regulated WNT5A as undesirable. We set the gene perturbation probability to 0.001. We assume that up-regulation of WNT5A is undesirable, and set ambiguous states in $\mathcal{D}$ to non-attractor states of the Boolean network, so that the control policy will not introduce new attractors. For the control genes, we choose each gene in the network, except WNT5A, as the control gene, in turn.

The aggregated probability mass of undesirable states for the uncontrolled network is $\pi_u(\bar{\mu}) = 0.1711$. Table 7 shows the shift made by all control policies based on different control genes. The maximum shift induced by μ_{UC} occurs when either MMP3 or PHOC are the control genes. However, the maximum shift for μ_{PC} occurs

Table 6. Boolean network of a metastatic melanoma network

Node Name	Gene Name	Predictor Function
v^{1}	WNT5A	$\left(\text{S100P } \textbf{AND } \text{MMP3 } \textbf{AND } \overline{\text{PHOC}}\right) \textbf{ OR } \left(\overline{\text{MMP3}} \textbf{ AND } \text{PHOC}\right)$
v^{2}	pirin	$\left(\overline{\text{WNT5A}} \textbf{ AND } \overline{\text{S100P}} \textbf{ AND } \text{MMP3}\right) \textbf{ OR } \left(\text{WNT5A } \textbf{AND } \overline{\text{S100P}} \textbf{ AND } \overline{\text{MMP3}}\right)$
v^{3}	S100P	MART1
v^{4}	RET1	$\left(\overline{\text{WNT5A}} \textbf{ AND } \text{pirin } \textbf{AND } \text{RET1}\right) \textbf{ OR } \left(\overline{\text{pirin}} \textbf{ AND } \text{RET1}\right)$
v^{5}	MMP3	$\left(\text{RET1 } \textbf{AND } \text{synuclein}\right) \textbf{ OR } \overline{\text{synuclein}}$
v^{6}	PHOC	$\left(\overline{\text{RET1}} \textbf{ AND } \overline{\text{MART1}}\right) \textbf{ OR } \left(\text{RET1 } \textbf{AND } \text{MART1 } \textbf{AND } \text{STC2}\right)$
v^{7}	MART1	MART1
v^{8}	HADHB	$\left(\text{WNT5A } \textbf{AND } \text{MMP3}\right)$ $\textbf{OR } \left(\overline{\text{MMP3}} \textbf{ AND } \overline{\text{synuclein}}\right) \textbf{ OR } \left(\text{WNT5A } \textbf{AND } \overline{\text{MMP3}} \textbf{ AND } \text{synuclein}\right)$
v^{9}	synuclein	$\left(\overline{\text{WNT5A}} \textbf{ AND } \overline{\text{MART1}} \textbf{ AND } \overline{\text{STC2}}\right)$ $\textbf{OR } \left(\text{RET1 } \textbf{AND } \overline{\text{MART1}} \textbf{ AND } \text{STC2}\right) \textbf{ OR } \text{MART1}$
v^{10}	STC2	$\overline{\text{S100P}}$

when MART1 is the control gene, which is an immunogen that the body sometimes recognizes and attacks via the immune system (Dissanayake et al., 2008; Qian & Dougherty, 2012).

CONCLUSION

Modeling tumor heterogeneity and the degree to which tumor cells respond to antitumor drugs are crucial to developing optimal cancer treatment. Using a recently proposed model for tumor growth inhibition, we have characterized uncertainties associated with drug response at different levels and built a probabilistic model for the entire intervention process. While accounting for practical effects arising from

Table 7. Amount of shift in the metastatic melanoma network for different control genes and policies

Gene \ Policy	MFPT	BOA	SSD	CSSD	Optimal UC	conSSD	conCSSD	Optimal PC
pirin	0.1632	0.1630	0.1630	0.1632	0.1632	0.0003	0.0043	0.0548
S100P	0.1679	0.1673	0.1679	0.1679	0.1681	0.1548	0.0080	0.0139
RET1	0.1681	0.1681	0.1646	0.1681	0.1681	0.0000	0.0005	0.1473
MMP3	0.1701	0.1701	0.1701	0.1701	0.1701	0.0054	0.0059	0.0076
PHOC	0.1701	0.1701	0.1701	0.1701	0.1701	0.0041	0.0057	0.0076
MART1	0.1684	0.1684	0.1684	0.1684	0.1684	0.0165	0.0016	0.1682
HADHB	0.0000	0.0000	0.0000	0.0000	0.0000	0.0000	0.0000	0.0000
synuclein	0.0029	0.0029	0.0029	0.0029	0.0029	0.0013	0.0029	0.0029
STC2	0.1671	0.1278	0.1674	0.1674	0.1674	0.1146	0.0040	0.0070

therapeutic constraints, we assumed PBNs as our model for GRNs and derived optimal control policies with random-length duration of treatment effectiveness. This was accomplished by taking a unified approach in the framework of finite-state MDPs with partially controlled decision intervals and infinite-horizon discounted cost. This framework is general enough to include several other, simple but practical, intervention models. We presented a simulation study on randomly generated instantaneously random PBNs and demonstrated that optimal intervention policies outperform suboptimal ones with a significant probability. We also showed the effectiveness of optimal intervention policies on mutated mammalian cell cycle regulation.

Although the control policies derived from this procedure are optimal, their optimality is relative to subjective cost functions defined by consulting physicians or biologists. We resolved this issue by addressing optimality directly relative to the long-run behavior of the underlying network and phenotypes of interest. We utilized a linear programming approach to optimally move the steady-state mass of undesirable states toward desirable states. We provided a simulation study on synthetic PBNs with different dynamical properties and a metastatic melanoma network, where we demonstrated the superiority of optimal methods over several suboptimal approaches.

Besides the approaches presented here, many additional methods have been proposed: (1) to constrain the intervention so as to limit the expected number of interventions (Ching et al., 2009; Faryabi et al., 2008); (2) to deal with biological model uncertainties via robust intervention policies (Pal, Datta, & Dougherty, 2008, 2009; Yousefi & Dougherty, 2014); (3) to address unpredictable effects of a drug on multiple off-target genes (Yousefi & Ivanov, 2013); (4) to address asynchronicity of updating rules in gene networks (Faryabi et al., 2008); and (5) to mitigate the computational burden from complex gene networks (Akutsu, Hayashida, Ching, & Ng, 2007; Faryabi, Datta, & Dougherty, 2007; Ivanov, Simeonov, Ghaffari, Qian, & Dougherty, 2010; Ng, Zhang, Ching, & Akutsu, 2006; Vahedi et al., 2008).

Although the applicability of many of the discussed approaches may appear to hinge on first demonstrating the validity of PBNs as a model for GRNs, these approaches are not restricted to PBNs. In fact, the main assumption underlying most of these approaches is that the dynamics of GRNs can be well-approximated by Markovian stochastic processes. This assumption is not restrictive once we realize that the class of Markovian models encompasses a large body of both deterministic and stochastic dynamical models of GRNs (Kim et al., 2002; Goutsias & Jenkinson, 2013). As long as we can characterize the joint probability distribution of the stochastic process of interest and have a well-defined objective function, with reasonable assumptions on the intervention model, we can derive either an analytical or an approximate solution to various intervention problems. The question now is whether inferring the joint probability distribution is possible, both computationally and experimentally. To answer this question, we must first answer two even more fundamental questions: (1) what assumptions are we willing to impose on the form of this joint probability distribution? and (2) what types of data, and how much, do we need to collect to achieve an accuracy, in terms of model prediction error, within a desirable range? Nonetheless, it is this author's belief that methods for optimal or approximate intervention will play an important role in the near future when the medical community is ready to adopt model-based design of intervention strategies for cancer treatment.

REFERENCES

Akutsu, T., Hayashida, M., Ching, W.-K., & Ng, M. K. (2007). Control of Boolean networks: Hardness results and algorithms for tree structured networks. *Journal of Theoretical Biology*, *244*(4), 670–679. doi:10.1016/j.jtbi.2006.09.023 PMID:17069859

Altman, E. (1999). *Constrained Markov Decision Processes* (Vol. 7). London, U.K.: Chapman & Hall/CRC.

Bertsekas, D. P. (2005). *Dynamic Programming and Optimal Control*. Belmont, MA: Athena Scientific.

Bittner, M., Meltzer, P., Chen, Y., Jiang, Y., Seftor, E., Hendrix, M., & Trent, J. et al. (2000). Molecular classification of cutaneous malignant melanoma by gene expression profiling. *Nature*, *406*(6795), 536–540. doi:10.1038/35020115 PMID:10952317

Brun, M., Dougherty, E. R., & Shmulevich, I. (2005). Steady-state probabilities for attractors in probabilistic Boolean networks. *Signal Processing*, *85*(10), 1993–2013. doi:10.1016/j.sigpro.2005.02.016

Brunton, L. L., Chabner, B., & Knollmann, B. C. (2010). *Goodman & Gilman's The Pharmacological Basis of Therapeutics, 12e*. New York, NY: McGraw-Hill Professional.

Cao, Y., & Samuels, D. C. (2009). Discrete stochastic simulation methods for chemically reacting systems. *Methods in Enzymology*, *454*, 115–140. doi:10.1016/S0076-6879(08)03805-6 PMID:19216925

Chaouiya, C., Remy, E., & Thieffry, D. (2008). Petri net modelling of biological regulatory networks. *Journal of Discrete Algorithms*, *6*(2), 165–177. doi:10.1016/j.jda.2007.06.003

Chen, K.-C., Wang, T.-Y., Tseng, H.-H., Huang, C.-Y. F., & Kao, C.-Y. (2005). A stochastic differential equation model for quantifying transcriptional regulatory network in Saccharomyces cerevisiae. *Bioinformatics (Oxford, England)*, *21*(12), 2883–2890. doi:10.1093/bioinformatics/bti415 PMID:15802287

Ching, W. K., Zhang, S. Q., Jiao, Y., Akutsu, T., Tsing, N. K., & Wong, A. S. (2009). Optimal control policy for probabilistic Boolean networks with hard constraints. *IET Systems Biology*, *3*(2), 90–99. doi:10.1049/iet-syb.2008.0120 PMID:19292563

Cojocaru, L., & Agur, Z. (1992). A theoretical analysis of interval drug dosing for cell-cycle-phase-specific drugs. *Mathematical Biosciences*, *109*(1), 85–97. doi:10.1016/0025-5564(92)90053-Y PMID:1591451

Datta, A., Choudhary, A., Bittner, M. L., & Dougherty, E. R. (2003). External Control in Markovian Genetic Regulatory Networks. *Machine Learning*, *52*(1), 169–191. doi:10.1023/A:1023909812213

Datta, A., & Dougherty, E. R. (2006). *Introduction to Genomic Signal Processing with Control*. Boca Raton, FL: CRC. doi:10.1201/9781420006674

Dayneka, N. L., Garg, V., & Jusko, W. J. (1993). Comparison of four basic models of indirect pharmacodynamic responses. *Journal of Pharmacokinetics and Pharmacodynamics*, *21*(4), 457–478. doi:10.1007/BF01061691 PMID:8133465

De Jong, H. (2002). Modeling and simulation of genetic regulatory systems: A literature review. *Journal of Computational Biology*, *9*(1), 67–103. doi:10.1089/10665270252833208 PMID:11911796

de Vladar, H. P., & González, J. A. (2004). Dynamic response of cancer under the influence of immunological activity and therapy. *Journal of Theoretical Biology*, *227*(3), 335–348. doi:10.1016/j.jtbi.2003.11.012 PMID:15019501

Derman, C. (1970). *Finite State Markovian Decision Processes* (Vol. 107). Orlando, FL: Academic Press, Inc.

Dissanayake, S. K., Olkhanud, P. B., O'Connell, M. P., Carter, A., French, A. D., Camilli, T. C., & Weeraratna, A. T. et al. (2008). Wnt5A regulates expression of tumor-associated antigens in melanoma via changes in signal transducers and activators of transcription 3 phosphorylation. *Cancer Research*, *68*(24), 10205–10214. doi:10.1158/0008-5472.CAN-08-2149 PMID:19074888

Dougherty, E. R., & Datta, A. (2005). Genomic signal processing: Diagnosis and therapy. *IEEE Signal Processing Magazine*, *22*(1), 107–112. doi:10.1109/MSP.2005.1407722

Dougherty, E. R., Pal, R., Qian, X., Bittner, M. L., & Datta, A. (2010). Stationary and structural control in gene regulatory networks: Basic concepts. *International Journal of Systems Science*, *41*(1), 5–16. doi:10.1080/00207720903144560

Dougherty, E. R., & Shmulevich, I. (2012). On the Limitations of Biological Knowledge. *Current Genomics*, *13*(7), 574–587. doi:10.2174/138920212803251445 PMID:23633917

El Samad, H., Khammash, M., Petzold, L., & Gillespie, D. (2005). Stochastic modelling of gene regulatory networks. *International Journal of Robust and Nonlinear Control*, *15*(15), 691–711. doi:10.1002/rnc.1018

Faryabi, B., Chamberland, J.-F., Vahedi, G., Datta, A., & Dougherty, E. R. (2008). Optimal intervention in asynchronous genetic regulatory networks. *IEEE Journal of Selected Topics in Signal Processing*, *2*(3), 412–423. doi:10.1109/JSTSP.2008.923853

Faryabi, B., Datta, A., & Dougherty, E. R. (2007). On approximate stochastic control in genetic regulatory networks. *IET Systems Biology*, *1*(6), 361–368. doi:10.1049/iet-syb:20070015 PMID:18203582

Faryabi, B., Vahedi, G., Chamberland, J.-F., Datta, A., & Dougherty, E. R. (2008). Optimal Constrained Stationary Intervention in Gene Regulatory Networks. *EURASIP Journal on Bioinformatics & Systems Biology*, (1): 620767. PMID:18516276

Faure, A., Naldi, A., Chaouiya, C., & Thieffry, D. (2006). Dynamical analysis of a generic Boolean model for the control of the mammalian cell cycle. *Bioinformatics (Oxford, England)*, *22*(14), e124–e131. doi:10.1093/bioinformatics/btl210 PMID:16873462

Friedman, N. (2004). Inferring cellular networks using probabilistic graphical models. *Science Signaling*, *303*(5659), 799–805. PMID:14764868

Giacinti, C., & Giordano, A. (2006). RB and cell cycle progression. *Oncogene*, *25*(38), 5220–5227. doi:10.1038/sj.onc.1209615 PMID:16936740

Goutsias, J., & Jenkinson, G. (2013). Markovian dynamics on complex reaction networks. *Physics Reports*, *529*(2), 199–264. doi:10.1016/j.physrep.2013.03.004

Heppner, G. H. (1984). Tumor Heterogeneity. *Cancer Research*, *44*(6), 2259–2265. PMID:6372991

Himmelstein, K. J., & Bischoff, K. B. (1973). Mathematical representations of cancer chemotherapy effects. *Journal of Pharmacokinetics and Pharmacodynamics*, *1*(1), 51–68. doi:10.1007/BF01060027

Huang, Y., Tienda-Luna, I., & Wang, Y. (2009). Reverse engineering gene regulatory networks. *IEEE Signal Processing Magazine*, *26*(1), 76–97. doi:10.1109/MSP.2008.930647 PMID:20046885

Iglesias, P. A., & Ingalls, B. P. (2010). *Control Theory and Systems Biology*. MIT Press.

Ivanov, I., Simeonov, P., Ghaffari, N., Qian, X., & Dougherty, E. R. (2010). Selection policy-induced reduction mappings for Boolean networks. *IEEE Transactions on Signal Processing*, *58*(9), 4871–4882. doi:10.1109/TSP.2010.2050314

Julius, A. A., Halasz, A., Sakar, M. S., Rubin, H., Kumar, V., & Pappas, G. J. (2008). Stochastic Modeling and Control of Biological Systems: The Lactose Regulation System of Escherichia Coli. *IEEE Transactions on Automatic Control*, *53*(Special Issue), 51–65. doi:10.1109/TAC.2007.911346

Kauffman, S. A. (1969). Metabolic stability and epigenesis in randomly constructed genetic nets. *Journal of Theoretical Biology*, *22*(3), 437–467. doi:10.1016/0022-5193(69)90015-0 PMID:5803332

Keméṅy, J. G., & Snell, J. L. (1960). *Finite Markov Chains*. Princeton, NJ: Van Nostrand.

Kim, S., Li, H., Dougherty, E. R., Cao, N., Chen, Y., Bittner, M., & Suh, E. B. (2002). Can Markov chain models mimic biological regulation? *Journal of Biological System*, *10*(04), 337–357. doi:10.1142/S0218339002000676

Kyoda, K., Baba, K., Onami, S., & Kitano, H. (2004). DBRF–MEGN method: An algorithm for deducing minimum equivalent gene networks from large-scale gene expression profiles of gene deletion mutants. *Bioinformatics (Oxford, England)*, *20*(16), 2662–2675. doi:10.1093/bioinformatics/bth306 PMID:15166016

Magni, P., Germani, M., De Nicolao, G., Bianchini, G., Simeoni, M., Poggesi, I., & Rocchetti, M. (2008). A Minimal Model of Tumor Growth Inhibition. *IEEE Transactions on Bio-Medical Engineering*, *55*(12), 2683–2690. doi:10.1109/TBME.2008.913420 PMID:19126447

Magni, P., Simeoni, M., Poggesi, I., Rocchetti, M., & De Nicolao, G. (2006). A mathematical model to study the effects of drugs administration on tumor growth dynamics. *Mathematical Biosciences*, *200*(2), 127–151. doi:10.1016/j.mbs.2005.12.028 PMID:16516246

Matsuno, H., Doi, A., Nagasaki, M., & Miyano, S. (2000). Hybrid Petri net representation of gene regulatory network.*Pacific Symposium on Biocomputing. Pacific Symposium on Biocomputing*, (pp. 341–352).

Ng, M., Zhang, S.-Q., Ching, W.-K., & Akutsu, T. (2006). A Control Model for Markovian Genetic Regulatory Networks. *Transactions on Computational Systems Biology V*, *4070*, 36–48. doi:10.1007/11790105_4

Pal, R., Datta, A., & Dougherty, E. R. (2006). Optimal infinite-horizon control for probabilistic Boolean networks. *IEEE Transactions on Signal Processing*, *54*(6), 2375–2387. doi:10.1109/TSP.2006.873740

Pal, R., Datta, A., & Dougherty, E. R. (2008). Robust intervention in probabilistic Boolean networks. *IEEE Transactions on Signal Processing*, *56*(3), 1280–1294. doi:10.1109/TSP.2007.908964

Pal, R., Datta, A., & Dougherty, E. R. (2009). Bayesian Robustness in the Control of Gene Regulatory Networks. *IEEE Transactions on Signal Processing*, *57*(9), 3667–3678. doi:10.1109/TSP.2009.2022872

Pe'er, D., & Hacohen, N. (2011). Principles and Strategies for Developing Network Models in Cancer. *Cell, 144*(6), 864–873. doi:10.1016/j.cell.2011.03.001 PMID:21414479

Qian, X., & Dougherty, E. R. (2008). Effect of Function Perturbation on the Steady-State Distribution of Genetic Regulatory Networks: Optimal Structural Intervention. *IEEE Transactions on Signal Processing*, *56*(10), 4966–4976. doi:10.1109/TSP.2008.928089

Qian, X., & Dougherty, E. R. (2012). Intervention in Gene Regulatory Networks via Phenotypically Constrained Control Policies Based on Long-Run Behavior. *IEEE/ACM Transactions on Computational Biology and Bioinformatics*, *9*(1), 123–136. doi:10.1109/TCBB.2011.107 PMID:21788674

Qian, X., Ivanov, I., Ghaffari, N., & Dougherty, E. R. (2009). Intervention in gene regulatory networks via greedy control policies based on long-run behavior. *BMC Systems Biology*, *3*(1), 61. doi:10.1186/1752-0509-3-61 PMID:19527511

Queinnec, I., Garcia, G., Niculescu, S.-I., & Tarbouriech, S. (2007). *Biology and control theory: current challenges* (Vol. 357). Springer. doi:10.1007/978-3-540-71988-5

Ross, K. W. (1989). Randomized and past-dependent policies for Markov decision processes with multiple constraints. *Operations Research*, *37*(3), 474–477. doi:10.1287/opre.37.3.474

Shmulevich, I., & Aitchison, J. D. (2009). Deterministic and stochastic models of genetic regulatory networks. *Methods in Enzymology*, *467*, 335–356. doi:10.1016/S0076-6879(09)67013-0 PMID:19897099

Shmulevich, I., Dougherty, E. R., Kim, S., & Zhang, W. (2002). Probabilistic Boolean networks: A rule-based uncertainty model for gene regulatory networks. *Bioinformatics (Oxford, England)*, *18*(2), 261–274. doi:10.1093/bioinformatics/18.2.261 PMID:11847074

Shmulevich, I., Dougherty, E. R., & Zhang, W. (2002). Control of stationary behavior in probabilistic Boolean networks by means of structural intervention. *Journal of Biological System*, *10*(04), 431–445. doi:10.1142/S0218339002000706

Shmulevich, I., Dougherty, E. R., & Zhang, W. (2002). Gene perturbation and intervention in probabilistic Boolean networks. *Bioinformatics (Oxford, England)*, *18*(10), 1319–1331. doi:10.1093/bioinformatics/18.10.1319 PMID:12376376

Simon, R., & Norton, L. (2006). The Norton-Simon hypothesis: Designing more effective and less toxic chemotherapeutic regimens. *Nature Clinical Practice. Oncology, 3*(8), 406–407. doi:10.1038/ncponc0560 PMID:16894366

Vahedi, G., Faryabi, B., Chamberland, J.-F., Datta, A., & Dougherty, E. R. (2008). Intervention in gene regulatory networks via a stationary mean-first-passage-time control policy. *IEEE Transactions on Bio-Medical Engineering, 55*(10), 2319–2331. doi:10.1109/TBME.2008.925677 PMID:18838357

Vahedi, G., Faryabi, B., Chamberland, J.-F., Datta, A., & Dougherty, E. R. (2009). Optimal intervention strategies for cyclic therapeutic methods. *IEEE Transactions on Bio-Medical Engineering, 56*(2), 281–291. doi:10.1109/TBME.2008.2003092 PMID:19272947

Weeraratna, A. T., Jiang, Y., Hostetter, G., Rosenblatt, K., Duray, P., Bittner, M., & Trent, J. M. (2002). Wnt5a signaling directly affects cell motility and invasion of metastatic melanoma. *Cancer Cell, 1*(3), 279–288. doi:10.1016/S1535-6108(02)00045-4 PMID:12086864

White, D. J. (1978). *Finite Dynamic Programming: An Approach to Finite Markov Decision Processes*. New York, NY: John Wiley & Sons, Inc.

Xiao, Y., & Dougherty, E. R. (2007). The impact of function perturbations in Boolean networks. *Bioinformatics (Oxford, England), 23*(10), 1265–1273. doi:10.1093/bioinformatics/btm093 PMID:17379691

Yousefi, M. R., Datta, A., & Dougherty, E. R. (2012). Optimal Intervention Strategies for Therapeutic Methods With Fixed-Length Duration of Drug Effectiveness. *IEEE Transactions on Signal Processing, 60*(9), 4930–4944. doi:10.1109/TSP.2012.2202114

Yousefi, M. R., Datta, A., & Dougherty, E. R. (2013). Optimal Intervention in Markovian Gene Regulatory Networks With Random-Length Therapeutic Response to Antitumor Drug. *IEEE Transactions on Bio-Medical Engineering, 60*(12), 3542–3552. doi:10.1109/TBME.2013.2272891 PMID:23864151

Yousefi, M. R., & Dougherty, E. R. (2013). Intervention in gene regulatory networks with maximal phenotype alteration. *Bioinformatics (Oxford, England), 29*(14), 1758–1767. doi:10.1093/bioinformatics/btt242 PMID:23630177

Yousefi, M. R., & Dougherty, E. R. (2014). A comparison study of optimal and suboptimal intervention policies for gene regulatory networks in the presence of uncertainty. *EURASIP Journal on Bioinformatics & Systems Biology*, (6): 1–13. PMID:24708650

Yousefi, M. R., & Ivanov, I. (2013). Optimal control of gene regulatory networks with uncertain intervention effects. In *Global Conference on Signal and Information Processing (GlobalSIP)*, (pp. 105–108). IEEE. doi:10.1109/GlobalSIP.2013.6736826

Zañudo, J. G. T., & Albert, R. (2015). Cell Fate Reprogramming by Control of Intracellular Network Dynamics. *PLoS Computational Biology*, *11*(4), e1004193. doi:10.1371/journal.pcbi.1004193 PMID:25849586

Chapter 6
Why Are Multiple Regulators Required for Transcription of Each Gene?

Hiroshi Kobayashi
Chiba University, Japan

ABSTRACT

The human genome has been proposed to contain numerous regions called "consensus sequences" that are recognized by DNA-binding proteins required for gene expression. Results obtained with a computer simulation showed that: (1) all random sequences consisting of less than 15 base pairs were present in the human genome, and (2) consensus sequences reported to date were found in an artificial genome created randomly by a computer. These results suggest that conserved sequences consisting of more than 15 base pairs are required for accurate gene expression. No consensus sequence consisting of more than 15 base pairs has been identified to date. Thus, the co-association of multiple proteins bound side by side is required for appropriate gene expression because the total number of conserved sequences can exceed 15, whereas the binding of a single protein has no physiological role because its consensus sequence is present randomly in the human genome.

DOI: 10.4018/978-1-5225-0353-8.ch006

INTRODUCTION

Reports of various regulatory networks for gene expression are now accumulating in mammals, demonstrating that branched complex networks are present. It has been suggested that mammals have a large number of DNA-binding proteins for gene expression. In contrast, one regulatory protein is generally involved in individual gene expression, and such expression is often mediated by only RNA polymerase without other DNA-binding proteins in bacteria. The basic question is why multiple DNA-binding proteins are required for the regulation of single gene expression in mammals. One explanation may be that mammals have more complex functions than bacteria. However, the functions operating in a single mammalian cell are less complex than those in a single bacterial cell, although the functions in the whole body are highly complex in total. It is indispensable for our improved understanding of gene-regulating networks to clarify why multiple regulators are required for each gene expression in mammalian cells.

The human genome has been almost completely analyzed (International Human Genome Sequencing Consortium, 2004), and numerous mechanisms for transcriptional regulation have been proposed (Abnizova & Gilks, 2006; Barrett, Fletcher, & Wilton, 2012). In the first step of transcription, various regulatory proteins are bound to their specific sites with a nucleotide sequence called a "consensus sequence". Experimental data have demonstrated that a given regulatory protein recognizes its consensus sequence to regulate transcription. The bioinformatics of mammalian genomes suggest numerous sites for the binding of regulatory proteins which cooperatively regulate gene expression (Abnizova, Subhankulova, & Gilks, 2007; Barrett, Fletcher, & Wilton, 2012).

The human genome has approximately 3 billion base pairs, and many consensus sequences reported to date have less than 16 base pairs (Barrett, Fletcher, & Wilton, 2012). Theoretically, a sequence consisting of less than 16 base pairs is present randomly in the human genome if its sequence is close to random. In contrast, if the human genome is not random, consensus sequences will be conserved only at sites required for gene expression. It is therefore indispensable to clarify whether or not the human genome is close to random.

To clarify this, the following were theoretically assessed using a computer in the present study: (1) whether or not sequences consisting of 11 to 16 base pairs created randomly are present in the human genome; (2) whether or not consensus sequences reported to date can be found in an artificial genome created by random numbers using a computer. The results suggested that the human genome has areas whose sequences are close to random and the areas account for more than 2.7x10^8 (4^14) base pairs.

In molecular biology, theoretical examinations have been few to date. Theoretical verification of biological phenomena is indispensable for our improved understanding of the regulation of gene expression to complement arguments based on experimental studies. In this chapter, we first discuss why multiple regulators are required for the expression of a single gene, then, we focus on the benefits of having multiple regulators, and finally, on the question why mammalian cells have complex systems for signal pathways. The discussion is based on theoretical results obtained with a computer simulation.

COMPUTER SIMULAION ANALYSIS OF CONSENSUS SEQUENCES

Random Sequences Existing in the Human Genome

It is accepted that areas participating in gene expression have been conserved, but other non-coding regions without a specific function may be randomized. Theoretically, any given sequence consisting of less than 16 base pairs can be present in a random sequence consisting of 1,073,741,824 (4^15) base pairs or more (Table 1). The human genome has approximately 3,000,000,000 base pairs. This means that the human genome contains any random sequences consisting of less than 16 base pairs if the sequence of the human genome is random.

In the first simulation, artificial sequences were constructed using random numbers generated by a computer, as described in Appendix 1, and it was examined

Table 1. Probability of a random sequence

Number of Base Pairs	Probability	Average Number Per 3x10^9 Base Pairs
7	6.10x10^-5 (1/4^7)	1.83x10^5
8	1.53x10^-5 (1/4^8)	4.58x10^4
9	3.81x10^-6 (1/4^9)	1.14x10^4
10	9.54x10^-7 (1/4^10)	2.86x10^3
11	2.38x10^-7 (1/4^11)	715
12	5.96x10^-8 (1/4^12)	179
13	1.49x10^-8 (1/4^13)	44.7
14	3.73x10^-9 (1/4^14)	11.2
15	9.31x10^-10 (1/4^15)	2.79
16	2.33x10^-10 (1/4^16)	0.698

whether or not the human genome contained such artificial random sequences, as described in Appendix 2. Three random sequences consisting of 15 base pairs were found one or more times in the human genome among 5 random sequences tested (Table 2). Additional 15 random sequences consisting of 15 base pairs were examined, and 8 sequences were present in the human genome (data not shown). Thus, the human genome contains approximately 50% random sequences consisting of 15 base pairs tested. All random sequences consisting of less than 15 base pairs tested were present in the human genome (Table 2).

Some artificial sequences were present at a high frequency in the human genome (indicated by asterisks in Table 2). These sequences may contain regions which are important for cell functions. For example, the No.6 sequence contains CAGG at the 3' end, and this sequence is required for splicing (Lou, Yang, Cote, Berget, & Gagel, 1995). The No.13 sequence contains AAA at the 5' end, and this is a part of the binding sites of TBP and E2F-4 (Table 3). The No. 20 and 25 sequences may contain a part of the sequences that are essential for unidentified cellular functions. Some sequences (No. 2, 7, 9, 11, 12, 15, and 17) contained translational stop codons presented by underlining (Table 2), and this may be the reason for their low frequency in the human genome. These results suggest that the human genome contains huge areas whose sequences are close to random and therefore contains any sequences consisting of less than 15 base pairs.

Consensus Sequences in Artificial Genomes Created by a Computer

Next, it was examined whether or not the consensus sequences proposed experimentally were present in an artificial genome consisting of 3 billion base pairs created using random numbers with a computer, as described in Appendix 2. All consensus sequences tested were found in the artificial genome (Table 3), and the numbers of consensus sequences presented in the artificial genome were close to those calculated using mathematical probability, shown in Table 1, suggesting that the consensus sequences reported to date are present in the areas whose sequences are close to random besides conserved sites. The same results were obtained even if a non-fixed base pair represented by N was inserted in the middle of the consensus sequence when the number of N was fixed (Table 3).

The frequencies of consensus sequences in the human genome were of the same order as those in the artificial genome except for EBS-1 and E2F-4 (Table 3). The human genome was reported to have TGACTCA, the binding site of AP-1, at a frequency of 8.08x10^-5 (Zhou et al., 2005), and this value corresponds to 2.4x10^5

Table 2. Random sequences in the human genome

Sequence Number	Random Sequence	Number of Base Pairs	Number in the Human Genome
1	AGACCTGCACG	11	376
2	GCGAAAATTAT	11	198
3	TTTGCCCGTTG	11	273
4	GCAAGGCGATA	11	83
5	AATGTTCCTCG	11	255
6	CAGCACTGCAGG	12	925*
7	CATCACAGCGGT	12	58
8	CTCCGGAAACCA	12	105
9	TAAGCTACGGCG	12	4
10	GGCGGAAAGCAG	12	116
11	ACGTGTGACCCTC	13	16
12	ACACTAGTTGTGT	13	58
13	AAAGATAAAGTAA	13	687*
14	TTTGGCTGTTGAC	13	110
15	GGGATTTAGCGAC	13	5
16	CTCAGCCCTACTAT	14	16
17	ACAAAGTTTAGCGC	14	1
18	CACCACGAGGAGTG	14	7
19	ACAAGTACCACATC	14	13
20	AAGAGAATCTAGTA	14	34*
21	CTCGGGATCCGACCT	15	0
22	GCCGACGCGTGGCCC	15	1
23	GTTAGCTCGATCCCT	15	0
24	AGGAGATGAGAGCCG	15	1
25	CACATATTCCCATTG	15	40*
26	TGTGCGAAATCATCCA	16	0
27	CGCGCCTGCCGCTCCT	16	0
28	CCTACAAACCCGTTGG	16	0
29	TCTGCTCTTAGATACA	16	1
30	CAAAGTCTCAGTCCCG	16	2

Table 3. Consensus sequences in an artificial genome

Regulatory Proteins	Consensus Sequence	Length	Number of Sequences	
			Artificial Genome	Human Genome
TBP (Maston, et al., 2006)	TATATAAA	8	(4.57±0.02)x10^4	NC
NF-κB (Hilton & Wang, 2003)	GGGGAGTTT	9	(1.14±0.01)x10^4	2.41x10^4
EBS-1 (Nair et al., 1997)	CGCCTACGC	9	(1.15±0.01)x10^4	5.67x10^2
IEF-1 (Ohlsson et al., 1988)	GCCAATCTG	9	(1.14±0.01)x10^4	1.63x10^4
SP-1 (Kim et al., 1992)	CCCCCACGCC	10	(2.84±0.04)x10^3	4.58x10^3
p53 (Shiio et al., 1992)	GGAAGTGACG	10	(2.88±0.06)x10^3	1.62x10^3
AP-1 (Lim & Garzino-Demo, 2000)	CTTGACTCCGC	11	728±30	127
E2F-4 (Goodet al., 1996)	TTTCGCGCCAAA	12	179±10	13
CREB (Spiro et al., 1995)	TGGCGTANNNNNTGCGTCA	14*	10.2±5.5	NC
Artificial	TGGCGTANNTGCGTCA	14*	11.2±2.5	NC
Artificial	TGGCGTATGCGTCA	14	10.2±2.2	NC
HOX4 (Zappavigna et al., 1991)	GACACATTAATCTAT	15	2.2±1.8	6

NC, not counted. *, N (any nucleotide) is not counted.

copies, being slightly higher than that calculated using mathematical probability (Table 1). This frequency was almost the same in all 24 chromosomes (Zhou et al., 2005).

The Human Genome May Have Areas Whose Sequences Are Close to Random

The results outlined in the previous section suggest that the human genome has areas whose sequences are close to random and those areas account for more than 2.7x10^8 (4^14) base pairs in total because all random sequences consisting of less than 15 base pairs were found in the human genome. Introns in the human genome account for 24% of the total genome, and 75% of it lies in intergenic DNA, whereas the region of exons is only 1.1% (Venter et al., 2001). A large part of the introns and intergenic regions may have been randomized during the evolution of human beings except for promoters and regulatory areas for gene expression. However, it has been reported that introns have some functions (Chorev & Carmel, 2012). If so, a part of introns may be conserved. Although a part of introns and intergenic regions are conserved besides exons, a large part of these regions may be close to random. Consequently, it is possible that any sequences consisting of less than 15 base pairs are present in the areas whose sequences are close to random, and the possibility increases as the number of base pairs decreases.

This is very important, because many questions raised in experimental studies reported to date can be reconsidered based on it. The following represent a list of such the questions:

1. There are many consensus sequences in the upstream region of the transcriptional initiation site (Hilton & Wang, 2003). Why is the binding of multiple regulatory proteins required?
2. Consensus sequences are often found in areas distant from the transcriptional initiation sites, sometimes over 1,000 base pairs away (Zhang et al., 2015). How does such a region far removed from the initiation sites regulate transcription?
3. Consensus sequences are found in the downstream region of the transcriptional termination site. Since experimental studies have often shown that binding to such sites stimulates gene expression, such a region is named an "enhancer" (Bhatia et al., 2013; Heinz et al., 2015), but the mechanism for enhancement is still unclear, although various mechanisms have been proposed. Does the binding really enhance transcription in vivo?
4. The consensus sequences are often found within an intron (Xue et al., 2004; Kesanakurti, et al., 2013). What is the physiological significance of consensus sequences in an intron?

PHYSIOLOGICAL FUNCTIONS OF CONSENSUS SEQUENCES

Consensus Sequences with Physiological Roles

Consensus sequences with physiological roles may be conserved at specific sites, but not be at random sites. What kinds of sequence are functional? The present simulation results suggest that functional sequences should consist of more than 15 base pairs. However, consensus sequences reported to date consist of less than 15 base pairs.

Multiple binding sites for regulatory proteins have been found in the upstream areas of coding regions (Dunham et al., 2012), and the sites are generally located side by side (for example, see Hilton & Wang, 2003). If multiple binding sites are present side by side and bound proteins associate with each other, the region will contain a conserved sequence consisting of more than 15 base pairs in total that can regulate the gene expression appropriately because such a sequence is not present randomly in the human genome. Consensus sequences do not need to adjoin in the absence of spaces and the number of base pairs between the consensus sequences should be fixed for the interaction of bound proteins. In fact, CCCCCACGCCNNNNNGGGAGTTT (N is any base pair) containing the binding sites of SP-1 and NF-κB was not found in the artificial genomes tested. The negative result was obtained again if the number of N was set at zero or 10. Thus, it can be argued that gene expression is regulated by the co-association of regulatory factors bound side by side, as proposed previously (Dunham et al., 2012).

The conserved sequence consisting of more than 15 base pairs is a minimum requirement for appropriate gene expression. In many cases, the gene expression is stimulated or inhibited by various factors, such as growth factors, cytokines, hormones, and external environmental factors. Cells may have specific regulatory proteins for regulation by these factors. The consensus sequences participating in the regulation of gene expression should be located side by side.

Consensus Sequences with No Physiological Role

Any consensus sequences consisting of less than 15 base pairs are suggested to be present in the areas whose sequences are close to random, and such consensus sequences located in the areas may have no physiological role. It has already been reported that there are non-functional consensus sequences (Daenen, van Roy, & De Bleser, 2008). Since genes whose function has been clarified number less than 30% of the total genes, it is still possible that unidentified consensus sequences are present next to identified ones. Except for this case, consensus sequences that are present alone are argued to be non-functional.

RNA Polymerase Does Not Work Without the Binding of Additional Regulatory Protein(s)

Transcription is initiated by the binding of RNA polymerase II with associated proteins, TFIIA, TFIIB, TFIID, TFIIE, TFIIF, and TFIIH. The binding site of the RNA polymerase II complex is called a "promoter", which consists of two sites, BRE and TATA box. The former and latter are the recognition sites of TFIIB and TFIID, respectively (Stephen & Kadonaga, 2003). The consensus sequences of BRE and TATA box are (G/C)(G/C)(G/A)CGCC and TATA(A/T)AA(G/A), respectively. The probability they are present is 2.98x10^-8 (1/2^3x4^4x4^6x2^2), which corresponds to 89.4 per 3x10^9 base pairs. Promoter-enhancing TATA box polymorphisms have been reported (Savinkova et al., 2009). Therefore, it is possible that the consensus sequence of the polymerase II complex is present in the areas whose sequences are close to random besides the promoter.

How is appropriate transcriptional initiation attained? When the binding site(s) of another DNA-binding protein(s) is located by the RNA polymerase-binding site, the region contains a conserved sequence of more than 15 base pairs in total, and the region can initiate transcription correctly because such a sequence is not present randomly in the human genome, consisting of 3 billion base pairs. Thus, the binding of the RNA polymerase II complex to the promoter with the co-binding of other proteins is essential for the high fidelity of transcriptional initiation.

Experimental analysis has revealed that the functional binding sites are located closer to transcriptional start sites (Whitfield et al., 2012). The association of CBP and USF2 with the LCR on the β-globin gene locus mediates recruitment of RNA polymerase II to the β-globin gene locus for expression at a high level (Rosenberg, Fan, Lin, Liang, & Bungert, 2013). These experimental data support the present assumption.

The Binding of Proteins to Non-Functional Consensus Sequences

Regulatory proteins can be equally bound to their consensus sequences, even if the sites are non-functional. When the binding of proteins occurs at non-functional sites in exons and introns, transcriptional elongation may be interrupted. When regulatory proteins to enhance gene expression are bound to non-functional sites in intergenic regions, the amount of free regulatory proteins that have to work at physiological sites decreases, resulting in the attenuation of gene expression.

Therefore, the binding of regulatory proteins to non-functional sites should be suppressed. It has been proposed that the dissociation of histones from the regulatory regions induced by histone modification promotes gene expression (Alexandrov et al., 2012; von Hippel, Johnson, & Marcus, 2013). The histones may prevent the binding of regulatory proteins to non-functional sites. Histones are generally modified simultaneously in the wide range of chromosomes. The modification may be dependent on cell cycle phases (Mews et al., 2014) or cell differentiation (Nair et al., 2014). Both induce changes in expression of a large number of genes. It is less likely that gene-specific histone modification occurs in a small number of genes.

An alternative possibility is that the binding affinity of regulatory proteins to their binding site is too low to form a stable complex, and consequently the binding of regulatory proteins to non-functional sites is minimized.

The Binding Affinity of Regulatory Proteins for Gene Expression

To confirm the above assumption, it would be essential to measure the binding kinetics of DNA-binding proteins to DNA. The dissociation constants of SPY-, HOX-, and NF-κB-DNA complexes were 57 nM (Sánchez-Moreno, Coral-Vázquez, Méndez, & Canto, 2008), 2.2 nM (Zhang, Larsen, Stadler, & Ames, 2011), and 2.7-8.2 nM (Alverdi, Hetrick, Joseph, & Komives, 2014), respectively. RFXANK, one subunit of RFX, was reported to be bound to X-box DNA at a dissociation constant of 200-850 nM that was dependent on the NaCl concentration and temperature (Chakraborty, Sengupta, Bhattacharya, Banerjee, & Chakrabarti, 2010). There are reports on ligand-receptor complex formation on the cell surface, and the dissociation constants have been reported to be 0.2 to 40 nM (Felder, LaVin, Ullrich, & Schlessinger, 1992; Wilkinson et al., 2001; Andrews, Holloway, Puddicombe, Holgate, & Davies, 2002).

Computer simulation studies showed that ligand-receptor complex formation requires an excess amount of receptors when the dissociation constant is higher than 1 nM (Kobayashi, Azuma, Konagaya, 2007a). In fact, cells have been reported to show an excess amount of receptors (Teramura et al., 2006). In the case of the binding of regulatory proteins to DNA, cells contain two copies of a given gene per cell. Therefore, some stabilizing mechanism is required for the binding of regulatory proteins to their functional consensus sequences.

Computer simulation results suggest that the ligand-receptor binary complex is not formed when the dissociation constant is higher than 1 nM (Kobayashi, Azuma, & Konagaya, 2007a). In contrast, the ligand-adaptor-receptor ternary complex can be formed when K_D (K_D1 x K_D2) is 0.1 nM or less (Kobayashi, Azuma, & Yasunaga, 2009), where K_D1 and K_D2 are dissociation constants of ligand-receptor and ligand-receptor-adaptor complexes, respectively. K_D is calculated to be 0.1 nM when K_D1 = K_D2 =10,000 nM. As described above, protein-DNA complex formation has similar kinetic parameters to ligand-receptor complex formation, and, hence, the kinetic data of ligand-receptor complex formation can be applicable to DNA-protein complex formation, supporting the idea that the ternary complex of multiple proteins and DNA is stable even if each dissociation constant (K_D1 and K_D2) is as high as 10,000 nM.

When the dissociation constant of the ternary complex (K_D1 x K_D2) is 0.1 nM or less, the dissociation of the ternary complex is very slow after the free ligand is removed (Kobayashi, Azuma, & Konagaya, 2007a). If the simulation results are applicable to protein-DNA complex formation, the dissociation of proteins from DNA has been argued to be very slow after free proteins are inactivated. One possible mechanism would be that the proteins bound to DNA are inactivated as well as the free proteins. The affinity of proteins for DNA is decreased by the inactivation, and the inactivated proteins promptly dissociate from the DNA. To the best of my knowledge, the inactivation of proteins bound to DNA has not been examined to date. Experimental studies to confirm this are essential.

Experimental Data against the Present Hypothesis

The present simulation results suggest that the binding of single DNA-binding protein has no physiological role. The experimental studies, however, demonstrated the binding of a single regulator without interaction with another protein (for example, see Zanotto et al., 2009). One possible explanation for the discrepancy is that the experimental evidence was obtained under conditions different from those present in normal physiology. The binding of proteins to their specific regions is dependent on the amounts of proteins and DNA. When an excess of proteins is used, binding to non-functional sites can be detected in vitro, even if the binding affinity is too low to form a stable complex with DNA in situ.

When the binding is at an equilibrium state between binding and dissociation at non-functional sites because of low affinity, the binding is temporal, resulting in low-level attenuation of transcriptional activity. Such low-level attenuation affects cell functions very little, because excess amounts of mRNA and proteins are generally synthesized.

In both cases, it is important to measure binding under conditions which are close to those in vivo. To the best of my knowledge, there has been no experimental report showing the binding of proteins to a single site under conditions which are similar to those in vivo, because the exact concentrations of DNA-binding proteins in the nucleus are difficult to measure.

BENEFITS OF GENE REGULATION BY MULTIPLE REGULATORS

Limitation of the Number of Nucleotides Recognized by a Single Protein

The next question is why no protein which recognizes more than 15 base pairs has evolved. Five nanometers (50Å) of the DNA-binding site is required for the binding of 15 base pairs of B-form DNA. Therefore, a protein with a binding pocket of more than 5 nm may be very large. Which is more advantageous, one big protein or a multimer complex of small proteins? The fact that no such large DNA-binding protein has been identified leads us to argue that a large protein has some disadvantages.

The possibility of making mistakes in transcription and translation may increase with a bigger protein. The fidelity of transcription mediated by RNA polymerase II has been poorly understood. An average error frequency of transcription by T7 RNA polymerase was estimated to be one in 2 x 10^4 (Huang, Brieba, & Sousa, 2000). Translational errors are caused mainly by the misacylation of a tRNA by its aminoacyl-tRNA synthetase and the selection of an incorrect tRNA during the elongation cycle, and an observed fidelity was one error in 10^3–10^4 polymerized amino acids. (Cochella & Green, 2005). It was reported that approximately 1% of methionine residues are aminoacylated to non-methionyl tRNAs in mammalian cells (Netzer et al., 2009). Thus, the error frequency during transcription and translation can be estimated to be one in several thousand codes. One thousand codes correspond to protein of approximately 100,000 daltons. This may be the reason why living organisms have few proteins over 100,000 daltons.

Regulation of Gene Expression by Heterologous Multimer is Advantageous

Which is better, a heterogeneous or homologous multimer? The data reported until now have suggested that both cases are feasible as clustering different consensus sequences and repeating the same consensus sequence have been found in the upstream region of the transcriptional initiation sites (Hilton & Wang, 2003). The binding of heterogeneous proteins to DNA has more advantage in gene regulation. Firstly, a common protein can be used for different gene expressions. Secondly, regulation by the binding of heterogeneous proteins may be advantageous for the complex regulation of gene expression, such as the induction of a given gene's expression by different stimuli and expression of multiple genes induced by a single stimulus.

COMPUTER ANALYSIS SUPPORTS COMPLEXITY OF SINGAL PATHWAYS

Why Are Complex Signal Pathways Required?

The necessity of multiple regulatory proteins for each gene expression has now emerged theoretically. In addition to minimum requirements obtained theoretically, many additional proteins may participate in the stimulation or inhibition of gene expression responding to various factors, such as growth factors, cytokines, hormones, and external environmental factors. A single external signal has often been reported to induce the expression of multiple genes. For example, a large number of proteins have to work to achieve cell proliferation, and the binding of a given ligand to its receptor initiates proliferation. Although it is possible that some proteins required for cell proliferation are already present in the resting cells, cells may have to initiate the transcription of a large number of genes after the binding of one growth factor to a receptor. Thus, a given ligand-receptor complex has to activate many kinds of DNA-binding protein.

It has been proposed that the signal transfer pathways consist of many signal transfer proteins between the receptor and DNA-binding proteins and that the pathways are generally branched, such as the MAP kinase pathway (Morrison, 2012, a011254), TLR pathway (O'Neill, Golenbock, & Bowie, 2013), and Akt pathway (Hemmings & Restuccia, 2012). Why is such a complex system required? It is clear that the ligand-receptor complex is unable to activate DNA-binding proteins directly, because the ligand-receptor complex is located on the cell surface and proteins regulating gene expression work in the nucleus. A possible mechanism is that one signal transfer protein travels to the nucleus after activation by a ligand-receptor complex on the cell surface, and then the activated signal protein activates all DNA-binding proteins participating in the expression of target genes in the nucleus, as illustrated in Figure 1a.

Why do mammalian cells not have such a system? Proteins are generally activated via the phosphorylation of amino acid residues mediated by enzymes called protein kinases. If a protein kinase can activate many proteins (Figure 1a), the catalytic center of such kinases can bind to many kinds of protein. In this case, the error frequency of binding to non-substrate proteins increases. For the fidelity of signal transduction, each protein kinase activates only one or a few proteins (Lee, Lin, Nam, Parang, & Sun, 2003). As illustrated in Figure 1b, even if one protein kinase activates only two proteins, 8 proteins can be activated after three steps. These may be the reasons why multiple steps are required.

Activation Level of Signal Proteins Required for Gene Expression

It has been reported that signal proteins are activated immediately after the binding of a ligand to its receptor, and the elevated level of activated signal proteins decreases to the steady state level without the release of ligands (Brightman & Fell, 2000; Yamada, Shiono, Joo, & Yoshimura, 2003; Kholodenko, 2006, pp.165-176). Computer simulation demonstrated that a high level of activation of signal proteins delays the signal cessation after removing the external ligand (Kobayashi, Azuma, & Yasunaga, 2009). The delay may be disadvantageous for cell functions. These data suggest that 100% activation of each protein is not necessary for signal transfer, while the efficiency of signal transfer and subsequent gene expression may be attenuated when the activated protein level is too low.

Figure 1. Models of signal pathways

L-R, ligand-receptor complex; A to C, signal transfer proteins; R1 to R8, DNA-binding proteins for regulation of gene expression. (a) Signal transfer protein A activates 8 proteins. (b) Each signal protein activates two proteins.

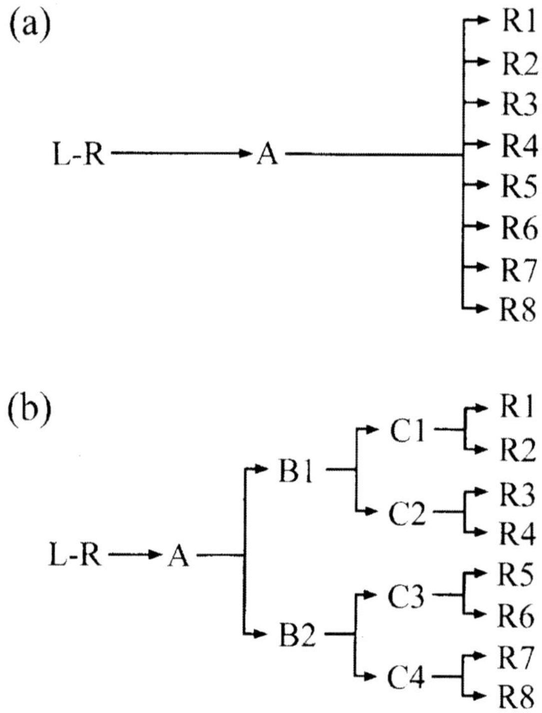

Many cross-talks have been reported (Morrison, 2012; O'Neill, Golenbock, & Bowie, 2013; Hemmings & Restuccia, 2012). Cross-talks are necessary for expression of the same gene due to different stimuli. It should be kept in mind that non-physiological cross-talk can be detected experimentally. For example, protein activation can be detected even if the activated protein level is less than that required for signal transduction.

Therefore, it is essential for our improved understanding of gene expression networks to clarify the activated protein level required for appropriate signal transfer. Even if the activation is detected experimentally, the level below the minimum requirement is not significant physiologically.

In this chapter, the regulation of gene expression has been discussed theoretically. Computer simulation is a useful method of theoretical examination. Many papers containing computer simulation results have already been published. In general, computer simulation analyses have been performed using special computer systems that are not familiar to general biologists. Our group has reported several computer simulation studies using a personal computer and sometimes a note-type computer (Kobayashi, 2003, pp.211-216; Kobayashi, Azuma, & Konagaya, 2007a; Kobayashi, Azuma, & Konagaya, 2007b; Kobayashi, Azuma, & Yasunaga, 2009). Platforms for programing are now available for a personal computer and are novice- friendly. The platform and source codes of the simulation program used in this study are appended in Appendix 3. The author is a biochemist and not a computer scientist, but some simulation analyses were carried out by himself. Computer simulation studies are now a common tool for general biologists. Source codes used in this study may be easily adapted to other simulation studies.

CONCLUSION

The simulation results presented in this chapter suggest that the human genome contains areas whose sequences are close to random, and any consensus sequences consisting of less than 15 base pairs are present in the areas. For the accurate regulation of gene expression, more than 15 base pairs should be conserved in the human genome. Big proteins which could recognize the wide range encompassing more than 15 base pairs may be undesirable because the transcriptional and translational errors increase in the big proteins, leading us to argue that the co-binding of multiple proteins in the upstream region of the transcriptional initiation site also participates in gene expression. In contrast, isolated consensus sequences may be non-functional, even if binding to such a sequence can be detected in vitro. The experimental studies on the functions of such consensus sequences are less significant physiologically, but the binding of a single protein to those consensus sequences needs further study. Future studies based on the assumption that each protein could perform its own role in gene expression in isolation should be avoided because the involvement of multiple proteins is essential for expression of any genes but not for complex regulation.

Only discussion presented in this chapter is based on simulation results, and suggests how future experimental studies that could promote our understanding of the regulation of gene expression could be designed. We claim that the most important experimental study aim is to clarify the number of DNA-binding proteins required for each gene expression and their binding affinity to DNA in vivo. The internal concentrations of the proteins are also critically important data. Future investigations of additional protein(s) are essential if only a single protein is currently identified for the specific gene expression. These studies are indispensable to avoid inaccurate results concerning gene expression regulation.

REFERENCES

Abnizova, I., & Gilks, W. R. (2006). Studying statistical properties of regulatory DNA sequences, and their use in predicting regulatory regions in the eukaryotic genomes. *Briefings in Bioinformatics*, *7*(1), 48–54. doi:10.1093/bib/bbk004 PMID:16761364

Abnizova, I., Subhankulova, T., & Gilks, W. (2007). Recent computational approaches to understand gene regulation: Mining gene regulation in silico. *Current Genetics*, *8*, 79–91. PMID:18660846

Alexandrov, B. S., Fukuyo, Y., Lange, M., Horikoshi, N., Gelev, V., Rasmussen, K. Ø., & Usheva, A. et al. (2012). DNA breathing dynamics distinguish binding from nonbinding consensus sites for transcription factor YY1 in cells. *Nucleic Acids Research*, *40*(20), 10116–10123. doi:10.1093/nar/gks758 PMID:22904068

Alverdi, V., Hetrick, B., Joseph, S., & Komives, E. A. (2014). Direct observation of a transient ternary complex during IκBα-mediated dissociation of NF-κB from DNA. *Proceedings of the National Academy of Sciences of the United States of America*, *111*(1), 225–230. doi:10.1073/pnas.1318115111 PMID:24367071

Andrews, A. L., Holloway, J. W., Puddicombe, S. M., Holgate, S. T., & Davies, D. E. (2002). Kinetic analysis of the interleukin-13 receptor complex. *The Journal of Biological Chemistry*, *277*(48), 46073–46078. doi:10.1074/jbc.M209560200 PMID:12354755

Barrett, L. W., Fletcher, S., & Wilton, S. D. (2012). Regulation of eukaryotic gene expression by the untranslated gene regions and other non-coding elements. *Cellular and Molecular Life Sciences: CMLS*, *69*(21), 3613–3634. doi:10.1007/s00018-012-0990-9 PMID:22538991

Bhatia, S., Bengani, H., Fish, M., Brown, A., Divizia, M. T., de Marco, R., & Kleinjan, D. A. et al. (2013). Disruption of Autoregulatory Feedback by a Mutation in a Remote, Ultraconserved PAX6 Enhancer Causes Aniridia. *American Journal of Human Genetics*, *93*(6), 1126–1134. doi:10.1016/j.ajhg.2013.10.028 PMID:24290376

Brightman, F. A., & Fell, D. A. (2000). Differential feedback regulation of the MAPK cascade underlies the quantitative differences in EGF and NGF signalling in PC12 cells. *FEBS Letters*, *482*(3), 169–174. doi:10.1016/S0014-5793(00)02037-8 PMID:11024454

Chakraborty, M., Sengupta, A., Bhattacharya, D., Banerjee, S., & Chakrabarti, A. (2010). DNA binding domain of RFX5: Interactions with X-box DNA and RFXANK. *Biochimica et Biophysica Acta*, *1804*(10), 2016–2024. doi:10.1016/j.bbapap.2010.07.009 PMID:20637319

Chorev, M., & Carmel, L. (2012). The function of introns. *Frontiers in Genetics*, *3*, 1–15. doi:10.3389/fgene.2012.00055 PMID:22518112

Cochella, L., & Green, R. (2005). Fidelity in protein synthesis. *Current Biology*, *15*(14), R536–R540. doi:10.1016/j.cub.2005.07.018 PMID:16051156

Daenen, F., van Roy, F., & De Bleser, P. J. (2008). Low nucleosome occupancy is encoded around functional human transcription factor binding sites. *BMC Genomics*, *9*(1), 332. doi:10.1186/1471-2164-9-332 PMID:18627598

Dunham, I., Kundaje, A., Aldred, S. F., Collins, P. J., Davis, C. A., & Doyle, F.,… Birney, E., ENCODE Project Consortium. (2012). An integrated encyclopedia of DNA elements in the human genome. *Nature*, *489*, 57–74. PMID:22955616

Felder, S., LaVin, J., Ullrich, A., & Schlessinger, J. (1992). Kinetics of binding, endocytosis, and recycling of EGF receptor mutants. *The Journal of Biological Chemistry*, *117*, 203–212. PMID:1556153

Good, L., Dimri, G. P., Campisi, J., & Chen, K. Y. (1996). Regulation of dihydrofolate reductase gene expression and E2F components in human diploid fibroblasts during growth and senescence. *Journal of Cellular Physiology*, *168*(3), 580–588. doi:10.1002/(SICI)1097-4652(199609)168:3<580::AID-JCP10>3.0.CO;2-3 PMID:8816912

Heinz, S., Romanoski, C. E., Benner, C., & Glass, C. K. (2015). The selection and function of cell type-specific enhancers. *Nature Reviews. Molecular Cell Biology*, *16*(3), 144–154. doi:10.1038/nrm3949 PMID:25650801

Hemmings, B. A., & Restuccia, D. F. (2012). PI3K-PKB/Akt pathway. *Cold Spring Harbor Perspectives in Biology*, *4*(9), a011189. doi:10.1101/cshperspect.a011189 PMID:22952397

Hilton, T. L., & Wang, E. H. (2003). Transcription factor IID recruitment and Sp1 activation. Dual function of TAF1 in cyclin D1 transcription. *The Journal of Biological Chemistry*, *278*(15), 12992–13002. doi:10.1074/jbc.M300412200 PMID:12569092

Huang, J., Brieba, L. G., & Sousa, R. (2000). Misincorporation by wild-type and mutant T7 RNA polymerases: Identification of interactions that reduce misincorporation rates by stabilizing the catalytically incompetent open conformation. *Biochemistry*, *39*(38), 11571–11580. doi:10.1021/bi000579d PMID:10995224

Human Genome Sequencing Consortium, I.International Human Genome Sequencing Consortium. (2004). Finishing the euchromatic sequence of the human genome. *Nature*, *431*(7011), 931–945. doi:10.1038/nature03001 PMID:15496913

Kesanakurti, D., Chetty, C., Rajasekhar Maddirela, D., Gujrati, M., & Rao, J. S. (2013). Essential Role of Cooperative NF-κB and Stat3 Recruitment to ICAM-1 Intronic Consensus Elements in the Regulation of Radiation-induced Invasion and Migration in Glioma. *Oncogene*, *32*(43), 5144–5155. doi:10.1038/onc.2012.546 PMID:23178493

Kholodenko, B. N. (2006). Cell-signalling dynamics in time and space. *Nature Reviews. Molecular Cell Biology*, *7*(3), 165–176. doi:10.1038/nrm1838 PMID:16482094

Kim, S. J., Onwuta, U. S., Lee, Y. I., Li, R., Botchan, M. R., & Robbins, P. D. (1992). The retinoblastoma gene product regulates Sp1-mediated transcription. *Molecular and Cellular Biology*, *12*(6), 2455–2463. doi:10.1128/MCB.12.6.2455 PMID:1588949

Kobayashi, H. (2003). Computer Simulation of Cytoplasmic pH Regulation Mediated by the F-type H^+-ATPase. *Biochimica et Biophysica Acta*, *1607*(2-3), 211–216. doi:10.1016/j.bbabio.2003.10.001 PMID:14670611

Kobayashi, H., Azuma, R., & Konagaya, A. (2007a). Kinetic Analysis of Ligand-Receptor Complex Formation with the Aid of Computer Simulation. In *AINAW, 2007, Advanced Information Networking and Applications Workshops, International Conference on, Advanced Information Networking and Applications Workshops, International Conference on 2007*, (vol. 1, pp. 727-732). The Institute of Electrical and Electronics Engineers.

Kobayashi, H., Azuma, R., & Konagaya, A. (2007b). Clustering of membrane proteins in the pre-stimulation stage is required for signal transduction: A computer analysis. *Signal Transduction*, *7*(4), 329–339. doi:10.1002/sita.200600126

Kobayashi, H., Azuma, R., & Yasunaga, T. (2009). Expression of excess receptors and negative feedback control of signal pathways are required for rapid activation and prompt cessation of signal transduction. *Cell Communication and Signaling*, *7*(1), 3. doi:10.1186/1478-811X-7-3 PMID:19254388

Lee, S., Lin, X., Nam, N. H., Parang, K., & Sun, G. (2003). Determination of the substrate-docking site of protein tyrosine kinase C-terminal Src kinase. *Proceedings of the National Academy of Sciences of the United States of America*, *100*(25), 14707–14712. doi:10.1073/pnas.2534493100 PMID:14657361

Lim, S. P., & Garzino-Demo, A. (2000). The human immunodeficiency virus type 1 Tat protein up-regulates the promoter activity of the beta-chemokine monocyte chemoattractant protein 1 in the human astrocytoma cell line U-87 MG: Role of SP-1, AP-1, and NF-kappaB consensus sites. *Journal of Virology*, *74*(4), 1632–1640. doi:10.1128/JVI.74.4.1632-1640.2000 PMID:10644332

Lou, H., Yang, Y., Cote, G. J., Berget, S. M., & Gagel, R. F. (1995). An intron enhancer containing a 5' splice site sequence in the human calcitonin/calcitonin gene-related peptide gene. *Molecular and Cellular Biology*, *15*(12), 7135–7142. doi:10.1128/MCB.15.12.7135 PMID:8524281

Maston, G. A., Evans, S. K., & Green, M. R. (2006). Transcriptional Regulatory Elements in the Human Genome. *Annual Review of Genomics and Human Genetics*, *7*(1), 29–59. doi:10.1146/annurev.genom.7.080505.115623 PMID:16719718

Matsumoto, M., & Nishimura, T. (1998). Mersenne twister: A 623-dimensionally equidistributed uniform pseudo-random number generator. *ACM Transactions on Modeling and Computer Simulation*, *8*(1), 3–30. doi:10.1145/272991.272995

Mews, P., Zee, B. M., Liu, S., Donahue, G., Garcia, B. A., & Berger, S. L. (2014). Histone methylation has dynamics distinct from those of histone acetylation in cell cycle reentry from quiescence. *Molecular and Cellular Biology*, *34*(21), 3968–3980. doi:10.1128/MCB.00763-14 PMID:25154414

Morrison, D. K. (2012). MAP kinase pathways. *Cold Spring Harbor Perspectives in Biology*, *4*(11), a011254. doi:10.1101/cshperspect.a011254 PMID:23125017

Nair, N. U., Lin, Y., Manasovska, A., Antic, J., Grnarova, P., Sahu, A. D., & Moret, B. M. E. et al. (2014). Study of cell differentiation by phylogenetic analysis using histone modification data. *BMC Bioinformatics*, *15*(1), 269. doi:10.1186/1471-2105-15-269 PMID:25104072

Nair, P., Muthukkumar, S., Sells, S. F., Han, S. S., Sukhatme, V. P., & Rangnekar, V. M. (1997). Early growth response-1-dependent apoptosis is mediated by p53. *The Journal of Biological Chemistry*, *272*(32), 20131–20138. doi:10.1074/jbc.272.32.20131 PMID:9242687

Netzer, N., Goodenbour, J. M., David, A., Kimberly, A., Dittmar, K. A., & Jones, R. B. (2009). Innate Immune and Chemically Triggered Oxidative Stress Modifies Translational Fidelity. *Nature*, *462*(7272), 522–526. doi:10.1038/nature08576 PMID:19940929

O'Neill, L. A., Golenbock, D., & Bowie, A. G. (2013). The history of Toll-like receptors - redefining innate immunity. *Nature Reviews. Immunology*, *13*(6), 453–460. doi:10.1038/nri3446 PMID:23681101

Ohlsson, H., Karlsson, O., & Edlund, T. (1988). A beta-cell-specific protein binds to the two major regulatory sequences of the insulin gene enhancer. *Proceedings of the National Academy of Sciences of the United States of America*, *85*(12), 4228–4231. doi:10.1073/pnas.85.12.4228 PMID:3288987

Rosenberg, M., Fan, A. X., Lin, I. J., Liang, S. Y., & Bungert, J. (2013). Cell-cycle specific association of transcription factors and RNA polymerase ii with the human β-globin gene locus. *Journal of Cellular Biochemistry*, *114*(9), 1997–2006. doi:10.1002/jcb.24542 PMID:23519692

Sánchez-Moreno, I., Coral-Vázquez, R., Méndez, J. P., & Canto, P. (2008). Full-length SRY protein is essential for DNA binding. *Molecular Human Reproduction*, *14*(6), 325–330. doi:10.1093/molehr/gan021 PMID:18453550

Savinkova, L. K., Ponomarenko, M. P., Ponomarenko, P. M., Drachkova, I. A., Lysova, M. V., Arshinova, T. V., & Kolchanov, N. A. (2009). TATA box polymorphisms in human gene promoters and associated hereditary pathologies. *Biochemistry (Mosc)*, *74*(2), 117–129. doi:10.1134/S0006297909020011 PMID:19267666

Shiio, Y., Yamamoto, T., & Yamaguchi, N. (1992). Negative regulation of Rb expression by the p53 gene product. *Proceedings of the National Academy of Sciences of the United States of America*, *89*(12), 5206–5210. doi:10.1073/pnas.89.12.5206 PMID:1608930

Spiro, C., Bazett-Jones, D. P., Wu, X., & McMurray, C. T. (1995). DNA structure determines protein binding and transcriptional efficiency of the proenkephalin cAMP-responsive enhancer. *The Journal of Biological Chemistry*, *270*(46), 27702–27710. doi:10.1074/jbc.270.46.27702 PMID:7499237

Stephen, T. S., & Kadonaga, J. T. (2003). The RNA polymerase II core promoter. *Annual Review of Biochemistry*, *72*(1), 449–479. doi:10.1146/annurev.biochem.72.121801.161520 PMID:12651739

Teramura, Y., Ichinose, J., Takagi, H., Nishida, K., Yanagida, T., & Sako, Y. (2006). Single-molecule analysis of epidermal growth factor binding on the surface of living cells. *The EMBO Journal*, *25*(18), 4215–4222. doi:10.1038/sj.emboj.7601308 PMID:16946702

Venter, J. C., Adams, M. D., Myers, E. W., Li, P. W., Mural, R. J., Sutton, G. G., & Zhu, X. et al. (2001). The Sequence of the Human Genome. *Science*, *291*(5507), 1304–1351. doi:10.1126/science.1058040 PMID:11181995

von Hippel, P. H., Johnson, N. P., & Marcus, A. H. (2013). Fifty years of DNA "Breathing": Reflections on old and new approaches. *Biopolymers*, *99*, 923–954. PMID:23840028

Whitfield, T. W., Wang, J., Collins, P. J., Partridge, E. C., Aldred, S. F., Trinklein, N. D., & Weng, Z. et al. (2012). Functional analysis of transcription factor binding sites in human promoters. *Genome Biology*, *13*, R50. PMID:22951020

Wilkinson, J. C., Stein, R. A., Guyer, C. A., Beechem, J. M., & Staros, J. V. (2001). Realtime kinetics of ligand/cell surface receptor interactions in living cells: Binding of epidermal growth factor to the epidermal growth factor receptor. *Biochemistry*, *40*(34), 10230–10242. doi:10.1021/bi010705t PMID:11513601

Xue, W., Wang, J., Shen, Z., & Zhu, H. (2004). Enrichment of transcriptional regulatory sites in non-coding genomic region. *Bioinformatics (Oxford, England)*, *20*(4), 569–575. doi:10.1093/bioinformatics/btg450 PMID:14990453

Yamada, S., Shiono, S., Joo, A., & Yoshimura, A. (2003). Control mechanism of JAK/STAT signal transduction pathway. *FEBS Letters*, *534*(1-3), 190–196. doi:10.1016/S0014-5793(02)03842-5 PMID:12527385

Zanotto, E., Häkkinen, A., Teku, G., Shen, B., Ribeiro, A. S., & Jacobs, H. T. (2009). NF-Y influences directionality of transcription from the bidirectional Mrps12/Sarsm promoter in both mouse and human cells. *Biochimica et Biophysica Acta*, *1789*(5), 432–442. doi:10.1016/j.bbagrm.2009.05.001 PMID:19439209

Zappavigna, V., Renucci, A., Izpisúa-Belmonte, J. C., Urier, G., Peschle, C., & Duboule, D. (1991). HOX4 genes encode transcription factors with potential auto- and cross-regulatory capacities. *The EMBO Journal*, *10*, 4177–4187. PMID:1756725

Zhang, X., Jin, J. Y., Wu, J., Qin, X., Streilein, R., Hall, R. P., & Zhang, J. Y. (2015). RNA-Seq and ChIP-Seq reveals SQSTM1/p62 as a key mediator of JunB suppression of NF-κB-dependent inflammation. *The Journal of Investigative Dermatology*, *135*(4), 1016–1024. doi:10.1038/jid.2014.519 PMID:25501661

Zhang, Y., Larsen, C. A., Stadler, H. S., & Ames, J. B. (2011). Structural basis for sequence specific DNA binding and protein dimerization of HOXA13. *PLoS ONE*, *6*(8), e23069. doi:10.1371/journal.pone.0023069 PMID:21829694

Zhou, H., Zarubin, T., Ji, Z., Min, Z., Zhu, W., Downey, J. S., & Han, J. (2005). Frequency and Distribution of AP-1 Sites in the Human Genome. *DNA Research*, *12*(2), 139–150. doi:10.1093/dnares/12.2.139 PMID:16303745

KEY TERMS AND DEFINITIONS

Computer Simulation: Analysis of biological phenomena using computer programs in this chapter.

Consensus Sequence: The region of DNA to which proteins are bound.

Dissociation Constant: The equilibrium constant of complex dissociation.

Gene Expression: The same meaning as transcription in a narrow sense. Transcription and the subsequent translation are sometimes referred to together as gene expression.

Ligand: External stimuli that induce or inhibit gene expression in this chapter.

Regulatory Protein: Proteins which are bound to DNA and regulate gene expression.

Regulatory Region: The region of DNA to which RNA polymerase and regulatory proteins are bound to regulate gene expression.

Target Gene: A gene whose expression is regulated by a given ligand.

Transcription: Synthesis of messenger RNA whose sequence is complementary to the sequence of a gene.

Translation: Synthesis of protein according to nucleotide sequences of messenger RNA.

APPENDIX 1: GENERATION OF RANDOM SEQUENCES

Real-type pseudo-uniform random numbers (N) with the range $0 \leq N < 1$ were generated as reported previously (Matsumoto and Nishimura, 1998). A, T, C, and G were set when $0 \leq N < 0.25$, $0.25 \leq N < 0.50$, $0.50 \leq N < 0.75$, and $0.75 \leq N < 1$, respectively.

APPENDIX 2: SEARCH FOR HOMOLOGOUS SEQUENCES

The existence of random sequences consisting of 11 to 16 base pairs in the human genome was investigated with BLAST of the National Library of Medicine, USA using the GENEBANK database. An artificial genome consisting of 3,000,000,000 bases was created as described above, and the existence of consensus sequences proposed to date was examined in the artificial genome using a simulation program made for this study. The average numbers and standard deviations were obtained from 5 simulations using artificial genomes created independently.

APPENDIX 3: THE SOURCE CODE OF THE PROGRAM USED IN THE SIMULATIONS

The source codes of the program were implemented with Visual Studio C++.net (Microsoft Co.), and the program was run on a personal computer with Windows XP (Microsoft Co.). Source codes supplied from Microsoft Co. are not presented.

```
<Mainroutine.h>
#include <io.h>
#include <sys/stat.h>
#include <fcntl.h>
#define MAX_SEQ_NUMBER 1000
#define Max_SEQ_Num 30
#define MODE 1
#define REPEAT 10
static int DISPLAY=0;
static unsigned long Process=0;
static int SeqNum=0;
static int SEQ_Num=0;
static char *DATA[6];
static char *Save_seq[REPEAT*Max_SEQ_Num][1];
```

```
class Simulation_Parameter {
        public:
                int Seq[MAX_SEQ_NUMBER];
                unsigned long Answer[MAX_SEQ_NUMBER];
}; Simulation_Parameter Simu;
class Random_Parameter {
        public:  unsigned SEED;
}; Random_Parameter Rand;
void CconsensusView::selection(int x, int y)
{
        CDC* pDC=GetDC();
        CString str;
        str.Format("Start");
        pDC->TextOut(50,100,str);
#if (MODE==0)
        int i;
        for (i=0; i<REPEAT*20; i++) *Save_seq[i]=" ";
        get_sequence();
        SaveSequence();
#endif
#if (MODE==1)
        Rand.SEED=(unsigned)time(NULL);
        sgenrand(Rand.SEED);
        display_seq(0, SEQ_Num);
        homologue(SEQ_Num);
        SaveData();
#endif
        ReleaseDC(pDC);
}
void CconsensusView::get_sequence()
{
        int i, n=SEQ_Num;
        Rand.SEED=(unsigned)time(NULL);
        sgenrand(Rand.SEED);
        for (i=0; i<REPEAT; i++)
        {
                sequence(i, n);
                display_seq(i, n);
        }
}
```

```
void CconsensusView::sequence(int k, int n)
{
        CDC* pDC=GetDC();
        CString str;        int i;
        for (i=0; i<n; i++)
        {
                Simu.Seq[i]=nucleotide();
        }
        for (i=0; i<n; i++)
        {
                if (Simu.Seq[i]==0) *Save_seq[k*Max_SEQ_
Num+i]="X";
                if (Simu.Seq[i]==1) *Save_seq[k*Max_SEQ_
Num+i]="A";
                if (Simu.Seq[i]==2) *Save_seq[k*Max_SEQ_
Num+i]="G";
                if (Simu.Seq[i]==3) *Save_seq[k*Max_SEQ_
Num+i]="T";
                if (Simu.Seq[i]==4) *Save_seq[k*Max_SEQ_
Num+i]="C";
        }
        *Save_seq[k*Max_SEQ_Num+Max_SEQ_Num-1]=" ";
                str.Format("k=%d, n=%d", k,n);
                pDC->TextOut(400,200+k*Max_SEQ_Num,str);
        for (i=0; i<n; i++)
        {
                str.Format("%s", Save_seq[k*Max_SEQ_Num+i]
[0]);
                pDC->TextOut(500+i*10,200+k*Max_SEQ_Num,str);
        }
        ReleaseDC(pDC);
}
int CconsensusView::nucleotide()
{
        double A; int N;
        A=genrand();
        if (A>0.75) N=1; //"A";
        if (A<=0.75 && A>0.5) N=2; //"G";
        if (A<=0.5 && A>0.25) N=3; //"T";
```

```
        if (A<=0.25) N=4; //"C";
        return N;
}
void CconsensusView::display_seq(int k, int n)
{
        CDC* pDC=GetDC();
        CString str;
        int i,x=50,y=150+50*k;
        str.Format(": %d", SEQ_Num);
                pDC->TextOut(300,50, str);
        for (i=0; i<n; i++)
        {
                if (Simu.Seq[i]==0) str.Format("%s", "X");
                if (Simu.Seq[i]==1) str.Format("%s", "A");
                if (Simu.Seq[i]==2) str.Format("%s", "G");
                if (Simu.Seq[i]==3) str.Format("%s", "T");
                if (Simu.Seq[i]==4) str.Format("%s", "C");
                pDC->TextOut(x+i*10,y,str);
                str.Format("%d", Simu.Seq[i]);
                pDC->TextOut(x+i*10,y-20,str);
        }
        ReleaseDC(pDC);
}
void CconsensusView::homologue(int n)
{
        CDC* pDC=GetDC();
        CString str;
        unsigned long Max=1000000, Trial=3000, k, m;
        int A, i, l, match, B, Sequence[MAX_SEQ_NUMBER];  //j,
h,
        for (i=0; i<n; i++)
        {
                Sequence[i]=nucleotide();
        }
        A=0;B=n;
        for (m=0; m<Trial; m++)
        {
                for (k=0; k<Max; k++)
                        {
                                i=0; match=0;
```

```
                                for (i=0; i<n; i++)
                                {
                                        if (Simu.Seq[i]!=0 &&
Simu.Seq[i]!=Sequence[i]) match=1;
                                }
                                if (match==0)  A++;
                                for  (l=0; l<n-1; l++)
                                        Sequence[l]=Sequence[
l+1];

                                Sequence[n-1]=nucleotide();
                        }
                        Process=m;
#if (MODE==1)
        str.Format("%d",m);
        pDC->TextOut(100,100,str);
#endif
        }
        DISPLAY=A+1;
        display_hom(pDC, DISPLAY-1);
        ReleaseDC(pDC);
}
void CconsensusView::display_hom(CDC *pDC, int A)
{
        CString str;
        str.Format("%d", A);
        pDC->TextOut(200,150,str);
}
void CconsensusView::display_process(CDC *pDC)
{
        CString str;
        str.Format("%s %d", "Process", Process);
        pDC->TextOut(100,300,str);
}
void CconsensusView::InputNewValues(int nChar)
{
        int N=0;
        if (nChar==65) N=1; //A
        if (nChar==71) N=2; //G
        if (nChar==84) N=3; //T
        if (nChar==67) N=4; //C
```

```
        if (nChar==88) N=0; //X
        Simu.Seq[SeqNum]=N;
        SeqNum++;
}
static int Number_Place=0;
void CconsensusView::InputNewNumber(int nChar)
{
        int KEY_FLAG=0;
        if (nChar==13) {KEY_FLAG=1;Number_Place=0;}
        if (KEY_FLAG==0 && nChar<13)
        {
                if (Number_Place==0) SEQ_Num=nChar;
                else SEQ_Num=SEQ_Num*10+nChar;
                Number_Place++;
        }
}
void CconsensusView::SaveSequence()
{
        int fh, byteread;
        int i, j;
        CString FileName, str;
        if (SEQ_Num>0 && SEQ_Num<Max_SEQ_Num)
        FileName.Format(".\DATA\%d.txt",SEQ_Num);
        else FileName.Format(".\DATA\%s.txt","0");
        fh=_creat(FileName, _S_IREAD|_S_IWRITE);
        for (i=0; i<REPEAT; i++)
        {
                for (j=0; j<Max_SEQ_Num; j++)
                {
                        str.Format ("%s", Save_seq[i*Max_SEQ_
Num+j][0]);
                 byteread=_write(_open(FileName, _O_RDWR|_O_
TEXT|_O_APPEND),str,1);
                }
                str.Format ("\n");
                 byteread=_write(_open(FileName, _O_RDWR|_O_
TEXT|_O_APPEND),str,2);
        }
}
void CconsensusView::SaveData()
```

```
{
        int fh, byteread;
        int j;
        CString FileName, str;
        if (SEQ_Num>0 && SEQ_Num<Max_SEQ_Num)
        FileName.Format(“./DATA/%s%d.txt”,”DATA”,SEQ_Num);
        else FileName.Format(“./DATA/%s%s.txt”,”DATA”,”0”);
        fh=_creat(FileName, _S_IREAD|_S_IWRITE);
        for (j=0; j<SEQ_Num; j++)
        {
                if (Simu.Seq[j]==0) str.Format (“X”);
                if (Simu.Seq[j]==1) str.Format (“A”);
                if (Simu.Seq[j]==2) str.Format (“G”);
                if (Simu.Seq[j]==3) str.Format (“T”);
                if (Simu.Seq[j]==4) str.Format (“C”);
                byteread=_write(_open(FileName, _O_RDWR|_O_
TEXT|_O_APPEND),str,1);
        }
        str.Format (“  ”);
        byteread=_write(_open(FileName, _O_RDWR|_O_TEXT|_O_
APPEND),str,2);
                str.Format (“%10d”, DISPLAY-1);
        byteread=_write(_open(FileName, _O_RDWR|_O_TEXT|_O_
APPEND),str,10);
}
<consensusView.cpp>
#include “stdafx.h”
#include “consensus.h”
#include “consensusDoc.h”
#include “consensusView.h”
#ifdef _DEBUG
#define new DEBUG_NEW
#endif
#include “mainroutine.h”
#include “random.h”
#include “.\consensusview.h”
IMPLEMENT_DYNCREATE(CconsensusView, CView)
BEGIN_MESSAGE_MAP(CconsensusView, CView)
        // Standard printing commands
        ON_COMMAND(ID_FILE_PRINT, CView::OnFilePrint)
```

```
        ON_COMMAND(ID_FILE_PRINT_DIRECT, CView::OnFilePrint)
        ON_COMMAND(ID_FILE_PRINT_PREVIEW,
CView::OnFilePrintPreview)
        ON_WM_LBUTTONDOWN()
        ON_WM_KEYDOWN()
END_MESSAGE_MAP()
// CconsensusView construction/destruction
CconsensusView::CconsensusView()
{
        // TODO: add construction code here
}
CconsensusView::~CconsensusView()
{
}
BOOL CconsensusView::PreCreateWindow(CREATESTRUCT& cs)
{
        // TODO: Modify the Window class or styles here by
modifying
        return CView::PreCreateWindow(cs);
}
// CconsensusView drawing
void CconsensusView::OnDraw(CDC* pDC)
{
        CconsensusDoc* pDoc = GetDocument();
        ASSERT_VALID(pDoc);
        if (!pDoc)
                return;
        // TODO: add draw code for native data here
        CString str;
        str.Format("MODE=%d    input sequence number", MODE);
                pDC->TextOut(50, 50, str);
}
// CconsensusView printing
BOOL CconsensusView::OnPreparePrinting(CPrintInfo* pInfo)
{
        // default preparation
        pInfo->SetMaxPage(5);
        return DoPreparePrinting(pInfo);
}
void CconsensusView::OnBeginPrinting(CDC* /*pDC*/, CPrintInfo*
```

```
/*pInfo*/)
{
        // TODO: add extra initialization before printing
}
void CconsensusView::OnEndPrinting(CDC* /*pDC*/, CPrintInfo*
/*pInfo*/)
{
        // TODO: add cleanup after printing
}
// CconsensusView diagnostics
#ifdef _DEBUG
void CconsensusView::AssertValid() const
{
        CView::AssertValid();
}
void CconsensusView::Dump(CDumpContext& dc) const
{
        CView::Dump(dc);
}
CconsensusDoc* CconsensusView::GetDocument() const // non-debug
version is inline
{
        ASSERT(m_pDocument->IsKindOf(RUNTIME_
CLASS(CconsensusDoc)));
        return (CconsensusDoc*)m_pDocument;
}
#endif //_DEBUG
// CconsensusView message handlers
void CconsensusView::OnLButtonDown(UINT nFlags, CPoint point)
{
        CView::OnLButtonDown(nFlags, point);
        int x=point.x, y=point.y;
                selection(x, y);
}
void CconsensusView::OnPrint(CDC* pDC, CPrintInfo* pInfo)
{
        int nPage; CString str; CFont font;
        nPage=pInfo->m_nCurPage;
}
void CconsensusView::OnKeyDown(UINT nChar, UINT nRepCnt, UINT
```

```
nFlags)
{
        CView::OnKeyDown(nChar, nRepCnt, nFlags);
        if (nChar>0 && nChar<65) InputNewNumber(nChar-48);
        if (nChar>96 && nChar<106) InputNewNumber(nChar-96);
        if (nChar==13) InputNewNumber(nChar);
        if (SeqNum<SEQ_Num)
        if (nChar==65 || nChar==67 || nChar==71 ||nChar==84
||nChar==88) InputNewValues(nChar);
        display_seq(0, SEQ_Num);}
```

Chapter 7
Computational Inference of Gene Regulation from Whole-Transcriptome Analysis of Early Embryos

Sung-Joon Park
The University of Tokyo, Japan

Kenta Nakai
The University of Tokyo, Japan

ABSTRACT

To ensure totipotency of new zygotes, immediately after fertilization, two gametes activate genetic and epigenetic processes that highly interplay and lead to sequential waves of gene expression. Since the first report of fertilization a hundred years ago, a long-standing question to what factors govern the dramatic transcriptional changes remains elusive. Recently, significant advances in biological experiment technology and computational methodology have taken place in understanding of the regulatory dynamics during fertilization. This chapter provides an overview of the recent progress in characterizing mammalian fertilization, and focuses on the computational methods applicable to the inference of regulation for understanding early embryo development. In particular, this chapter introduces the linear regression modeling and log-linear graphical modeling to identify potential key regulators in the higher-order conditional distribution where statistical paradox often occurs.

DOI: 10.4018/978-1-5225-0353-8.ch007

INTRODUCTION

This chapter will provide a broad overview of biological and computational efforts that are undertaken to elucidate the mechanism underlying mammalian fertilization. Since the first report of fertilization a hundred years ago, great progress has been achieved through cellular and molecular analyses. The first part of this chapter will focus on the progress. Recent advances in computational methodology offer new insights into understanding of embryo development. The second part of this chapter will summarize computational methods to infer gene regulatory networks. In particular, a regression-based method and log-linear graphical model are explained with examples of model organisms. In the last part, recent studies that have applied the computational methods with a large-scale whole-transcriptome data are described. This chapter will also discuss methodological limitations to be overcome in the future.

BACKGROUND

Fertilization is the developmental process that is responsible for the maintenance of the species. Fertilization is also to be an error-prone process; in human, 10–30% of fertilized oocytes are aneuploidy causative of birth defects such as Down syndrome (Hassold & Hunt, 2001). Spermatozoa also show abnormalities during cell divisions frequently, which is a reason of infertile male (Egozcue et al., 2005). Understanding why such abnormalities occur and how we discriminate them are needed to prevent diseases and to improve therapeutic applications. Moreover, identifying key molecules of fertilization enhances the efficiency of reprogramming of somatic cells into pluripotent stem cells (Maekawa et al., 2011; Shinagawa et al., 2014; Xu et al., 2015).

Recent technical advances in developmental biology established the global gene expression profile that is often referred to as "*programmed waves*" (Hamatani, Carter, Sharov, & Ko, 2004; Ko et al., 2000; S. J. Park et al., 2013). Speculations on the gene expression waves using database systems, such as DBTMEE (S. J. Park, Shirahige, Ohsugi, & Nakai, 2015), unveiled the existence of maternal-to-zygotic transition (MZT) that consists of the degradation of maternal factors and the onset of zygotic gene activation (ZGA) (Schultz, 2002). It is thought that maternal proteins and RNAs are necessary for the oocyte maturation and chromatin remodeling, but become harmful to development (M. T. Lee, Bonneau, & Giraldez, 2014; H. Wang & Dey, 2006). Sperm contribution is indispensable to achieve the proper and timely removal and replacement of maternal factors (S. J. Park et al., 2013).

A long-standing question on embryo development is what factors govern the dramatic transcriptional changes immediately after fertilization. Even though available information increased, the regulatory mechanism underlying the sequential waves of gene expression still remains unknown. In this regard, significant recent advances in biological experiment technology and computational methodology have taken place in understanding gene regulatory dynamics. These include high-throughput sequencing for transcriptomes and high-performance reverse engineering algorithms. There is no doubt that these advances help establishing the profile of regulatory dynamics, which gives a clue to the answer.

In the remainder of this chapter, we start by reviewing the latest progress in the contexts of mammalian embryo development. One may be surprised about the current state of huge quantity of knowledge regarding fertilization. We then review the computational methods to infer gene regulatory networks, which mainly focus on the linear regression approach coupled with log-linear graphical model to identify potential key regulators in the higher-order conditional distribution. This chapter will explore the usage of computational inferences that gives shed light on the regulatory dynamics during fertilization.

PREIMPLANTATION EMBRYONIC DEVELOPMENT

Characteristics of Sequential Set of Cellular Events

In mammalian reproduction, the gametes that are highly specialized haploid cells physically contact each other and lead to fertilization that confers the ability to produce all the cells of the body, which is referred to as totipotency (Condic, 2014). This cell fusion is the first step of beginning of life, and also is the successful termination of the sophisticated maturation of germ cells, which is called gametogenesis. During gametogenesis, spermatogonia and oogonia become fully matured gametes through a dramatic set of cellular processes (Evans, Hardy, & Storey, 2001; Guan, Boavida, & McCormick, 2001; Handel & Schimenti, 2010). In particular, meiotic cell division that occurs differently in the gametes is a crucial stage to assure the passage of genetic information to subsequent generations; it is known that meiotic deficiency is associated with infertility or developmental abnormality in offspring (Egozcue et al., 2005; Handel & Schimenti, 2010; Sarrate, Vidal, & Blanco, 2014).

In oogenesis, a primary oocyte developed from diploid oogonium through mitosis undergoes the first meiotic division, forming a haploid oocyte and a polar body. The haploid oocyte starts the second meiotic division that causes metaphase II (MII) arrest. This MII oocyte is fulfilled by oocyte-specific materials that are required for further development (Gurdon, 1962), and awaits fertilization in a transcriptionally inactive state (Clift & Schuh, 2013; R. Li & Albertini, 2013; H. Wang & Dey, 2006). In spermatogenesis, diploid spermatogonium develops a primary spermatocyte through mitosis, and the primary spermatocyte goes through the first meiotic division to give rise to two haploid secondary spermatocytes. Each of the secondary spermatocytes goes through the second meiotic division to produce two spermatids that undergo differentiation and maturation without further cell division. The spermatids finally develop spermatozoa with morphological changes (Guan et al., 2001; Krawetz, 2005). Spermatozoa acquire capacitation, the capacity of fertilization, by residing several hours in the female genital system, where the structural and functional changes of spermatozoa occur (Evans et al., 2001; Ferramosca & Zara, 2014).

In sperm-oocyte fusion, the membrane of capacitated sperm has been destabilized in the female genital system, so it can bind to the extracellular coat surrounding oocytes. The coat is called zona pellucida (P. Wassarman et al., 1999). Sperm's acrosome reacts to the zona pellucida, accompanied by such as calcium signaling and phosphorylation to ensure the penetration of sperm through the coat. Once the penetration has been completed, the structure of zona pellucida is changed to prevent the risk of polyspermy, fertilizing by more than one sperm (Hoodbhoy et al., 2005; Sato, 1979; P. M. Wassarman, 1999). The sperm penetrated through the zona pellucida starts to fuse with the plasma membrane of the oocyte, where the entire contents of the sperm including highly compacted nucleus are eventually released into the oocyte cytoplasm. This sperm-oocyte fusion triggers fertilization forming a zygote. The zygote that includes maternal and paternal pronuclei undergoes a sequential set of cellular processes that transforms it to a totipotent embryo (Clift & Schuh, 2013; Schultz, 2002; H. Wang & Dey, 2006).

Immediately after fertilization, the zygote undergoes asymmetric cellular processes at one-cell stage, i.e. resumption of the meiotic cell division in the female nucleus and decompaction in the male nucleus. The one-cell embryo undergoes a series of cell divisions up to the morula stage that develops the blastocyst formation. During these cell divisions, the cellular machinery involving various key molecules that are inherited from the gametes or newly synthesized governs the developmental competence of the embryo in a timely manner in order to avoid abortive development (Bui et al., 2011; S. J. Park et al., 2013). The principle of reproduction is highly conserved during evolution in most mammals. On the other hand, errors in the reproduction are surprisingly frequent; in human oocytes, 10–30% of fertilized oocytes are aneuploidy (Hassold & Hunt, 2001) causative of birth defects such as Down syndrome. Though the estimated frequency is ~3 times lower than in oocytes, the meiotic abnormality is also observed from spermatogenesis, which cause infertile male (Egozcue et al., 2005).

Recent reviews have provided more details on the germ cells development and preimplantation embryogenesis (Evans et al., 2001; Guan et al., 2001; Iwafuchi-Doi & Zaret, 2014; Krawetz, 2005; Lesch & Page, 2012; Simpson, Caballero, Jungbluth, Chen, & Old, 2005).

Molecular Aspects during Fertilization

The gametes maintain genomes in an asymmetric configuration of chromatin; that is, female with nucleus arrested in MII and male with super-compacted nucleus. Simultaneously, they contain specific proteins and RNAs to be necessary for fertilization followed by the transition of the asymmetry to symmetry. For example, to prevent from the risk of polyspermy that may cause embryonic lethality, zona pellucida sperm-binding proteins (ZPs) alter their three-dimensional structure after binding by one sperm (Dean, 2004). In mice, a key player in such structural change of ZPs is ovastacin, Astl; astacin-like metalloendopeptidase (M12 family); see its expression from http://dbtmee.hgc.jp/gene_card.php?id=2002875. Ovastactin exhibits restricted expression in oocytes, and cleaves ZP2 to prevent postfertilization sperm binding (Burkart, Xiong, Baibakov, Jimenez-Movilla, & Dean, 2012; Clift & Schuh, 2013).

In regard to sperm contribution, sperm delivers key molecules into the oocyte cytoplasm, and triggers the onset of early embryonic development. For example, the fusion of the sperm with the plasma membrane of the oocyte induces Ca^{2+} oscillations in which phospholipase C zeta initiates (Ramadan, Kashir, Jones, & Coward, 2012). It is believed that Phospholipase C zeta (Plcz1, http://dbtmee.hgc.jp/gene_card.php?id=2016893) exists only in sperm. Interestingly, mRNA of Plcz1 has also been found from parthenogenetic one-cell embryo in which sperm effects are absent (S. J. Park et al., 2013), which suggests an intriguing possibility of ovarian Plcz1. The increase of Ca^{2+} levels upon fertilization triggers the resumption of meiotic cell division of MII oocytes. In this division, oocyte-specific cohesin complexes stored in MII oocytes play a crucial role in the rapid and precise segregation of sister chromatids, forming the female pronucleus and the second polar body (Tachibana-Konwalski et al., 2010; Weng, Jeffreys, & Bickel, 2014).

In the sperm nucleus, protamines compact DNA at least six-fold more than in somatic cells. Since the protamines are sperm specific, Prm1, for example, is highly expressed only at sperm (http://dbtmee.hgc.jp/gene_card.php?id=2017413). Importantly, some of sperm nucleosomes escape from this compaction to immediately contribute chromatin remodeling in the zygote after fertilization (Aoshima, Inoue, Sawa, & Okada, 2015; Erkek et al., 2013; Hammoud et al., 2009; Miller, Brinkworth, & Iles, 2010; van der Heijden et al., 2008). The protamines are rapidly replaced with maternally hyperacetylated histones, including histone H3 variant H3.3, an active nucleosome marker (Torres-Padilla, Bannister, Hurd, Kouzarides, & Zernicka-Goetz, 2006; van der Heijden et al., 2005). Again, this process requires the recruitment of maternal factors and newly synthesized proteins (Jullien et al., 2014; Krawetz, 2005; Ward & Coffey, 1991).

After fertilization, approximately 90% of maternal mRNAs are rapidly degraded through two-cell stage (Piko & Clegg, 1982; Schultz, 2002), whereas a number of coding and non-coding genes are transiently expressed during early embryo development (Hamatani et al., 2004; S. J. Park et al., 2013; Zeng, Baldwin, & Schultz, 2004). Although many mechanisms remain elusive, these molecular features have been extensively studied, suggesting selective nature rather than permissive. Recently, mimicking the contribution of molecules important in the embryonic stages enhances the efficiency in various biological applications; for example, reprogramming of somatic cells into pluripotent stem cells by Glis1 (Maekawa et al., 2011), by Hist1h2aa and Hist1h2ba (Shinagawa et al., 2014), and by Dppa3 (Xu et al., 2015).

Programmed Gene Expression Waves

Recent technical advances in cell biology have characterized the programmed waves of gene expression during embryonic development (Figure 1). These studies were based on expressed sequence tags (Ko et al., 2000), microarrays (Hamatani et al., 2004; Vassena et al., 2011; Q. T. Wang et al., 2004; Zeng et al., 2004), and high-throughput RNA sequencing (RNA-seq) (Macfarlan et al., 2012; S. J. Park et al., 2013; Tang et al., 2009; Xue et al., 2013). The gene expression profiles unveiled the existence of maternal-to-zygotic transition (MZT) that consists of the degradation of maternal factors and the onset of zygotic gene activation (ZGA) (Schultz, 2002). Since parthenogenetic embryos exhibit a high level of mRNA production that leads to developmental arrest (S. J. Park et al., 2013), the presence of sperm effects that associate with stored maternal RNAs is required to achieve the proper and timely transition.

Although the gametes are transcriptionally inert, fertilization immediately turns on the switch of transcription, the first wave referred to as the minor ZGA. This transition has been confirmed by transcriptional and translational inhibitions in mammalian embryos (Hamatani et al., 2004; S. J. Park et al., 2013; Vassena et al., 2011). In mice, embryos upon transcriptional inhibition arrest development as two-cell embryos, suggesting the importance of minor ZGA to fuse pronuclei into zygotic nucleus that requires chromatin remodeling, such as the replacement of protamines with histones that gives accessibility to reprogramming factors (M. T. Lee et al., 2014). The minor ZGA occurs at one-cell stage in mice (S. J. Park et al., 2013) and human (Xue et al., 2013). Then, it globally shifts to the most dramatic transition of gene expression accompanied with burst of *de novo* zygotic transcripts, which is referred to as major ZGA; in mice, over 9,000 genes are differentially expressed in this transition (S. J. Park et al., 2013). It is now widely accepted that the timing of ZGA is species specific; at two-cell stage in mice, at four-cell stage

Figure 1. Schematic representation of programmed gene expression waves and epigenetic changes during mouse early embryo development

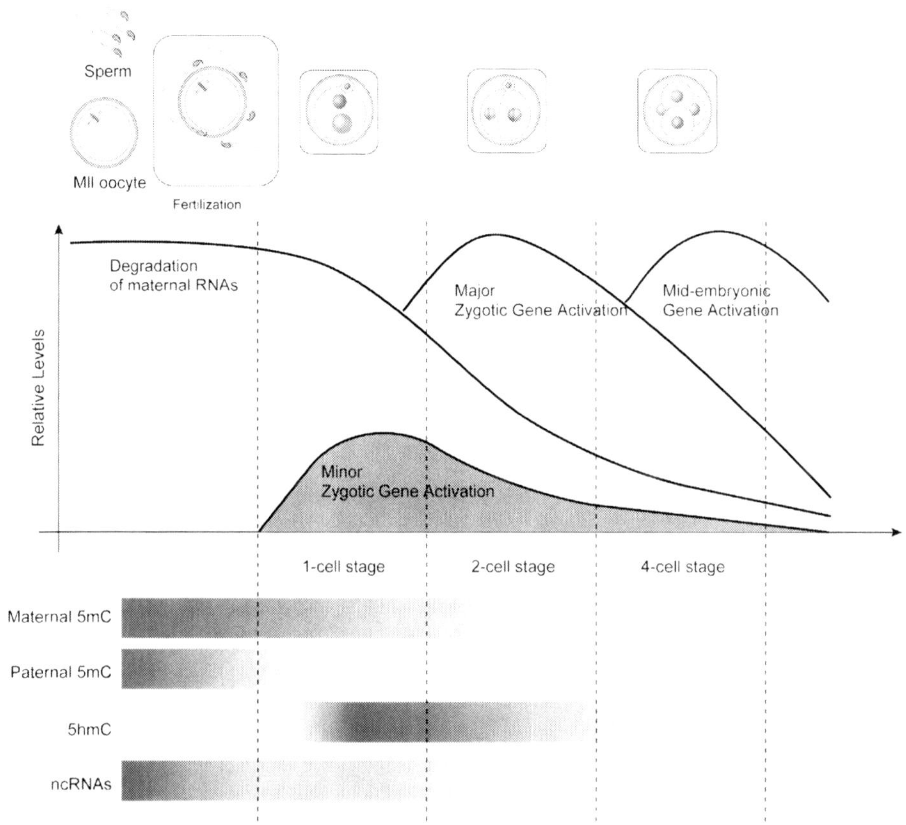

in pigs, at eight-cell stage in human, cat and horse, at sixteen-cell stage in cow, sheep, rabbit, and monkey (Brunet-Simon, Henrion, Renard, & Duranthon, 2001; Cao et al., 2014; M. T. Lee et al., 2014; S. J. Park et al., 2013; Vassena et al., 2011).

It is noteworthy that proteins translated from maternal mRNAs play an indispensable role in the transition; several proteomic approaches have demonstrated the requirement of protein synthesis in oocyte maturation (J. Chen et al., 2011), as well as the presence of specific proteins in zygote (S. Wang et al., 2010; Yurttas, Morency, & Coonrod, 2010). These studies suggest that specific maternal mRNAs escaped from the degradation are recruited to polysomes after fertilization. Importantly, such proteins are highly involved in the biological processes of *de novo* transcription and protein degradation (L. Li, Lu, & Dean, 2013; S. Wang et al., 2010). Therefore, it is thought that maternal proteins and RNAs are necessary for MII oocyte maturation and chromatin remodeling at one-cell stage, but become harmful to zygotic development (M. T. Lee et al., 2014; H. Wang & Dey, 2006).

Unfortunately, it is still hard to distinguish maternal proteins and RNAs from *de novo* zygotic proteins transcripts due in part to the uncoupled nature of transcription and translation during fertilization (Nothias, Miranda, & DePamphilis, 1996; Schultz, 2002). This makes it difficult to connect the programmed waves with the translational activity, and needs multi-omics profiling. In this regard, recent efforts applying state-of-the-art techniques that enables the profiling at single-cell level (Hou et al., 2013; Pan, 2014) have a potential to provide the clearer picture of parental roles (Deng, Ramskold, Reinius, & Sandberg, 2014; Xue et al., 2013). On the other hand, the gene expression patterns highly coordinate with epigenetic modifiers (Aoshima et al., 2015; Erkek et al., 2013; Kobayashi et al., 2012). Therefore, interrogating the interplay between genetic and epigenetic effects on early embryonic development is becoming recognized as important for understanding the programmed waves.

Epigenetic Dynamics

Fertilization activates epigenetic reprogramming to ensure totipotency of newly formed embryos. Histone replacement and modifications, DNA (de)methylation, and non-coding RNAs (ncRNAs) are known to play a role in the epigenetic reprogramming. Remarkably, this reprogramming is achieved by asymmetric processes in parental genomes (Figure 1).

In sperm genome, approximately 90% of genomic regions are hypermethylated through the modification of cytosine nucleotide by adding a methyl group to 5th carbon (5mC). Nonetheless, 4−15% of the sperm genomic regions are retained in nucleosomes where the localizations are highly correlated with protamine-free regions and hypomethylated promoter regions (Erkek et al., 2013; Hammoud et al., 2009; Kobayashi et al., 2012). These nucleosomes predominantly contain histone variants, including TH2B, H2BFWT, H2A.X, H2A.Z (Gatewood, Cook, Balhorn, Schmid, & Bradbury, 1990; Talbert & Henikoff, 2010). Intriguingly, the retention of H3.3 in sperm necessary for the male pronucleus formation (Wen et al., 2014) is co-enriched with an active histone maker H3K4me3, trimethylated at Lys4 of histone H3 (Erkek et al., 2013). This sperm-specific H3.3 is rapidly replaced by maternally provided H3.3 after fertilization, which influences the ZGA (Aoshima et al., 2015). In addition, H3K27me3, a modification of H3.1 and H3.2, associated with silencing, and bivalent histone marks (i.e. H3K4me3-H3K27me3) in sperm are also associated with the ZGA. This distinctive configuration of paternal chromatin is thought to be a template for the chromatin remodeling of zygote, suggesting a potential way to transmit epigenome information to zygote which is called transgenerational epigenome inheritance (Brunner, Nanni, & Mansuy, 2014; Dias & Ressler, 2014; Heard & Martienssen, 2014; Talbert & Henikoff, 2010; van der Heijden et al., 2008).

On the other hand, oocytes retain histones that are required not only for male pronucleus formation but also oocyte maturation. For example, H1foo, oocyte-specific histone H1 (http://dbtmee.hgc.jp/gene_card.php?id=2009673), is expressed during oogenesis, but is rapidly degraded until two-cell stage. It has been revealed that H1foo incorporates into sperm chromatin (Mizusawa et al., 2010) and that its pervasive binding to linker DNAs destabilizes the chromatin structure to give accessibility to the reprogramming factors (Jullien et al., 2014). As another example, MacroH2A (http://dbtmee.hgc.jp/gene_card.php?id=2009725) that is expressed at the gametes associates with transcriptional repression, such as X chromosome inactivation and inactive autosomal genes (Chang et al., 2005; Gamble, Frizzell, Yang, Krishnakumar, & Kraus, 2010). Like H1foo, macroH2A is removed by two-cell stage and appeared again at the later embryonic stage. Thus, macroH2A is thought to be an epigenetic barrier to the reprogramming of somatic cells into pluripotent stem cells (Gaspar-Maia et al., 2013). Interestingly, at one-cell stage, some of histone modifications are exclusively observed from parental nuclei, though their functions are largely unknown; e.g. H4K20me3, H3K9me3, H3K27me2, H3K27me3 in female (Cowell et al., 2002), hyperacetylated H4 in male (Adenot, Mercier, Renard, & Thompson, 1997).

DNA methylation and demethylation mechanisms involve various enzymes (Messerschmidt, Knowles, & Solter, 2014), including DNA methyltransferase family for methylation maintenance, *de novo* methylation, and passive/active demethylation. Recently, it has been found that the active demethylation includes 5-hydroxymethylcytosine (5hmC) by ten-eleven translocation (TET) family of dioxygenases (Tahiliani et al., 2009). Although recent investigations based on sensitive genomic high-throughput technologies offer new insight into the function of TET family (Peat et al., 2014), the role of hydroxylation is still unclear. Immediately after fertilization, genomic methylation in sperm is globally lost by active demethylation. Simultaneously, the level of 5hmC is increased. Oocyte genome is also passively demethylated, but it is rather protected compared to the sperm. These 5mC and 5hmC in zygote are continually decreased until blastocyst stage where remethylation occurs. Importantly, DNA methylation of imprinted gene regions is maintained during the global demethylation process, ensuring parent-of-origin-specific gene expression (Cantone & Fisher, 2013; Kobayashi et al., 2012; Messerschmidt et al., 2014; Smith et al., 2012).

Concomitant with the gene expression waves and DNA methylations, ncRNAs are highly enriched during development (S. J. Park et al., 2013). Among them, maternally provided micro RNAs (miRNAs), which is shorter than ~20 nucleotides, mediate the degradation of maternal RNAs (M. T. Lee et al., 2014; Tang et al., 2007). The levels of maternal miRNAs, including let-7 family and miR-30 family, rapidly decline probably by adenylation of miRNA 3'-end (M. Lee et al., 2014), and are replaced with zygotic miRNAs (Svoboda & Flemr, 2010)(Svoboda & Flemr, 2010). Recently, the ncRNA-mediated early embryo development has been further expanded with long non-coding RNAs (lncRNAs) that is longer than ~200 nucleotides (Fatica & Bozzoni, 2014). In the near future, their functional importance will attract attention as a key regulator for genomic imprinting, maintenance of totipotency, and acquisition of developmental potency (Hamazaki, Uesaka, Nakashima, Agata, & Imamura, 2015; Necsulea et al., 2014).

There is no doubt that the interplay of genetic and epigenetic effects exhibiting asymmetric features in parental genomes is indispensable to tune gene expression levels and patterns during the sequential set of cellular and molecular events at very early embryonic stages. This remarkable crosstalk involves highly complex kinetics of various molecules and needs to be described in more detail.

COMPUTATIONAL INFERENCE OF GENE REGULATION

Difficulties in Characterizing Transcriptional Regulation during Fertilization

We have accumulated pieces of a puzzle that help assembling the landscape showing what events happen at cellular and molecular levels during embryonic development. Fundamentally, upon intricate epigenetic modifications after fertilization, chromatin remodeling mediates promoter regions where transcription by RNA polymerase commences. Then, it provides an opportunity to the transcriptional machinery regulating the initiation of MZT.

In *Drosophila*, maternal Zelda that binds a *cis*-regulatory heptamer motif TAGteam is required for activating the first zygotic genes (Liang et al., 2008). In zebrafish, Nanog, Pou5f1 and SoxB1 have been reported as the activators for MZT (M. T. Lee et al., 2013). These transcription factors (TFs) likely recognize the binding-positions and timings through distinctive promoter shapes (Haberle et al., 2014; Hoskins et al., 2011). In mouse and human, specific TF-binding sites (TFBSs), such as TATA box and SP-1 (Gazdag, Rajkovic, Torres-Padilla, & Tora, 2007; Vassena et al., 2011;

Worrad, Ram, & Schultz, 1994), have been found from maternal and zygotic genes activated immediately after fertilization. These TFBSs are differently enriched in parthenogenetic embryos (S. J. Park et al., 2013). In mice, Foxd1, Nkx2-5, Sox18, Myod1, and Runx1 have been annotated as *de novo* and sperm-driven factors for fine-tuning the ZGA (S. J. Park et al., 2013).

Despite the presence of important factors for the onset of MZT (Iwafuchi-Doi & Zaret, 2014), a long-standing question to how such factors contribute to the timely regulation of the programmed waves remains elusive. This is due in part to the scarcity of genome-wide *bona fide* TF occupancy data. High-throughput sequencing of TF-DNA interactions, i.e. chromatin immunoprecipitation coupled with massive parallel DNA sequencing (ChIP-seq), is promising to overcome this problem, but to prepare successful ChIP-seq libraries for multiple TFs from homogeneous embryo samples is extremely difficult. Even if the identification of genome-wide TF-bindings succeeded, to infer their activities and regulatory networks requires incorporating multilayer evidence, such as expression profiles, protein-protein interactions, and nucleosome occupancy.

For these reasons, *in silico* methods should be performed taking into account the affordable information. For instance, gene expression profiles during fertilization will be incorporated into reverse engineering algorithms to infer regulatory networks. Also, specific gene promoters that lack TF occupancy data in embryos can be modeled by motif enrichment analysis (S. J. Park et al., 2013), and putative key regulators for such genes can be identified by regression models for example (S. J. Park et al., 2014).

Reverse Engineering of Regulatory Networks

As learnt from the tremendous community efforts (Marbach et al., 2012), we can infer regulatory networks through reverse engineering algorithms that convert gene expression profiles into functionally meaningful genetic information. Due to the scarcity of *in vivo* regulatory information during fertilization, this approach holds a promise to provide a comprehensive window into the programmed gene expression waves.

The reverse engineering algorithms successfully applied in broad biological research are based on statistical approaches that test significance of pairwise gene expression correlation in steady-state perturbation response data (Basso et al., 2005; Cahan et al., 2014; Faith et al., 2007), learning stochastic features from expression profiles and inferring causality (Friedman, 2004; Friedman, Linial, Nachman, & Pe'er, 2000; Santra, 2014; Zou & Hastie, 2005), regression methods to fit expression data to a prior model (Holter, Maritan, Cieplak, Fedoroff, & Banavar, 2001; Yeung, Tegner, & Collins, 2002; Zhang, Ching, Tsing, Leung, & Guo, 2010), and graphical probabilistic models with a framework for dimensionality reduction (Friedman, 2004; S. J. Park et al., 2014; Z. Wang, Xu, San Lucas, & Liu, 2013). Recent studies have expanded with time-course expression data to investigate network dynamics. These include methods based on B-spline basis with the coefficients estimated by the least squares procedure (Storey, Xiao, Leek, Tompkins, & Davis, 2005), probabilistic Boolean and Bayesian networks coupled with optimization approaches (Angelini, De Canditiis, Mutarelli, & Pensky, 2007; Martin, Zhang, Martino, & Faulon, 2007), and network flow optimization with given interaction networks (Patil, Kumagai, Liang, Suzuki, & Nakai, 2013), and multiple regression models (Zhang et al., 2010). The aforementioned methods have resolved broad theoretical and practical issues, including error-prone features of biological dataset and limited number of experiments. One can further find a list of affordable methods from Marbach et al., 2012.

To simplify it to a network of transcriptional regulations we are interested in here, although all of these algorithms are applicable with slight modifications, the regression-based modeling that explains gene expression by estimating TF activities is a much more straightforward approach (Beer & Tavazoie, 2004; Bussemaker, Foat, & Ward, 2007; Das, Pellegrini, & Gray, 2009; Gao, Foat, & Bussemaker, 2004). Importantly, the modeling methods need to consider the combinatorial nature of TFs without being caught by spurious associations that lead to misinterpretations. As a solution to this, log-linear graphical model (LGM) is a unique probabilistic approach (Datta & Zhao, 2008; Edwards, 2000; Lauritzen, 1996; Morita, Nakamura, Hamada, & Takahashi, 2011; S. J. Park et al., 2014). Next, we take a close look at the principles of LGM and regression model. Of note, these are useful not only for understanding fertilization, but also for a wide range of biological applications.

Inference of Combinatorial TF-Bindings

Log-Linear Graphical Model

A ChIP-seq experiment with a specific antibody for a TF provides the TF-binding instances in a genome scale. After integrating independent ChIP-seq datasets of multiple TFs, if we found that a TF-TF pair is inclusively or exclusively binding (Figure 2A), then we will conclude that the TFs are co-associated. However, there is a chance that the observation has been affected by other TF-bindings (i.e. higher-order conditional dependency). This phenomenon is known as Simpson's paradox (Heydtmann, 2002). To unveil such conditional (in)dependency, LGM tests statistical significance of TF-binding frequency in the higher-order conditional distribution, and presents the result as a simplified undirected graph. Its web-based installation is available at http://dbtmee.hgc.jp/tools/llm/.

Figure 2. Schematic representation of analysis procedure for LGM and its results in the modeling of yeast TFs

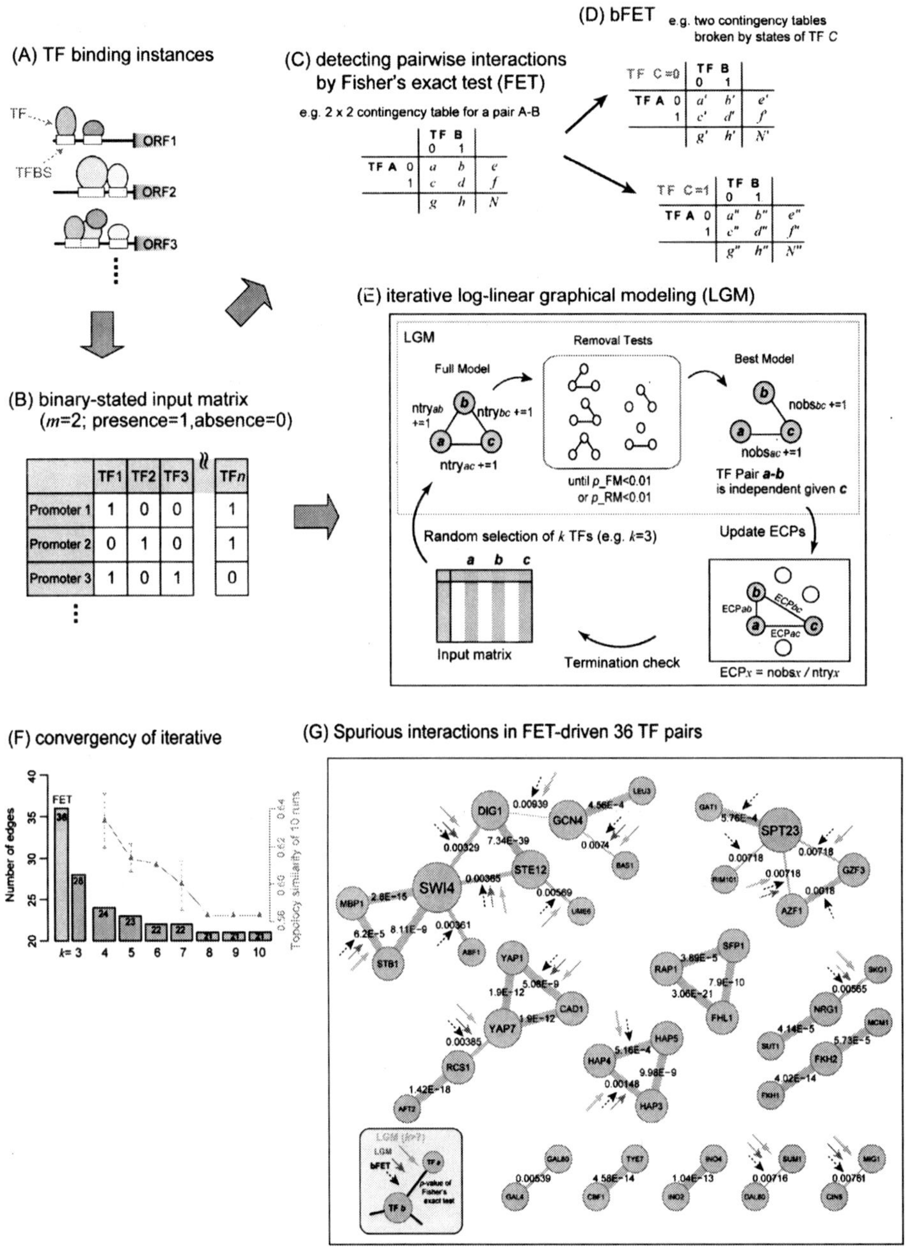

LGM is originated from log-linear model (LM). LM has been applied for analyzing the association among *n* variables categorized into *m* states, where *n* and *m* stand for the number of TFs and the number of discretized binding instances for a TF, respectively (Figure 2B). Suppose that LM considers binary-stated variables *A*, *B*, and *C* (i.g. *m*=2 and *n*=3). The observed counts fall into 2^3 dimensional contingency table by cross-classifying the variable states. Full model (FM), so called since it contains all the possible effects, writes the logarithm of cell probabilities as

$$\log p_{ijk} = \lambda + \lambda_i^A + \lambda_j^B + \lambda_k^C + \lambda_{ij}^{AB} + \lambda_{ik}^{AC} + \lambda_{jk}^{BC} + \lambda_{ijk}^{ABC},$$

where *i*, *j* and *k* are the state indices of *A*, *B*, and *C*, λs are unknown parameters, and parameters $\lambda_{ij}^{AB}, \lambda_{ik}^{AC}, \lambda_{jk}^{BC}$ stand for the interaction effects among indexed variables. Particularly, λ_{ijk}^{ABC} represents the maximal interaction effect in the LM. Following the model hierarchy, when we removed λ_{ij}^{AB} for example, we also remove all the higher-order relatives of this term. Therefore, the reduced model (RM) by this removal test is written as

$$\log p_{ijk} = \lambda + \lambda_i^A + \lambda_j^B + \lambda_k^C + \lambda_{ik}^{AC} + \lambda_{jk}^{BC}.$$

The hierarchy yields the unique combination of interaction effects, and thus a LM corresponds to a certain conditional independency. That is, the above RM can be reformulated as

$$p_{ijk} = (p_{i+k} \cdot p_{+jk} \,/\, p_{++k}),$$

where '+' denotes summation over the corresponding index. This formula is equivalent to Pr(*A*=*i*, *B*=*j* | *C*=*k*) Pr(*B*=*j* | *C*=*k*), meaning that *A* and *B* are independent in the conditional distribution given *C*.

A likelihood ratio G^2 statistics is used to measure model adequacy (Lauritzen, 1996). Deviance G^2 under a model M_t at step *t* of the removal test is measured by likelihood ratio test of M_t versus FM written as

$$G_t^2 = 2\sum_l O_l \log \frac{O_l}{E_l^t},$$

where the vector l stands for the state combination of variables, O_l is the observation, and E_l is the expectation under M_t. Difference of G^2 among two nested models, $M_t \supseteq M_{t+1}$, is given as $d_{t,t+1} = G_{t+1}^2 - G_t^2$. M_{t+1} is the RM caused by a removal of interactions. Since these deviances follow asymptotic χ_{df}^2 distribution where df is the degree of freedom, two p-values of χ^2 test (i.e. G_t^2 and $d_{t,t+1}$), which referred to as p_FM and p_RM respectively, provide the statistical significance of M_t.

Theoretically, the maximal interaction effects in hierarchical LMs correspond to cliques in an undirected graph. This guarantees that LMs can be read off undirected graphs, so-called log-linear graphical model. Therefore, graphical models can be interpreted solely in terms of conditional independence. In other words, the graphical models visually simplify the probabilistic conditional independence structure specified in the complex higher-order interactions. Decomposability in the graphical model, i.e. to contain no cycles of length greater or equal to four without a chord, ensures efficient parameter estimations and closed form expressions for approximating statistics. For this reason, LGM constrains the removal tests; a removal is accepted if it causes the largest p_RM (≥ 0.01) under the current model (i.e. parsimonious), but we seek another eligible edge if this removal yields a non-decomposable graph. This procedure is repeated until p_FM and/or any p_RM are less than certain thresholds (e.g. <0.01).

Iterative Random Sampling

As a practical issue, a large number of TFs reach the astronomical dimension in LGM. To find the nearly optimal parsimonious RM even in such higher dimensional space, an iterative sampling scheme has been proposed (Figure 2E) (S. J. Park et al., 2014). Let $G=(V,E)$ is an undirected graph, where V is a finite set of vertices (i.e. TFs) and E is a set of edges representing the associations of vertex pairs. The iterative LGM is given as follows;

1. A TF set $S=(s_1,\ldots,s_k)$, a non-redundant combination of k TFs, is randomly chosen from n TFs and used to build a FM.
2. For all possible vertex pairs (s_i,s_j), the trial number $ntry_{ij}$ of an edge between s_i and s_j is counted up (i.e. full model of k variables).
3. LGM infers the best graphical model $g=(v,e)$, where e is the set of edges.
4. For all possible vertex pairs (s_i,s_j), if an edge in e links a pair (s_i,s_j), the observed edge frequency $nobs_{ij}$ is counted up.
5. For all possible vertex pairs (s_i,s_j), edge conservation probability (ECP) for a pair (s_i,s_j) is calculated by $nobs_{ij} / ntry_{ij}$.

6. If the global graph $G(V,E)$, where E is a set of edges whose ECPs are 1.0, is not altered during large number of samplings, this procedure is terminated. Otherwise, step1-step5 is performed again.

Under the assumption that satisfying samplings have been done, if an edge linking TF A and TF B retains ECP=1.0, this means that none of k–2 variables block A-B interaction. Importantly, the converged graph shows the independence structure expected to observe in the entire interactions.

A Case Study

There are great attempts to decipher the regulatory codes governing the transcriptional machinery of human and mouse (L. Chen et al., 2014; Stergachis et al., 2014). As a model system for those, budding yeast *Saccharomyces cerevisiae* that has a relatively compact genome and has been addressed by TF ChIP experiments provides an opportunity for readily interpretable TF-TF associations (Harbison et al., 2004; T. I. Lee et al., 2002; MacIsaac et al., 2006).

For example, Park et al. have studied the conditional dependency among 61 yeast TFs using LGM (S.J. Park, Ichinose, & Yada, 2008). The authors compiled 557 Promoters × 61 TFs matrix and used it as an input to LGM. Each vector for a promoter represented the binary states of TFs; that is 1 for presence or 0 for absence of a TF-binding ($p < 0.001$, probe bound statistics in ChIP experiments). These instances investigated within 1kbp upstream regions from translation start sites were associated with 763 yeast genes. In general, Fisher's exact test (FET) is widely used to detect interactive TF pairs in the hypergeometric distribution given by fixed marginal (not conditional) frequencies as a contingency table (Figure 2C). FET inferred that the matrix involved 36 interactive TF pairs at 1% error level (Figure 2G). A marginal contingency table can be broken into two conditional tables by the states of a stratum TF, reflecting the conditional interactions among three TFs (Figure 2D). All the possible combinations reach 35,990 ($={}_{61}C_3$). In this situation, FET said that 19 out of 36 pairs were conditionally independent (denoted as bFET in Figure 2G). Similarly, LGM with these broken tables suggested that 11 of the FET-driven pairs were spurious (Figure 2G). The difference between bFET and LGM are due to inherent methodological features. In much higher-order conditional distribution, the iterative random sample of LGM (k>7) showed a converged graph with 21 edges among 34 TFs (Figure 2F), suggesting that 15 of the FET-driven pairs were spurious (Figure 2G). These results underscore the importance of handling the paradoxical feature of statistics to model the combinatorial nature of TFs.

To predict key regulators in the MZT, the iterative random sampling of LGM has been incorporated into the regression modeling described below (S. J. Park et al., 2014). This application has a great potential to forge a way in the development of new inference strategies that reduce the possibility of false discovery.

Identification of Potential Key Regulators

Linear Regression Model

The reconstruction of transcriptional regulatory networks aims for identifying key activators and repressors. For this purpose, many studies have adopted the linear regression modeling (Bussemaker, Li, & Siggia, 2001; Gao et al., 2004; Ouyang, Zhou, & Wong, 2009; S. J. Park & Nakai, 2011; Suzuki et al., 2009). A linear regression model is written as

$$Y_i = \sum_j w_j X_{ij} + e_i,$$

where Y_i is the expression level of gene *i* in logarithm, X_{ij} is the *j*th explanatory variable for gene *i*, w_j is the regression coefficient (RC) of the *j*th variable, and e_i is the error term. In the case that ChIP-seq data is available, *X*s can be represented by scores with weighted ChIP-seq signals (Ouyang et al., 2009; S. J. Park & Nakai, 2011). If not available, *X*s are approximated by computational approaches, including methods to search potential TFBSs (Irie et al., 2011) and to weight them by various features such as inferences of LGM (S. J. Park et al., 2014). If the RC for an *X* is positive, then this model implies that *X* has positive effect on gene expression (i.e. activator). Conversely, if the coefficient is negative, *X* may act as a repressor (Das et al., 2009).

To give an example, the linear regression model predicts the synthetic promoter activities (Gertz, Siggia, & Cohen, 2009); L2 library in the original paper includes random combinations of 8 TFBSs (Reb1, Reb1r, Gcr1, Gcr1r, Mig, Migr, Rap1, Rap1r, where 'r' represents the reverse orientation) on 471 yeast promoters integrated into TRP1 locus. The gene expression level was quantified by the intensity of yellow fluorescent protein and transformed into base-10 logarithm. By using the frequency of TFBS on a synthetic promoter as the explanatory variable, the regression model explained 76% of the variance in expression (Figure 3A). Moreover, the model implies that Mig has repressive activity which is consistent with the results in Gertz's study and that Gcr1r has relatively less effect (Figure 3B).

Figure 3. Estimation of TFBS activities in yeast synthetic promoter by regression modelling

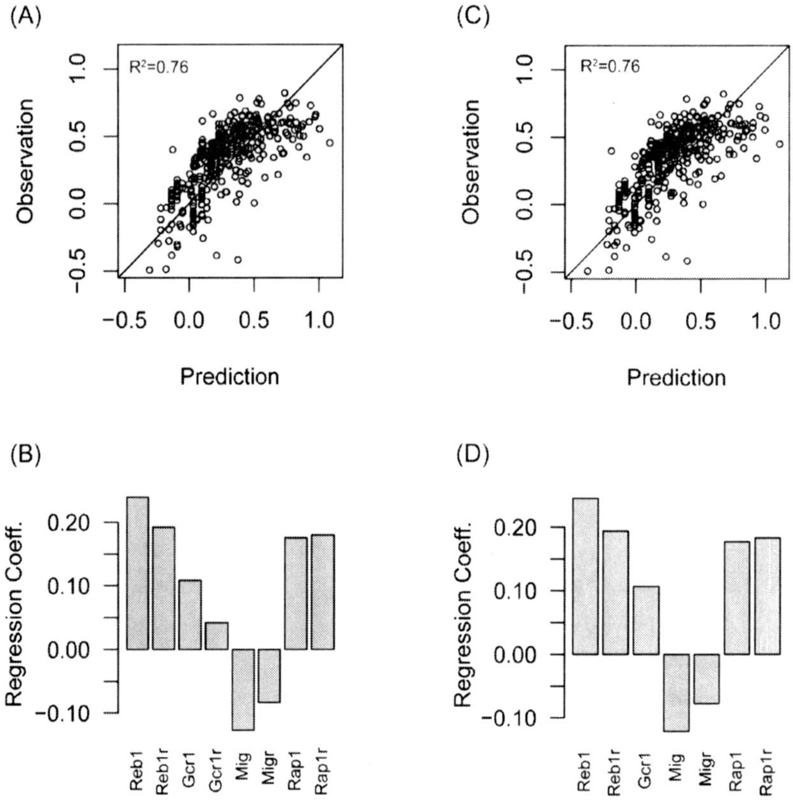

Feature Selection

To choose a subset of informative features from the input data, the feature selection process is performed by assessing redundancy and irrelevancy of variables. For example, in the yeast synthetic promoter modeling, even if Gcr1r that has the weakest RC is removed from the input data, the performance of regression model is not decreased (Figure 3C, 3D), suggesting that Gcr1r is irrelevant to explaining gene expression. Note that the simplest model without loss of predictive power is the first choice.

To find irrelevant explanatory variables, AIC (Akaike's information criterion) and *k*-fold cross-validation (CV) are widely used (Irie et al., 2011; O'Connor & Bailey, 2014; Ouyang et al., 2009). AIC is given as $AIC=-2(\ln L-K)$, where L is the likelihood of the input data on given model and K is the degree of freedom of the model. This equation implies that a good-fitted candidate model which minimizes the Kullback-Leibler distance should have fewer parameters. If the removal of an explanatory variable reduces AIC, then this candidate model is accepted. Otherwise, another variable is tested. This stepwise AIC is repeated until any removal does not minimize AIC.

CV can be used to test the predictive power of models. That is, after splitting the input data into k blocks, the RC *ws* are fitted with $k-1$ blocks. Then, the fitted model predicts *Y*s of the remaining part of the data using its *X*s of the remaining part. Since the stepwise AIC process is neither exhaustive nor rigorous, variables that are overlooked by AIC may further improve the predictive power. To rescue such variables, Park et al. have tested iterative procedures coupled with 5-fold CV (S. J. Park & Nakai, 2011; S. J. Park et al., 2014).

INFERENCE OF TRANSCRIPTIONAL REGULATION IN ZGA

Who's Talking First?

The earlier study conducting somatic cell nuclei transfer (SCNT) revealed that oocytes can reprogram somatic nuclei into zygotic nuclei (Gurdon, 1962) by a deterministic reprogramming process that is closely similar to fertilization (Jullien et al., 2014; Jullien, Pasque, Halley-Stott, Miyamoto, & Gurdon, 2011). This finding has been further extended to the production of pluripotent stem cells using defined factors (Takahashi et al., 2007) and zygote SCNT (Egli et al., 2011). However, most studies have found abnormalities and developmental arrest from the induced stem cells (Egli et al., 2011; Ma et al., 2014), suggesting the existence of reprogramming barriers. During natural fertilization, parental factors can overcome these barriers by communicating with each other. So, who will begin to talk first? The identification of triggers and their functions is a long-standing question.

Over last few decades several studies have been conducted to understand the transcriptome dynamics during fertilization by RNA-seq analysis. These help to identify the origin of potential TFs, which can be further used for inferring transcriptional regulation. For example, Park et al. have performed a large-scale whole-transcriptome analysis with mouse one-cell normal and parthenogenetic embryos, sperm, and transcriptional inhibition (S. J. Park et al., 2013). They then categorized genes into six groups according to the expression states among embryos and sperm. These categories include sperm-driven mRNAs (expressed at sperm but not at oocyte) and newly transcribed mRNAs (i.e. *de novo* transcript, expressed at one-cell embryo but not at sperm and oocyte) at one-cell stage. Moreover, they analyzed the enrichment of JASPAR motifs (Vlieghe et al., 2006) in the promoters of each group, finding 103 significant motifs. The results revealed that the genes in *de novo* transcript group at one-cell stage possess well-conserved TFBSs distinctive from genes in other groups.

Furthermore, by utilizing TRANSFAC (Wingender et al., 2000) that provides the list of TFs binding to the enriched JASPAR motifs, they retrieved 108 TFs whose mRNAs are in the six groups; 97 maternal and 11 non-maternal mRNAs (3 *de novo*, 2 sperm-driven, and 6 promiscuous factors). These TF-coding genes (browsable at http://dbtmee.hgc.jp/regulation.php) are candidates to turn on the regulatory switches of enriched motifs. Interestingly, through a network analysis, they revealed that 5 *de novo* and sperm-driven TFs (Foxd1, Nkx2-5, Myod1, Runx1, and Sox18) potentially involve self-regulation, i.e. a TF-coding gene requires the bindings of itself to its promoter. For example, Sox18 targeting 286 genes that are transcribed at one-cell stage but not transcribed at oocyte (http://dbtmee.hgc.jp/network/tf-gene.php?TF=Sox18&CNT=286&CATE=Nascent-1C) is absent in oocyte, but it is detected at one-cell stage and in sperm. Sox18 is potentially regulated by 41 TFs including itself and 3 other sperm-driven TFs Myod1, Runx1, and Sox9 (http://dbtmee.hgc.jp/network/gene-tf.php?ID=DBTMEE:0019624&Gene=Sox18&RefSeq=NM_009236&CC=0.0520459&Category=Spr1C).

Therefore, it is thought that sperm-driven mRNAs of TFs prime themselves for rapid transcription immediately after fertilization. Indeed, parthenogenetic embryos in which sperm makes no genetic contribution exhibit delayed onset of the gene activation (Bui et al., 2011). By coordinating with maternal TFs, these sperm-driven primers may give the competence to overcome the reprogramming barriers.

What Comes Next?

Since a general difficulty in dealing with early embryos for TF ChIP-seq experiments, the characterization of global regulatory networks during fertilization has been conducted by integrative and comparative approaches that utilize genome-wide expression data. For example, Vassena et al. mapped their embryo expression data to the pluripotency network reported previously (Vassena et al., 2011). Cao et al. built the co-expression network for pig preimplantation development with orthologous genes to mouse (Cao et al., 2014). Xue et al. identified gene modules co-expressed during human preimplantation development and found that some of the stage-specific modules are preserved in mouse (Xue et al., 2013). Although these studies may help to deepen our understanding of the regulations, more efforts are required to answer what factors govern the transcriptional changes during development, or more generally how *cis*- and *trans*-regulatory dynamics relate to the cellular and molecular features that define mammal early embryogenesis.

In particular, the MZT that comes after fertilization is one of the most dramatic events, accompanying the first major transcriptional change; approximately 90% of maternal mRNAs undergo degradation by two-cell stage, and the major ZGA commences at this stage. To elicit key regulators in the MZT, Park et al. have proposed a computational promoter modeling based on the linear regression model and LGM (S. J. Park et al., 2014). They originally used this approach for studying key TFs during hematopoiesis.

The modeling using the dataset Park et al. 2013 starts by detecting differentially expressed genes (DEGs); 878 up- (Class B) and 947 down-regulated genes (Class A) at two-cell stage (2C) compared to at the oocyte (Oo) (Figure 4A). Because LGM requires a longer runtime for testing the conditional independency among larger number of explanatory variables, they first picked up 400 genes that were most expressed in each class. After preparing DNA sequences from promoter regions spanning ±5kbp from transcription start sites of the genes, they searched TRANSFAC TFBSs by MATCH tools (Kel et al., 2003) and prepared TFs that are known to bind to the TFBSs. Then, they built four regression models, where response variables are gene expression of Class A in oocyte and at two-cell stage, and gene expression of Class B in oocyte and at two-cell stage (Figure 4B).

The explanatory variables in each model were designed by incorporating the similarity score of TFBSs weighted by positional bias, the expression fold change of bound TFs, and their ECPs of the iterative random sampling of LGM (see the definition of TFBS-Gene association score (TGAS) V in the original paper). Each model was tested by 100 runs of the procedure proposed in Park et al. 2014 (Figure 4B). The procedure comprises the stepwise AIC process and the iterative regression procedure coupled with random variable selection and 5-fold CV (S. J. Park et al., 2014). To assess the statistical significance of 100 RCs for an explanatory variable, they performed single-sample t-test using the RCs, which rejects the null hypothesis that the mean value of RCs is equal to zero ($p < 0.05$). The final RCs were assigned by the mean values (Figure 4C) that reflect the higher-order interaction of TFs and are necessary for explaining gene expression in each model.

One can speculate on the change of TFBS activities in the models, which may define the MZT. For example, FXR/RXR-alpha, the leftmost-TFBS in Figure 4C and targeted by nuclear receptor Nr1h4 (http://dbtmee.hgc.jp/gene_card.php?id=2014619), shows the most significant repressive activity for Class A genes. This repressive activity is not changed between the oocyte and two-cell stage. However, this activity becomes

Figure 4. Identification of potential key regulators in MZT

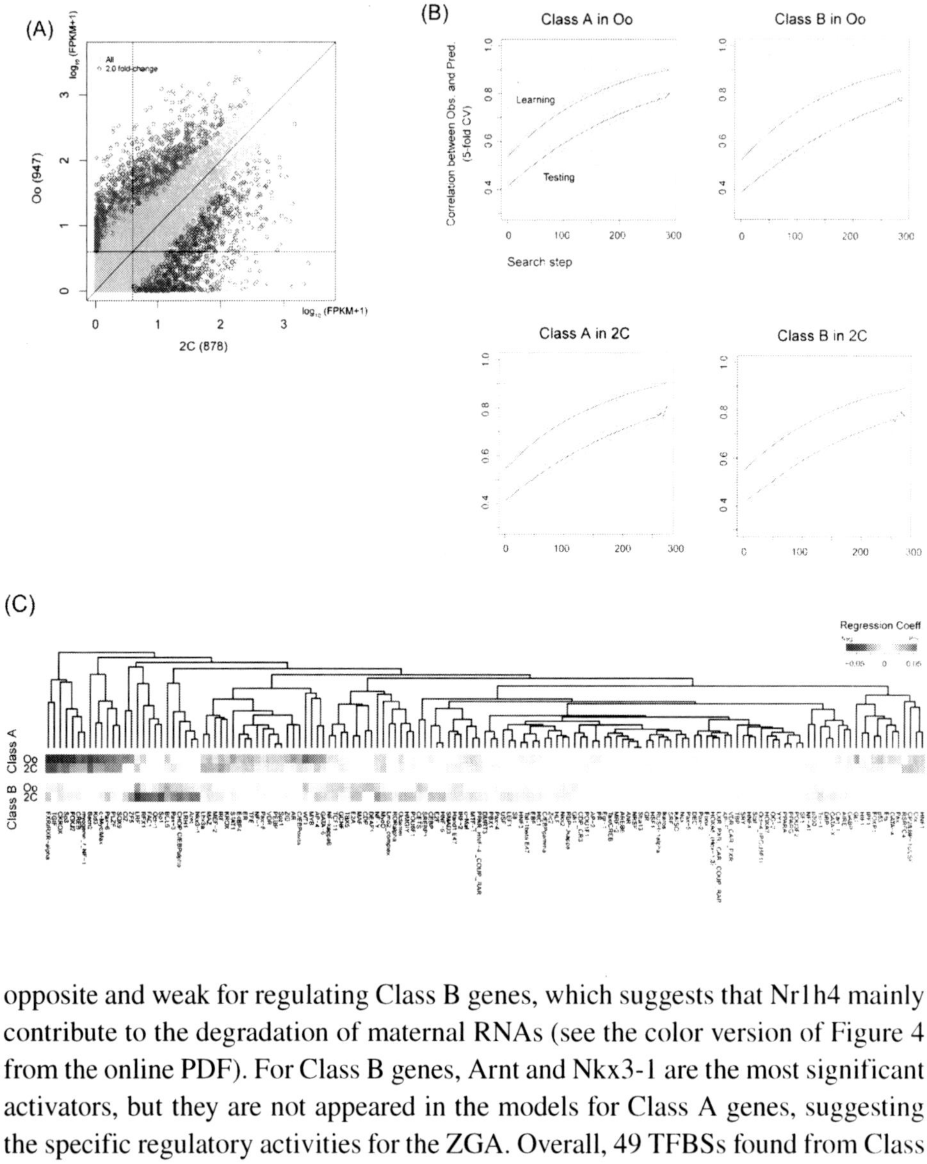

opposite and weak for regulating Class B genes, which suggests that Nr1h4 mainly contribute to the degradation of maternal RNAs (see the color version of Figure 4 from the online PDF). For Class B genes, Arnt and Nkx3-1 are the most significant activators, but they are not appeared in the models for Class A genes, suggesting the specific regulatory activities for the ZGA. Overall, 49 TFBSs found from Class A and 42 TFBSs found from Class B exhibited no changes in the mode of their activities between the oocyte and two-cell stage. In addition, 16% (8/49) for Class A and 83% (35/42) for Class B more significantly contributed to the regulation at two-cell stage compared with at the oocyte. This propensity is distinguishable from what observed from the lineage commitment in blood cells (S. J. Park et al., 2014).

FUTURE RESEARCH DIRECTIONS

In spite of the rising influx of data, many questions still remain unclear, which include what factors govern the MZT, or more generally how *cis-* and *trans-*regulatory dynamics relate to the features that define mammal early embryogenesis. By giving examples with model organisms, we reviewed reverse engineering algorithms that convert gene expression profiles into functionally meaningful genetic information. It is important to know that these methods applicable for a wide range of biological interest have resolved broad theoretical and practical issues, including noisy and error-prone features of biological datasets and limited number of experiments.

In particular, we introduced the linear regression modeling and log-linear graphical modeling to identify potential key regulators in the higher-order conditional distribution where statistical paradox often occurs. The key points in the regression approach are the design of explanatory variable and the method of finding parsimonious models. In addition, to generalize the regression model by avoiding model over-fitting is a difficult challenge. It will be interesting to investigate whether the over-fitting relates to biological aspects. The log-linear graphical model is a unique approach to detect the dependency of discretized variables without being caught by spuriousness, which enhances the investigation of the combinatorial nature of TFs. A problem is that the log-linear graphical model becomes infeasible when a large number of variables are given, which needs practical and theoretical studies to overcome this limitation.

In studying fertilization, due to the paucity of regulatory information from experiments, *in silico* approaches are indispensable to integrate and to extend expression datasets for a deeper understanding of regulatory dynamics. That is, differentially expressed TFs at a certain stage can be identified from the whole-transcriptome analysis, and such expression information can help to examine the explanatory variable in modeling regulatory dynamics along with predicted features including the enrichment of TF-binding motifs. Moreover, as shown in the published study (S. J. Park & Nakai, 2011), to incorporate the interplay between genetic and epigenetic effects into the explanatory variable gives new insights. In this regard, many studies have recently generated epigenetic profiles during fertilization including DNA methylation (Kobayashi et al., 2012) and histone modifications (Erkek et al., 2013), and further works need to take into account the way that efficiently merges such epigenetic profile with genetic profile. As a particular one of several possibilities, since the presence of DNA methylation and specific histone marks in TFBSs may affect the affinity of cognate TFs, their dependency that can be inferred by LGM will allow to assess what type of explanatory variables improves predictability and why.

Regulatory dynamics exerted by the transcriptional machinery underlay the programmed waves of gene expression at early embryos. It will be exciting to establish the profile of regulatory dynamics and to see how they are switching from one to another. To this end, more attempts in biological experiments are required to validate TF-target gene relationships that are challenging in general. The attempts will include high-throughput sequencing technologies that capture allele-specific gene expression and protein-protein interactions, and transcriptional and translation inhibitions for whole or specific molecules in gametes and embryos. Further computational studies incorporating such attempts will offer new insight into the regulatory dynamics.

CONCLUSION

Early embryo development is unique in the way that it exhibits the sequential set of gene expression patterns strictly regulated by the highly complex interplay between genetic and epigenetic effects. In spite of the progress achieved through the genome-wide analysis of landmark events, its applications in clinical therapy and stem cell biology still remain low in success rates. The improvement can be achieved by highly reliable biological experiment technology and computational methodology. In the near future, cellular and molecular characteristics of embryo development will be significantly upgraded. Thereby, computational methods will increase the predictive power and become applicable to the modeling of regulatory dynamics, which makes it possible to take a closer look at how species are maintained.

REFERENCES

Adenot, P. G., Mercier, Y., Renard, J. P., & Thompson, E. M. (1997). Differential H4 acetylation of paternal and maternal chromatin precedes DNA replication and differential transcriptional activity in pronuclei of 1-cell mouse embryos. *Development*, *124*(22), 4615–4625. Retrieved from http://www.ncbi.nlm.nih.gov/pubmed/9409678 PMID:9409678

Angelini, C., De Canditiis, D., Mutarelli, M., & Pensky, M. (2007). A Bayesian approach to estimation and testing in time-course microarray experiments. *Stat Appl Genet Mol Biol, 6*, Article24. doi:10.2202/1544-6115.1299

Aoshima, K., Inoue, E., Sawa, H., & Okada, Y. (2015). Paternal H3K4 methylation is required for minor zygotic gene activation and early mouse embryonic development. *EMBO Reports*, *16*(7), 803–812. doi:10.15252/embr.201439700 PMID:25925669

Basso, K., Margolin, A. A., Stolovitzky, G., Klein, U., Dalla-Favera, R., & Califano, A. (2005). Reverse engineering of regulatory networks in human B cells. *Nature Genetics*, *37*(4), 382–390. doi:10.1038/ng1532 PMID:15778709

Beer, M. A., & Tavazoie, S. (2004). Predicting gene expression from sequence. *Cell*, *117*(2), 185–198. doi:10.1016/S0092-8674(04)00304-6 PMID:15084257

Brunet-Simon, A., Henrion, G., Renard, J. P., & Duranthon, V. (2001). Onset of zygotic transcription and maternal transcript legacy in the rabbit embryo. *Molecular Reproduction and Development*, *58*(2), 127–136. doi:10.1002/1098-2795(200102)58:2<127::AID-MRD1>3.0.CO;2-A PMID:11139224

Brunner, A. M., Nanni, P., & Mansuy, I. M. (2014). Epigenetic marking of sperm by post-translational modification of histones and protamines. *Epigenetics Chromatin*, *7*(1), 2. doi:10.1186/1756-8935-7-2 PMID:24443974

Bui, H. T., Wakayama, S., Mizutani, E., Park, K. K., Kim, J. H., Van Thuan, N., & Wakayama, T. (2011). Essential role of paternal chromatin in the regulation of transcriptional activity during mouse preimplantation development. *Reproduction (Cambridge, England)*, *141*(1), 67–77. doi:10.1530/REP-10-0109 PMID:20974742

Burkart, A. D., Xiong, B., Baibakov, B., Jimenez-Movilla, M., & Dean, J. (2012). Ovastacin, a cortical granule protease, cleaves ZP2 in the zona pellucida to prevent polyspermy. *The Journal of Cell Biology*, *197*(1), 37–44. doi:10.1083/jcb.201112094 PMID:22472438

Bussemaker, H. J., Foat, B. C., & Ward, L. D. (2007). Predictive modeling of genome-wide mRNA expression: From modules to molecules. *Annual Review of Biophysics and Biomolecular Structure*, *36*(1), 329–347. doi:10.1146/annurev.biophys.36.040306.132725 PMID:17311525

Bussemaker, H. J., Li, H., & Siggia, E. D. (2001). Regulatory element detection using correlation with expression. *Nature Genetics*, *27*(2), 167–171. doi:10.1038/84792 PMID:11175784

Cahan, P., Li, H., Morris, S. A., Lummertz da Rocha, E., Daley, G. Q., & Collins, J. J. (2014). CellNet: Network biology applied to stem cell engineering. *Cell*, *158*(4), 903–915. doi:10.1016/j.cell.2014.07.020 PMID:25126793

Cantone, I., & Fisher, A. G. (2013). Epigenetic programming and reprogramming during development. *Nature Structural & Molecular Biology*, *20*(3), 282–289. doi:10.1038/nsmb.2489 PMID:23463313

Cao, S., Han, J., Wu, J., Li, Q., Liu, S., Zhang, W., & Li, N. et al. (2014). Specific gene-regulation networks during the pre-implantation development of the pig embryo as revealed by deep sequencing. *BMC Genomics*, *15*(1), 4. doi:10.1186/1471-2164-15-4 PMID:24383959

Chang, C. C., Ma, Y., Jacobs, S., Tian, X. C., Yang, X., & Rasmussen, T. P. (2005). A maternal store of macroH2A is removed from pronuclei prior to onset of somatic macroH2A expression in preimplantation embryos. *Developmental Biology*, *278*(2), 367–380. doi:10.1016/j.ydbio.2004.11.032 PMID:15680357

Chen, J., Melton, C., Suh, N., Oh, J. S., Horner, K., Xie, F., & Conti, M. et al. (2011). Genome-wide analysis of translation reveals a critical role for deleted in azoospermia-like (Dazl) at the oocyte-to-zygote transition. *Genes & Development*, *25*(7), 755–766. doi:10.1101/gad.2028911 PMID:21460039

Chen, L., Kostadima, M., Martens, J. H., Canu, G., Garcia, S. P., Turro, E., & Rendon, A. et al. (2014). Transcriptional diversity during lineage commitment of human blood progenitors. *Science*, *345*(6204), 1251033. doi:10.1126/science.1251033 PMID:25258084

Clift, D., & Schuh, M. (2013). Restarting life: Fertilization and the transition from meiosis to mitosis. *Nature Reviews. Molecular Cell Biology*, *14*(9), 549–562. doi:10.1038/nrm3643 PMID:23942453

Condic, M. L. (2014). Totipotency: What it is and what it is not. *Stem Cells and Development*, *23*(8), 796–812. doi:10.1089/scd.2013.0364 PMID:24368070

Cowell, I. G., Aucott, R., Mahadevaiah, S. K., Burgoyne, P. S., Huskisson, N., Bongiorni, S., & Singh, P. B. et al. (2002). Heterochromatin, HP1 and methylation at lysine 9 of histone H3 in animals. *Chromosoma*, *111*(1), 22–36. doi:10.1007/s00412-002-0182-8 PMID:12068920

Das, D., Pellegrini, M., & Gray, J. W. (2009). A primer on regression methods for decoding cis-regulatory logic. *PLoS Computational Biology*, *5*(1), e1000269. doi:10.1371/journal.pcbi.1000269 PMID:19180174

Datta, D., & Zhao, H. (2008). Statistical methods to infer cooperative binding among transcription factors in Saccharomyces cerevisiae. *Bioinformatics (Oxford, England)*, *24*(4), 545–552. doi:10.1093/bioinformatics/btm523 PMID:17989095

Dean, J. (2004). Reassessing the molecular biology of sperm-egg recognition with mouse genetics. *BioEssays*, *26*(1), 29–38. doi:10.1002/bies.10412 PMID:14696038

Deng, Q., Ramskold, D., Reinius, B., & Sandberg, R. (2014). Single-cell RNA-seq reveals dynamic, random monoallelic gene expression in mammalian cells. *Science*, *343*(6167), 193–196. doi:10.1126/science.1245316 PMID:24408435

Dias, B. G., & Ressler, K. J. (2014). Parental olfactory experience influences behavior and neural structure in subsequent generations. *Nature Neuroscience*, *17*(1), 89–96. doi:10.1038/nn.3594 PMID:24292232

Edwards, D. (2000). *Introduction to Graphical Modelling*. Springer. doi:10.1007/978-1-4612-0493-0

Egli, D., Chen, A. E., Saphier, G., Ichida, J., Fitzgerald, C., Go, K. J., & Eggan, K. et al. (2011). Reprogramming within hours following nuclear transfer into mouse but not human zygotes. *Nature Communications*, *2*, 488. doi:10.1038/ncomms1503 PMID:21971503

Egozcue, J., Sarrate, Z., Codina-Pascual, M., Egozcue, S., Oliver-Bonet, M., Blanco, J., & Vidal, F. et al. (2005). Meiotic abnormalities in infertile males. *Cytogenetic and Genome Research*, *111*(3-4), 337–342. doi:10.1159/000086907 PMID:16192712

Erkek, S., Hisano, M., Liang, C. Y., Gill, M., Murr, R., Dieker, J., & Peters, A. H. et al. (2013). Molecular determinants of nucleosome retention at CpG-rich sequences in mouse spermatozoa. *Nature Structural & Molecular Biology*, *20*(7), 868–875. doi:10.1038/nsmb.2599 PMID:23770822

Evans, J. P., Hardy, D. M., & Storey, B. T. (2001). *Sperm–Egg Interactions: Sperm–Egg Binding in Mammals eLS*. John Wiley & Sons, Ltd.

Faith, J. J., Hayete, B., Thaden, J. T., Mogno, I., Wierzbowski, J., Cottarel, G., & Gardner, T. S. et al. (2007). Large-scale mapping and validation of Escherichia coli transcriptional regulation from a compendium of expression profiles. *PLoS Biology*, *5*(1), e8. doi:10.1371/journal.pbio.0050008 PMID:17214507

Fatica, A., & Bozzoni, I. (2014). Long non-coding RNAs: New players in cell differentiation and development. *Nature Reviews. Genetics*, *15*(1), 7–21. doi:10.1038/nrg3606 PMID:24296535

Ferramosca, A., & Zara, V. (2014). Bioenergetics of mammalian sperm capacitation. *Biomed Res Int*, *902953*. doi:10.1155/2014/902953 PMID:24791005

Friedman, N. (2004). Inferring cellular networks using probabilistic graphical models. *Science*, *303*(5659), 799–805. doi:10.1126/science.1094068 PMID:14764868

Friedman, N., Linial, M., Nachman, I., & Pe'er, D. (2000). Using Bayesian networks to analyze expression data. *Journal of Computational Biology*, *7*(3-4), 601–620. doi:10.1089/106652700750050961 PMID:11108481

Gamble, M. J., Frizzell, K. M., Yang, C., Krishnakumar, R., & Kraus, W. L. (2010). The histone variant macroH2A1 marks repressed autosomal chromatin, but protects a subset of its target genes from silencing. *Genes & Development*, *24*(1), 21–32. doi:10.1101/gad.1876110 PMID:20008927

Gao, F., Foat, B. C., & Bussemaker, H. J. (2004). Defining transcriptional networks through integrative modeling of mRNA expression and transcription factor binding data. *BMC Bioinformatics*, *5*(1), 31. doi:10.1186/1471-2105-5-31 PMID:15113405

Gaspar-Maia, A., Qadeer, Z. A., Hasson, D., Ratnakumar, K., Leu, N. A., Leroy, G., & Bernstein, E. et al. (2013). MacroH2A histone variants act as a barrier upon reprogramming towards pluripotency. *Nature Communications*, *4*, 1565. doi:10.1038/ncomms2582 PMID:23463008

Gatewood, J. M., Cook, G. R., Balhorn, R., Schmid, C. W., & Bradbury, E. M. (1990). Isolation of four core histones from human sperm chromatin representing a minor subset of somatic histones. *The Journal of Biological Chemistry*, *265*(33), 20662–20666. Retrieved from http://www.ncbi.nlm.nih.gov/pubmed/2243112 PMID:2243112

Gazdag, E., Rajkovic, A., Torres-Padilla, M. E., & Tora, L. (2007). Analysis of TATA-binding protein 2 (TBP2) and TBP expression suggests different roles for the two proteins in regulation of gene expression during oogenesis and early mouse development. *Reproduction (Cambridge, England)*, *134*(1), 51–62. doi:10.1530/REP-06-0337 PMID:17641088

Gertz, J., Siggia, E. D., & Cohen, B. A. (2009). Analysis of combinatorial cis-regulation in synthetic and genomic promoters. *Nature*, *457*(7226), 215–218. doi:10.1038/nature07521 PMID:19029883

Guan, Y., Boavida, L., & McCormick, S. (2001). *Gametogenesis eLS*. John Wiley & Sons, Ltd.

Gurdon, J. B. (1962). The developmental capacity of nuclei taken from intestinal epithelium cells of feeding tadpoles. *Journal of Embryology and Experimental Morphology*, *10*, 622–640. Retrieved from http://www.ncbi.nlm.nih.gov/pubmed/13951335 PMID:13951335

Haberle, V., Li, N., Hadzhiev, Y., Plessy, C., Previti, C., Nepal, C., & Lenhard, B. et al. (2014). Two independent transcription initiation codes overlap on vertebrate core promoters. *Nature, 507*(7492), 381–385. doi:10.1038/nature12974 PMID:24531765

Hamatani, T., Carter, M. G., Sharov, A. A., & Ko, M. S. (2004). Dynamics of global gene expression changes during mouse preimplantation development. *Developmental Cell, 6*(1), 117–131. doi:10.1016/S1534-5807(03)00373-3 PMID:14723852

Hamazaki, N., Uesaka, M., Nakashima, K., Agata, K., & Imamura, T. (2015). Gene activation-associated long noncoding RNAs function in mouse preimplantation development. *Development, 142*(5), 910–920. doi:10.1242/dev.116996 PMID:25633350

Hammoud, S. S., Nix, D. A., Zhang, H., Purwar, J., Carrell, D. T., & Cairns, B. R. (2009). Distinctive chromatin in human sperm packages genes for embryo development. *Nature, 460*(7254), 473–478. doi:10.1038/nature08162 PMID:19525931

Handel, M. A., & Schimenti, J. C. (2010). Genetics of mammalian meiosis: Regulation, dynamics and impact on fertility. *Nature Reviews. Genetics, 11*(2), 124–136. doi:10.1038/nrg2723 PMID:20051984

Harbison, C. T., Gordon, D. B., Lee, T. I., Rinaldi, N. J., Macisaac, K. D., Danford, T. W., & Young, R. A. et al. (2004). Transcriptional regulatory code of a eukaryotic genome. *Nature, 431*(7004), 99–104. doi:10.1038/nature02800 PMID:15343339

Hassold, T., & Hunt, P. (2001). To err (meiotically) is human: The genesis of human aneuploidy. *Nature Reviews. Genetics, 2*(4), 280–291. doi:10.1038/35066065 PMID:11283700

Heard, E., & Martienssen, R. A. (2014). Transgenerational epigenetic inheritance: Myths and mechanisms. *Cell, 157*(1), 95–109. doi:10.1016/j.cell.2014.02.045 PMID:24679529

Heydtmann, M. (2002). The nature of truth: Simpson's Paradox and the limits of statistical data. *QJM, 95*(4), 247–249. doi:10.1093/qjmed/95.4.247 PMID:11937652

Holter, N. S., Maritan, A., Cieplak, M., Fedoroff, N. V., & Banavar, J. R. (2001). Dynamic modeling of gene expression data. *Proceedings of the National Academy of Sciences of the United States of America, 98*(4), 1693–1698. doi:10.1073/pnas.98.4.1693 PMID:11172013

Hoodbhoy, T., Joshi, S., Boja, E. S., Williams, S. A., Stanley, P., & Dean, J. (2005). Human sperm do not bind to rat zonae pellucidae despite the presence of four homologous glycoproteins. *The Journal of Biological Chemistry, 280*(13), 12721–12731. doi:10.1074/jbc.M413569200 PMID:15677449

Hoskins, R. A., Landolin, J. M., Brown, J. B., Sandler, J. E., Takahashi, H., Lassmann, T., & Celniker, S. E. et al. (2011). Genome-wide analysis of promoter architecture in Drosophila melanogaster. *Genome Research, 21*(2), 182–192. doi:10.1101/gr.112466.110 PMID:21177961

Hou, Y., Fan, W., Yan, L., Li, R., Lian, Y., Huang, J., & Qiao, J. et al. (2013). Genome analyses of single human oocytes. *Cell, 155*(7), 1492–1506. doi:10.1016/j.cell.2013.11.040 PMID:24360273

Irie, T., Park, S. J., Yamashita, R., Seki, M., Yada, T., Sugano, S., & Suzuki, Y. et al. (2011). Predicting promoter activities of primary human DNA sequences. *Nucleic Acids Research, 39*(11), e75. doi:10.1093/nar/gkr173 PMID:21486745

Iwafuchi-Doi, M., & Zaret, K. S. (2014). Pioneer transcription factors in cell reprogramming. *Genes & Development, 28*(24), 2679–2692. doi:10.1101/gad.253443.114 PMID:25512556

Jullien, J., Miyamoto, K., Pasque, V., Allen, G. E., Bradshaw, C. R., Garrett, N. J., & Gurdon, J. B. et al. (2014). Hierarchical molecular events driven by oocyte-specific factors lead to rapid and extensive reprogramming. *Molecular Cell, 55*(4), 524–536. doi:10.1016/j.molcel.2014.06.024 PMID:25066233

Jullien, J., Pasque, V., Halley-Stott, R. P., Miyamoto, K., & Gurdon, J. B. (2011). Mechanisms of nuclear reprogramming by eggs and oocytes: A deterministic process? *Nature Reviews. Molecular Cell Biology, 12*(7), 453–459. doi:10.1038/nrm3140 PMID:21697902

Kel, A. E., Gossling, E., Reuter, I., Cheremushkin, E., Kel-Margoulis, O. V., & Wingender, E. (2003). MATCH: A tool for searching transcription factor binding sites in DNA sequences. *Nucleic Acids Research, 31*(13), 3576–3579. doi:10.1093/nar/gkg585 PMID:12824369

Ko, M. S., Kitchen, J. R., Wang, X., Threat, T. A., Hasegawa, A., Sun, T., & Doi, H. et al. (2000). Large-scale cDNA analysis reveals phased gene expression patterns during preimplantation mouse development. *Development, 127*(8), 1737–1749. Retrieved from http://www.ncbi.nlm.nih.gov/entrez/query.fcgi?cmd=Retrieve&db=PubMed&dopt=Citation&list_uids=10725249 PMID:10725249

Kobayashi, H., Sakurai, T., Imai, M., Takahashi, N., Fukuda, A., Yayoi, O., & Kono, T. et al. (2012). Contribution of intragenic DNA methylation in mouse gametic DNA methylomes to establish oocyte-specific heritable marks. *PLOS Genetics, 8*(1), e1002440. doi:10.1371/journal.pgen.1002440 PMID:22242016

Krawetz, S. A. (2005). Paternal contribution: New insights and future challenges. *Nature Reviews. Genetics*, *6*(8), 633–642. doi:10.1038/nrg1654 PMID:16136654

Lauritzen, S. (1996). *Graphical Models (Oxford Statistical Science Series)*. Oxford University Press.

Lee, M., Choi, Y., Kim, K., Jin, H., Lim, J., Nguyen, T. A., & Kim, V. N. et al. (2014). Adenylation of maternally inherited microRNAs by Wispy. *Molecular Cell*, *56*(5), 696–707. doi:10.1016/j.molcel.2014.10.011 PMID:25454948

Lee, M. T., Bonneau, A. R., & Giraldez, A. J. (2014). Zygotic Genome Activation During the Maternal-to-Zygotic Transition. *Annual Review of Cell and Developmental Biology*, *30*(1), 581–613. doi:10.1146/annurev-cellbio-100913-013027 PMID:25150012

Lee, M. T., Bonneau, A. R., Takacs, C. M., Bazzini, A. A., DiVito, K. R., Fleming, E. S., & Giraldez, A. J. (2013). Nanog, Pou5f1 and SoxB1 activate zygotic gene expression during the maternal-to-zygotic transition. *Nature*, *503*(7476), 360–364. doi:10.1038/nature12632 PMID:24056933

Lee, T. I., Rinaldi, N. J., Robert, F., Odom, D. T., Bar-Joseph, Z., Gerber, G. K., & Young, R. A. et al. (2002). Transcriptional regulatory networks in Saccharomyces cerevisiae. *Science*, *298*(5594), 799–804. doi:10.1126/science.1075090 PMID:12399584

Lesch, B. J., & Page, D. C. (2012). Genetics of germ cell development. *Nature Reviews. Genetics*, *13*(11), 781–794. doi:10.1038/nrg3294 PMID:23044825

Li, L., Lu, X., & Dean, J. (2013). The maternal to zygotic transition in mammals. *Molecular Aspects of Medicine*, *34*(5), 919–938. doi:10.1016/j.mam.2013.01.003 PMID:23352575

Li, R., & Albertini, D. F. (2013). The road to maturation: Somatic cell interaction and self-organization of the mammalian oocyte. *Nature Reviews. Molecular Cell Biology*, *14*(3), 141–152. doi:10.1038/nrm3531 PMID:23429793

Liang, H. L., Nien, C. Y., Liu, H. Y., Metzstein, M. M., Kirov, N., & Rushlow, C. (2008). The zinc-finger protein Zelda is a key activator of the early zygotic genome in Drosophila. *Nature*, *456*(7220), 400–403. doi:10.1038/nature07388 PMID:18931655

Ma, H., Morey, R., O'Neil, R. C., He, Y., Daughtry, B., Schultz, M. D., & Mitalipov, S. et al. (2014). Abnormalities in human pluripotent cells due to reprogramming mechanisms. *Nature*, *511*(7508), 177–183. doi:10.1038/nature13551 PMID:25008523

Macfarlan, T. S., Gifford, W. D., Driscoll, S., Lettieri, K., Rowe, H. M., Bonanomi, D., & Pfaff, S. L. et al. (2012). Embryonic stem cell potency fluctuates with endogenous retrovirus activity. *Nature*, *487*(7405), 57–63. doi:10.1038/nature11244 PMID:22722858

MacIsaac, K. D., Wang, T., Gordon, D. B., Gifford, D. K., Stormo, G. D., & Fraenkel, E. (2006). An improved map of conserved regulatory sites for Saccharomyces cerevisiae. *BMC Bioinformatics*, *7*(1), 113. doi:10.1186/1471-2105-7-113 PMID:16522208

Maekawa, M., Yamaguchi, K., Nakamura, T., Shibukawa, R., Kodanaka, I., Ichisaka, T., & Yamanaka, S. et al. (2011). Direct reprogramming of somatic cells is promoted by maternal transcription factor Glis1. *Nature*, *474*(7350), 225–229. doi:10.1038/nature10106 PMID:21654807

Marbach, D., Costello, J. C., Kuffner, R., Vega, N. M., Prill, R. J., Camacho, D. M., & Stolovitzky, G. et al. (2012). Wisdom of crowds for robust gene network inference. *Nature Methods*, *9*(8), 796–804. doi:10.1038/nmeth.2016 PMID:22796662

Martin, S., Zhang, Z., Martino, A., & Faulon, J. L. (2007). Boolean dynamics of genetic regulatory networks inferred from microarray time series data. *Bioinformatics (Oxford, England)*, *23*(7), 866–874. doi:10.1093/bioinformatics/btm021 PMID:17267426

Messerschmidt, D. M., Knowles, B. B., & Solter, D. (2014). DNA methylation dynamics during epigenetic reprogramming in the germline and preimplantation embryos. *Genes & Development*, *28*(8), 812–828. doi:10.1101/gad.234294.113 PMID:24736841

Miller, D., Brinkworth, M., & Iles, D. (2010). Paternal DNA packaging in spermatozoa: More than the sum of its parts? DNA, histones, protamines and epigenetics. *Reproduction (Cambridge, England)*, *139*(2), 287–301. doi:10.1530/REP-09-0281 PMID:19759174

Mizusawa, Y., Kuji, N., Tanaka, Y., Tanaka, M., Ikeda, E., Komatsu, S., & Yoshimura, Y. et al. (2010). Expression of human oocyte-specific linker histone protein and its incorporation into sperm chromatin during fertilization. *Fertility and Sterility*, *93*(4), 1134–1141. doi:10.1016/j.fertnstert.2008.11.028 PMID:19147139

Morita, M., Nakamura, M., Hamada, M., & Takahashi, S. (2011). Combinatorial motif analysis of regulatory gene expression in Mafb deficient macrophages. *BMC Systems Biology*, *5*(Suppl 2), S7. doi:10.1186/1752-0509-5-S2-S7 PMID:22784578

Necsulea, A., Soumillon, M., Warnefors, M., Liechti, A., Daish, T., Zeller, U., & Kaessmann, H. et al. (2014). The evolution of lncRNA repertoires and expression patterns in tetrapods. *Nature*, *505*(7485), 635–640. doi:10.1038/nature12943 PMID:24463510

Nothias, J. Y., Miranda, M., & DePamphilis, M. L. (1996). Uncoupling of transcription and translation during zygotic gene activation in the mouse. *The EMBO Journal*, *15*(20), 5715–5725. Retrieved from http://www.ncbi.nlm.nih.gov/pubmed/8896464 PMID:8896464

O'Connor, T. R., & Bailey, T. L. (2014). Creating and validating cis-regulatory maps of tissue-specific gene expression regulation. *Nucleic Acids Research*, *42*(17), 11000–11010. doi:10.1093/nar/gku801 PMID:25200088

Ouyang, Z., Zhou, Q., & Wong, W. H. (2009). ChIP-Seq of transcription factors predicts absolute and differential gene expression in embryonic stem cells. *Proceedings of the National Academy of Sciences of the United States of America*, *106*(51), 21521–21526. doi:10.1073/pnas.0904863106 PMID:19995984

Pan, X. (2014). Single Cell Analysis: From Technology to Biology and Medicine. *Single Cell Biol*, *3*(1). doi:10.4172/2168-9431.1000106 PMID:25177539

Park, S. J., Ichinose, N., & Yada, T. (2008). Probabilistic Graphical Modeling for Large-scale Combinatorial Regulation of Transcription Factors. In *Proc. the workshop on Knowledge, Language, and Learning in Bioinformatics* (KLLBI) (pp. 72-86). Retrieved from http://www.hgc.jp/~park/research/data/kllbi2008/park_kllbi2008.pdf

Park, S. J., Komata, M., Inoue, F., Yamada, K., Nakai, K., Ohsugi, M., & Shirahige, K. (2013). Inferring the choreography of parental genomes during fertilization from ultralarge-scale whole-transcriptome analysis. *Genes & Development*, *27*(24), 2736–2748. doi:10.1101/gad.227926.113 PMID:24352427

Park, S. J., & Nakai, K. (2011). A regression analysis of gene expression in ES cells reveals two gene classes that are significantly different in epigenetic patterns. *BMC Bioinformatics*, *12*(Suppl 1), S50. doi:10.1186/1471-2105-12-S1-S50 PMID:21342583

Park, S. J., Shirahige, K., Ohsugi, M., & Nakai, K. (2015). DBTMEE: A database of transcriptome in mouse early embryos. *Nucleic Acids Research*, *43*, D771–D776. doi:10.1093/nar/gku1001 PMID:25336621

Park, S. J., Umemoto, T., Saito-Adachi, M., Shiratsuchi, Y., Yamato, M., & Nakai, K. (2014). Computational promoter modeling identifies the modes of transcriptional regulation in hematopoietic stem cells. *PLoS ONE*, *9*(4), e93853. doi:10.1371/journal.pone.0093853 PMID:24710559

Patil, A., Kumagai, Y., Liang, K. C., Suzuki, Y., & Nakai, K. (2013). Linking transcriptional changes over time in stimulated dendritic cells to identify gene networks activated during the innate immune response. *PLoS Computational Biology*, *9*(11), e1003323. doi:10.1371/journal.pcbi.1003323 PMID:24244133

Peat, J. R., Dean, W., Clark, S. J., Krueger, F., Smallwood, S. A., Ficz, G., & Reik, W. et al. (2014). Genome-wide Bisulfite Sequencing in Zygotes Identifies Demethylation Targets and Maps the Contribution of TET3 Oxidation. *Cell Reports*, *9*(6), 1990–2000. doi:10.1016/j.celrep.2014.11.034 PMID:25497087

Piko, L., & Clegg, K. B. (1982). Quantitative changes in total RNA, total poly(A), and ribosomes in early mouse embryos. *Dev Biol*, *89*(2), 362-378. doi:0012-1606(82)90325-6

Ramadan, W. M., Kashir, J., Jones, C., & Coward, K. (2012). Oocyte activation and phospholipase C zeta (PLCzeta): Diagnostic and therapeutic implications for assisted reproductive technology. *Cell Communication and Signaling*, *10*(1), 12. doi:10.1186/1478-811X-10-12 PMID:22591604

Santra, T. (2014). A bayesian framework that integrates heterogeneous data for inferring gene regulatory networks. *Front Bioeng Biotechnol*, *2*, 13. doi:10.3389/fbioe.2014.00013 PMID:25152886

Sarrate, Z., Vidal, F., & Blanco, J. (2014). Meiotic abnormalities in metaphase I human spermatocytes from infertile males: Frequencies, chromosomes involved, and the relationships with polymorphic karyotype and seminal parameters. *Asian Journal of Andrology*, *16*(6), 838–844. doi:10.4103/1008-682X.135126 PMID:25080930

Sato, K. (1979). Polyspermy-preventing mechanisms in mouse eggs fertilized in vitro. *The Journal of Experimental Zoology*, *210*(2), 353–359. doi:10.1002/jez.1402100219 PMID:541603

Schultz, R. M. (2002). The molecular foundations of the maternal to zygotic transition in the preimplantation embryo. *Human Reproduction Update*, *8*(4), 323–331. doi:10.1093/humupd/8.4.323 PMID:12206467

Shinagawa, T., Takagi, T., Tsukamoto, D., Tomaru, C., Huynh, L. M., Sivaraman, P., & Ishii, S. et al. (2014). Histone variants enriched in oocytes enhance reprogramming to induced pluripotent stem cells. *Cell Stem Cell, 14*(2), 217–227. doi:10.1016/j.stem.2013.12.015 PMID:24506885

Simpson, A. J., Caballero, O. L., Jungbluth, A., Chen, Y. T., & Old, L. J. (2005). Cancer/testis antigens, gametogenesis and cancer. *Nature Reviews. Cancer, 5*(8), 615–625. doi:10.1038/nrc1669 PMID:16034368

Smith, Z. D., Chan, M. M., Mikkelsen, T. S., Gu, H., Gnirke, A., Regev, A., & Meissner, A. (2012). A unique regulatory phase of DNA methylation in the early mammalian embryo. *Nature, 484*(7394), 339–344. doi:10.1038/nature10960 PMID:22456710

Stergachis, A. B., Neph, S., Sandstrom, R., Haugen, E., Reynolds, A. P., Zhang, M., & Stamatoyannopoulos, J. A. et al. (2014). Conservation of trans-acting circuitry during mammalian regulatory evolution. *Nature, 515*(7527), 365–370. doi:10.1038/nature13972 PMID:25409825

Storey, J. D., Xiao, W., Leek, J. T., Tompkins, R. G., & Davis, R. W. (2005). Significance analysis of time course microarray experiments. *Proceedings of the National Academy of Sciences of the United States of America, 102*(36), 12837–12842. doi:10.1073/pnas.0504609102 PMID:16141318

Suzuki, H., Forrest, A. R., van Nimwegen, E., Daub, C. O., Balwierz, P. J., Irvine, K. M., & Hayashizaki, Y. et al. (2009). The transcriptional network that controls growth arrest and differentiation in a human myeloid leukemia cell line. *Nature Genetics, 41*(5), 553–562. doi:10.1038/ng.375 PMID:19377474

Svoboda, P., & Flemr, M. (2010). The role of miRNAs and endogenous siRNAs in maternal-to-zygotic reprogramming and the establishment of pluripotency. *EMBO Reports, 11*(8), 590–597. doi:10.1038/embor.2010.102 PMID:20651740

Tachibana-Konwalski, K., Godwin, J., van der Weyden, L., Champion, L., Kudo, N. R., Adams, D. J., & Nasmyth, K. (2010). Rec8-containing cohesin maintains bivalents without turnover during the growing phase of mouse oocytes. *Genes & Development, 24*(22), 2505–2516. doi:10.1101/gad.605910 PMID:20971813

Tahiliani, M., Koh, K. P., Shen, Y., Pastor, W. A., Bandukwala, H., Brudno, Y., & Rao, A. et al. (2009). Conversion of 5-methylcytosine to 5-hydroxymethylcytosine in mammalian DNA by MLL partner TET1. *Science, 324*(5929), 930–935. doi:10.1126/science.1170116 PMID:19372391

Takahashi, K., Tanabe, K., Ohnuki, M., Narita, M., Ichisaka, T., Tomoda, K., & Yamanaka, S. (2007). Induction of pluripotent stem cells from adult human fibroblasts by defined factors. *Cell*, *131*(5), 861–872. doi:10.1016/j.cell.2007.11.019 PMID:18035408

Talbert, P. B., & Henikoff, S. (2010). Histone variants--ancient wrap artists of the epigenome. *Nature Reviews. Molecular Cell Biology*, *11*(4), 264–275. doi:10.1038/nrm2861 PMID:20197778

Tang, F., Barbacioru, C., Wang, Y., Nordman, E., Lee, C., Xu, N., & Surani, M. A. et al. (2009). mRNA-Seq whole-transcriptome analysis of a single cell. *Nature Methods*, *6*(5), 377–382. doi:10.1038/nmeth.1315 PMID:19349980

Tang, F., Kaneda, M., O'Carroll, D., Hajkova, P., Barton, S. C., Sun, Y. A., & Surani, M. A. et al. (2007). Maternal microRNAs are essential for mouse zygotic development. *Genes & Development*, *21*(6), 644–648. doi:10.1101/gad.418707 PMID:17369397

Torres-Padilla, M. E., Bannister, A. J., Hurd, P. J., Kouzarides, T., & Zernicka-Goetz, M. (2006). Dynamic distribution of the replacement histone variant H3.3 in the mouse oocyte and preimplantation embryos. *The International Journal of Developmental Biology*, *50*(5), 455–461. doi:10.1387/ijdb.052073mt PMID:16586346

van der Heijden, G. W., Dieker, J. W., Derijck, A. A., Muller, S., Berden, J. H., Braat, D. D., & de Boer, P. et al. (2005). Asymmetry in histone H3 variants and lysine methylation between paternal and maternal chromatin of the early mouse zygote. *Mechanisms of Development*, *122*(9), 1008–1022. doi:10.1016/j.mod.2005.04.009 PMID:15922569

van der Heijden, G. W., Ramos, L., Baart, E. B., van den Berg, I. M., Derijck, A. A., van der Vlag, J., & de Boer, P. et al. (2008). Sperm-derived histones contribute to zygotic chromatin in humans. *BMC Developmental Biology*, *8*(1), 34. doi:10.1186/1471-213X-8-34 PMID:18377649

Vassena, R., Boue, S., Gonzalez-Roca, E., Aran, B., Auer, H., Veiga, A., & Izpisua Belmonte, J. C. (2011). Waves of early transcriptional activation and pluripotency program initiation during human preimplantation development. *Development*, *138*(17), 3699–3709. doi:10.1242/dev.064741 PMID:21775417

Vlieghe, D., Sandelin, A., De Bleser, P. J., Vleminckx, K., Wasserman, W. W., van Roy, F., & Lenhard, B. (2006). A new generation of JASPAR, the open-access repository for transcription factor binding site profiles. *Nucleic Acids Research*, *34*, D95–D97. doi:10.1093/nar/gkj115 PMID:16381983

Wang, H., & Dey, S. K. (2006). Roadmap to embryo implantation: Clues from mouse models. *Nature Reviews. Genetics*, *7*(3), 185–199. doi:10.1038/nrg1808 PMID:16485018

Wang, Q. T., Piotrowska, K., Ciemerych, M. A., Milenkovic, L., Scott, M. P., Davis, R. W., & Zernicka-Goetz, M. (2004). A genome-wide study of gene activity reveals developmental signaling pathways in the preimplantation mouse embryo. *Dev Cell, 6*(1), 133-144. doi:S1534580703004040

Wang, S., Kou, Z., Jing, Z., Zhang, Y., Guo, X., Dong, M., . . . Gao, S. (2010). Proteome of mouse oocytes at different developmental stages. *Proc Natl Acad Sci U S A, 107*(41), 17639-17644. doi: 10.1073/pnas.1013185107

Wang, Z., Xu, W., San Lucas, F. A., & Liu, Y. (2013). Incorporating prior knowledge into Gene Network Study. *Bioinformatics (Oxford, England)*, *29*(20), 2633–2640. doi:10.1093/bioinformatics/btt443 PMID:23956306

Ward, W. S., & Coffey, D. S. (1991). DNA packaging and organization in mammalian spermatozoa: Comparison with somatic cells. *Biology of Reproduction*, *44*(4), 569–574. doi:10.1095/biolreprod44.4.569 PMID:2043729

Wassarman, P., Chen, J., Cohen, N., Litscher, E., Liu, C., Qi, H., & Williams, Z. (1999). Structure and function of the mammalian egg zona pellucida. *The Journal of Experimental Zoology*, *285*(3), 251–258. doi:10.1002/(SICI)1097-010X(19991015)285:3<251::AID-JEZ8>3.0.CO;2-1 PMID:10497324

Wassarman, P. M. (1999). Mammalian fertilization: Molecular aspects of gamete adhesion, exocytosis, and fusion. *Cell*, *96*(2), 175–183. doi:10.1016/S0092-8674(00)80558-9 PMID:9988213

Wen, D., Banaszynski, L. A., Liu, Y., Geng, F., Noh, K. M., Xiang, J., & Rafii, S. et al. (2014). Histone variant H3.3 is an essential maternal factor for oocyte reprogramming. *Proceedings of the National Academy of Sciences of the United States of America*, *111*(20), 7325–7330. doi:10.1073/pnas.1406389111 PMID:24799717

Weng, K. A., Jeffreys, C. A., & Bickel, S. E. (2014). Rejuvenation of meiotic cohesion in oocytes during prophase I is required for chiasma maintenance and accurate chromosome segregation. *PLOS Genetics*, *10*(9), e1004607. doi:10.1371/journal.pgen.1004607 PMID:25211017

Wingender, E., Chen, X., Hehl, R., Karas, H., Liebich, I., Matys, V., & Schacherer, F. et al. (2000). TRANSFAC: An integrated system for gene expression regulation. *Nucleic Acids Research*, *28*(1), 316–319. doi:10.1093/nar/28.1.316 PMID:10592259

Worrad, D. M., Ram, P. T., & Schultz, R. M. (1994). Regulation of gene expression in the mouse oocyte and early preimplantation embryo: Developmental changes in Sp1 and TATA box-binding protein, TBP. *Development, 120*(8), 2347–2357. Retrieved from http://www.ncbi.nlm.nih.gov/pubmed/7925035 PMID:7925035

Xu, X., Smorag, L., Nakamura, T., Kimura, T., Dressel, R., Fitzner, A., & Krishna Pantakani, D. V. et al. (2015). Dppa3 expression is critical for generation of fully reprogrammed iPS cells and maintenance of Dlk1-Dio3 imprinting. *Nature Communications, 6*, 6008. doi:10.1038/ncomms7008 PMID:25613421

Xue, Z., Huang, K., Cai, C., Cai, L., Jiang, C. Y., Feng, Y., & Fan, G. et al. (2013). Genetic programs in human and mouse early embryos revealed by single-cell RNA sequencing. *Nature, 500*(7464), 593–597. doi:10.1038/nature12364 PMID:23892778

Yeung, M. K., Tegner, J., & Collins, J. J. (2002). Reverse engineering gene networks using singular value decomposition and robust regression. *Proceedings of the National Academy of Sciences of the United States of America, 99*(9), 6163–6168. doi:10.1073/pnas.092576199 PMID:11983907

Yurttas, P., Morency, E., & Coonrod, S. A. (2010). Use of proteomics to identify highly abundant maternal factors that drive the egg-to-embryo transition. *Reproduction (Cambridge, England), 139*(5), 809–823. doi:10.1530/REP-09-0538 PMID:20106898

Zeng, F., Baldwin, D. A., & Schultz, R. M. (2004). Transcript profiling during preimplantation mouse development. *Developmental Biology, 272*(2), 483–496. doi:10.1016/j.ydbio.2004.05.018 PMID:15282163

Zhang, S. Q., Ching, W. K., Tsing, N. K., Leung, H. Y., & Guo, D. (2010). A new multiple regression approach for the construction of genetic regulatory networks. *Artificial Intelligence in Medicine, 48*(2-3), 153–160. doi:10.1016/j.artmed.2009.11.001 PMID:19963359

Zou, H., & Hastie, T. (2005). Regularization and variable selection via the elastic net. *Journal of the Royal Statistical Society. Series B, Statistical Methodology, 67*(2), 301–320. doi:10.1111/j.1467-9868.2005.00503.x

KEY TERMS AND DEFINITIONS

High-Throughput Sequencing: A high-speed and low-cost technique that deciphers DNA sequences; e.g. RNA-seq analyzing RNAs, ChIP-seq analyzing DNAs that interacted with a protein.

Linear Regression Model: A statistical method to explain the relationship between response variables (i.e. gene expression) and explanatory variables (i.e. promoter features).

Log-Linear Graphical Model (LGM): A probabilistic method to detect the dependency of pairwise associations in higher-order conditional distribution.

Maternal-to-Zygotic Transition (MZT): A developmental stage where embryos become under the control of zygotic genome.

MII Oocyte: An unfertilized female gamete that is arrested at the second metaphase.

Parthenogenesis: A process of embryo development in which sperm makes no genetic contribution.

Somatic Cell Nuclei Transfer (SCNT): An experimental technique in which the nucleus extracted from a somatic cell is transferred into the cytoplasm of oocyte.

Transcription Factor (TF): Proteins that bind to specific DNA sequences and regulate gene transcription.

Zygotic Gene Activation (ZGA): A component of MZT which accompanies *de novo* zygotic transcripts.

Chapter 8
Prioritize Transcription Factor Binding Sites for Multiple Co-Expressed Gene Sets Based on Lasso Multinomial Regression Models

Hong Hu
University of Illinois – Chicago, USA

Yang Dai
University of Illinois – Chicago, USA

ABSTRACT

Computational prediction of cis-regulatory elements for a set of co-expressed genes based on sequence analysis provides an overwhelming volume of potential transcription factor binding sites. It presents a challenge to prioritize a set of functional transcription factors and their binding sites on the regulatory regions of the target genes that are relevant to the gene expression study. A novel approach based on the use of lasso multinomial regression models is proposed to address this problem. We examine the ability of the lasso models using a time-course microarray data obtained from a comprehensive study of gene expression profiles in skin and mucosal in mouse over all stages of wound healing.

DOI: 10.4018/978-1-5225-0353-8.ch008

INTRODUCTION

Transcription is a key component in the flow of the genetic information within the living organisms by replicating the information into messenger RNA (mRNA) from a section of the DNA (Levine & Tjian, 2003). Transcription factors (TFs) are a class of proteins that bind to transcription factor binding sites (TFBSs) on DNA and regulate the gene expression (Latchman, 1997) in a context-specific manner. The transcription regulation enables cells to respond to intra- and extra-cellular signals to maintain the cellular activities. Previous studies have estimated that in human genome there are about 20,000~25,000 genes, which cover only~5% of the DNA sequences. The rest of the genome associates with non-coding RNA molecules, regulatory DNA sequences, short or long interspersed elements, introns, and sequences currently with unknown functions (Lander et al., 2001). Due to the high complexity of regulatory relationship within living organisms, the identification of TFs and their TFBSs in a context-specific manner is an important but unsolved problem in genomic study (Chai et al., 2014; Collins, Green, Guttmacher, Guyer, & Institute, 2003; W. P. Lee & Tzou, 2009; Spellman et al., 1998).

EXPERIMENTAL IDENTIFICATION OF TRANSCRIPTION FACTOR BINDING SITES

Large scale experimental methods have been proposed for the identification of TFBSs on human genome for a given TF. The recently developed technologies of chromatin-immunoprecipitation followed by massively sequencing (ChIP-seq) or microarray (ChIP-array) allow researchers to capture the genome-wide binding profile of a TF under a given experimental condition (Iyer et al., 2001; Johnson, Mortazavi, Myers, & Wold, 2007; Landt et al., 2012; Robertson et al., 2007). The Encyclopedia of DNA Elements (ENCODE) Consortium, which aims to build a comprehensive list of annotation for functional elements in human genome, has performed a large number of ChIP-seq experiments (Dunham et al., 2012). However, this technology is limited by the availability of the TF-specific antibodies and the high experimental cost. Even with the consortium effort, the ChIP-seq data generated in the ENCODE project only covers a limited number of TFs for a few cell types (Dunham et al., 2012). Although the cost for ChIP-seq study is reducing with the rapid development of the next generation sequencing technology, it is not always possible to use this experimental approach. This is because there are situations where the key set of TFs that regulate the underlying process are unknown *a prior*.

Sequence-Based Computational Identification of Transcription Factor Binding Sites from Co-Expressed Genes

Studies have shown that TFs bind to the DNA sequences in a sequence-specific manner. That is, each TF binds to the specific DNA regions to control the rate of transcription (Latchman, 1997). Therefore, using the prior knowledge of sequence features as well as the TF binding preference, the TF binding sites can be predicted from the sequence information. This in turn provides clue on potential involving TFs.

Measuring temporal gene expression profiles upon certain stimulus to a specific type of cells or tissues have been used to interrogate the transcription regulatory mechanism (Hirose et al., 2008; Y. Liu, Jiang, & Zhang, 2009; Schena, Shalon, Davis, & Brown, 1995; Spellman et al., 1998). Gene expression time-course data capture the dynamics of transcription regulation over time, thus providing valuable information for the inference of the transcriptional regulatory relationship between TFs and their target genes (L. Chen et al., 2010; D'Haeseleer, Liang, & Somogyi, 2000; Ljung, 1987; Wahde & Hertz, 2000; Yeung, Tegner, & Collins, 2002). From the analysis of a temporal gene expression profile, genes can be divided into groups based on their co-expression patterns (Eisen, Spellman, Brown, & Botstein, 1998). Some groups comprise of genes that respond immediately upon the stimulus, representing either early up-regulated or early down-regulated expression changes. These genes are likely to be the direct targets of some key TFs that regulate the cellular activities responding to the stimulation. It is extremely important to know what TFs are involved, what genes are being targeted and where are the binding sites of the TFs on the regulatory regions of the target genes. Besides the promoter regions, TFs may also bind to the distal enhancer regions which are far away from the transcription starting site of a target gene (Burger, Gaidatzis, Schubeler, & Stadler, 2013; H. Hu & Dai, 2014; Marsman & Horsfield, 2012). However, there's little location information is available regarding the distal regulatory regions. Therefore, most transcriptional regulation studies still focus on TF binding to the promoter sequence of a gene.

Under the common hypothesis that genes with similar expression profiles are likely to be regulated by a common set of TFs (Altman & Raychaudhuri, 2001), various methods have been proposed to analyze the promoter sequences of a set of co-expressed genes based on the detection of over-representing TFBSs for certain TFs. These methods can be generally categorized into two types: *de novo* identification of overrepresented motifs and TFBS prediction using Position Weight Matrix (PWM) scanning.

de novo Identification of Overrepresented Motifs

The *de novo* identification methods are mostly use for searching the motifs without prior knowledge. The assumption underlying is that the regulatory elements are enriched in the promoters and are evolutionary conserved. Therefore, searching for the overrepresented motifs across regulatory regions may discover the novel regulatory elements (Bailey et al., 2009; Huber & Bulyk, 2006; Smith, Sumazin, Xuan, & Zhang, 2006). The common approaches for the *de novo* motif discovery start with a set of the sequences extracted from the genome. These sequences are typically of a fixed length and reside in the promoter regions of a set of genes either co-expressed or with similar functional annotations. Then an algorithm is deployed for the discovery of the most overrepresented motifs according to the proposed motif model and statistical measure. To increase the sensitivity of the motif discovery, additional prior knowledge about gene regulation can be integrated. The prior knowledge generally includes the context of the putative regulatory elements, such as interactions with other nearby regulatory elements, nearby CpG islands, and overall positions of the putative regulatory elements in the DNA structure. Additionally, individual genes in a set of co-expressed genes may display different levels of influences in gene regulation. This prior knowledge can also be used as a weight function to increase the influence of the potential important genes in *de novo* motif discovery. Other information such as regulatory regions of orthologous genes can also be utilized as the prior knowledge.

The *de novo* motif discovery methods can be summarized into 4 levels (Sandve & Drablos, 2006). In first level, short contiguous sequences that are likely to be bounded by TFs are modeled by single motif models. Each sequence segment in a regulatory region is assigned a distinct score based on the matching score that describes the similarity between the subsequence start at a position and an underlying consensus model on the gene. The models for obtaining matching score can be generally classified into two types: 1) deterministic models with binary scores (Bussemaker, Li, & Siggia, 2000; L. J. Jensen & Knudsen, 2000; van Helden, Andre, & Collado-Vides, 1998) and 2) probabilistic models with weighted scores (Ben-Gal et al., 2005; Berg & von Hippel, 1987; Kechris, van Zwet, Bickel, & Eisen, 2004; O'Flanagan, Paillard, Lavery, & Sengupta, 2005; Zhao, Huang, & Speed, 2005). In addition, an occurrence prior probability could be assigned, representing prior belief (probability) of the occurrence of a regulatory element at the position on a gene. The prior probability can be obtained based on influence of genetic context on the activity of regulatory elements (S. T. Jensen, Liu, Zhou, & Liu, 2004; X. Liu, Brutlag, & Liu, 2001; McCue et al., 2001), including the distance of the regulatory elements to the transcription start site, sequence conservation in orthologous genes, DNA structure and the presence of CpG islands (Pudimat, Schukat-Talamazzini, &

Backofen, 2004). In the second level, the composite motif model is utilized which contains a set of single motifs. Clusters of binding sites for cooperating TFs, also known as modules, are believed to be essential building blocks of the regulatory machinery (Werner, 1999). For a set of positions, each represents a single motif, score for a composite motif is a function of scores for each single motif and the scores for the distances between motifs. Several methods have been proposed to measure the inter-motif distance within a module by either restricting the distance between the motifs (Y. J. Hu, 2003; X. Liu et al., 2001) or all single motifs to be within a window of given length (GuhaThakurta & Stormo, 2001; Rebeiz, Reeves, & Posakony, 2002). The conservation of inter-motif distance across modules is also used for scoring distance (Wagner, 1999). On the other hand, multiple schemes have been proposed to use a single measure to represent a set of single motifs and distance scores. For methods that use deterministic match models with binary component scores, the composite motif score is calculated as the count of the binary single motif values (Brazma, Vilo, Ukkonen, & Valtonen, 1997; Scherf, Klingenhoff, & Werner, 2000). For methods using non-binary single motif scores, the composite score is calculated as the sum of single motif scores and distance scores (Frech & Werner, 1997). In the third level, the methods focus on the presence of motifs across the whole regulatory regions of a gene (Bussemaker, Li, & Siggia, 2001; Conlon, Liu, Lieb, & Liu, 2003; Wang & Stormo, 2003; Xing, Wu, Jordan, & Karp, 2004). These methods are generally applied to the prediction of which genes are regulated by a TF or module, or evaluation for the significance of a *de novo* discovered motif. In the fourth level, the methods focus on the action of several sets of modules on a set of genes (Jonassen, 1997; Ren et al., 2000; Rigoutsos & Floratos, 1998; Zhu, Pilpel, & Church, 2002). The scores are based on overrepresentation of motifs or on correspondence between motif scores and experimental data.

The *de novo* motif discovery is usually used for novel TFBSs prediction without prior knowledge and generally involves extensive computation (Simcha, Price, & Geman, 2012). Therefore, the TFBSs prediction with prior knowledge, such as position weight matrices scanning, is widely used.

The Prediction of TFBSs Using Position Weight Matrices (PWMs) Scanning

The second type of TFBSs prediction approaches are more widely applied in transcriptional regulation studies. This type of approaches relies on the prediction of TFBSs using position weight matrices (PWMs) of known TFs on the promoter sequences of a set of co-expressed genes (Stormo, 2000; Wasserman & Sandelin, 2004). PWMs, also known as position-specific scoring matrices (PSSMs), are commonly used to represent the motifs (patterns) in biological sequences. For example, the TF FOXD1 belongs to the forkhead family, which is characterized by a distinct forkhead domain. It functions in kidney development by promoting nephron progenitor differentiation, and in the development of the retina and optic chiasm. It may also regulate inflammatory reactions and prevent autoimmunity (Thackray, 2014). In previous experiments, the binding sites of FOXD1 with 8 bp long are identified 20 times. In the matrix below, the binding sites are aligned, each column represents frequency of each base appear at each genomic position (Mathelier et al., 2014). The DNA sequences with the similar content have big preference to be bound by FOXD1.

$$\begin{matrix} A \\ C \\ G \\ T \end{matrix}\begin{bmatrix} 1 & 0 & 19 & 20 & 18 & 1 & 20 & 7 \\ 1 & 0 & 1 & 0 & 1 & 18 & 0 & 2 \\ 17 & 0 & 0 & 0 & 1 & 0 & 0 & 3 \\ 1 & 20 & 0 & 0 & 0 & 1 & 0 & 8 \end{bmatrix}$$

PWMs are compiled from the TF-bound DNA sequences that are discovered from experiments and are aligned based on multiple sequence alignment (for more details of PWMs construction, please refer to METHOD section). The curated PWMs have been collected in databases such as JASPAR and TRANSFAC (Matys et al., 2003; Sandelin, Alkema, Engstrom, Wasserman, & Lenhard, 2004; Wingender, 2008). One can use a PWM to scan the sequences of interest by the sliding window method to obtain the probability (or similarity score) of the TF binding at each genomic position (Bailey et al., 2009; Stormo, Schneider, Gold, & Ehrenfeucht, 1982). The positions with probability or log transformed probability (similarity score) beyond a given significance threshold are taken as predicted TFBSs. The threshold is usually determined from an empirical background probability distribution of similarity scores that are computed from properly constructed random sequences. However, due to the short length of the binding motifs represented by PWMs, the motif scanning procedure usually results in an overwhelming number of the predicted TFBSs on a specified regulatory region. Since TFBSs are predicted based on the sequence information along, it is impossible to discriminate which of the predicted TFBSs may be occupied by the TFs under the experimental condition. Therefore, other functional genomic data, such as gene expression, need be integrated into the inference of the TFs and corresponding TFBSs, to identify the functional TFBSs out of numerous predicted binding sites.

Numerous computational models have been proposed for learning TF-target relationship from gene expression data (Budden, Hurley, & Crampin, 2014; D. Das, Nahle, & Zhang, 2006; Natarajan, Yardimci, Sheffield, Crawford, & Ohler, 2012). For example, researchers have studied TF-tissue association between a limited number of TFs and the proximal promoter regions of genes that are specifically expressed in certain types of tissues based on the analysis of microarray or expressed tags (EST) data (Qian et al., 2005; Yu, Lin, Zack, & Qian, 2006). Previously, the linear regression model was applied to model the abundance of motif in promoter of the target gene vs. the expression level of the target gene (Bussemaker et al., 2001)

$$Y_g = \sum_m X_{g,m}\beta_m + \varepsilon_g$$

where $X_{g,m}$ is the abundance of motif m in promoter of gene g, Y_g is the expression level of gene g. The summation is over all the motifs which are believed to contribute to the expression of gene g. This basic model has since been expanded to model interactions between motifs that appear together in the same promoter sequence (R. Das & Gerstein, 2004; Keles, van der Laan, & Eisen, 2002). However, the methods so far mentioned have attained some degree of success with data from yeast, the applicability of these methods in higher order organisms is still unproven.

The sequence-bases approaches, including both *de novo* motif discovery and PWMs scanning, generally only consider the sequence information. This does not fit into the real biological circumstance that the transcription regulation is condition specific. That is, under differential biological condition, in different tissues, the TFs have different binding profiles therefore regulate the transcription, which determine the cell development, differentiation and responses to the external stimulation (Altman & Raychaudhuri, 2001). Therefore, other information, especially the gene expression profiles, which is the direct consequences of transcription regulation, need to be integrated to identify the condition-specific TF binding profiles.

THE APPROACH PROPOSED IN THIS CHAPTER

The TFBSs prediction methods, especially PWMs scanning, generate a huge amount of predicted TFBSs, in which only a small subset of TFBSs of a limited number of TFs are functional under certain biological circumstance. On the other hand, the sequence-based PWMs scanning may generate a large amount of false-positives in TFBSs prediction results. Therefore, for each set of co-expressed genes identified from a time course gene expression study, we want identify a set of TFs and associated TFBSs out of the numerous predicted candidates from PWMs. This method does not intend to discover all TFs involved. Instead, it aims to detect those distinctively regulate the individual co-expressed gene sets. As the number of candidate TFs/TFBSs is generally far beyond the number of the sets of co-expressed genes, feature selection procedure is required.

In this chapter, an approach based on lasso penalized multinomial logistic regression is presented to prioritize a set of TFs and their TFBSs from numerous candidates for each of the multiple sets of co-expressed genes identified from a time-course gene expression study. The proposed lasso models overcome the difficulty of dealing with an overwhelming number of TFBSs by seeking the tradeoff between the fitness and the number of variables to be included in the models. To further effectively model the TFs and their specific binding sites, each promoter region is divided into multiple bins and the information of the predicted TFBSs in each bin are incorporated to model the binding of TFs on the target genes. The significance of overrepresentation of the TFBSs in the gene set is also taken into account through weighting on similarity scores of TFBSs. We further propose two sub-models to consider the binding sites of a single TF in a single bin and the joint effect of binding of two TFs, each TF may have binding sites in a single bin or in two bins.

Method

Our goal is to identify multiple sets of TFs, each of which may regulate a set of co-expressed genes that responded to certain external stimulus in specific cell or tissue. Starting with a time-course gene expression profile, a clustering analysis will be performed first to divide genes into groups where genes in each group exhibit a similar temporal gene expression profile. For each co-expressed gene set, the promoter regions are scanned using PWMs scanning to predict TFBSs that are significantly over-represented. The TFBSs and the corresponding TFs are used to form putative TF-gene interactions with weighted affinities, considering both the regulatory effect of single TFs and joint effect of two TFs, are used as input to the proposed lasso-penalized multinomial logistic regression models. Our framework is summarized in Figure 1.

PWM and Similarity Score

Here we first provide an explanation of the form of PWM and show how to calculate the similarity score for PWM scanning (Sinha, 2006; Stormo, 2000). Generally, a PWM representing a nucleotide sequence motif is a matrix with 4 rows, each row for each nucleotide base (e.g. A, T, G and C). The number of columns of the PWM depends on the length of the motif (e.g. the length of the sequence). Suppose there's a set of n aligned nucleotide sequences W with length L and they are bound by a specific TF, which is known from experiments. To compile a PWM from this set of sequences, we first calculate a position frequency matrix (PFM) that counts the frequency of each nucleotide base at each position. Then the position probability

Figure 1. The overall work flow of the proposed method to prioritize a set of TFs using lasso-penalized multinomial regression for each of the multiple sets co-expressed genes

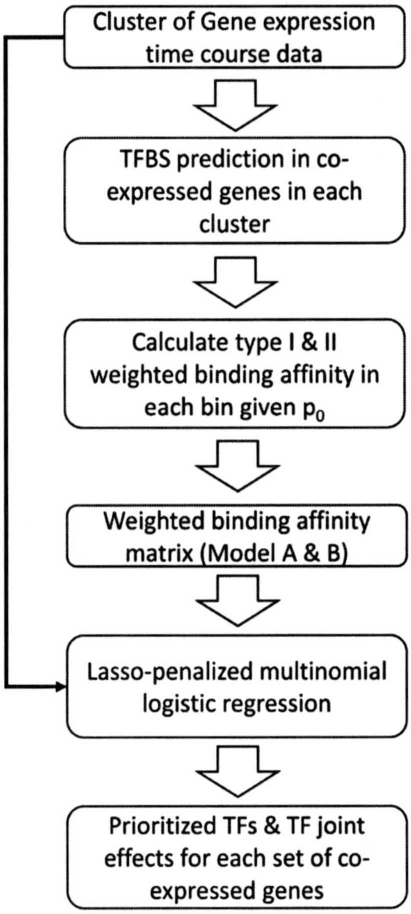

matrix (PPM) *M*, which represents the probability of each nucleotide at each position, is calculated by dividing the PFM by the length of the sequences *n* at each position. Therefore, the probability that a base *k* ($k = A, C, G, T$) appears at position *l* ($l=1, \ldots, L$) can be calculated as

$$M_{k,l} = \frac{1}{n}\sum_{i=1}^{n} I(W_{il} = k)$$

where $I(\bullet)$is the indicator function.

Here we give an example. A set of nucleotide sequences *W* have 5 sequences, each is of 9 bases long.

$$W = \begin{matrix} TAGGTCATT \\ CAGGTATAC \\ TGTGTGAGT \\ ACAGTCAGT \\ ACAGTCAGT \end{matrix}$$

The corresponding PFM is

$$PFM = \begin{matrix} A \\ C \\ G \\ T \end{matrix} \begin{bmatrix} 2 & 2 & 1 & 0 & 0 & 2 & 4 & 1 & 0 \\ 1 & 1 & 0 & 0 & 0 & 2 & 0 & 1 & 1 \\ 0 & 1 & 3 & 5 & 0 & 1 & 0 & 2 & 0 \\ 2 & 1 & 1 & 0 & 5 & 0 & 1 & 1 & 4 \end{bmatrix}$$

Therefore, the corresponding PPM *M* can be calculated as

$$M = \begin{matrix} A \\ C \\ G \\ T \end{matrix} \begin{bmatrix} 0.4 & 0.4 & 0.2 & 0 & 0 & 0.4 & 0.8 & 0.2 & 0 \\ 0.2 & 0.2 & 0 & 0 & 0 & 0.4 & 0 & 0.2 & 0.2 \\ 0 & 0.2 & 0.6 & 1.0 & 0 & 0.2 & 0 & 0.4 & 0 \\ 0.4 & 0.2 & 0.2 & 0 & 1.0 & 0 & 0.2 & 0.2 & 0.8 \end{bmatrix}$$

Both PFM and PPM assume the statistical independence between the positions of within a motif. That is, the probability of a nucleotide base appearing at one position is independent of the probability of a nucleotide base appearing at another position. We can use PPM to calculate the probability of a given sequence to be bound by the specific TF. For example, if the nucleotide sequence *S=ACGGTCAGC*, the probability of this sequence to be bound by the TF represented by the above PPM is:

$$p(S|M) = 0.4 \times 0.2 \times 0.6 \times 1.0 \times 1.0 \times 0.4 \times 0.8 \times 0.4 \times 0.2 = 0.0012288.$$

The PWM is calculated as the log likelihood of the PPM. Each element in PPM is log transformed and weighted by a background model b_k which represents the weight for each base, that is, $M'_{k,l} = \ln(M_{k,l} / b_k)$. The simplest background model assumes b_k=0.25 at each genomic base. Therefore, the corresponding PWM M' can be calculated as,

$$M' = \begin{matrix} A \\ C \\ G \\ T \end{matrix}\begin{bmatrix} 0.47 & 0.47 & -0.22 & -\infty & -\infty & 0.47 & 1.16 & -0.22 & -\infty \\ -0.22 & -0.22 & -\infty & -\infty & -\infty & 0.47 & -\infty & -0.22 & -0.22 \\ -\infty & -0.22 & 0.88 & 1.39 & -\infty & -0.22 & -\infty & 0.47 & -\infty \\ 0.47 & -0.22 & -0.22 & -\infty & 1.39 & -\infty & -0.22 & -0.22 & 1.16 \end{bmatrix}$$

The similarity score for sequence S is calculated as the sum of log likelihoods at each position.

Similarity Score = 0.47 – 0.22 + 0.88 + 1.39 + 1.39 + 0.47 + 1.16 + 0.47 – 0.22 = 5.79

The 0 count in PFM introduces $-\infty$in the PWM. This inconvenience can be eliminated by introducing the pseudocounts (Nishida, Frith, & Nakai, 2009). In practice, the background model can be determined from random sequences that are unlikely bound by the TF.

The following information on each predicted TFBS of a specific TF in the promoter of a given gene is used for the subsequent modeling.

1. The genomic position of each TFBS.
2. The similarity score of each TFBS.
3. The enrichment fold *p*-value for the TFBSs of the specific TF in the set of co-expressed genes.

This *p*-value represents the enrichment of the number of predicted TFBSs in the set compared to those generated from a randomly generated sequences background. In this paper, we utilized the cREMaG (*cis*-Regulatory Elements in the Mammalian Genome) database for the PWM scanning (Piechota, Korostynski, & Przewlocki, 2010). The random background sequences used in the cREMaG database were pre-generated and scanned with all the PWMs collected by the authors for the prediction of TFBSs. For a given PWM and a query set of co-expressed genes, the TFBSs for each sequence in the set are predicted first. Then the fold-difference in TFBS frequency is computed by dividing the observed TFBS frequency with the background frequency and normalized to a z-score based on the fold-difference distribution for all PWMs. The *p*-value is defined as the probability of obtaining a specific range of z-scores using the standard normal distribution.

Proposed Multinomial Models

Suppose there are a total M sets of co-expressed genes from the clustering analysis. Each gene is assigned with a set label $y_i=m$, $m=1,\ldots,M$. To further model the TFBSs of a TF in the specific genomic region on the promoter of a gene, we divide the promoter into multiple bins of certain length. For a gene i in the co-expressed gene set m, its promoter is divided into K bins in order to control the size of independent variables in our model. We model the set label of each gene using the multinomial logistic regression (Hastie, Tibshirani, & Friedman, 2009).

$$\log\frac{p(y_i = m \mid X_i)}{p(y_i = M \mid X_i)} = \begin{bmatrix} x_{i11} & \cdots & \cdots & \cdots & x_{i1K} \\ \vdots & \ddots & & & \vdots \\ \vdots & & x_{ijk} & & \vdots \\ \vdots & & & \ddots & \vdots \\ x_{iJ1} & \cdots & \cdots & \cdots & x_{iJK} \end{bmatrix} \begin{bmatrix} \vdots \\ \vdots \\ \beta^m_{jk} \\ \vdots \\ \vdots \end{bmatrix} + \beta^m_0 \text{ for } m = 1,\ldots,M-1 \tag{1}$$

where x_{ijk} describes the weighted binding affinity (defined below) of TF j on all its TFBSs in bin k in the promoter of gene i in co-expressed gene set m, $i=1,\ldots,N$; $j=1,\ldots,J$; $k=1,\ldots,K$. Here J is the total number of TFs of interest. Through the regression model, we aim to learn the coefficients $\beta^m = \begin{bmatrix} \beta^m_{11} & \cdots & \beta^m_{jk} & \cdots & \beta^m_{Jk} \end{bmatrix}^T$. A non-zero coefficient β^m_{jk} implies that the model selects the TFBSs of TF j in bin k in gene set m. By the nature of multinomial logistic regression, we are able to obtain simultaneously a set of TFs and their corresponding binding sites (bins) for each set of co-expressed genes.

Two types of weighted binding affinities are considered. The **type I** weighted binding affinity is defined as

$$x_{ijk} = -\bar{s}_{ijk} \ln p_{m_j} \tag{2}$$

where $\bar{s}_{ijk}$ is the average similarity score of all the predicted TFBSs of TF j in bin k in the promoter region of gene i; p_{m_j} is the enrichment fold p-value of TFBSs of TF j in co-expressed gene set m. If TF j has no predicted TFBS in bin k in the promoter region of gene i, $x_{ijk}=0$.

The type II weighted binding affinity is designed to describe the joint effects of TFBSs of TF *j* in two different bins *k* and *k'* ($k \neq k'$), or joint effects of TFBSs of TF *j* and TF *j'* ($j \neq j'$) in bins *k* and *k'* (*k* and *k'* can be either different or identical) in the promoter of gene *i*:

$$x_{ij''k''} = \sqrt{xi_{jk} \times x_{ij'k'}} \tag{3}$$

To improve the ability of our model in selecting the TFs that best classify the co-expressed gene sets, we propose to prioritize the TFs with highly significant enrichment fold *p*-values. By the definition, a TF with a smaller enrichment fold *p*-value indicates a higher significance of the TFBS overrepresentation in a set of co-expressed genes. Therefore, we want to assign a higher weight on TFs with more stringent enrichment fold *p*-values. More specifically, given a pre-specify threshold $0<p_0<1$, we modify x_{ijk} for type I affinity in Formula (2) as:

$$x_{ijk} = \begin{cases} -\bar{s}_{ijk} \bullet \ln p_{m_j} & p_{m_j} > p_0 \\ \bar{s}_{ijk} \bullet \left(\ln p_{m_j}\right)^2 & p_{m_j} < p_0 \end{cases} \tag{4}$$

Note $-\ln(1/e)=1$. Therefore, if $0 < p_{m_j} < 1/e$, $-\ln p_{m_j} > 1$; and if $1/e < p_{m_j} < 1$, $0 < -\ln p_{m_j} < 1$.

Based on this relationship, we redefined the weighting by further modifying the Formula (4).

If $p_0<1/e$

$$x_{ijk} = \begin{cases} -\bar{s}_{ijk} \bullet \ln p_{m_j} & p_{m_j} > p_0 \\ \bar{s}_{ijk} \bullet \left(\ln p_{m_j}\right)^2 & p_{m_j} < p_0 < 1/e \end{cases} \tag{5}$$

If $1/e<p_0<1$

$$x_{ijk} = \begin{cases} -\overline{s}_{ijk} \bullet \ln p_{m_j} & p_{m_j} > p_0 \\ \overline{s}_{ijk} \bullet \sqrt{-\ln p_{m_j}} & 1/e < p_{m_j} < p_0 \\ \overline{s}_{ijk} \bullet (\ln p_{m_j})^2 & p_{m_j} < 1/e < p_0 \end{cases} \quad (6)$$

Using this scheme, we hope that the model can prioritize the TFs depending on the level of the significant enrichment of TFBSs. Likewise, $x_{i''k''}$ for type II weighted binding affinity are modified. In summary, the type I and type II in our work are calculated using Formula (3), (5) and (6).

Based on the above definitions, we set up the multinomial logistic regression model as in Formula (1). Since the model has a sheer number of the independent variables that exceeds the number of the observations, especially when the type II variables are considered, it cannot be solved properly. To resolve this difficulty, we utilized the least absolute shrinkage and selection operator (LASSO) penalty in the multinomial logistic regression for model selection. The use of lasso type penalty for model selection is justified by the fact that the regulatory relationship between TFs and target genes is sparse. That is, only a limited number of TFBSs from the numerous candidates are sufficient for the best discrimination of the co-expressed genes.

The lasso regression is a penalized regression model that uses a penalty term to achieve a sparse solution, especially for the problems where the number of predictors far beyond the number of the observations (Krishnapuram, Carin, Figueiredo, & Hartemink, 2005; Tibshirani, 1996). We add the *l2* penalty term in our model as defined in Formula (1) with the objective function written as:

$$\min\{L(\beta,\beta_0,\lambda)\} = \min\{-l(\beta,\beta_0) + \lambda \sum_{jkm} |\beta_{jk}^m|\} \quad (5)$$

where $l(\boldsymbol{\beta},\boldsymbol{\beta}_0)$ is the multinomial likelihood function for Formula (1), $(\boldsymbol{\beta},\boldsymbol{\beta}_0)$ is the vector which combines all (β^m, β_0^m), $m = 1,\ldots,M$; λ is a penalty parameter with positive value. In this work the 10-fold cross validation was used to choose the optimal value of λ based on the minimum cross-validated error rate. With the optimally chosen λ value, the best model can be determined at a balance point between the maximum likelihood estimator and the number of independent variables (the number of non-zeros in estimated β).

RESULTS

Dataset

We used a published temporal microarray gene expression dataset to demonstrate the utility of our proposed models (L. Chen et al., 2010). This dataset includes the systematic and comprehensive study of gene express in skin and mucosal in a mouse model over the key stages of wound healing process. Using microarray techniques, the gene expression was measured in skin and tongue tissues at 0, 6, 12 and 24 hours as well as 3, 5, 7 and 10 days after wound respectively. The gene expression in normal skin and tongue tissues from the same mouse were also profiled at the same time points as control. Significantly differentially expressed (DE) genes in skin and tongue wounds were identified respectively using the one-way ANOVA with the corrected p-value < 1e-5 for control of the false discovery rate (FDR) and the absolute fold change ≥2 in at least one time point in relation to the normal tissue.

Using the K-means clustering method, the DE genes were clustered in each tissue data to multiple sets of co-expressed genes with similar expression profiles. The number of the clusters was verified using the principal component analysis method. In each tissue, genes are classified into five clusters and labeled as: *early down*; *early up high*, *early up medium*; *early up low* and *late up* according to the average temporal gene expression profile in each cluster (Chen et al., 2010). In skin wounds, genes showing rapid early-up regulation at 12~24 hours and rapid drop after day 3 were classified as early up high, early up medium and early up low clusters respectively, according to their degree of up regulation at the early stage. Genes in early down group in skin showed early down regulation and reach the dew point at 12~24 hours then gradually returned back to normal by day 10. Genes in late up group in skin showed slightly down-regulation in first 24 hours and then underwent a slowly up regulation starting at day 3. In tongue wounds, 3 groups of genes also showed up regulation, and were classified into high, medium and low groups by the level of their up regulation level respectively. However, the early up response came faster, started at 6 hours and got plateaued at 6~24 hours, comparing to 12~24 hours in skin wounds. Genes in early down group in tongue started down-regulation at 6 hours, reached the dew point between 12 hours to days, and returned back to normal by day 7, which had a longer down-regulation times comparing to the analogous groups in skin. The late up group in tongue started slightly down-regulation in first 24 hours and started the up-regulation slowly at day 3 until day 10. The numbers of genes in each cluster identified for each tissue are summarized in Table 1.

Table 1. Number of genes in each co-expressed set in skin and tongue

Co-Expressed Set	Skin	Tongue
Early down	236	47
Early up high	54	18
Early up low	376	119
Early up medium	158	72
Late up	130	88

Genes in clusters *early down* and *early up* are likely to be immediate regulatory targets of a key set of TFs that respond to the wound healing respectively. Genes in the *late up* cluster could be the secondary targets, which are likely to be regulated by the products of the target genes of the key TFs. Therefore, the analysis of TFBSs in the promoters of these genes can facilitate the identification of key TFs that play an important role in wound healing.

The promoter region of each gene was specified as 10,000bp upstream of the transcription start site (TSS) and 1,000bp downstream of the TSS. The prediction of TFBSs was carried out using the cREMaG database (Piechota et al., 2010) which was built based on the PWMs included in TRANSFAC and JASPAR databases (Matys et al., 2003; Sandelin et al., 2004). The cREMaG database precompiled the predicted TFBSs for each gene for each PWM in various random background sequences based on which it calculates the enrichment fold p-value for a set of query genes. The parameters used in our queries are listed in Table 2.

As described in the previous section, the query results, including the similarity scores of TFBSs and the enrichment fold p-values, were used to calculate the input matrices for the lasso multinomial regression models. The upstream of the TSS in

Table 2. Parameter settings for cREMaG database query

Parameter	Selection
Conservation threshold	70%
Top percent of conserved region	100%
Max number of most conserved TFBSs	Top 100
Coding/Non-coding sequence	Non-coding
Length of upstream segment	10,000bp
Length of downstream segment	1,000bp
Random TFBS occurrence	2TFBS/10,000bp

each promoter was partitioned into 5 bins each with length of 2,000bp while the 1000bp downstream of the TSS was designated as one bin. Therefore, there were 6 bins in total covering the promoter region of each gene.

Evaluation Procedure

Two lasso multinomial logistic regression models including different types of independent variables were considered. Our models analyze multiple sets of co-expressed genes simultaneously and identify the set of TFs and their binding regions (at the resolution of a bin) that are best associated to each set of the co-expressed genes.

Model A: Use *X* matrix of Type I variables to represent the average effect of all TFBSs of individual TFs in a bin on a promoter.

Model B: Use *X* matrices both Type I and Type II variables to represent the effects explained by 1) single TF and 2) either the effects of TFBSs of a single TF in 2 bins or the effects of TFBSs of 2 different TFs in one bin or two bins on a promoter.

The Bioconductor package *glmnet* (J. Friedman, Hastie, & Tibshirani, 2010) was used to solve the lasso multinomial logistic regression with the 10-fold cross-validation procedure. To avoid the randomness involved in training, we ran the 10-fold cross-validation for 100 times. The cross-validated minimum error rates and corresponding optimal λs from 100 runs were averaged into a single value respectively. The average minimum error rates and average optimal λ are summarized in Table 3.

From Table 3 we can see that the cross-validated minimum error rate depends not only on λ but also on the fold enrichment threshold p_0. To examine the influence of the threshold on the performance of the models, we varied the value of p_0 to impose different weights on TFBSs. We observe that the lowest cross-validated

Table 3. Optimal λ and minimum error rate

		Skin		Tongue	
	Fold p-Value Threshold (p_0)	λ	Min Error Rate	λ	Min Error Rate
Model A	0.01	0.018	0.464	0.018	0.455
	0.05	0.009	0.302	0.014	0.425
Model B	0.01	0.037	0.538	0.067	0.616
	0.05	0.027	0.404	0.078	0.501

minimum error rate was reached at certain threshold p_0; an example in skin when using the Model A with restricted number of independent variables (explained below) is given in Figure 2.

As we stated in Method, the enrichment fold *p-value* p_0 represents the enrichment of TFBS for each TF in a set of the co-expressed genes against the random background. A smaller p_0 represents a TF with significant more TFBSs. In our method, we aim to put more weight on the TFs with smaller p_0 to be prioritized. From Figure 2 we can tell that when p_0 is chosen 0.02 ~ 0.08, the min error rate from lasso regression is stable, when p_0 is chosen as 0.1, the performance dropped. When p_0 keeps increasing, too many TFs will be assigned higher weights thus over-whelming independent variables will be prioritized, as in the construction of weighted binding affinity matrix in Formula (6). When $p_0<0.05$, too few TFs will be assigned higher weights to be prioritized. Together with the results listed in Table 3, we choose $p_0=0.05$ as the threshold for optimal performance. Furthermore, the performance of Model A was better than Model B in data from both tissue types, especially in Skin. This might because the number of independent variables in Model B was far larger that in Model A and the number of observations was not sufficient for the lasso regression to select the most informative variables for Model B. In the subsequent analysis, we only report the results from Model A in both tissues and from Model B in Skin only at $p_0=0.05$.

Figure 2. The minimum error rate reached at an optimal value when different enrichment fold p-value threshold p_0 was used in Model A, in skin, with $Df_{max}=100$

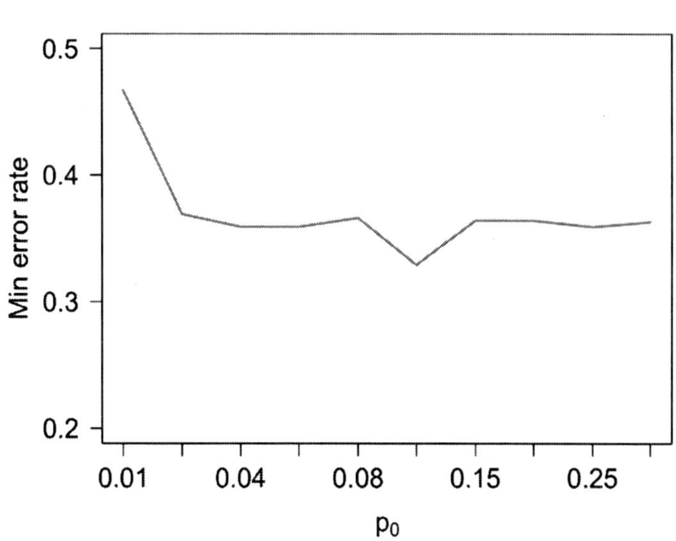

To further control the number of variables selected in the models, we restricted the maximum number of the independent variables (Df_{max}) that are allowed to be included in each model. The results are summarized in Table 4. Within each tissue, we observe that in either Model A or Model B, when the upper bound of Df_{max} was reduced to 100, there was no significant increase in the cross-validated minimum error rate comparing to the models without the Df_{max} restriction. Therefore, the final results on the selected TFs were determined with $Df_{max}=100$ as in Table 5.

To further evaluate the significance of the method, we asked whether our methods have selected the TFs that bind to the specific genomic regions in the gene sets by chance. To investigate this, we compared the performance of the models on data sets in which the TFs were randomly selected to bind to the promoter regions of the co-expressed genes. To construct the random datasets, the elements in the weighted binding affinity matrices in Model A and Model B were randomly shuffled. As the TFBS enrichment fold *p-values* were calculated within individual co-expressed gene sets, the TFBS affinity weights within each gene set were shuffled. That is, the weighted binding affinity scores for a TF in a specific bin in the promoter region of a given gene were randomly assigned to a different bin of another gene's promoter for a different TF in the same co-expressed gene set. In this way we maintained the enrichment fold *p-value* unchanged.

For Model B, we added the random joint regulatory effects for the binding of one TF in different bins or the binding of two TFs in a single bin or different bins on a promoter region. We randomly shuffled the elements in weighted binding affinity matrix in Model B within each set of co-expressed genes based on the assumptions that the TF binding sites were randomly distributed across the promoter region and the interactions between the TF binding sites were randomly distributed.

For each tissue data, we generated 1000 random weighted binding affinity matrices for Model A and Model B respectively under different p_0, while keeping the class labels for the genes unchanged. The lasso multinomial logistic regression models were applied to each random data set, using the same procedure as we used on the original dataset. The minimum error rates from 1000 shuffling datasets were used to generate an empirical background distribution for each original matrix X.

Table 4. Minimum error rate with maximum number of variables

		Skin Min Error Rate		**Tongue Min Error Rate**	
	$\mathbf{P_0}$	**No Df_{max}**	**Df_{max} = 100**	**No Df_{max}**	**Df_{max} = 100**
Model A	0.05	0.302	0.359	0.425	0.480
Model B	0.05	0.404	0.420	-	-

Table 5. Some identified effects in different models

	Co-Expressed Set	TF	TF Class
Skin Model A	Early down	Lhx3	HOMEO
	Early up high	RELA	REL
		NF-κB	REL
	Early up low2	CREB1	bZIP
		MZF1_5_13	ZN-FINGER
		ELK1	ETS
		ELK4	ETS
	Early up medium	ELF5	ETS
		SPIB	ETS
	Late up	Evi1	ZN-FINGER
		Nobox	HOMEO
Skin Model B	Early down	Lhx3	HOMEO
		*CREB1	bZIP
	Early up high	RELA	REL
		NF-κB	REL
	Early up low	ELK1	ETS
	Early up medium	RELA	REL
		*RELA	REL
		FOS	bZIP
		*SPIB	ETS
	Late up	Evi1	ZN-FINGER
Tongue Model A	Early down	MZF_5-13	ZN_FINGER
		Pax4	PAIRED-HOMEO
		NR1H2-RXRA	NUCLEAR RECEPTOR
		ELK1	ETS
	Early up high	Cebpa	bZIP
		NF-κB	REL
	Early up low	Foxq1	FORKHEAD
		IRF2	TRP-CLUSTER
	Early up medium	IRF1	TRP-CLUSTER
		Foxd3	FORKHEAD
	Late up	MEF2A	MADS

* This TF has been predicted to have joint effect with the TF in the previous row

The *p-values* for the minimum error rates obtained from the original matrices were obtained based on the distribution. We observed that the minimum error rates for both models are significant (p–value<0.01). This evaluation further supports the conclusion that our method selected a set of informative TFs and their binding sites may contain useful biologically relevant TFs.

The TFBSs of different TFs in different bins correspond to Type I and Type II matrices were selected from the final lasso penalized multinomial logistic regression using the following criteria. For each model, we repeated 100 runs with the best penalty parameter determined with the cross-validation procedure using p_0=0.05. The coefficient for each independent variable was calculated as the averaged value from the 100 models. To properly select the variables with non-zero coefficient, a threshold needs to be specified. We selected the variables with an absolute value larger than the threshold. Obviously, a large threshold cutoff value would filter out most variables. On the other side, a small threshold would retain almost all variables. By selecting proper threshold, variables with small coefficients that only appeared in a limited number of the 100 models were filtered out. The relationship between the threshold and the number of variables are shown in Figure 3 for Model A in both tissues and Model B in skin only.

From Figure 3, we found that if the threshold 0.001 was used, then the variables with non-zero coefficients appeared in at least 50 models out of 100 mentioned above. That is, when using 0.001 as threshold, the median of the coefficients of the selected variables are bigger than the threshold. Therefore, we used this threshold and retained the variables whose absolute values of the averaged coefficients were greater than 0.001. Table 5 provides a list of TFs that were found by the models in data of each tissue.

Several prioritized TFs can be verified from the published literatures. For example, NF-κB was identified in both Model A or B for both tissue data in *early up high* group. NF-κB is a ubiquitous TF found in almost all animal cell types and it involves in cellular responses to stimuli and plays a key role in regulating the immune response to infection (Gilmore, 2006). Other tissue-specific TFs were also identified. Lhx3, which is a TF required for pituitary development and motor neuron specification, was only identified in skin models (Sloop, Showalter, Von Kap-Herr, Pettenati, & Rhodes, 2000). Several TFs with joint effects were also found, including RELA and NF-κB. Researchers have revealed that NF-κB binds to RELA to form a complex that was activated, and then translocated into the nucleus to bind to the gene regulatory region (Nolan, Ghosh, Liou, Tempst, & Baltimore, 1991). Another example is the identified joint effects between RELA binding to different bins on promoters, in *early up medium* group in Model B in skin. Previous experimental

Figure 3. The number of selected variables using different threshold: (a) Skin, Model A; (b) Skin, Model B; (c) Tongue, Model A

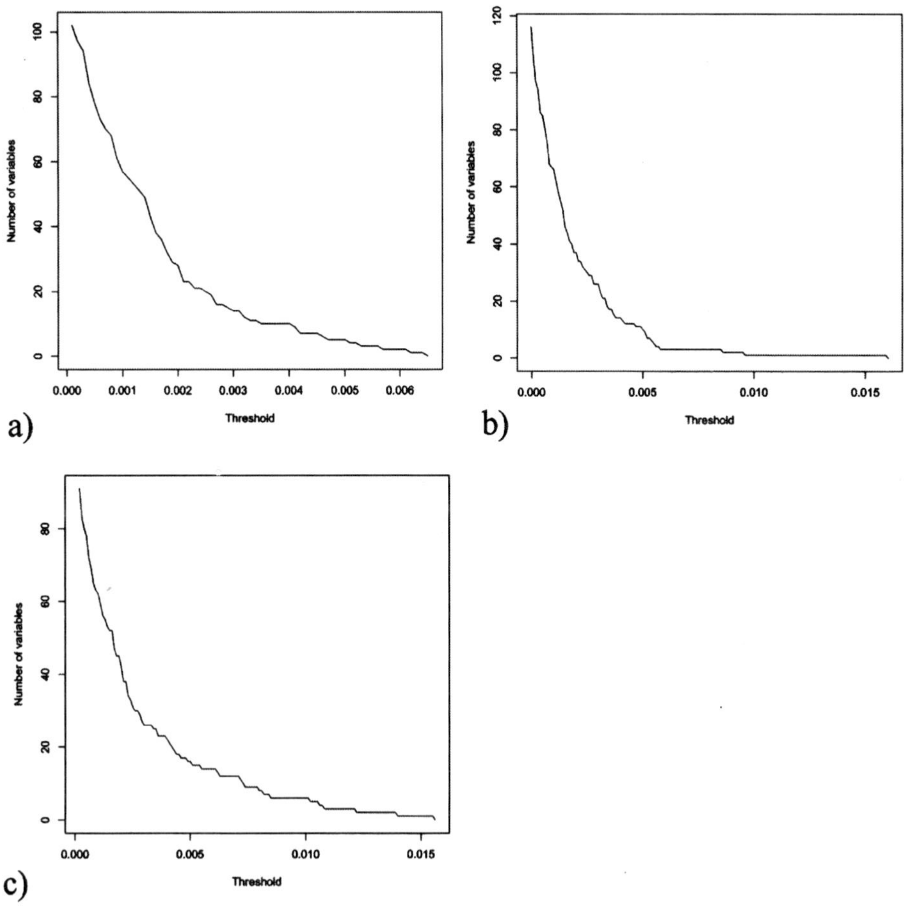

studies have revealed the interaction of RELA and therefore the joint effects can be explained as the TF dimmer formed by RELA bind to the promoter region of the target gene and regulate the gene expression (Hassa, Covic, Bedford, & Hottiger, 2008; Veena et al., 2014; Yang, Hori, Sanda, & Okamoto, 1999).

DISCUSSION

Transcription factors bind to their regulatory sites on the DNA sequences to determine the biological process in cells and tissues. Time-course gene expression data are excellent resource for the inference of TFs and their regulating targets based on the prediction of over-represented TFBSs on the promoters of co-expressed genes detected from a specific experiment. However, the overwhelming number of the predicted TFBSs from the sequence analysis presents an obstacle in determining the functional regulatory relationship between TFs 2and their target genes. Furthermore, the identification of combinatorial patterns of two TFs is not trivial from the predicted results. We have proposed two lasso multinomial logistic regression models to learn a set of strongly associated TFs and their TFBSs in promoters of a set of co-expressed genes. Our models integrate the binding similarity, the statistical significance of over-representation and the position of the TFBSs. To take into account the multiple numbers of TFBSs in a small region on promoter, a promoter is divided into 6 bins and the binding similarity scores of the TFBSs of a TF in the same bin was aggregated into a single value, which was further weighted by the significance of overrepresentation of the TFBSs in the co-expressed gene set. In this way, we were able to determine the importance of multiple TFBSs that may be occupied by the same TF or different TFs at different locations on promoters. The preliminary evaluation of the proposed lasso multinomial logistic regression models based on the dataset of wound healing in skin and tongue has demonstrated the potential ability of this approach to identify functional transcriptional regulatory elements. We note, however, the resolution of the TF binding sites dealt in the proposed models was determined by the bin size and is still relatively low.

The Underlying Assumption of Co-Expression and Co-Regulation

In introduction, we have stated that the underlying assumption for transcriptional regulation study using gene expression time course data is that a set of co-expressed genes showing similar expression profiles are likely to be regulated by a common set of the TFs (Altman & Raychaudhuri, 2001). In our approach, the lasso-penalized multinomial logistic regression is applied to regress the weighted binding affinity representing the TF bindings on the promoter region of each gene against the class of each gene through clustering. On the other hand, this lasso-penalized regression model can also be applied to the gene clustering result not obtained through expression time course data, the underlying assumption may be violated. Therefore, the prioritized TFs for each cluster obtained from non-time course data may not regulate the set of the gene simultaneously.

Re-Construction of Gene Regulatory Network

Transcriptional regulation is one of the key mechanisms in regulation in living organisms. By learning the interactions between TFs and target genes, researchers can rebuild the complex relationship between TF-coding genes and target genes. In a bigger picture, transcriptional regulation study is an essential part in reconstructing of gene regulatory networks (GRNs). A GRN describes the complex regulatory relationship between genes as well as their products in cells or tissues (Alon, 2007; W. P. Lee & Tzou, 2009). These approaches for GRNs reconstruction can be generally classified into two types, depending on whether discrete variables or continuous variables are involved. The models using discrete variables include *Boolean network models*, *probabilistic Boolean network models* and *Bayesian network models*.

Boolean network models are the simplest form of the GRN in which genes are assumed to only exist in discrete states (C. C. Chen & Zhong, 2008). The states of genes are represented by Boolean variables as either on (active or expressed) or off (inactive or unexpressed). The arc between the nodes represent the regulation activity between the genes. The simplest method to construct a Boolean network is to directly infer the network from literatures. When experimental data are available, a Boolean network can be inferred using time course gene expression data based on a correlation measurement between genes (Laubenbacher & Stigler, 2004) or machine learning algorithms (Michalewicz, 1996). In the latter case the genetic algorithm is commonly utilized to obtain the optimal network structure where the genes (nodes) and regulations (edges) are coded as strings and adaptive operators (such as crossover and mutation operations). Although the algorithms to learn a Boolean network is easy to implement, its underlying assumption is over simplified and cannot capture the system dynamics.

In a learning framework using *probabilistic Boolean network*, each network node is assigned multiple regulatory functions, each of which is assigned a pre-defined probability. The formed probabilistic Boolean networks are stochastic. At each time point, the regulatory function of a node is random selected according to a given probability. To construct the probabilistic Boolean network, candidate Boolean networks need to be identified in advance using the methods mentioned above, as in simple Boolean network models. Then the candidate Boolean networks are compiled and probabilities are assigned to each node. Details on the construction of probabilistic Boolean networks can be found in literature (Shmulevich, Dougherty, Kim, & Zhang, 2002). Notably, learning a probabilistic Boolean network significantly increases the computational complexity comparing to the Boolean network. Therefore this approach is not applicable in learning of large scale networks.

Bayesian network models are another type of popular methods for discrete variables. A Bayesian network is a directed acyclic graph that describes probabilistic relationships between network nodes (N. Friedman, Linial, Nachman, & Pe'er, 2000; Kim, Imoto, & Miyano, 2003; Zou & Conzen, 2005). The arcs between nodes describe both regulatory functions and conditional dependencies between the nodes. The construction of a Bayesian network model requires network structure learning and parameter learning. Using the Bayesian scoring metric, each possible network structure model can be assigned a score as the logarithm of the probability of the network structure that correctly describes the given set of experimental or literature data. However, the Bayesian network approach is not practical to be used in the context where multiple TFBSs are considered for a single TF due to the exponentially large number of variables involved.

The second type of GRN construction methods utilizes continuous variables, including *differential equations* and *neural networks*. The models based on *differential equations* includes those using linear ordinary differential equations (ODEs) or nonlinear power law differential equations (Cinquemani, Milias-Argeitis, Summers, & Lygeros, 2008; de Jong, 2002). In these models, the change in the expression level of a gene at a certain time (discrete or continuous) is characterized by a function that takes the regulatory influence (activation or inhibition) of other genes into account. Comparing to the discrete variable models, the differential equations are capable of characterizing the system dynamic of a GRN in a more accurate manner by modeling the experimental variables continuously. The application of differential equations also agrees with the theory of system analysis and control design in dynamic system. However, the differential equations, especially the nonlinear differential equations, are difficult to solve. This fact makes them hard to be applied to infer large scale GRNs. Neural network models are another type of methods that incorporate continuous variables (W.-P. Lee & Yang, 2008; Vohradsky, 2001). In a neural network method, each node represents a particular gene and the edges connecting the nodes are defined as the regulatory interactions. Each node has a connection to every node in the network including itself. The expression level of a gene at a time point can be assessed using all other genes (nodes). The regulatory effect on a gene is derived as a linear or nonlinear function of expression levels of all other genes that have regulatory connection to it. A scoring function is used to evaluate the network for the fitness to the given experimental data. By maximizing the score (the network performance) or minimizing the system error, the parameters can be estimated. However, due the computational complexity introduced by the network size, the neural network method is still difficult to be applied to large scale networks.

In summary, the methods mentioned above for the GRN construction using discrete or continuous variables require heavy computational loads. Thus they cannot be used to infer large scale TF regulation networks (W. P. Lee & Tzou, 2009). In addition, these methods are proposed for modeling the TF-target relationship; they do not explicitly predict the region on promoters where a TF may bind.

FURTHER DEVELOPMENT

To further expand our model, several possible directions can be explored. First, the TFBS prediction can be improved by using the more up-to-date PWM database such as JASPAR 2014 (Mathelier et al., 2014). Using different nucleotide frequencies for different genomic regions can refine the background sequences for the TFBS enrichment analysis. Second, our models had a strong tendency to choose a variable corresponding to x_{ijk} for just one cluster due to the nature of multi-class classification. This may not be always biologically correct because one TF may have similar binding patterns in different sets of genes. Further development of a more accurate mathematical model is required to allow the flexibility to capture this behavior. Third, in our model, independent variables represent the binding strength of different TFs in different promoter regions. They may not fully meet the orthogonal assumption in lasso regression. To address this issue, i.e., when strong correlations exist between the independent variables, additional modifications can be made to identify explanatory TFs in a group manner. This can be achieved through group lasso regression or other novel methods (Yuan & Lin, 2006).

REFERENCES

Alon, U. (2007). Network motifs: Theory and experimental approaches. *Nature Reviews. Genetics*, *8*(6), 450–461. doi:10.1038/nrg2102 PMID:17510665

Altman, R. B., & Raychaudhuri, S. (2001). Whole-genome expression analysis: Challenges beyond clustering. *Current Opinion in Structural Biology*, *11*(3), 340–347. doi:10.1016/S0959-440X(00)00212-8 PMID:11406385

Bailey, T. L., Boden, M., Buske, F. A., Frith, M., Grant, C. E., Clementi, L., . . . Noble, W. S. (2009). MEME SUITE: tools for motif discovery and searching. *Nucleic Acids Research, 37*, W202-208. doi: 10.1093/nar/gkp335

Ben-Gal, I., Shani, A., Gohr, A., Grau, J., Arviv, S., Shmilovici, A., & Grosse, I. et al. (2005). Identification of transcription factor binding sites with variable-order Bayesian networks. *Bioinformatics (Oxford, England)*, *21*(11), 2657–2666. doi:10.1093/bioinformatics/bti410 PMID:15797905

Berg, O. G., & von Hippel, P. H. (1987). Selection of DNA binding sites by regulatory proteins. Statistical-mechanical theory and application to operators and promoters. *Journal of Molecular Biology*, *193*(4), 723–750. doi:10.1016/0022-2836(87)90354-8 PMID:3612791

Brazma, A., Vilo, J., Ukkonen, E., & Valtonen, K. (1997). Data mining for regulatory elements in yeast genome. *Proceedings of the International Conference on Intelligent Systems for Molecular Biology*, *5*, 65–74. PMID:9322017

Budden, D. M., Hurley, D. G., & Crampin, E. J. (2014). Predictive modelling of gene expression from transcriptional regulatory elements. *Briefings in Bioinformatics*. doi:10.1093/bib/bbu034 PMID:25231769

Burger, L., Gaidatzis, D., Schubeler, D., & Stadler, M. B. (2013). Identification of active regulatory regions from DNA methylation data. *Nucleic Acids Research*, *41*(16), e155. doi:10.1093/nar/gkt599 PMID:23828043

Bussemaker, H. J., Li, H., & Siggia, E. D. (2000). Building a dictionary for genomes: Identification of presumptive regulatory sites by statistical analysis. *Proceedings of the National Academy of Sciences of the United States of America*, *97*(18), 10096–10100. doi:10.1073/pnas.180265397 PMID:10944202

Bussemaker, H. J., Li, H., & Siggia, E. D. (2001). Regulatory element detection using correlation with expression. *Nature Genetics*, *27*(2), 167–171. doi:10.1038/84792 PMID:11175784

Chai, L. E., Loh, S. K., Low, S. T., Mohamad, M. S., Deris, S., & Zakaria, Z. (2014). A review on the computational approaches for gene regulatory network construction. *Computers in Biology and Medicine*, *48*, 55–65. doi:10.1016/j.compbiomed.2014.02.011 PMID:24637147

Chen, C. C., & Zhong, S. (2008). Inferring gene regulatory networks by thermodynamic modeling. *BMC Genomics*, *9*(Suppl 2), S19. doi:10.1186/1471-2164-9-S2-S19 PMID:18831784

Chen, L., Arbieva, Z. H., Guo, S. J., Marucha, P. T., Mustoe, T. A., & DiPietro, L. A. (2010). Positional differences in the wound transcriptome of skin and oral mucosa. *BMC Genomics*, *11*, 15. doi: 10.1186/1471-2164-11-471

Cinquemani, E., Milias-Argeitis, A., Summers, S., & Lygeros, J. (2008). Stochastic dynamics of genetic networks: Modelling and parameter identification. *Bioinformatics (Oxford, England)*, *24*(23), 2748–2754. doi:10.1093/bioinformatics/btn527 PMID:18845579

Collins, F. S., Green, E. D., Guttmacher, A. E., & Guyer, M. S. (2003). A vision for the future of genomics research. *Nature*, *422*(6934), 835–847. doi:10.1038/nature01626 PMID:12695777

Conlon, E. M., Liu, X. S., Lieb, J. D., & Liu, J. S. (2003). Integrating regulatory motif discovery and genome-wide expression analysis. *Proceedings of the National Academy of Sciences of the United States of America*, *100*(6), 3339–3344. doi:10.1073/pnas.0630591100 PMID:12626739

D'Haeseleer, P., Liang, S., & Somogyi, R. (2000). Genetic network inference: From co-expression clustering to reverse engineering. *Bioinformatics (Oxford, England)*, *16*(8), 707–726. doi:10.1093/bioinformatics/16.8.707 PMID:11099257

Das, D., Nahle, Z., & Zhang, M. Q. (2006). Adaptively inferring human transcriptional subnetworks. *Molecular Systems Biology*, *2*. doi: 10.1038/msb4100067

Das, R., & Gerstein, M. (2004). A method using active-site sequence conservation to find functional shifts in protein families: Application to the enzymes of central metabolism, leading to the identification of an anomalous isocitrate dehydrogenase in pathogens. *Proteins-Structure Function and Bioinformatics*, *55*(2), 455–463. doi:10.1002/prot.10639 PMID:15048835

de Jong, H. (2002). Modeling and simulation of genetic regulatory systems: A literature review. *Journal of Computational Biology*, *9*(1), 67–103. doi:10.1089/10665270252833208 PMID:11911796

Dunham, I., Kundaje, A., Aldred, S. F., Collins, P. J., Davis, C., Doyle, F., & Consortium, E. P. et al. (2012). An integrated encyclopedia of DNA elements in the human genome. *Nature*, *489*(7414), 57–74. doi:10.1038/Nature11247 PMID:22955616

Eisen, M. B., Spellman, P. T., Brown, P. O., & Botstein, D. (1998). Cluster analysis and display of genome-wide expression patterns. *Proceedings of the National Academy of Sciences of the United States of America*, *95*(25), 14863–14868. doi:10.1073/pnas.95.25.14863 PMID:9843981

Frech, K., & Werner, T. (1997). Specific modelling of regulatory units in DNA sequences.*Pacific Symposium on Biocomputing*, (pp. 151-162).

Friedman, J., Hastie, T., & Tibshirani, R. (2010). Regularization Paths for Generalized Linear Models via Coordinate Descent. *Journal of Statistical Software*, *33*(1), 1–22. doi:10.18637/jss.v033.i01 PMID:20808728

Friedman, N., Linial, M., Nachman, I., & Pe'er, D. (2000). Using Bayesian networks to analyze expression data. *Journal of Computational Biology*, *7*(3-4), 601–620. doi:10.1089/106652700750050961 PMID:11108481

GuhaThakurta, D., & Stormo, G. D. (2001). Identifying target sites for cooperatively binding factors. *Bioinformatics (Oxford, England)*, *17*(7), 608–621. doi:10.1093/bioinformatics/17.7.608 PMID:11448879

Hassa, P. O., Covic, M., Bedford, M. T., & Hottiger, M. O. (2008). Protein arginine methyltransferase 1 coactivates NF-kappaB-dependent gene expression synergistically with CARM1 and PARP1. *Journal of Molecular Biology*, *377*(3), 668–678. doi:10.1016/j.jmb.2008.01.044 PMID:18280497

Hastie, T., Tibshirani, R., & Friedman, J. H. (2009). *The elements of statistical learning: data mining, inference, and prediction* (2nd ed.). New York, NY: Springer. doi:10.1007/978-0-387-84858-7

Hirose, O., Yoshida, R., Imoto, S., Yamaguchi, R., Higuchi, T., Charnock-Jones, D. S., & Miyano, S. et al. (2008). Statistical inference of transcriptional module-based gene networks from time course gene expression profiles by using state space models. *Bioinformatics (Oxford, England)*, *24*(7), 932–942. doi:10.1093/bioinformatics/btm639 PMID:18292116

Hu, H., & Dai, Y. (2014). A Model-based approach to transcription regulatory network reconstruction from time-course gene expression data. *Conference Proceedings of the Annual International Conference of the IEEE Engineering in Medicine and Biology Society*. doi:10.1109/EMBC.2014.6944690

Hu, Y. J. (2003). Finding subtle motifs with variable gaps in unaligned DNA sequences. *Computer Methods and Programs in Biomedicine*, *70*(1), 11–20. doi:10.1016/S0169-2607(01)00198-5 PMID:12468123

Huber, B. R., & Bulyk, M. L. (2006). Meta-analysis discovery of tissue-specific DNA sequence motifs from mammalian gene expression data. *BMC Bioinformatics*, *7*(1), 229. doi:10.1186/1471-2105-7-229 PMID:16643658

Iyer, V. R., Horak, C. E., Scafe, C. S., Botstein, D., Snyder, M., & Brown, P. O. (2001). Genomic binding sites of the yeast cell-cycle transcription factors SBF and MBF. *Nature*, *409*(6819), 533–538. doi:10.1038/35054095 PMID:11206552

Jensen, L. J., & Knudsen, S. (2000). Automatic discovery of regulatory patterns in promoter regions based on whole cell expression data and functional annotation. *Bioinformatics (Oxford, England)*, *16*(4), 326–333. doi:10.1093/bioinformatics/16.4.326 PMID:10869030

Jensen, S. T., Liu, X. S., Zhou, Q., & Liu, J. S. (2004). Computational discovery of gene regulatory binding motifs: A Bayesian perspective. *Statistical Science*, 188–204.

Johnson, D. S., Mortazavi, A., Myers, R. M., & Wold, B. (2007). Genome-wide mapping of in vivo protein-DNA interactions. *Science*, *316*(5830), 1497–1502. doi:10.1126/science.1141319 PMID:17540862

Jonassen, I. (1997). Efficient discovery of conserved patterns using a pattern graph. *Computer Applications in the Biosciences*, *13*(5), 509–522. PMID:9367124

Kechris, K. J., van Zwet, E., Bickel, P. J., & Eisen, M. B. (2004). Detecting DNA regulatory motifs by incorporating positional trends in information content. *Genome Biology*, *5*(7), R50. doi:10.1186/gb-2004-5-7-r50 PMID:15239835

Keles, S., van der Laan, M., & Eisen, M. B. (2002). Identification of regulatory elements using a feature selection method. *Bioinformatics (Oxford, England)*, *18*(9), 1167–1175. doi:10.1093/bioinformatics/18.9.1167 PMID:12217908

Kim, S. Y., Imoto, S., & Miyano, S. (2003). Inferring gene networks from time series microarray data using dynamic Bayesian networks. *Briefings in Bioinformatics*, *4*(3), 228–235. doi:10.1093/bib/4.3.228 PMID:14582517

Krishnapuram, B., Carin, L., Figueiredo, M. A., & Hartemink, A. J. (2005). Sparse multinomial logistic regression: Fast algorithms and generalization bounds. *IEEE Transactions on Pattern Analysis and Machine Intelligence*, *27*(6), 957–968. doi:10.1109/TPAMI.2005.127 PMID:15943426

Lander, E. S., Linton, L. M., Birren, B., Nusbaum, C., Zody, M. C., Baldwin, J., & Morgan, M. J. et al. (2001). Initial sequencing and analysis of the human genome. *Nature*, *409*(6822), 860–921. doi:10.1038/35057062 PMID:11237011

Landt, S. G., Marinov, G. K., Kundaje, A., Kheradpour, P., Pauli, F., Batzoglou, S., & Snyder, M. et al. (2012). ChIP-seq guidelines and practices of the ENCODE and modENCODE consortia. *Genome Research*, *22*(9), 1813–1831. doi:10.1101/gr.136184.111 PMID:22955991

Latchman, D. S. (1997). Transcription factors: An overview. *International Journal of Biochemistry & Cell Biology, 29*(12), 1305-1312. doi: 10.1016/s1357-2725(97)00085-x

Laubenbacher, R., & Stigler, B. (2004). A computational algebra approach to the reverse engineering of gene regulatory networks. *Journal of Theoretical Biology*, *229*(4), 523–537. doi:10.1016/j.jtbi.2004.04.037 PMID:15246788

Lee, W. P., & Tzou, W. S. (2009). Computational methods for discovering gene networks from expression data. *Briefings in Bioinformatics*, *10*(4), 408–423. doi:10.1093/bib/bbp028 PMID:19505889

Lee, W.-P., & Yang, K.-C. (2008). A clustering-based approach for inferring recurrent neural networks as gene regulatory networks. *Neurocomputing*, *71*(4), 600–610. doi:10.1016/j.neucom.2007.07.023

Levine, M., & Tjian, R. (2003). Transcription regulation and animal diversity. *Nature*, *424*(6945), 147–151. doi:10.1038/nature01763 PMID:12853946

Liu, X., Brutlag, D. L., & Liu, J. S. (2001). BioProspector: discovering conserved DNA motifs in upstream regulatory regions of co-expressed genes.*Pacific Symposium on Biocomputing*, (pp. 127-138).

Liu, Y., Jiang, B., & Zhang, X. (2009). Gene-set analysis identifies master transcription factors in developmental courses. *Genomics*, *94*(1), 1–10. doi:10.1016/j.ygeno.2009.02.005 PMID:19272436

Ljung, L. (1987). *System identification: Theory for the user* (Vol. 198). Academic Press.

Marsman, J., & Horsfield, J. A. (2012). Long distance relationships: Enhancer-promoter communication and dynamic gene transcription. *Biochimica et Biophysica Acta (BBA) - Bioenergetics*, *1819*(11-12), 1217–1227. doi:10.1016/j.bbagrm.2012.10.008

Mathelier, A., Zhao, X., Zhang, A. W., Parcy, F., Worsley-Hunt, R., Arenillas, D. J., & Wasserman, W. W. et al. (2014). JASPAR 2014: An extensively expanded and updated open-access database of transcription factor binding profiles. *Nucleic Acids Research*, *42*, D142–D147. doi:10.1093/nar/gkt997 PMID:24194598

Matys, V., Fricke, E., Geffers, R., Gossling, E., Haubrock, M., Hehl, R., & Wingender, E. et al. (2003). TRANSFAC(R): Transcriptional regulation, from patterns to profiles. *Nucleic Acids Research*, *31*(1), 374–378. doi:10.1093/nar/gkg108 PMID:12520026

McCue, L., Thompson, W., Carmack, C., Ryan, M. P., Liu, J. S., Derbyshire, V., & Lawrence, C. E. (2001). Phylogenetic footprinting of transcription factor binding sites in proteobacterial genomes. *Nucleic Acids Research*, *29*(3), 774–782. doi:10.1093/nar/29.3.774 PMID:11160901

Michalewicz, Z. (1996). Genetic algorithms + data structures = evolution programs (3rd rev. ed.). Berlin: Springer-Verlag.

Natarajan, A., Yardimci, G. G., Sheffield, N. C., Crawford, G. E., & Ohler, U. (2012). Predicting cell-type-specific gene expression from regions of open chromatin. *Genome Research*, *22*(9), 1711–1722. doi:10.1101/gr.135129.111 PMID:22955983

Nishida, K., Frith, M. C., & Nakai, K. (2009). Pseudocounts for transcription factor binding sites. *Nucleic Acids Research*, *37*(3), 939–944. doi:10.1093/nar/gkn1019 PMID:19106141

Nolan, G. P., Ghosh, S., Liou, H. C., Tempst, P., & Baltimore, D. (1991). DNA binding and I kappa B inhibition of the cloned p65 subunit of NF-kappa B, a rel-related polypeptide. *Cell*, *64*(5), 961–969. doi:10.1016/0092-8674(91)90320-X PMID:2001591

O'Flanagan, R. A., Paillard, G., Lavery, R., & Sengupta, A. M. (2005). Non-additivity in protein-DNA binding. *Bioinformatics (Oxford, England)*, *21*(10), 2254–2263. doi:10.1093/bioinformatics/bti361 PMID:15746285

Piechota, M., Korostynski, M., & Przewlocki, R. (2010). Identification of cis-regulatory elements in the mammalian genome: The cREMaG database. *PLoS ONE*, *5*(8), e12465. doi:10.1371/journal.pone.0012465 PMID:20824209

Pudimat, R., Schukat-Talamazzini, E. G., & Backofen, R. (2004). *Feature Based Representation and Detection of Transcription Factor Binding Sites*. Paper presented at the German Conference on Bioinformatics.

Qian, J., Esumi, N., Chen, Y., Wang, Q., Chowers, I., & Zack, D. J. (2005). Identification of regulatory targets of tissue-specific transcription factors: Application to retina-specific gene regulation. *Nucleic Acids Research*, *33*(11), 3479–3491. doi:10.1093/nar/gki658 PMID:15967807

Rebeiz, M., Reeves, N. L., & Posakony, J. W. (2002). SCORE: A computational approach to the identification of cis-regulatory modules and target genes in whole-genome sequence data. Site clustering over random expectation. *Proceedings of the National Academy of Sciences of the United States of America*, *99*(15), 9888–9893. doi:10.1073/pnas.152320899 PMID:12107285

Ren, B., Robert, F., Wyrick, J. J., Aparicio, O., Jennings, E. G., Simon, I., & Young, R. A. et al. (2000). Genome-wide location and function of DNA binding proteins. *Science*, *290*(5500), 2306–2309. doi:10.1126/science.290.5500.2306 PMID:11125145

Rigoutsos, I., & Floratos, A. (1998). Combinatorial pattern discovery in biological sequences: The TEIRESIAS algorithm. *Bioinformatics (Oxford, England)*, *14*(1), 55–67. doi:10.1093/bioinformatics/14.1.55 PMID:9520502

Robertson, G., Hirst, M., Bainbridge, M., Bilenky, M., Zhao, Y. J., Zeng, T., & Jones, S. et al. (2007). Genome-wide profiles of STAT1 DNA association using chromatin immunoprecipitation and massively parallel sequencing. *Nature Methods*, *4*(8), 651–657. doi:10.1038/nmeth1068 PMID:17558387

Sandelin, A., Alkema, W., Engstrom, P., Wasserman, W. W., & Lenhard, B. (2004). JASPAR: An open-access database for eukaryotic transcription factor binding profiles. *Nucleic Acids Research, 32*(90001), D91–D94. doi:10.1093/nar/gkh012 PMID:14681366

Sandve, G. K., & Drablos, F. (2006). A survey of motif discovery methods in an integrated framework. *Biology Direct, 1*(1), 11. doi:10.1186/1745-6150-1-11 PMID:16600018

Schena, M., Shalon, D., Davis, R. W., & Brown, P. O. (1995). Quantitative monitoring of gene expression patterns with a complementary DNA microarray. *Science, 270*(5235), 467–470. doi:10.1126/science.270.5235.467 PMID:7569999

Scherf, M., Klingenhoff, A., & Werner, T. (2000). Highly specific localization of promoter regions in large genomic sequences by PromoterInspector: A novel context analysis approach. *Journal of Molecular Biology, 297*(3), 599–606. doi:10.1006/jmbi.2000.3589 PMID:10731414

Shmulevich, I., Dougherty, E. R., Kim, S., & Zhang, W. (2002). Probabilistic Boolean Networks: A rule-based uncertainty model for gene regulatory networks. *Bioinformatics (Oxford, England), 18*(2), 261–274. doi:10.1093/bioinformatics/18.2.261 PMID:11847074

Simcha, D., Price, N. D., & Geman, D. (2012). The limits of de novo DNA motif discovery. *PLoS ONE, 7*(11), e47836. doi:10.1371/journal.pone.0047836 PMID:23144830

Sinha, S. (2006). On counting position weight matrix matches in a sequence, with application to discriminative motif finding. *Bioinformatics (Oxford, England), 22*(14), e454–e463. doi:10.1093/bioinformatics/btl227 PMID:16873507

Sloop, K. W., Showalter, A. D., Von Kap-Herr, C., Pettenati, M. J., & Rhodes, S. J. (2000). Analysis of the human LHX3 neuroendocrine transcription factor gene and mapping to the subtelomeric region of chromosome 9. *Gene, 245*(2), 237–243. doi:10.1016/S0378-1119(00)00025-1 PMID:10717474

Smith, A. D., Sumazin, P., Xuan, Z., & Zhang, M. Q. (2006). DNA motifs in human and mouse proximal promoters predict tissue-specific expression. *Proceedings of the National Academy of Sciences of the United States of America, 103*(16), 6275–6280. doi:10.1073/pnas.0508169103 PMID:16606849

Spellman, P. T., Sherlock, G., Zhang, M. Q., Iyer, V. R., Anders, K., Eisen, M. B., & Futcher, B. et al. (1998). Comprehensive identification of cell cycle-regulated genes of the yeast Saccharomyces cerevisiae by microarray hybridization. *Molecular Biology of the Cell*, *9*(12), 3273–3297. doi:10.1091/mbc.9.12.3273 PMID:9843569

Stormo, G. D. (2000). DNA binding sites: Representation and discovery. *Bioinformatics (Oxford, England)*, *16*(1), 16–23. doi:10.1093/bioinformatics/16.1.16 PMID:10812473

Stormo, G. D., Schneider, T. D., Gold, L., & Ehrenfeucht, A. (1982). Use of the 'Perceptron' algorithm to distinguish translational initiation sites in E. coli. *Nucleic Acids Research*, *10*(9), 2997–3011. doi:10.1093/nar/10.9.2997 PMID:7048259

Thackray, V. G. (2014). Fox tales: Regulation of gonadotropin gene expression by forkhead transcription factors. *Molecular and Cellular Endocrinology*, *385*(1-2), 62–70. doi:10.1016/j.mce.2013.09.034 PMID:24099863

Tibshirani, R. (1996). Regression shrinkage and selection via the Lasso. *Journal of the Royal Statistical Society. Series B. Methodological*, *58*(1), 267–288.

van Helden, J., Andre, B., & Collado-Vides, J. (1998). Extracting regulatory sites from the upstream region of yeast genes by computational analysis of oligonucleotide frequencies. *Journal of Molecular Biology*, *281*(5), 827–842. doi:10.1006/jmbi.1998.1947 PMID:9719638

Veena, M. S., Wilken, R., Zheng, J. Y., Gholkar, A., Venkatesan, N., Vira, D., & Srivatsan, E. S. et al. (2014). p16 Protein and gigaxonin are associated with the ubiquitination of NFkappaB in cisplatin-induced senescence of cancer cells. *The Journal of Biological Chemistry*, *289*(50), 34921–34937. doi:10.1074/jbc.M114.568543 PMID:25331947

Vohradsky, J. (2001). Neural network model of gene expression. *The FASEB Journal*, *15*(3), 846–854. doi:10.1096/fj.00-0361com PMID:11259403

Wagner, A. (1999). Genes regulated cooperatively by one or more transcription factors and their identification in whole eukaryotic genomes. *Bioinformatics (Oxford, England)*, *15*(10), 776–784. doi:10.1093/bioinformatics/15.10.776 PMID:10705431

Wahde, M., & Hertz, J. (2000). Coarse-grained reverse engineering of genetic regulatory networks. *Bio Systems*, *55*(1-3), 129–136. doi:10.1016/S0303-2647(99)00090-8 PMID:10745116

Wang, T., & Stormo, G. D. (2003). Combining phylogenetic data with co-regulated genes to identify regulatory motifs. *Bioinformatics (Oxford, England)*, *19*(18), 2369–2380. doi:10.1093/bioinformatics/btg329 PMID:14668220

Wasserman, W. W., & Sandelin, A. (2004). Applied bioinformatics for the identification of regulatory elements. *Nature Reviews. Genetics*, *5*(4), 276–287. doi:10.1038/nrg1315 PMID:15131651

Werner, T. (1999). Models for prediction and recognition of eukaryotic promoters. *Mammalian Genome*, *10*(2), 168–175. doi:10.1007/s003359900963 PMID:9922398

Wingender, E. (2008). The TRANSFAC project as an example of framework technology that supports the analysis of genomic regulation. *Briefings in Bioinformatics*, *9*(4), 326–332. doi:10.1093/bib/bbn016 PMID:18436575

Xing, E. P., Wu, W., Jordan, M. I., & Karp, R. M. (2004). Logos: A modular bayesian model for de novo motif detection. *Journal of Bioinformatics and Computational Biology*, *2*(1), 127–154. doi:10.1142/S0219720004000508 PMID:15272436

Yang, J. P., Hori, M., Sanda, T., & Okamoto, T. (1999). Identification of a novel inhibitor of nuclear factor-kappaB, RelA-associated inhibitor. *The Journal of Biological Chemistry*, *274*(22), 15662–15670. doi:10.1074/jbc.274.22.15662 PMID:10336463

Yeung, M. K., Tegner, J., & Collins, J. J. (2002). Reverse engineering gene networks using singular value decomposition and robust regression. *Proceedings of the National Academy of Sciences of the United States of America*, *99*(9), 6163–6168. doi:10.1073/pnas.092576199 PMID:11983907

Yu, X., Lin, J., Zack, D. J., & Qian, J. (2006). Computational analysis of tissue-specific combinatorial gene regulation: Predicting interaction between transcription factors in human tissues. *Nucleic Acids Research*, *34*(17), 4925–4936. doi:10.1093/nar/gkl595 PMID:16982645

Yuan, M., & Lin, Y. (2006). Model selection and estimation in regression with grouped variables. *Journal of the Royal Statistical Society. Series B, Statistical Methodology*, *68*(1), 49–67. doi:10.1111/j.1467-9868.2005.00532.x

Zhao, X., Huang, H., & Speed, T. P. (2005). Finding short DNA motifs using permuted Markov models. *Journal of Computational Biology*, *12*(6), 894–906. doi:10.1089/cmb.2005.12.894 PMID:16108724

Zhu, Z., Pilpel, Y., & Church, G. M. (2002). Computational identification of transcription factor binding sites via a transcription-factor-centric clustering (TFCC) algorithm. *Journal of Molecular Biology*, *318*(1), 71–81. doi:10.1016/S0022-2836(02)00026-8 PMID:12054769

Zou, M., & Conzen, S. D. (2005). A new dynamic Bayesian network (DBN) approach for identifying gene regulatory networks from time course microarray data. *Bioinformatics (Oxford, England)*, *21*(1), 71–79. doi:10.1093/bioinformatics/bth463 PMID:15308537

KEY TERMS AND DEFINITIONS

Least Absolute Shrinkage and Selection Operator (LASSO): A shrinkage and selection method for linear regression. It minimizes the usual sum of squared errors, with a bound on the sum of the absolute values of the coefficients.

Multinomial Logistic Regression Model: A statistical classification method generalize logistic regression to multiclass problem, i.e., a model that is used to predict the probabilities of the different possible outcomes of a categorically distributed dependent variable (i.e. the set id of co-expressed genes), given a set of independent variables (i.e. the TF binding information on each gene).

Position Weight Matrix (PWM): A matrix that represents the sequence motif (i.e. for the TFBS) compiled from a set of aligned sequences (i.e. the aligned sequences are verified to be bound by a TF).

Time-Course Gene Expression: The gene expression level of a specific sample or tissue across multiple time points, typically the transcription level measured by microarray or RNA-seq techniques.

Transcription Factor (TF): Proteins that bind to specific DNA sequences and regulate gene transcription.

Transcription Factor Binding Site (TFBS): Genomic locations on the DNA sequences that the transcription factors bind to regulate gene transcription.

Chapter 9
Dynamical Analysis of Drug Efficacy and Mechanism of Action Using GFP Reporters

Jianping Hua
Texas A&M University, USA

Chao Sima
Texas A&M University, USA

Milana Cypert
Texas A&M University, USA

Edward R. Dougherty
Texas A&M University, USA

Jeffery M. Trent
Translational Genomics Research Institute, USA

Michael L. Bittner
Translational Genomics Research Institute, USA

ABSTRACT

To the development of effective cancer drug, it is necessary to, first, identify drugs and their possible combinations that could exert desired control over the type of cancer being considered; second, have a drug testing method that allows one to assess the variety of responses that can be provoked by drugs. To facilitate such an experiment-modeling-experiment cycle for drug development, a method based on the dynamical systems of pathways is presented. It involves a three-state experimental design: (1) formulate an oncologic pathway model of relevant cancer; (2) perturb the pathways with the drugs of known effects on components of the pathways of interest; and (3) measure process activity indicators at various points on cell populations. To evaluate the drug response in a high-throughput manner, a green fluorescent protein reporter-based technology has been developed. The authors apply the dynamical approach to several issues in the context of colon cancer cell lines.

DOI: 10.4018/978-1-5225-0353-8.ch009

INTRODUCTION

Target identification and validation are critical components of the targeted drug discovery (Bleicher2003, Hoelder 2012, Dahlin 2015). Predicting drug effectiveness through potential target is typically via a bottom-up conceptualization of tumor function. It assumes that sufficient knowledge regarding cellular processes can be gathered through genomic, proteomic, and metabolic profiling projects. Such knowledge can then be used to recognize tumor vulnerabilities, screen potential drugs and construct effective targeted therapies (Lee 2009, T.C.G.A. Network 2008). For this constructivist approach to succeed, the vulnerabilities must be based on simple linear relationships, in which inactivating a particular target will be sufficient to control the tumor.

To illustrate the kind of special situation that contributes to a direct correlation between therapy and disease control, we consider two cancer types, chronic myelogenous leukemia (CML) and acute promyelocytic leukemia (APML), in which substantial improvements in patient outcomes have been achieved by targeted therapy. In these cancers, a single type of translocation fusing parts of two genes is a pathognomonic indicator of the disease and an excellent biochemical target for therapeutic targeting (Chen 1991, Druker 2001, Nowell 1960, Rowley 1977). These cases seem to be fairly extreme outliers from the bulk of cancer types in two ways.

The first is the striking uniformity of CML and APML tumors. Significant homogeneity of particular, discriminating patterns of gene expression in a particular subtype of tumor is rare, yet both of these tumors show exactly this characteristic (Dong 2010, Nowiki 2003). It appears that such tumors arise from a specific partially differentiated precursor cell that exhibits most of the characteristics of a cancer and that those missing cancer characteristics occur as a consequence of a single genomic alteration (Pear 1998). This behavior is in stark contrast to the considerable heterogeneity of molecular pathologies that appear to arise from various differentiated precursors in many cancers, such as breast cancer (Perou 2000). Also consider melanoma, which originates from a single, highly-differentiated cell, the melanocyte, and has yet to be sub-classified into clinically useful groups on the basis of molecular pathology. The second way in which CML and APML differ from most cancers is that, as opposed to CML and APML, tumors arising from highly differentiated cell types appear to have to breach more types of controls opposing oncogenesis. Even when starting from a single differentiated type, they appear to break these controls in a wider variety of ways.

These differences in tumor-type homogeneity and underlying molecular pathology are inversely correlated to the efficacy of the drugs available for the differing tumor types, i.e., high efficacy with low tumor complexity and low efficacy with high tumor complexity. Due to the manifold regulatory structures in complex systems, multiple inputs affect each output and the control structure includes parallel, redundant, and feedback loop processes. Thus if the tumor itself is not actively responding to the drug, it is very likely that many operations susceptible to a drug will not even be operating. Even for the cases the drug does work, like the hits found through drug screening, the effect could be due to mechanism other than what is originally designed. So it's very important to identify the exact response the drug induces.

Traditionally, this is done by protein assays, like western blot (Brough 2009, Geisler 2015). The assay should examine the downstream of the desired targets, like the phosphorylation status of certain proteins. More recently, high-throughput technique like microarray and RNA-Seq are also available, which measure the transcriptional activity (Smith 2006, Sirota 2011). However, these methods are often time consuming and costly. Furthermore, it takes time for any drug to drive out desired effects, and the response time varies for different drugs. Yet the traditional approaches often only take one time point for examination, and often miss critical information (Orth 2008).

In this chapter we discuss a model-based dynamical experimental method to facilitate better pathology identification and better matching of tumors with targeted drugs. The method could potentially address many central questions relevant to the mechanics of drug function: Are the responses consistent with the intended targeting? Is a single intervention at a given point successful in stopping the altered function? Do combinations directed at a particular pair of proliferation and survival generators produce synergy in provoking cell death? If not, at what point or points in the relevant processes do the responses fail to push the cell in the intended direction?

The methodology presented in this chapter provides a paradigm for answering these questions. It utilizes the dynamic evolution of drug/cell interactions, as opposed to non-dynamic techniques incapable of ordering tumor regulatory decisions during successful and failed drug interventions. Dynamical methods also better exploit the ability to gain information relevant to tumor vulnerabilities by probing the same cells with different drugs, whose targets are in various positions of the regulatory pathways. This allows comparison of signaling changes introduced at different points in pathways affected by the drug. With such information, it's possible to understand the mechanism that leads to the success or failure of the drug, and help the researchers to come up with new hypotheses.

EXPERIMENTAL APPARATUS

Characterizing drug response requires experimental designs that ask specific questions about cell behavior in the absence of a drug and how cell activity changes when the drug is present. Design and execution of experiments aimed at understanding mechanism in terms of drug effects on its target(s) is carried out in three stages: (1) formulate a model of pathways in the processes potentially affected; (2) perturb the pathways with the test drug and drugs with known effects on components of the pathways of interest; and (3) measure process activity indicators at various points on cell populations.

A model of oncologic processes represents existing expectations based on currently available, experimentally derived information. These expectations could be the effect a drug on a particular component in a biological pathway. And it could be the effects induced by the drug on its target, like the various processes whose behavior is being driven by the target. For the majority of tumors, where a variety of processes are involved in tumor maintenance, complex modeling is required to identify the different processes and process components maintaining the tumor. The construction of typical pathway graphs depicting functional connections for which there is existing evidence. It provides a hypothesized wiring diagram to use in considering what kinds of reporting would be indicative of process changes consistent with current expectations.

A diagram showing a set of pathway connections expected to be affected by several different drugs is shown in Figure 1. This figure depicts possible ways that proliferative and survival signals generated by EGFR and ERBB2 could be transmitted through cellular signaling pathways. The produced changes in the expression status of various necessary intermediaries ultimately lead to further changes in genes carrying the functions required for proliferation and survival. It shows points along the pathways where characterized "reference" drugs that interfere with these processes act. A judicious choice of reporters, "reference" drugs, and cell lines with known alterations facilitates examination of whether many of the possible processes that an oncology drug could disrupt are in fact being disrupted in a cell line bearing interesting sets of alterations.

A critical step that is essential to the success of the described scheme is to measure the cellular activity in an accurate yet efficient manner through real experiments. Here we introduce a high-content fluorescent protein reporter-based experimental protocol, where the transcriptional activity of cells under given perturbation can be traced through the fluorescent reporters through automatic imaging and image processing.

Experimental Protocol

Fluorescent reporter technology has been used to study a range of individual and collective cell activities, for instance, to address questions concerning the transcriptional activity of individual cells (Chalfie 1994). Developmental questions

Figure 1. A wiring diagram showing proliferation and survival pathway elements whose transcriptional states could be altered in a cell exposed to various drugs. Nodes in shade are ones for which a reporter would be used to assess transcription for that gene. The potential drugs are indicated by the asymmetric bounding box and light gray labels.

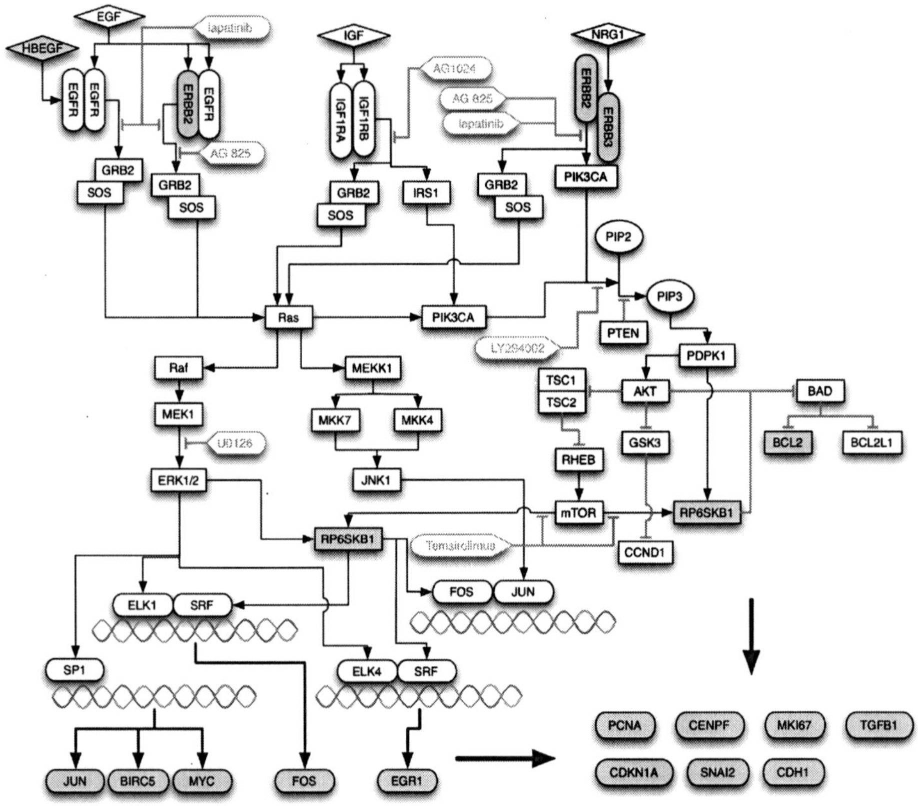

such as when and in which cells particular genes are expressed and where in the embryo a series of genes is expressed have also been addressed (Hao 2006). Such questions about the interaction of processes over time are closely related in form to drug response questions, in which the abundance of a fluorescent protein is used to measure cellular transcriptional responses to drugs. There are two basic types of fluorescent reporters based on the molecular structure: 1) Promoter-reporters that indicate the activity of the promoter for a particular gene. A fluorescent reporter to serve this purpose can be constructed by fusing the promoter region of a gene of interest with the coding sequence of a fluorescent protein, most commonly a green fluorescent protein (GFP). Then the abundance of the reporter can be used as an indicator of the activity of the promoter. 2) Fusion-reporters that indicate the cellular location of a protein, the stability of a protein, the status of large architectural elements in the cell, or the positioning of small molecules in a cell. This type of fluorescent reporter is made by an in-frame fusion of the coding sequence of either a whole protein or a protein domain to the coding sequence of a fluorescent protein. Then the coding sequence under the control of a promoter will produce that protein at a sufficient rate, resulting in constant detectable level of the fusion protein. In the proposed protocol, we will deal with the first type of fluorescent reporter only.

The presence of fluorescent molecules can be detected by imaging with an epifluorescent microscope. With fluorescent molecules in the very small volume of a cell, one can make quantitative measurements using only light to probe the cells. Cells can be exposed to the exciting light hourly over 40 hours without injury and many different fields of view in the same culture well can be sampled. Repeated serial imaging allows the response dynamics of a cell population to be followed over a time course. These measurements utilize an automated epifluorescent microscope. Reporters (molecular constructs) are employed that indicate activity levels for promoters of particular genes. A single assay is carried out by epifluorescent imaging of a portion of the bottom of a well in a 384 well plate, producing an image of the cells in that region (~200-300) bearing fluorescent reporters.

The fluorescent protein reporter approach has several advantages over standard biochemical analysis. In the latter, one examines a portion of the treated cell population at each time point. The cells in that portion are lysed and extracted to obtain the analyte of interest, irreversibly committing the investigator to measuring the average response of a large number of cells. The frequency and length of sampling is usually limited by the ability to handle the number of cells required per assay and by the cost of the test applied. For many types of measurements, the consequence is either taking a single pre- and post-drug exposure sample or a sparse sampling over time. Using fluorescent reporters allows the same cell population to be sampled many times, since the quantification can be done by microscopic scanning of cells in culture. Current automated scanners, e.g., ImageXpress Micro by Molecular Devices, are reasonably fast and can do a two-color scan on a 384 well plate residing in a cell culture chamber in 15 minutes to 1 hour, depending on the number of sites scanned per well. This produces data on 200-300 individual cells per site that can be analyzed both individually, as a distribution, or in aggregate, as an average. The capability to observe the distributions of cell response behavior facilitates detection of subpopulations with different susceptibilities to the drug. It also allows the determination of whether the response is continuous and graded, where the level of the measured function changes in increments in all cells simultaneously, or the response is binary, where the level switches sharply from one level to another (Biggar 2001).

In a properly designed plate, each condition will be tested in multiple wells to ensure both adequate coverage and replicates to assess technical variation. In our experiments, every condition is duplicated in three or more wells. There will also be media-only, unperturbed cell population to serve as negative controls and cell-lines with a highly expressed reporter to serve as positive controls. These control wells can serve as a final safe guard to detect any unexpected effects.

Figure 2 shows two typical fluorescent images sampled for the same site in a 48-hour drug treatment study based on the diagram shown in Figure 1. The original images are 16-bit images and are normalized to 8-bit for proper viewing. The colon cancer cell-line HCT116 is under observation here. The reporter is driven by a 1kb promoter fragment from MKI67, a nuclear protein whose abundance is tightly correlated with proliferation (Spyratos 2002). No drug was added for the first 5 hours, then lapatinib, a drug used to treat breast cancer, was added and imaged for 43 hours. In Figure 2, the left image was taken at the beginning of the experiment when no drug has been added, while the right image was taken 43 hours after lapatinib was added. There are about 300 cells in the image. The nuclear channel (blue) images the live cell DNA fluorophore, Vybrant Violet, and the promoter reporter activity channel (green) images the GFP reporter protein. For the case shown here, the change in transcriptional activity is significant. In the beginning, gene MKI67 is highly transcribed, indicating strong proliferation activity. At the end of the experiment, most of the GFP fluorescence was reduced to almost unobservable, a likely indicator of reduced transcription followed by apoptotic proteolysis.

Figure 2 shows that fluorescent reporters can help researchers track the cellular activity of the same cohort of cells, in this case, by direct visual evaluation; however, the volume of data that can be handled by visual inspection in a timely manner is very limited, while the volume of data generated by the current approach is always huge. With a 384-well plate there will be 384 videos for evaluation and the number can be much higher if the experiment requires multiple sites being chosen in each well or multiple plates to cover all experimental conditions. Furthermore, visual evaluation is unreliable when one needs to aggregate the results from duplicated wells or when one needs to quantitatively compare different conditions. Hence the high-content nature of the GFP reporter approach calls for a more automatic and quantitative solution to efficiently extract and summarize the captured transcriptional activity.

Figure 2. Two fluorescent images of the same imaging site for cell-line HCT116 with a promoter reporter for the gene MKI67 taken in the time course of one experiment: a) before any drug is applied; b) 43 hours after a drug lapatinib was added. *Tracking transcriptional activities with high-content epifluorescent imaging, J Hua et al., Journal of Biomedical Optics, vol. 17 (Copyright @ 2012), World Scientific Publishing Co. Used with permission.*

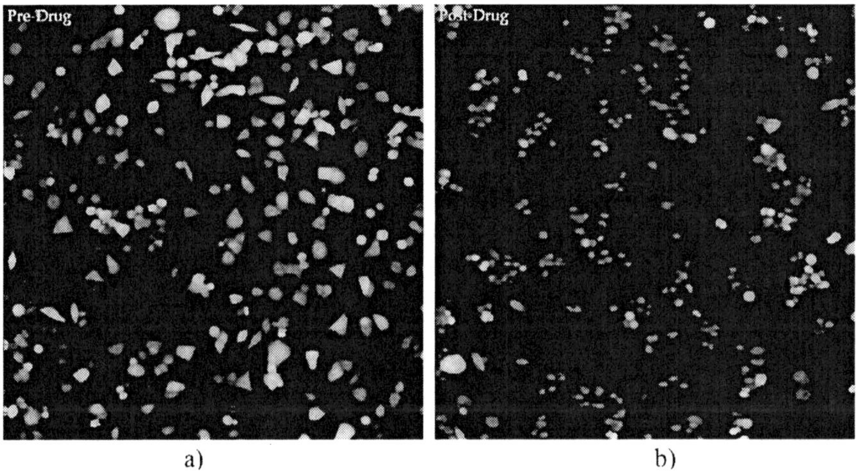

Image Processing

To facilitate automatic processing of the experiment results, the first step is image processing, where the transcriptional levels of the fluorescent images will be properly extracted, quantized, and saved. When designing the image processing method, several characteristics of cell imaging specific to the experimental protocol must be considered.

- **Cell-Level Activity:** In GFP reporter-based imaging, the transcriptional activity can be extracted at the single cell level. As shown in Figure 2b, although the proliferation activity of most cells has been turned off, there is still a small portion of cells whose transcriptional activity remains unchanged, indicating a possible sub-population of cells that is relatively resistant to the applied drug. Thus, a proper image processing method must be able to identify individual cells and extract the transcriptional level of each cell.
- **Robustness:** The demand for robust performance comes from both the time and space domains. First, it should be noted that each site is being imaged for a period of time and the extracted data will be studied as time-course

data rather than independent data. To further complicate the issue, a cell-line could have significantly different responses to different perturbations, so the imaging quality can also vary from well to well and time point to time point. Actually a well-designed experiment should be able to induce a variety of responses that allows the exploration of certain genetic regulation mechanisms that are only responding to specific stimuli. Ideally, if one can consider consecutive images in time and track the activity of individual cells, the results will be most informative. However, a considerable number of cells will be moving in or out of the imaging area during the experiment and the imaging site itself is not always fixed. These, plus the situations just described, make cell tracking challenging and limit its usefulness. Thus, we utilize single image processing to pursue the activity at a population scale, while making sure that the method is sufficiently robust to produce quality results over all conditions for the whole time course.

Achieving all of these objectives requires a fast image processing algorithm with good balance between speed and the ability to pick up almost all objects. The most common segmentation methods involve an adaptive-threshold-based approach like the one based on the top-hat transform (Dougherty 2003). Top-hat transform is a morphological processing operator. Like most image processing algorithm, the target objects, i.e., the stained nuclei in the nuclear channel and cell cytoplasm regions containing GFP protein in the GFP reporter channel are foreground to be identified and segmented, while the remaining parts of the images are deemed as background. The top-hat transform first applies an opening operator to the original image to remove the objects, which is the cells, in the foreground, and obtain a smooth background image. Then this opened image is subtracted from the original image, resulting a clean foreground image with background variation removed. We have found this approach, with a properly chosen threshold, to be fast and able to identify the foreground objects in general for almost all cases. However, it is difficult to find the optimal threshold for such an approach, which makes it hard to further identify individual nuclei. We decided to build our method using another morphological process, the watershed transformation (Vincent 1991), which is a popular and proven approach in cellular image processing. The time-course images are processed independently to maintain process speed and minimize the consequence of unexpected outliers: only one pair of images will be loaded and processed each time, and the results of outlier images can later be discarded without affecting other results.

Figure 3 provides the general flowchart on how a fluorescent image pair is processed: the nuclear channel is first processed to identify all nuclei (top part of

Figure 3. Flowchart of the image processing procedure
Tracking transcriptional activities with high-content epifluorescent imaging, J Hua et al., Journal of Biomedical Optics, vol. 17 (Copyright @2012), World Scientific Publishing Co. Used with permission.

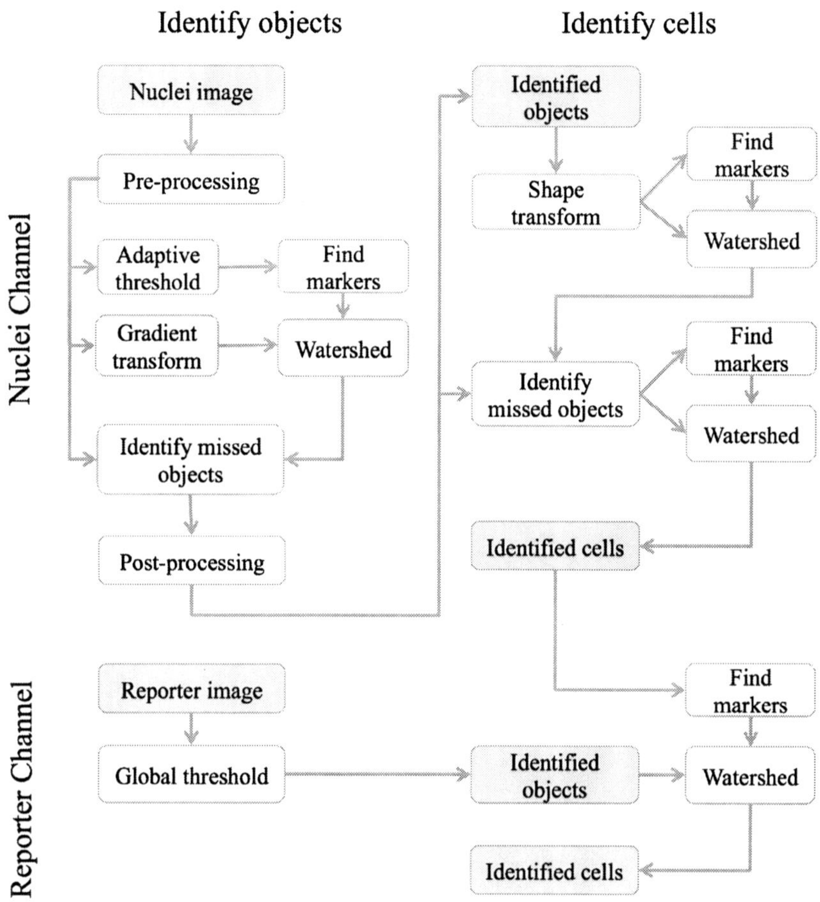

Figure 3) and the reporter channel is then analyzed to extract an individual cell's expression activity (bottom part of Figure 3). As a practical example, the critical steps in processing the image in Figure 2a are shown in Figure 4. In order to show the segmentation details, only a portion of the full image is shown here.

Nuclear channel processing contains two major steps. First, the foreground objects are identified, where an object is either an isolated nucleus or a cluster of nuclei (top left part of Figure 3). Before segmentation, the nuclear channel image is first pre-processed to remove tiny additive and subtractive noise via a single-stage open-close alternating sequential filter. We apply watershed-based segmentation on the gradient image. It is known that the accuracy of flood holes is critical to the

Figure 4. The segmentation of Figure 2a: a) row image of nuclear channel; b) markers for gradient-image-based watershed segmentation, where the white dots are the foreground markers, while the white lines are the boundary of background markers; c) identified foreground region; d) markers for nuclei de-clumping, the white dots are found by shape information, while the relatively large grey objects are markers found by intensity; the de-clumping is limited to the foreground region defined by the white boundary lines; e) final segmentation of the nuclei channel; f) reporter channel, where the white lines are the identified cell boundaries while the grey lines identify the nuclei used as markers.

Tracking transcriptional activities with high-content epifluorescent imaging, J Hua et al., Journal of Biomedical Optics, vol. 17 (Copyright @ 2012), World Scientific Publishing Co. Used with permission.

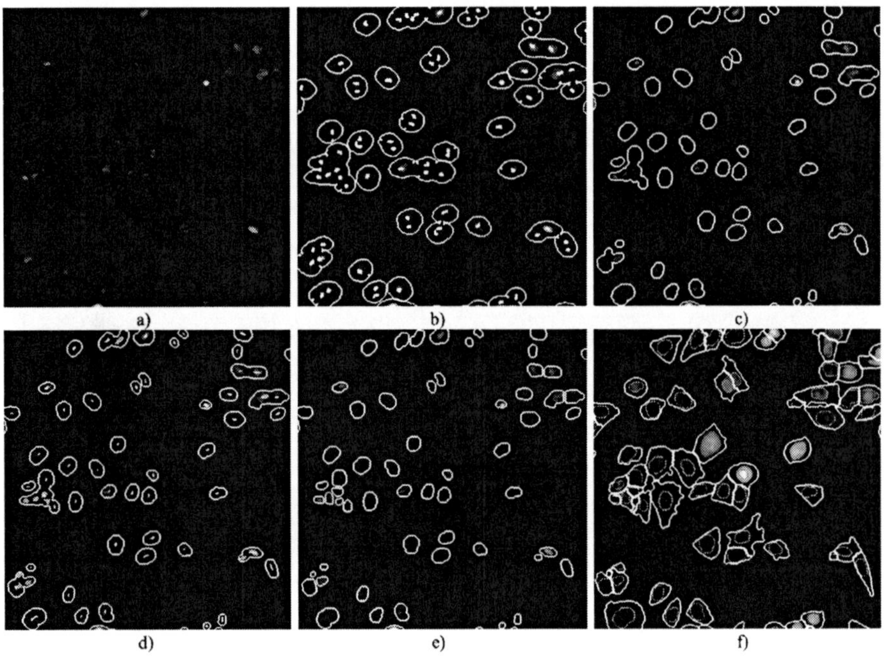

success of the watershed procedure. As is commonly done, we depend on the top-hat transform for marker selection. The local maxima inside the top-hat-identified foreground are filtered by the area open operation to form the foreground markers of size no smaller than a designated value, which is based on common nuclei sizes (Vincent 1993). Similarly, the background markers are found by morphologically thinning the top-hat-identified background. Thinning, rather than erosion, is chosen to avoid small and/or narrow passage regions between nuclei being lost. An example of the watershed markers can be viewed in Figure 4b.

Overall, the watershed on a gradient image gives a tight boundary on most nuclei; however, if a dim nucleus is attached to a bright nucleus or part of a nucleus is much brighter than other parts, the high gradient of the bright one might overshadow the dim one and the watershed segmentation could miss the dim region. To address this problem, the difference between top-hat transform and watershed results is used to identify and recover large missing objects. To obtain the final identified foreground objects, the segmentation results are sent for post-processing to remove tiny objects that can be artifacts or nuclear debris. An example of the final foreground can be found in Figure 4c.

The next step is to identify individual nuclei by de-clumping the clustered nuclei (top right part of Figure 3). In most cases, nuclei have circular or near-circular shapes, and the tight segmentation boundaries secured by the previous step make it a natural choice to use a shape-based de-clumping method, where we apply watershed segmentation on distance transformed foreground images. The local maxima found through morphological opening are set as markers for flooding holes. For the highly clustered regions, where the boundary shape is not informative enough, shape-based de-clumping might fail to separate some or all of the nuclei. These missed objects usually have a much larger area size and irregular shape than can be detected by the circularity shape factor. Intensity-based watershed segmentation is applied to further de-clump these missed objects, where the markers are found through the area open operator. An example of the watershed markers can be found in Figure 4d. To save space, we put both shape-based and intensity-based de-clumping markers in one image. Before passing the identified nuclei to the GFP channel processor, a final post-processing is done by removing the tiny objects generated by the de-clumping procedure. An example of the final segmentation results can be found in Figure 4e.

The parameters associated with image processing must be adjusted for each cell-line and experiment because critical factors associated with segmentation, like nuclei size, shape, and intensity range, can vary significantly due to experiment design, biological diversity, or imaging setting. In some extreme cases, the whole de-clumping part might be omitted if the nuclei do not have a round shape. Indeed, for certain cancer cell-lines, it is common to see large cells with large and/or irregularly shaped nuclei. For example, the breakage-fusion-bridge cycles can create very large nuclei or connected nuclei linked through anaphase or chromatin bridges. These irregular nuclei cannot be separated by standard criteria like shape or intensity, and can be ambiguous to determine even by experienced biologists. Hence, for such cases, to avoid further complication, it is better to skip the de-clumping step and treat the identified objects as nuclei.

As with the nuclear channel, the processing of the GFP reporter channel can be viewed as a two-step procedure: first, identify the objects; second, identify the cells. Here, both steps are much simpler than their counterparts in nuclear channel. Since the reporters are not bound to any other molecule, one would expect that the fluorescent reporter protein, once produced, should be all over the cytoplasm and quite evenly distributed. Thus, a global threshold can be used to find the objects that have fluorescent intensity over a designated value. The threshold is a fixed value offset by the background intensity, which is largely additive noise that should not be measured as part of the transcriptional activity. The background intensity can be retrieved by finding the mode in the lower part of the intensity histogram. If the cells are at a density where they essentially cover the entire bottom surface of the well, then the mode estimate of the background intensity could be false. In such a case, the imaging site will either be discarded during the quality check or an alternative approach will be used. The reference sites, which have the same media but no cells and are in the same plate, can be used to find the background intensity. To de-clump the objects into cells, intensity-based watershed segmentation was used. Since the nuclei must be contained in the cells, it is natural to use the identified nuclei as the markers of the presence of a cell for the watershed segmentation. Thanks to the success of nuclei channel segmentation, the most difficult part of GFP channel segmentation, identify the seeds for watershed, it now easy and straightforward. An example of the final GFP reporter channel segmentation results can be found in Figure 4f. It should be noted that although we treat the background noise as additive, there exist some multiplicative noise types. Such noise is only prominent when the overall intensity distribution range changes significantly, which can be detected when one observes a large background intensity change. In such cases, rather than trying to normalize such a variance, we discard the site and omit it in the following analysis.

Once the individual cells are identified, the transcriptional activity represented by the reporter is extracted for every cell by summing up the background subtracted pixel intensity of the whole cell area and taking a $\log_2$ transform before being exported. The $\log_2$ transform is taken because the changes of the chemical reactions prevalent in the cells are multiplicative rather than additive and the $\log_2$-transformed intensity can therefore better represent the dynamics of reporter abundance changes in the cell. The results are exported as a tabular data file for further processing. It should be noted that in our study, the transcription activity is collected from the reporter channel, while the nuclei channel only serves an auxiliary role for better segmentation. However, since the morphological changes of the nuclei can also reveal important information on cellular response, we are working on incorporating such information into our protocol.

The whole image processing procedure is done with MATLAB, with most morphological operators done by the SDC Morphology Toolbox for MATLAB (Dougherty 2003). For a standard image pair of size 1024 x 1024, with about 300 cells, it takes about 5 minutes to process a 48-time-point series, which is about 6s per image pair for a standard Windows-based desktop computer (3.3GHz processor, 4G memory). By splitting the plate into several blocks and running the program in parallel, a single plate can be processed overnight by a multi-core desktop computer.

Data Representation

Once the images are processed, transcriptional activity extracted, and the results exported, organized data representation is required to help researchers evaluate the quality of the experiment, investigate the outcome, and identify the biologically meaningful relationships.

As discussed in the last subsection, even under the same condition cells can be at different states with different expression levels due to various factors, such as cell cycle status, drug efficiency, heterogeneity of the cell population, etc. Hence, rather than obtain an average expression level, one should aggregate the cell-wise expression levels from all duplicate wells to obtain the expression profile across the whole population. If, for a given gene, the expression state of a cell at a certain time point is viewed as a random distribution, then the expression profile is a density function depending on the cell line, gene, external stimulant, and time. Although histograms are widely used to visualize the density function, it is cumbersome to put different profiles of different time points and conditions in one plot for comparison, which is critical to the time-course case-control study. Thus, we use one-dimension kernel density estimation with Gaussian kernel to obtain a density function, which has the properties of being nonnegative and having integral one (Parzen 1962, Bowman 1997). The smoothing parameter is selected by the Sheather-Jones method from Sheather (1991) implemented in the Matlab toolbox 'Smoothing, Functional Data Analysis, and Distance Weighted Discrimination Software,' by Marron (2004).

Figure 5 a shows expression profiles in the form of density functions obtained for the cell population imaged in Figure 2. The profiles of all 48 time points are shown. The lines are color-coded by grey-scale to indicate time. The two time points corresponding to the two images shown in Figure 2 are shown in bold lines, with the before-the-drug time point in black, and 43-hour time point in light-grey. As a comparison, Figure 5b shows the expression profiles of the un-drugged population of the same reporter and cell line. Note that in this example, since both plots are based on a single imaging site, the density curves are actually not as smooth as pooled results. The x-axis is the transcriptional activity level measured by the total fluorescent intensity in $\log_2$ scale. The figure clearly shows that the transcription level of target gene MKI67 is relatively high at the beginning of the experiment, with the peak at around 2^{17} (arbitrary camera intensity units). At the end of the

Figure 5. a) The evolution of the expression level histogram for the imaging site used in Figure 1 during the experiment where an MKI67 reporter in cell line HCT116 is responding to the drug, lapatinib is shown. The intensity of profile indicates time, starting with black at the beginning, and gradually changing to light grey at the end. The profiles at the beginning and ending time points are shown with bold lines; b) Expression profile for a un-drugged population of same reporter and cell line. *Tracking transcriptional activities with high-content epifluorescent imaging, J Hua et al., Journal of Biomedical Optics, vol. 17 (Copyright @2012), World Scientific Publishing Co. Used with permission.*

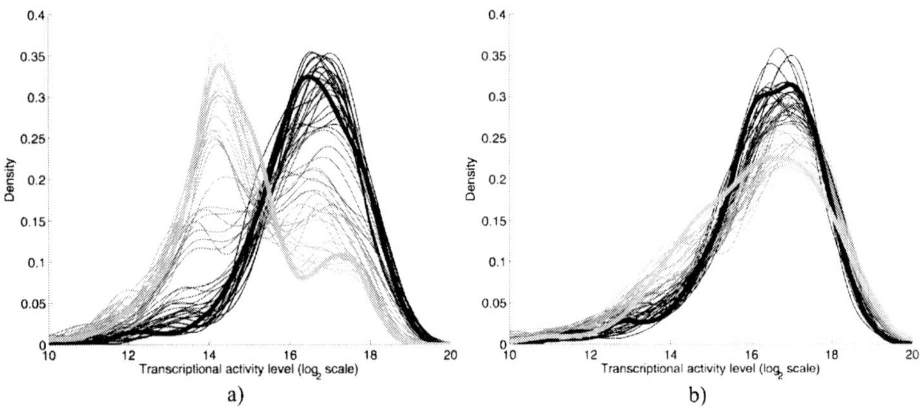

experiment, the transcription level of MKI68 has been greatly reduced to around 2^{14}, which is a decrease of roughly 8-fold, and is close to the fluorescence level that a cell without GFP would exhibit. The transcription profile also shows that there might be a small portion of cells that remain in a highly proliferative state as their MKI68 levels show no sign of decrease. The control population shows that without the influence of the drug, the transcription activities remain high in most of cells and only start to decrease in a small portion of cells at the end of the experiment, probably due to the depletion of growth factors in the media. This also confirms that there is little photobleaching in unaffected cells. More revealingly, the bi-modal expression profiles demonstrate beyond any doubt that when turning off, the state transition of MKI67 transcription activity is more of a switch-like procedure, rather than a gradual decrease of transcription level over the whole cell population. Furthermore, the switch-off of the transcription activity among responsive cells is not well synchronized, indicating different latency periods for cells to responding to the drug or even the existence of resistant sub-populations in the cell culture. Such a response dynamic is only observable when individual cell transcription activities, rather than the overall activity, are identifiable.

Viewing expression profiles as density functions allows researchers to compare transcription activities under different conditions in a single plot. In Figure 5, the transcription activities at different time points are compared. One can also compare the transcription activities under multiple conditions, such as different drugs or cell-lines. One can also put different plots side-by-side and make larger-scale comparisons. While providing a means for direct comparison, a density function does not quantize observed differences. Thus, while this density-based approach can provide straightforward comparison for small-scale studies, where only a few factors are involved and there are few conditions to track and compare, it is less suitable for studies of complex diseases like cancer, where one would like to perturb the cell-line of interest with multiple drugs aiming at different targets and track multiple transcription signatures representing several critically related pathways. To represent the cellular activities at such a large scale in a manageable manner, one needs to further quantize the expression profile while keeping the meaningful responses observable.

From the example shown Figure 5 and other observations, we notice that, for a cell culture at any given time, the distribution of transcription activity is mostly either uni-modal or bi-modal, indicating the cells are in either one or two dominating states most of the time. In the beginning of the experiment, owing to careful preparation, the transcription activity is highly homogeneous and the density function is always uni-modal with a sharp peak. Moreover, when the cell culture reacts to certain stimuli, the response is mostly uni-directional, which can lead to bi-modal distribution during state transition. If one could identify the valid states which the transcription activity can occupy and the percentage of cells in those states for any give time point, then the most informative part of the cellular dynamical system could be captured; however, we have found that such an absolute measure of cellular response is not practical on account of the following reasons: (A) It is hard to determine whether the distribution is uni-modal or bi-modal in many cases, especially when the estimated density functions is not smooth or during the transient period when cell response commences but is not prominent. (B) Assuming the nature of the distribution is correctly identified, then in a bi-modal case, if the two peaks are too close or the minor peak is too small, the state position of the minor peak could be hard to identify. (C) The absolute values of the transcription activity of the case population are of little meaning unless they are benchmarked against a control population. To ensure robustness of the quantized results, rather than using absolute measures, we adopt a relative measure that quantizes the transcription activity changes with respect to a control population.

As a relative measure, our quantization procedure intends to determine the size of the case population that behaves differently from the control population and the magnitude of the difference, which we denoted as *pop-shift* (PS) and *fold-change*

(FC), respectively. We denote the distributions of the case and control populations as F and G, respectively, and their corresponding density functions as $f(x)$ and $g(x)$, respectively. The control population can represent any condition that makes the comparison meaningful. The most common choice is the unperturbed population at the same time point, although it is also common to choose the same population at the beginning of the experiment. The intuitive idea for PS and FC is shown in Figure 6. Basically PS measures the non-overlapped areas of the two distributions, as indicated by either grey area in Figure 6, left and right representing the difference between $f(x)$ and $g(x)$ where $f(x) > g(x)$ and where $g(x) > f(x)$, respectively. Because $f(x)$ and $g(x)$ are density functions, the two grey areas are of same size, PS can be conveniently computed by using the total variation distance:

$$ps = ½ \int | f(x) - g(x) | \, dx.$$

For two identical distributions, $ps = 0$, while for two totally disjoint distributions, $ps = 1$. To evaluate FC, we measure the distance between the mass centers of the two non-overlapped areas, which indicate the extent of the population shifts:

$$fc = \int_{f(x)>g(x)} x\{f(x) - g(x)\}dx \,/ \int_{f(x)>g(x)} \{f(x) - g(x)\}dx -$$

$$\int_{f(x)\leq g(x)} x\{g(x) - f(x)\}dx \,/ \int_{f(x)\leq g(x)} \{g(x) - f(x)\}dx$$

Note that the FC can be conveniently computed as

Figure 6. Quantification of the relative transcription level shift and the percent of the population of cells showing altered transcription levels are carried out as shown to produce pop-shift and fold-change estimates

Tracking transcriptional activities with high-content epifluorescent imaging, J Hua et al., Journal of Biomedical Optics, vol. 17 (Copyright @ 2012), World Scientific Publishing Co. Used with permission.

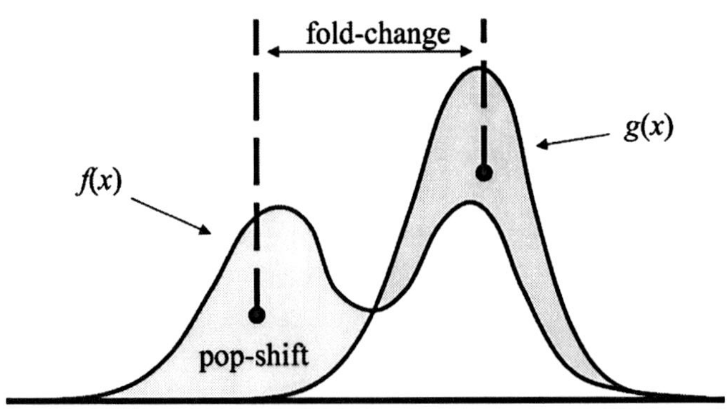

$$fc = \int x\{f(x) - g(x)\}\, dx \,/\, ps,$$

which is essentially the distance between the means of the two distributions, normalized by PS. Fold-change can be either positive or negative, indicating whether the transcription activity is increasing or decreasing with respect to the reference case. The product of PS and FC is given by the difference of mean cell-expression levels,

$$ps \cdot fc = \int xf(x)\, dx - \int xg(x)\, dx.$$

Although average expression levels or similar ideas are commonplace in many high-throughput genomic data collection techniques, such as microarrays, the term used here is actually different. The critical difference is that the average here is taken at the cell level over the log-transformed values, whereas the microarray expression level is averaged in the assay during sample preparation and the log transform is taken after the average expression level is retrieved. Thus, the measure used here can improve detection sensitivity by mitigating the distorting impact of over-expressing outliers, which can have concentration levels hundreds of times higher than most of the other cells.

Again note that the PS and FC measures assume case population shift in one direction with respect to the control population and are most reliable when the assumption holds. There are cases this assumption is violated. However, such cases are very rare and usually happen in the cell status transition, e.g., from high expression to inhibition. Once the PS and FC values are calculated for all time points of a certain condition, the PS and FC time sequences are separately smoothed with a median filter to remove small irregularities.

To show the transcription activity changes summarized by the PS and FC pair, we use a bar-plot pair as shown in Figure 7, which refers to the same drugged cell population imaged and processed in Figs. 2 and 5a. In Figure 7a, the initial state of the drugged population was chosen as the control, so for each time point, its current expression profile was compared to its own expression profile at the beginning of the experiment to obtain the PS and FC pair. In Figure 7b the un-drugged population shown in Figure 5b was chosen as the control, so for each time point, drugged population's current expression profile was compared to the expression profile of the same time point of the control population. In each plot, two bar-plots are put in parallel pairs with common x-axis aligned with experiment time points. The top bar-plot represents PS, while the bottom bar-plot, which is drawn downwardly, represents FC. In order to make the whole plot compact for better comparison, the absolute value of the change is shown and different filling colors are used to indicate up- or down-regulation: white for increased activity and black for reduced activity. Each tick in the y-axis of the PS plots corresponds to a 10% shift. The y-axis for FC

Figure 7. The bar-plots show the relative transcription activity of the cell population of Figure 2 throughout the experiment with the control population set to: a) the initial population distribution at the beginning of the experiment; b) the untreated population at the same time point. The drug was added after 5th hour. The top bar-plots show the pop-shift, while the bottom ones show the fold-change. Each tick upward on the y-axis of the upper, population-shift plots, corresponds to a 10% shift in population, while each tick downward on the y-axis of the lower, fold-change, plots indicate a 2-fold difference in expression level. The white bars indicate up-regulation while the black bars indicate down-regulation. The expression level of the initial state for both case and control are shown at the left of each plot as the log_2 *of the green fluorescent intensity.*

Tracking transcriptional activities with high-content epifluorescent imaging, J Hua et al., Journal of Biomedical Optics, vol. 17 (Copyright @ 2012), World Scientific Publishing Co. Used with permission.

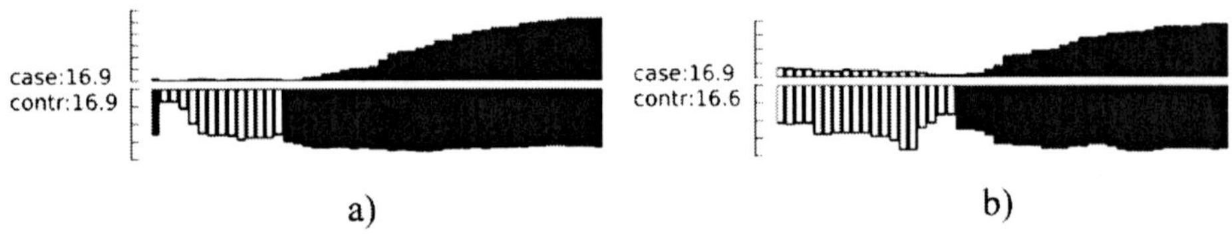

plots is shown in log_2 scale, i.e., two consecutive ticks indicate a 2-fold difference in concentration. Finally, although the PS and FC pair can reveal most of the cellular response activities, sometimes it's still helpful to have a measure on the absolute expression levels. Here we chose the initial state, represented by the peak positions in the density functions at the beginning of the experiment, to provide a general idea on the absolute transcriptional level of the target gene. The initial states of both case and control populations are shown at the left of the plot.

Figure 7 shows that under the influence of the drug lapatinib, the transcription of gene MKI67 is turned off, starting from about 10-15 hours after the drug is added. At the end of the experiment, transcription activities of about 40% of the cells have decreased to 1/8-1/16 of the no-drug control activity level.

RESULTS

Image Processing Performance

The essential performance measure for image processing is nuclei segmentation accuracy. To evaluate nuclei segmentation accuracy, we have randomly selected

4 fluorescent images from 4 different imaging sites that cover over 1000 nuclei. Besides the proposed method, two other methods are used: segmentation by the software CellProfiler and manual segmentation. CellProfiler is a free open-source cell image analysis software developed by the Broad Institute (Carpenter 2006, Lamprecht 2007) and we have used Cell Profiler 2.0 (r10997). The nuclei segmentation is conducted through the module IdentifyPrimaryObjects, which uses a thresholding-based approach followed by shape-based de-clumping. Since the size of the foreground region in our cell image can vary substantially from image to image and from time to time, based on the suggestions from the software, we choose the two-class Ostu Adaptive segmentation as the thresholding method. The default threshold is too lenient for the cell density observed in our experiment, thus a threshold correction factor of 10 is set to achieve good segmentation of the selected images. For the proposed method, no specific individual parameters are selected for the four images; instead, we used a global parameter setting, which is actually used for all images processed in this chapter.

Since there is no ground truth for the segmentation, we use manual segmentation by co-author M. L. Bittner, who has three decades experience with cell and nuclear morphology, as a benchmark and compare the other approaches to this manual benchmark for accuracy counts. We also included another independent manual setting by another experienced biologist. The inconsistency between the manual segmentations provides an idea of how well one could hope to do in comparison to the chosen standard. If two experienced biologists can only agree on 90% of the segmentation results, then it is likely that any claim of above 90% for automatic segmentation is just accidental.

Accuracy counts are divided into three categories: consistent nuclei, under-/over-segmentation, and missed nuclei. Segmentation results are shown in Table 1a. Performance is comparable for CellProfiler and the proposed method at about 85% accuracy rate. CellProfiler's performance decreases slightly when the cell density is high or the number of cells is large (over 300). Although the manual segmentation performance is better (relative to the benchmark manual segmentation), it is still below 90% and only 4% better than automatic segmentation methods, reflecting the overall difficulty of the problem and indicating good performance for the automatic methods. It should be noted that the total number of nuclei identified by each method are not the same. For computer-based approach, the most common errors are the over-/under-segmentation, while for manual segmentation, the discrepancy on whether to identify a certain object as cell play a much larger role in accuracy rate.

To check the robustness of the image processing, we examine images of the same sites of the four images evaluated in Table 1a but at different time points of the experiment. In general, both automatic segmentation methods work pretty well,

Table 1. Accuracy of nuclei segmentation: A, 4 random selected images from 4 different sites; B, an image later in the experiment from a site used in part A

	Approaches	Consistent	Over-/Under-Segmentation	Missed	Accuracy
a)	CellProfiler	1179	179	37	84.5%
	Manual	1263	93	62	89.1%
	Proposed	1183	166	39	85.2%
b)	CellProfiler	139	170	45	39.3%
	Manual	258	82	53	65.6%
	Proposed	250	85	27	69.1%

Tracking transcriptional activities with high-content epifluorescent imaging, J Hua et al., Journal of Biomedical Optics, vol. 17 (Copyright @2012), World Scientific Publishing Co. Used with permission.

except for one site, where the nuclei intensity has dropped significantly at the later stage of the experiment. To illuminate the problem, another sample image of that site has been picked out for accuracy counting and the results are listed in Table 1b, where we see that the segmentation performance has decreased significantly for all methods, with CellProfiler at a paltry 39.3%. The proposed method's performance at 69% is even better than the manual segmentation, reflecting the great uncertainty even for the experienced biologists. This large discrepancy is due to the extreme condition experienced by the cells under great pressure. Although the dye on the nuclei has become dimmer, the chromosomes have begun to shrink into small yet bright nodes, making it much harder to tell whether an object contains single or multiple nuclei. Moreover, some nuclei have started to break up, a sign of apoptosis, thereby creating small particles that further confuse the segmentation procedure. Overall, the proposed image processing protocol can provide fast and robust image segmentation performance with adequate accuracy for the purpose of the experiment. It is not because of CellProfiler's general performance that we did not use it, but because of its lack of robustness relative to the kind of image degradation that can occur with our technology - for normal images, both CellProfiler and our approach provide segmentation accuracy close to human performance.

Drug Sensitivity

In this section we consider two issues: (1) the variability of duplicated cases, (2) identification of cell subpopulations possessing differing degrees of drug sensitivity, (3) the consequences of different drug dosing strategies on cellular processes.

Reproducibility

To compare gene expression profiles obtained under the same condition and derive an empirical distribution of profile difference under the identical condition, we measure the variability of duplicated cases. If the experiment and data analysis protocol are highly reproducible, then there should be little difference. In one experiment, we have collected 56 wells of identical condition: cell-line HCT116, reporter for gene MKI67, with no drug applied. Each well contains one imaging site. The experiment lasts 48 hours, with images taken at every hour. After image processing and quality check, three time points were removed due to low imaging quality in significant number of imaging sites. The results of the remaining 45 time points are used to compute the profile difference distribution by emulating the standard experimental design, where each condition is covered by three sites. A scheme based on random sampling is used:

1. For a given time point, randomly find three sites to represent the target condition, and obtain the aggregated expression profile via the kernel-density distribution described in Data Representation subsection.
2. Randomly find three different sites at the same time point to represent a second sample of the same target condition, and obtain the aggregated expression profile.
3. Compute the pop-shift value between the two profiles.
4. Repeat step 1-3 for 1,000 times, save the results for this time point.
5. Repeat step 1-4 for every time point.

Thus by aggregating all PS values, we have an empirical distribution for profile variation of identical condition based on PS, which is equivalent to total variation, an important distance measure for probability distributions. Taking identical condition as null hypothesis, the empirical distribution indicates that a PS value over 0.072 will lead to the rejection of the null hypothesis at significance level 0.05, and a threshold of 0.09 corresponds to the significance level 0.01. For most case pairs, the reproducibility is very high, with PS values between 0.03 and 0.06, and very few PS values over 0.1. Thus if the PS value between the expression profiles of target case and control case is larger than 0.1, one can confidently claim there is a significant difference in the transcriptional response that is worth further investigation.

Identification of Cell Subpopulation

A key question regards the types and levels of heterogeneity shown by tumors in response to drugs. Averaging measurements do not provide a detailed assessment of either types or frequencies of heterogeneity. Distributional measurements provide more detailed information for detecting subpopulations. To illustrate, we consider an experiment in which mixtures having different levels of the colon cancer cell lines HT29 (lapatinib resistant) and HCT116 (lapatinib sensitive) are tested for response to lapatinib using an EGR1 promoter reporter. The mixtures of HT29 to HCT116 are 100:0, 85:15, 70:30, 40:60 and 0:100. Dynamical observations provide a sampling of the time-series evolution of the populations. The bar plots are shown in Figure 8. In it, the peaks of expression for the bottom three mixtures, dominated by HT29, are at lower starting values prior to drug addiction than the top two mixtures, dominated by HCT116, so that even prior to drug addiction there is indication of population differences. By the end of the drug exposure, the HCT116-dominating mixtures show shifted populations having less EGR1 promoter activity. The fractions of cells shifted are relatively proportional to the percentage of sensitive cells in each mix. The averaged result would not shift for the fully resistant population and would show increasing shift in proportion to the number of sensitive cells. There would be no way to know whether these shifts represented different populations

Figure 8. The bar plots for different mixtures of HCT116 and HT29 bearing a reporter for the EGR1 gene as response to Lapatinib

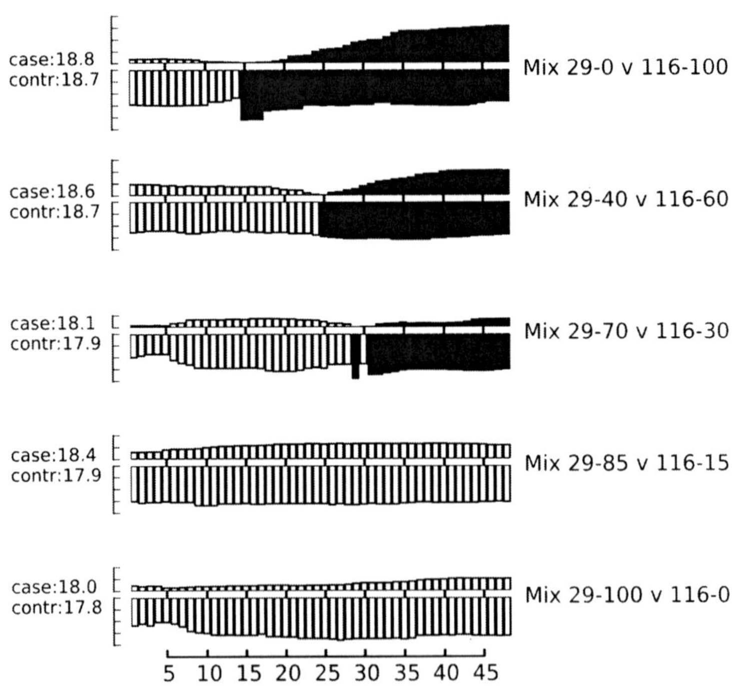

or homogeneous ones that are more or less responsive. Our new method facilitates measuring the change associated with each population, thereby allowing assessment of heterogeneous sensitive and resistant fractions in a tumor sample with a heterogeneous population. Such measurements can and have been made by flow cytometry for a long time, but only by taking samples out of the population at different times. We want to highlight the continuousness of our sampling on the same population.

Dosage Effects

Dosing studies focus on how the strength and timing of doses affect the target cells. The reporter assay allows one to determine dose-related alterations in the dynamics of drug response. Figure 9 shows the bar plots that depict the MKI67 response to lapatinib doses ranging from 1 to 32 micromolar. The experiment reveals that the cells are shifting at different speeds, although density distribution plots (not shown) indicate that they are still changing in a switch manner. At increased doses they are transitioning at a rate high enough that they look nearly continuous with a 1 per hour sampling rate. The main differences between the doses are the time at which shifting begins and the number of cells shifted in the time allowed. This kind of testing can facilitate studies of whether short pulses of higher dose or persistent dosing with lower doses provides the most favorable outcomes, a question attracting active investigation (Hsu 2010, Shah 2008).

Application to Cell Process Analysis: Cross-Talk in the HCT116 Cell Line

Thus far we have focused on straightforward properties of the technology; however, the kind of stochastic time-series analysis being carried out can be useful in a deeper way by facilitating the determination of possible mechanisms of action. In this section, using a specific example, we illustrate how knowledge of process dynamics can answer the central questions in drug effectiveness validation that have been raised in the introduction section, including the consistency/inconsistency of drug effect with intended targeting, and the extent of such consistency/inconsistency. With the help of the proposed dynamical measures, one can have a deeper understanding on the tumor vulnerabilities and the mechanism of drugs, which ultimately leads to new hypotheses concerning mechanism.

We have screened a set of colon cancer cell lines (HT29, SW620, SW480 and HCT116) against a variety of drugs. The cell lines have a number of activity levels and alterations in key components of their proliferation and survival pathways, involving the genes BRAF, KRAS and PIK3CA, as shown in Table 2. Referring

Figure 9. MKI67 gene as response to Lapatinib doses ranging from 1 to 32 micromolar, shown in bar plots
Tracking transcriptional activities with high-content epifluorescent imaging, J Hua et al., Journal of Biomedical Optics, vol. 17 (Copyright @2012), World Scientific Publishing Co. Used with permission.

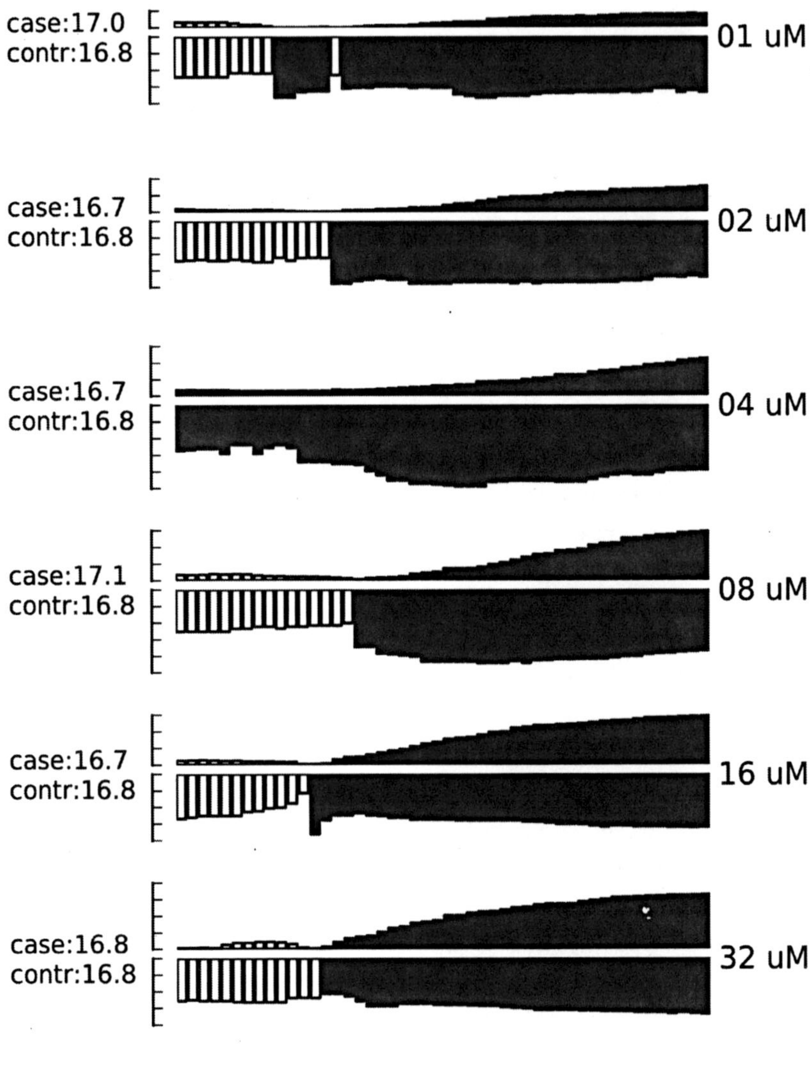

to the wiring diagram in Figure 1, it can be seen that activating mutations in the genes shown in Table 2 should readily compensate for loss of signaling from the upstream receptors. This is observed for the cell lines' responses to drugs acting on these receptors except in one case, lapatinib. As shown in Figure 1, lapatinib is a dual EGFR/ERBB2 kinase inhibitor. Histograms of the responses over time of

Table 2. Molecular characterizations of four colon cancer cell lines and predictions of their susceptibility to proliferation impairment by inhibitors of EGFR family and IGF1R receptors.

	Observed					Predicted	
	EGFR (IHC)	**BRAF Activating Mutation**	**KRAS Activating Mutation**	**PIK3CA Activating Mutation**	**PTEN Inactivation**	**FOS Promoter Down-regulated**	**MKI67 Promoter Down-regulated**
HT29	Weak	Yes	No	No	No	Unsure	Unsure
SW620	Negative	No	Yes	No	No	No	No
SW480	Strong	No	Yes	No	No	No	No
HCT116	Strong	No	Yes	Yes	No	No	No

Dyanmical Analysis of Drug Efficacy and Mechanism of Action Using GFP Reporters, J Hua et al., Journal of Biological Systems, vol. 20 (Copyright @2012), World Scientific Publishing Co. Used with permission.

the reporters for the genes FOS and MKI67 in each of the colon lines to lapatinib are shown in Figure 10. No response is observed in the lines HT29, SW480 and SW620, but a strong reduction in transcriptional activity is achieved in HCT116. This capability against HCT116 is unexpected in the light of what is known about these cell lines. If anything, HCT116 would appear to have the strongest set of

Figure 10. Distributional plots of the responses of the FOS and MKI67 promoter reporters in four colon cancer cell lines. The intensity of profile indicates time, starting with black at the beginning, and gradually changing to light grey at the end.
Dyanmical Analysis of Drug Efficacy and Mechanism of Action Using GFP Reporters, J Hua et al., Journal of Biological Systems, vol. 20 (Copyright @2012), World Scientific Publishing Co. Used with permission.

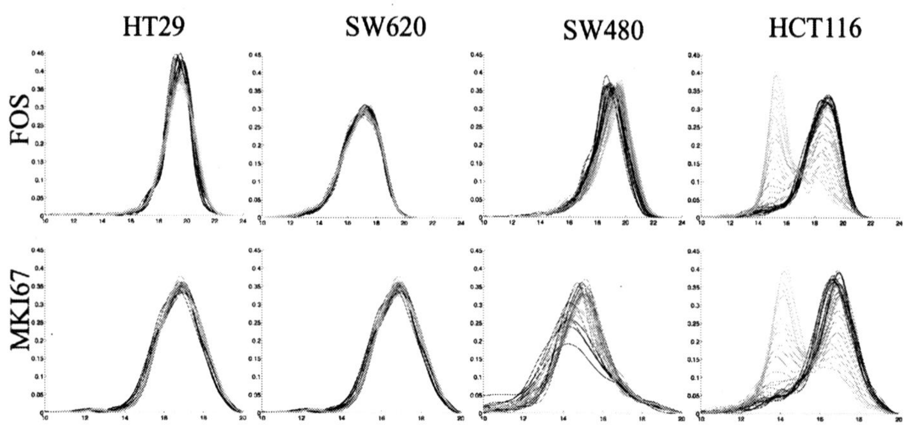

alterations to insure that lapatinib would fail to halt proliferation. In this cell line there are activating mutations in both survival and proliferation pathways, activated PIK3CA and activated KRAS (see Table 2).

To gain insight into how lapatinib overcomes these handicaps, we compare the information about the effects of lapatinib to that of five other drugs on the full set of reporters shown in Figure 1 for HTC116. In particular, responses can be categorized in terms of the extent of change in expression and percentage of the population effected. Figure 11 depicts the bar plots for the HCT116 for four different drugs.

Figure 11. The bar-plots of relative transcriptional activity in HCT116 for four different drugs are presented
Tracking transcriptional activities with high-content epifluorescent imaging, J Hua et al., Journal of Biomedical Optics, vol. 17 (Copyright @ 2012), World Scientific Publishing Co. Used with permission.

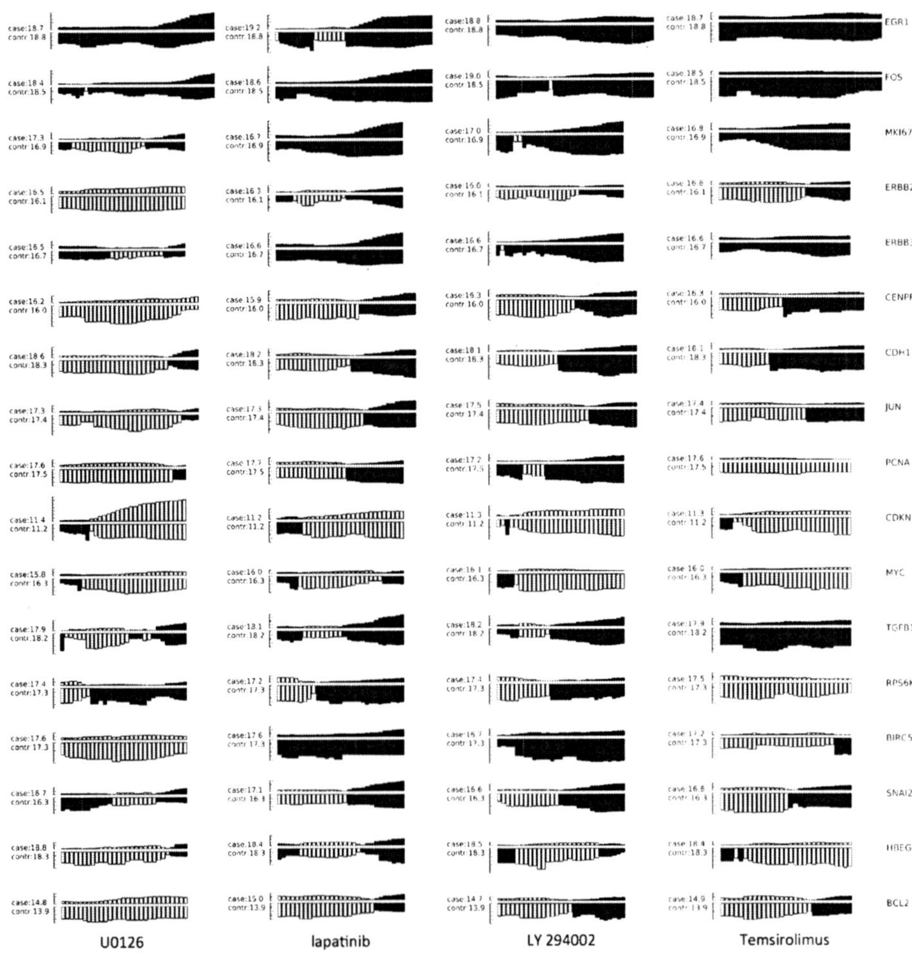

Again the percentage of cells shifted at each time point is determined by comparing cells with a reporter that are responding to an added drug to cells with the same reporter that are not exposed to the drug, removing any changes that are due only to the culture conditions.

Overall, HCT116 shows considerable transcription reduction over many reporters for a variety of drugs. Ranking drug responsiveness across the drugs for HCT116, one observes responses over most reporters but with decreasing percentages of cells shifted for lapatinib (EGFR/ERBB2), LY29004 (PI3K) and Temsirolimus (mTOR). This similarity of action with decreasing efficacy falls directly along a survival signaling pathway headed by an activated receptor heterodimer, ERBB2/ERBB3, and then proceeds along the canonical PI3K/AKT/mTOR pathway. The remaining drugs' inhibitory powers would be ranked U0126 (MEK1/2), AG1024 (IGF1R) and AG825 (ERBB2) (Due to the limitation of space, only the response of U0126 is shown in Figure 11). All of these second tier drugs deliver very low reductions of transcription of MKI67, the current "gold standard" in tumor pathology for determining the proliferative state of a tumor (Gerdes 1990, Scholzen 2000).

The engineering problem is to interpret the mechanistic features indicated by the data and formulate plausible hypotheses. For this purpose, a simplification of the original wiring diagram is shown in Figure 12. It depicts proliferation (left)

Figure 12. Relative strengths of drugs versus position in pathways and inferred crosstalk between survival and proliferative signal channels

Dyanmical Analysis of Drug Efficacy and Mechanism of Action Using GFP Reporters, J Hua et al., Journal of Biological Systems, vol. 20 (Copyright @2012), World Scientific Publishing Co. Used with permission.

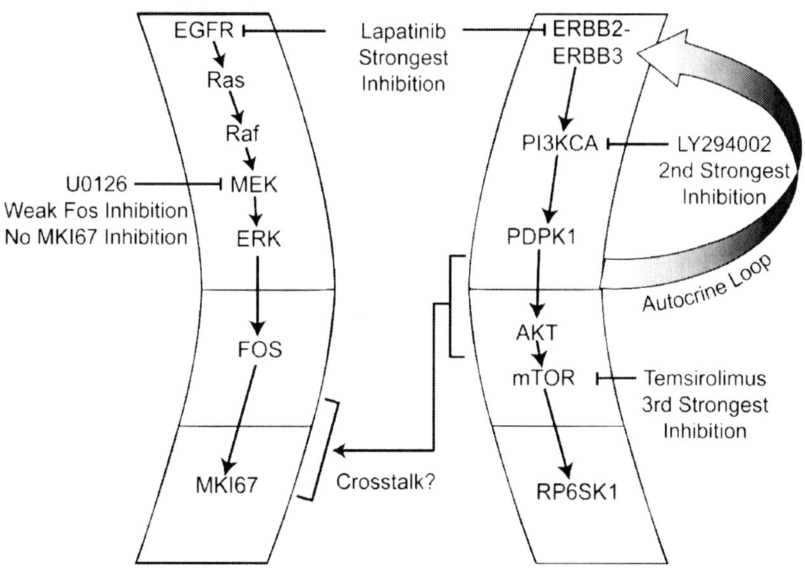

and survival (right) pathways and annotated with the relative strengths of effects of the drugs. In particular, it illustrates the relative ineffectiveness of U0126 on the proliferation pathway and the diminishing effect as intervention moves from the top downward on the survival signaling pathway (autocrine loop to be explained later). The relatively low effectiveness of inhibition at MEK and the higher inhibition at ERBB2/3 or PI3K suggests that the primary driver for proliferation is not achieved through the typical MEK dependent pathway. If the bulk of the signaling is being driven by ERBB2/3 machinery, then some form of hypothesized crosstalk between the survival and proliferative pathways is warranted. This is shown as an influence extending from the survival arm to the proliferation arm. The strong effect of laptanib on ERBB2/3 in conjunction with the hypothesized crosstalk would explain the observed ability of laptinib to stop both proliferative and survival signaling.

To consider the problem further, it is important to order the events that occur during induction, with the expectation that earlier events may be required for the occurrence of later events. This is only possible with dynamic response data. Figure 13a shows a plot of the expression levels and fractions of cells shifted relative to time for five reporters, TGFB1, ERBB3, EGR1, MKI67, and FOS. The variance in sequential change along the time course can be used to estimate the time point at which sustained unidirectional change in the direction of the final, largest shift (a commitment to the change) reaches a statistically significant level. The percentage of cells with altered transcriptional activity provides a measure to determine groupings that may be driven by the same set of processes: if changes must occur to allow a further set of changes to occur, then the initial set of changes has to occur in at least as large an initial cell population as that which changes as a consequence of the initial changes (see Figure 13b).

In this study, only the five reporters in Figure 13 show an early, consistently high, level of change in response to lapatinib. Early, temporally differentiated changes could not have arisen from high levels of proteolytic degradation due to induced apoptosis or autophagy, as these processes reduce all reporter levels simultaneously. The ordering of commitments suggests that action on ERBB2 reduces the activity of an autocrine loop keeping the level of expression of ERBB3 high. This conclusion is reached since lapatinib is most likely exerting its effect through either EGFR or the ERBB2 kinase. Since signaling through MEK is low and is high through PI3K, ERBB2 is the best candidate. Because activated ERBB3 is able to strongly stimulate the ERBB2 kinase, the reduction of ERBB3 expression would be expected to greatly diminish the observed signaling (Scholzen 2000). ERBB3's transcription drops early, while ERBB2's transcription does not, reinforcing the notion that ERBB3 loss is key to lapatinib's action. Receptors frequently are involved in autocrine loops, where their own activity directs their continuing expression, so it is plausible that this mechanism is in play. As TGFB1 and ERBB3 begin to show

Figure 13. Temporal changes, shown in hourly increments, based on the pop-shift part of the box plots: (a) temporal changes in expression level induced by lapatinib for five reporters; (b) the minimum percentage of cells required to drive a later change dependent on the earlier change is the highest percent of cells changed of the later event

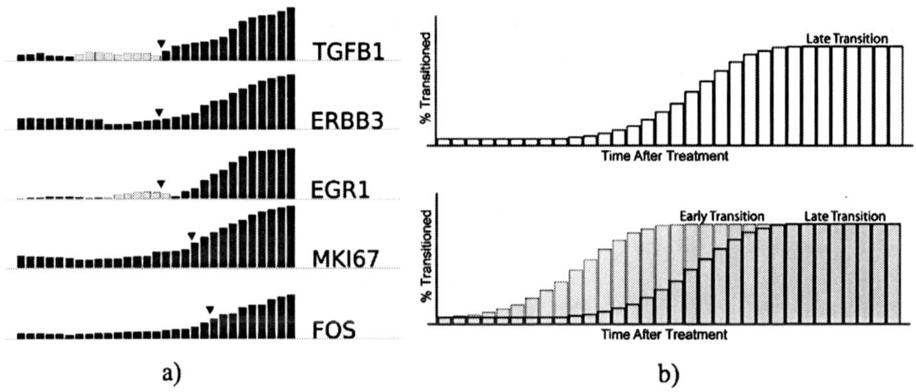

declining expression levels at the same time, it may be that TGFB1 is also involved in the autocrine loop. As LY294002 and Temsirolimus are also capable of reducing ERBB3 expression, though less and less effectively, one would expect that the reductions in EGR1, MKI67 and FOS would be similarly lowered if they were ERBB3 expression-dependent, and this is true. The decay in ability to reduce ERBB3 expression as drug interventions move down the survival signaling pathway suggests that the most effective way of influencing the loop is likely to be an offshoot of the mainline near the top of the chain (Figure 12).

Following these initial declines in Figure 13, EGR1 expression declines, followed by MKI67 and then FOS. This is consistent with the observation that U0126 is only partially effective at blocking FOS transcription and totally ineffective at blocking MKI67 transcription. So loss of FOS transcription needs not fall ahead of MKI67 transcription, since it appears not to be supporting MKI67 transcription. EGR1's decline prior to MKI67 is an indication that it may function to support MKI67 transcription in unperturbed HCT116 cells.

By combining the above chains of reasoning with extant genomic knowledge of HCT116 and biochemical knowledge about the portions of the pathway that seem to be key to lapatinib's success in reducing proliferation in HCT116, it is possible to propose a set of biomarkers that may be sufficient to guide the use of the drug in colon and perhaps other cancers. It is known that HCT116 does not have copy number increases of ERBB2, so that it is possible that ligand stimulated ERBB3's ability to greatly increase ERBB2's stimulation of survival signaling is key to proliferative and survival signaling as well as an autocrine loop supporting ERBB3 expression in HCT116 (Hellyer 2001, Soltoff 1994). If true, this would suggest that looking for high-level expression of ERBB3, and expression of NRG1, an ERBB3 activating ligand, combined with the absence of ERBB2 copy number increase. This could identify colon cancers responsive to lapatinib, and perhaps extremely sensitive to a combination therapy including lapatinib and an inhibitor of ERBB3 activation of ERBB2.

CONCLUSION

At early 2000s, a single paper had been published showing lapatinib's activity against the colon cancer cell line GEO, and no clinical trial of lapatinib in colon cancer has shown positive results (Zhou 2006). On the other hand, a considerable amount of work has been published on ERBB3's role in colon cancer. For example, a detailed demonstration of an ERBB3 autocrine loop was shown to generate growth factor independence (Venkateswarlu 2002). Another study demonstrated that when an ERRB3 colon-specific ablation mechanism is operating in ApcMin mice, the high level of intestinal adenomas expected in ApcMin mice is not observed, although regions of localized apoptosis were common (Lee 2009). Later work further showed that ERBB2/3 activity can produce signaling far exceeding the typically expected levels and generate a quantitatively distinct type of signaling (Amin 2010). Taken together, such publications could be seen as describing a set of conditions that indicate a possible niche for the use of lapatinib, namely, a pairing of lapatinib with a drug inhibiting ERBB3's activation to achieve a synergistic disruption of ERBB2/3 autocrine loops in some types of colon cancer. A proof of concept of the utility of this approach in breast cancer has recently been published (Cook 2011).

Whenever drug performance depends on conditions beyond the presence and activity of a specific target, the choice of biomarkers for patient stratification needs to be expanded. Producing more definite criteria that describe a cellular context where one or more drugs can produce strong responses spanning a number of proliferative and survival processes will definitely help. EGFR family members certainly play a role in many cancers. Yet the low success rate of drugs targeting them indicates that their involvement is heavily conditioned by other activities in the cell. Stratifying patients to match them with drugs having the best chance of providing a substantial benefit requires not only patient profiling, but also drug profiling. The profile of what cells a drug will impact is, in most cases, more complicated than simply requiring the drug's biochemical target to be present and active. Hence, a complete description of the conditions under which a drug will succeed must be developed. It would allow sufficiently detailed patient profiling to be carried out to identify those that have the characteristics required by the drug.

The lapatinib study of the preceding section illustrates how a standard engineering approach, based on three core steps, can be applied to complex biological systems to accomplish these ends. To wit: (1) build a model network of interactions the drug can be expected to perturb; (2) apply the perturbations for both the candidate drug and other drugs shown to be capable of perturbing parts of the hypothesized network; and (3) measure responses to the perturbations and use these to determine which expectations were correct and which were not. Typically, the knowledge associated with a screened drug is limited to knowing that the compound can greatly reduce the activity of the protein it has been designed against and that it has some level of activity against related proteins. Explicit testing of hypothesized mechanisms in a system framework extends such knowledge.

Step 1 is critical and perhaps the most difficult because it depends on the extent of our prior pathway knowledge. Because some relevant pathways will be unknown our models will be incomplete and therefore unfulfilled expectations in step 3 could arise from this incompleteness. This is a difficult problem because the number of potential augmentations is astronomical. Hence, efficient methods need to be developed based on using the literature in conjunction with objective-based search constraints to find potentially relevant pathways. Our efforts are not merely theoretical. We are already using the methods developed in this study in our drug discovery work.

Finally we would like to point out that the proposed protocol is based on in vitro cell-line-based experiments. The cell-line cultured in the lab environment is much homogeneous in the genetic profile than actual tumor, where the heterogeneity is overwhelming (Kandoth 2013), even inside the same tumor tissue (Gerlinger 2012). Besides heterogeneity, the drug effects on real tumor will also be affected by many other factors, like the drug uptake, penetration and depletion, plus various uncertainties either in real tissue environment or introduced through the drug delivery. Thus how to best extrapolate the knowledge gained from the cell-line-based study to animal model, ultimately the human model, is another difficult question that need to be addressed in the future.

REFERENCES

Amin, D. N., Sergina, N., Ahuja, D., Mcmahon, M., Blair, J. A., Wang, D., & Moasser, M. M. et al. (2010). Resiliency and vulnerability in the HER2-HER3 tumorigenic driver. *Science Translational Medicine*, *2*(16), 16ra7. doi:10.1126/scitranslmed.3000389 PMID:20371474

Biggar, S., & Crabtree, G. (2001). Cell signaling can direct either binary or graded transcriptional responses. *The EMBO Journal*, *20*(12), 3167–3176. doi:10.1093/emboj/20.12.3167 PMID:11406593

Bleicher, K. H., Böhm, H.-J., Müller, K., & Alanine, A. I. (2003). Hit and lead generation: Beyond high-throughput screening. *Nature Reviews. Drug Discovery*, *2*(5), 369–378. doi:10.1038/nrd1086 PMID:12750740

Bowman, A., & Azzalini, A. (1997). *Applied smoothing techniques for data analysis the kernel approach with S-Plus illustrations*. Oxford, UK: Clarendon Press.

Brough, P. A., Barril, X., Borgognoni, J., Chene, P., Davies, N. G. M., Davis, B., & Wright, L. et al. (2009). Combining hit identification strategies: Fragment-based and in silico approaches to orally active 2-aminothieno[2,3-d]pyrimidine inhibitors of the Hsp90 molecular chaperone. *Journal of Medicinal Chemistry*, *52*(15), 4794–4809. doi:10.1021/jm900357y PMID:19610616

Carpenter, A. E., Jones, T. R., Lamprecht, M. R., Clarke, C., Kang, I. H., Friman, O., & Sabatini, D. M. et al. (2006). CellProfiler: Image analysis software for identifying and quantifying cell phenotypes. *Genome Biology*, *7*(10), R100. doi:10.1186/gb-2006-7-10-r100 PMID:17076895

Chalfie, M., Tu, Y., Euskirchen, G., Ward, W., & Prasher, D. (1994). Green fluorescent protein as a marker for gene expression. *Science*, *263*(5148), 802–805. doi:10.1126/science.8303295 PMID:8303295

Chen, S. J., Zhu, Y. J., Tong, J. H., Dong, S., Huang, W., Chen, Y., & Qian, G. Q. et al. (1991). Rearrangements in the second intron of the RARA gene are present in a large majority of patients with acute promyelocytic leukemia and are used as molecular marker for retinoic acid-induced leukemic cell differentiation. *Blood*, *78*(10), 2696–2701. PMID:1668609

Cook, R. S., Garrett, J. T., Sanchez, V., Stanford, J. C., Young, C., Chakrabarty, A., & Arteaga, C. L. et al. (2011). ErbB3 ablation impairs PI3K/Akt-dependent mammary tumorigenesis. *Cancer Research*, *71*(11), 3941–3951. doi:10.1158/0008-5472.CAN-10-3775 PMID:21482676

Dahlin, J. L., Inglese, J., & Walters, M. A. (2015). Mitigating risk in academic preclinical drug discovery. *Nature Reviews. Drug Discovery*, *14*(4), 279–294. doi:10.1038/nrd4578 PMID:25829283

Dong, H., Kung, J., Bhardwaj, V., & Mcgill, J. (2010). Flow cytometry rapidly identifies all acute promyelocytic leukemias with high specificity independent of underlying cytogenetic abnormalities. *American Journal of Clinical Pathology*, *135*(1), 76–84. doi:10.1309/AJCPW9TSLQNCZAVT PMID:21173127

Dougherty, E., & Lotufo, R. (2003). *Hands-on morphological image processing*. Bellingham, WA: SPIE. doi:10.1117/3.501104

Druker, B. J., Sawyers, C. L., Kantarjian, H., Resta, D. J., Reese, S. F., Ford, J. M., & Talpaz, M. et al. (2001). Activity of a specific inhibitor of the BCR-ABL tyrosine kinase in the blast crisis of chronic myeloid leukemia and acute lymphoblastic leukemia with the philadelphia chromosome. *The New England Journal of Medicine*, *344*(14), 1038–1042. doi:10.1056/NEJM200104053441402 PMID:11287973

Geisler, C., Gaisa, N. T., Pfister, D., Fuessel, S., Kristiansen, G., Braunschweig, T., & Henkel, C. et al. (2015). Identification and validation of potential new biomarkers for prostate cancer diagnosis and prognosis using 2D-DIGE and MS. *BioMed Research International*, *2015*, 454256. doi:10.1155/2015/454256 PMID:25667921

Gerdes, J. (1990). Ki-67 and other proliferation markers useful for immunohistological diagnostic and prognostic evaluations in human malignancies. *Seminars in Cancer Biology*, *1*(3), 199–206. PMID:2103495

Gerlinger, M., Rowan, A. J., Horswell, S., Larkin, J., Endesfelder, D., Gronroos, E., & Swanton, C. et al. (2012). Intratumor heterogeneity and branched evolution revealed by multiregion sequencing. *The New England Journal of Medicine*, *366*(10), 883–892. doi:10.1056/NEJMoa1113205 PMID:22397650

Hao, L., Johnsen, R., Lauter, G., Baillie, D., & Bürglin, T. R. (2006). Comprehensive analysis of gene expression patterns of hedgehog-related genes. *BMC Genomics*, *7*(1), 280. doi:10.1186/1471-2164-7-280 PMID:17076889

Hellyer, N. J., Kim, M. S., & Koland, J. G. (2001). Heregulin-dependent Activation of Phosphoinositide 3-Kinase and Akt via the ErbB2/ErbB3 Co-receptor. *The Journal of Biological Chemistry*, *276*(45), 42153–42161. doi:10.1074/jbc.M102079200 PMID:11546794

Hoelder, S., Clarke, P. A., & Workman, P. (2012). Discovery of small molecule cancer drugs: Successes, challenges and opportunities. *Molecular Oncology*, *6*(2), 155–176. doi:10.1016/j.molonc.2012.02.004 PMID:22440008

Hsu, C. H., Shen, Y. C., Lin, Z. Z., Chen, P. J., Shao, Y. Y., Ding, Y. H., & Cheng, A. L. (2010). Phase II study of combining sorafenib with metronomic tegafur/uracil for advanced hepatocellular carcinoma. *Journal of Hepatology*, *53*(1), 126–131. doi:10.1016/j.jhep.2010.01.035 PMID:20416968

Kandoth, C., McLellan, M. D., Vandin, F., Ye, K., Niu, B., Lu, C., & Ding, L. et al. (2013). Mutational landscape and significance across 12 major cancer types. *Nature*, *502*(7471), 333–339. doi:10.1038/nature12634 PMID:24132290

Lamprecht, M. R., Sabatini, D. M., & Carpenter, A. E. (2007). CellProfiler™: Free, versatile software for automated biological image analysis. *BioTechniques*, *42*(1), 71–75. doi:10.2144/000112257 PMID:17269487

Lee, D., Yu, M., Lee, E., Kim, H., Yang, Y., Kim, K., & Threadgill, D. et al. (2009). Tumor-specific apoptosis caused by deletion of the ERBB3 pseudo-kinase in mouse intestinal epithelium. *The Journal of Clinical Investigation*, *119*(9), 2702–2713. doi:10.1172/JCI36435 PMID:19690388

Marron, J. S. (2004). *Smoothing, functional data analysis, and distance weighted discrimination software*. Retrieved from http://www.unc.edu/~marron/marron_software.html

Network, T. C. G. A. (2008). Comprehensive genomic characterization defines human glioblastoma genes and core pathways. *Nature*, *455*, 1061–1068. doi:10.1038/nature07385 PMID:18772890

Nowell, P. C., & Hungerford, D. A. (1960). A minute chromosome in human chronic granulocytic leukemia. *Science*, *132*, 1497.

Nowicki, M., Pawlowski, P., Fischer, T., Hess, G., Pawlowski, T., & Skorski, T. (2003). Chronic myelogenous leukemia molecular signature. *Oncogene*, *22*(25), 3952–3963. doi:10.1038/sj.onc.1206620 PMID:12813469

Orth, J. D., Tang, Y., Shi, J., Loy, C. T., Amendt, C., Wilm, C., & Mitchison, T. J. et al. (2008). Quantitative live imaging of cancer and normal cells treated with Kinesin-5 inhibitors indicates significant differences in phenotypic responses and cell fate. *Molecular Cancer Therapeutics*, *7*(11), 3480–3489. doi:10.1158/1535-7163.MCT-08-0684 PMID:18974392

Parzen, E. (1962). On Estimation of a Probability Density Function and Mode. *Annals of Mathematical Statistics*, *33*(3), 1065–1076. doi:10.1214/aoms/1177704472

Pear, W. S., Miller, J. P., Xu, L., Pui, J. C., Soffer, B., Quackenbush, R. C., & Baltimore, D. et al. (1998). Efficient and rapid induction of a chronic myelogenous leukemia-like myeloproliferative disease in mice receiving P210 bcr/abl-transduced bone marrow. *Blood*, *92*(10), 3780–3792. PMID:9808572

Perou, C. M., Sørlie, T., Eisen, M. B., van de Rijn, M., Jeffrey, S. S., Rees, C. A., & Botstein, D. et al. (2000). Molecular portraits of human breast tumours. *Nature*, *406*(6797), 747–752. doi:10.1038/35021093 PMID:10963602

Rowley, J. D., Golomb, H. M., & Dougherty, C. (1977). 15/17 translocation, a consistent chromosomal change in acute promyelocytic leukaemia. *Lancet*, *1*(8010), 549–550. doi:10.1016/S0140-6736(77)91415-5 PMID:65649

Scholzen, T., & Gerdes, J. (2000). The Ki-67 protein: From the known and the unknown. *Journal of Cellular Physiology*, *182*(3), 311–322. doi:10.1002/(SICI)1097-4652(200003)182:3<311::AID-JCP1>3.0.CO;2-9 PMID:10653597

Shah, N. P., Kasap, C., Weier, C., Balbas, M., Nicoll, J. M., Bleickardt, E., & Sawyers, C. L. et al. (2008). Transient potent BCR-ABL inhibition is sufficient to commit chronic myeloid leukemia cells irreversibly to apoptosis. *Cancer Cell*, *14*(6), 485–493. doi:10.1016/j.ccr.2008.11.001 PMID:19061839

Sheather, S. J., & Jones, M. C. (1991). A reliable data-based bandwidth selection method for Kernel density estimation. *Journal of the Royal Statistical Society. Series B. Methodological*, *53*(3), 683–690.

Sirota, M., Dudley, J. T., Kim, J., Chiang, A. P., Morgan, A. A., Sweet-Cordero, A., & Butte, A. J. et al. (2011). Discovery and preclinical validation of drug indications using compendia of public gene expression data. *Science Translational Medicine*, *3*(96), 96ra77. doi:10.1126/scitranslmed.3001318 PMID:21849665

Smith, L., Watson, M. B., O'Kane, S. L., Drew, P. J., Lind, M. J., & Cawkwell, L. (2006). The analysis of doxorubicin resistance in human breast cancer cells using antibody microarrays. *Molecular Cancer Therapeutics*, *5*(8), 2115–2120. doi:10.1158/1535-7163.MCT-06-0190 PMID:16928833

Soltoff, S. P., Carraway, K. L. III, Prigent, S. A., Gullick, W. G., & Cantley, L. C. (1994). ErbB3 is involved in activation of phosphatidylinositol 3-kinase by epidermal growth factor. *Molecular and Cellular Biology*, *14*(6), 3550–3558. doi:10.1128/MCB.14.6.3550 PMID:7515147

Spyratos, F., Ferrero-Poüs, M., Trassard, M., Hacène, K., Phillips, E., Tubiana-Hulin, M., & Le Doussal, V. (2002). Correlation between MIB-1 and other proliferation markers: Clinical implications of the MIB-1 cutoff value. *Cancer*, *94*(8), 2151–2159. doi:10.1002/cncr.10458 PMID:12001111

Venkateswarlu, S., Dawson, D. M., St Clair, P., Gupta, A., Willson, J. K., & Brattain, M. G. (2002). Autocrine heregulin generates growth factor independence and blocks apoptosis in colon cancer cells. *Oncogene*, *21*(1), 78–86. doi:10.1038/sj.onc.1205011 PMID:11791178

Vincent, L. (1993). Grayscale area opening and closing, their efficient implementation and applications. In *Proceedings in Eurasip Workshop on Mathematical Morphology and its Applications to Signal Processing*. Barcelona, Spain: Academic Press.

Vincent, L., & Soille, P. (1991). Watersheds in digital spaces: An efficient algorithm based on immersion simulations. *IEEE Transactions on Pattern Analysis and Machine Intelligence*, *13*(6), 583–598. doi:10.1109/34.87344

Zhou, Y., Li, S., Hu, Y. P., Wang, J., Hauser, J., Conway, A. N., & Brattain, M. G. et al. (2006). Blockade of EGFR and ErbB2 by the novel dual EGFR and ErbB2 tyrosine kinase inhibitor GW572016 sensitizes human colon carcinoma GEO cells to apoptosis. *Cancer Research*, *66*(1), 404–411. doi:10.1158/0008-5472.CAN-05-2506 PMID:16397255

KEY TERMS AND DEFINITIONS

Cell-Line: A cell culture that extracted from live tissue, usually a tumor, and cultivated under controlled conditions.

Cell Segmentation: The image processing procedure that identifies individual cells in microscopic images of cell culture.

Drug Response: How the cells react to drugs applied at different concentration levels, including apoptosis, survival, proliferation and others.

Fluorescent Protein Reporter: A fluorescent protein gene that shares the same regulatory sequence with the target gene.

Gene Network: A collection of physical entities in a cell that interact with each other.

Mathematical Morphology: A theory on signal processing, especially image processing, using geometrical structures.

Signaling Pathway: A series of molecular activities that transfer an external stimulant into the cell, and induce proper response.

Compilation of References

Abnizova, I., & Gilks, W. R. (2006). Studying statistical properties of regulatory DNA sequences, and their use in predicting regulatory regions in the eukaryotic genomes. *Briefings in Bioinformatics*, *7*(1), 48–54. doi:10.1093/bib/bbk004 PMID:16761364

Abnizova, I., Subhankulova, T., & Gilks, W. (2007). Recent computational approaches to understand gene regulation: Mining gene regulation in silico. *Current Genetics*, *8*, 79–91. PMID:18660846

Adalsteinsson, D., McMillen, D., & Elston, T. C. (2004). Biochemical Network Stochastic Simulator (BioNetS): Software for stochastic modeling of biochemical networks. *BMC Bioinformatics*, *5*(1), 24. doi:10.1186/1471-2105-5-24 PMID:15113411

Adenot, P. G., Mercier, Y., Renard, J. P., & Thompson, E. M. (1997). Differential H4 acetylation of paternal and maternal chromatin precedes DNA replication and differential transcriptional activity in pronuclei of 1-cell mouse embryos. *Development*, *124*(22), 4615–4625. Retrieved from http://www.ncbi.nlm.nih.gov/pubmed/9409678 PMID:9409678

Aittokallio, T., & Schwikowski, B. (2006). Graph-based methods for analysing networks in cell biology. *Briefings in Bioinformatics*, *73*(3), 243–255. doi:10.1093/bib/bbl022 PMID:16880171

Akutsu, T., Hayashida, M., Ching, W.-K., & Ng, M. K. (2007). Control of Boolean networks: Hardness results and algorithms for tree structured networks. *Journal of Theoretical Biology*, *244*(4), 670–679. doi:10.1016/j.jtbi.2006.09.023 PMID:17069859

Akutsu, T., Miyano, S., & Kuhara, S. (2000). Inferring qualitative relations in genetic networks and metabolic pathways. *Bioinformatics (Oxford, England)*, *16*(8), 727–734. doi:10.1093/bioinformatics/16.8.727 PMID:11099258

Albert, R. (2005). Scale-free networks in cell biology. *Journal of Cell Science*, *118*(21), 4947–4957. doi:10.1242/jcs.02714 PMID:16254242

Albert, R., & Barabasi, A. L. (2002). Statistical mechanics of complex networks. *Reviews of Modern Physics*, *74*(1), 47–97. doi:10.1103/RevModPhys.74.47

Albert, R., Jeong, H., & Barabási, A. L. (2001). Error and attack tolerance of complex networks. *Nature*, *409*(6819), 542. doi:10.1038/35054111 PMID:10935628

Alexandrov, B. S., Fukuyo, Y., Lange, M., Horikoshi, N., Gelev, V., Rasmussen, K. Ø., & Usheva, A. et al. (2012). DNA breathing dynamics distinguish binding from nonbinding consensus sites for transcription factor YY1 in cells. *Nucleic Acids Research*, *40*(20), 10116–10123. doi:10.1093/nar/gks758 PMID:22904068

Alon, U. (2007). Network motifs: Theory and experimental approaches. *Nature Reviews. Genetics*, *8*(6), 450–461. doi:10.1038/nrg2102 PMID:17510665

Altman, E. (1999). *Constrained Markov Decision Processes* (Vol. 7). London, U.K.: Chapman & Hall/CRC.

Altman, R. B., & Raychaudhuri, S. (2001). Whole-genome expression analysis: Challenges beyond clustering. *Current Opinion in Structural Biology*, *11*(3), 340–347. doi:10.1016/S0959-440X(00)00212-8 PMID:11406385

Altschul, S. F., Gish, W., Miller, W., Myers, E. W., & Lipman, D. J. (1990). Basic local alignment search tool. *Journal of Molecular Biology*, *215*(3), 403–410. doi:10.1016/S0022-2836(05)80360-2 PMID:2231712

Alverdi, V., Hetrick, B., Joseph, S., & Komives, E. A. (2014). Direct observation of a transient ternary complex during IκBα-mediated dissociation of NF-κB from DNA. *Proceedings of the National Academy of Sciences of the United States of America*, *111*(1), 225–230. doi:10.1073/pnas.1318115111 PMID:24367071

Amin, D. N., Sergina, N., Ahuja, D., Mcmahon, M., Blair, J. A., Wang, D., & Moasser, M. M. et al. (2010). Resiliency and vulnerability in the HER2-HER3 tumorigenic driver. *Science Translational Medicine*, *2*(16), 16ra7. doi:10.1126/scitranslmed.3000389 PMID:20371474

Anastassiou, D. (2007). Computational analysis of the synergy among multiple interacting genes. *Molecular Systems Biology*, *3*, 83. doi:10.1038/msb4100124 PMID:17299419

Anderson, L. M., & Yang, H. (2008). DNA looping can enhance lysogenic CI transcription in phage lambda. *Proceedings of the National Academy of Sciences of the United States of America*, *105*(15), 5827–5832. doi:10.1073/pnas.0705570105 PMID:18391225

Andrecut, M., Huang, S., & Kauffman, S. (2008). Heuristic approach to sparse approximation of gene regulatory networks. *Journal of Computational Biology*, *15*(9), 1173–1186. doi:10.1089/cmb.2008.0087 PMID:18844584

Andrecut, M., & Kauffman, S. (2008). On the sparse reconstruction of gene networks. *Journal of Computational Biology*, *15*(1), 21–30. doi:10.1089/cmb.2007.0185 PMID:18257675

Andrews, A. L., Holloway, J. W., Puddicombe, S. M., Holgate, S. T., & Davies, D. E. (2002). Kinetic analysis of the interleukin-13 receptor complex. *The Journal of Biological Chemistry*, *277*(48), 46073–46078. doi:10.1074/jbc.M209560200 PMID:12354755

Angelini, C., De Canditiis, D., Mutarelli, M., & Pensky, M. (2007). A Bayesian approach to estimation and testing in time-course microarray experiments. *Stat Appl Genet Mol Biol*, *6*, Article24. doi:10.2202/1544-6115.1299

Ao, P., Kown, C., & Qian, H. (2007). On the existence of potential landscape in the evolution of complex systems. *Complexity*, *12*(4), 19–27. doi:10.1002/cplx.20171

Aoshima, K., Inoue, E., Sawa, H., & Okada, Y. (2015). Paternal H3K4 methylation is required for minor zygotic gene activation and early mouse embryonic development. *EMBO Reports*, *16*(7), 803–812. doi:10.15252/embr.201439700 PMID:25925669

Arkin, A., Ross, J., & McAdams, H. H. (1998). Stochastic kinetic analysis of developmental pathway bifurcation in phage lambda-infected Escherichia coli cells. *Genetics*, *149*(4), 1633–1648. PMID:9691025

Ashburner, M., Ball, C. A., Blake, J. A., Botstein, D., Butler, H., Cherry, J. M., & Sherlock, G. (2000). Gene Ontology: Tool for the unification of biology. *Nature Genetics*, *25*(1), 25–29. doi:10.1038/75556 PMID:10802651

Aurell, E., Brown, S., Johanson, J., & Sneppen, K. (2002). Stability puzzles in phage lambda. *Physical Review E: Statistical, Nonlinear, and Soft Matter Physics*, *65*(5), 051914. doi:10.1103/PhysRevE.65.051914 PMID:12059600

Bailey, T. L., Boden, M., Buske, F. A., Frith, M., Grant, C. E., Clementi, L., . . . Noble, W. S. (2009). MEME SUITE: tools for motif discovery and searching. *Nucleic Acids Research, 37*, W202-208. doi: 10.1093/nar/gkp335

Bairoch, A., Apweiler, R., Wu, C. H., Barker, W. C., Boeckmann, B., Ferro, S., & Yeh, L. S. et al. (2005). The universal protein resource (UniProt). *Nucleic Acids Research*, *33*(Database issue), 154–159. doi:10.1093/nar/gki070 PMID:15608167

Barabasi, A. L. (2002). *Linked: How Everything Is Connected to Everything Else and What It Means*. Perseus.

Barabasi, A. L. (2009). Scale-Free Networks: A Decade and Beyond. *Science*, *325*(5939), 412–413. doi:10.1126/science.1173299 PMID:19628854

Barabasi, A. L., & Oltvai, Z. (2004). Network biology: Understanding the cell's functional organization. *Nature Reviews. Genetics*, *5*(2), 101–113. doi:10.1038/nrg1272 PMID:14735121

Barrera, J., Cesar-Jr, R. M., Martins-Jr, D. C., Vencio, R. Z. N., Merino, E. F., Yamamoto, ... Portillo, H. A. (2007). Constructing probabilistic genetic networks of Plasmodium falciparum, from dynamical expression signals of the intraerythrocytic development cycle. In P. McConnell, S. M. Lin, & P. Hurban (Eds.), Methods of Microarray Data Analysis V (pp. 11-26). Springer.

Barrett, L. W., Fletcher, S., & Wilton, S. D. (2012). Regulation of eukaryotic gene expression by the untranslated gene regions and other non-coding elements. *Cellular and Molecular Life Sciences: CMLS*, *69*(21), 3613–3634. doi:10.1007/s00018-012-0990-9 PMID:22538991

Basso, K., Margolin, A. A., Stolovitzky, G., Klein, U., Dalla-Favera, R., & Califano, A. (2005). Reverse engineering of regulatory networks in human B cells. *Nature Genetics*, *37*(4), 382–390. doi:10.1038/ng1532 PMID:15778709

Baumbach, J., Tauch, A., & Rahmann, S. (2009). Towards the integrated analysis, visualization and reconstruction of microbial gene regulatory networks. *Briefings in Bioinformatics*, *10*(1), 75–83. doi:10.1093/bib/bbn055 PMID:19074493

Beard, D. A., & Qian, H. (2008). *Chemical biophysics: quantitative analysis of cellular systems*. Cambridge University Press. doi:10.1017/CBO9780511803345

Beer, M. A., & Tavazoie, S. (2004). Predicting gene expression from sequence. *Cell*, *117*(2), 185–198. doi:10.1016/S0092-8674(04)00304-6 PMID:15084257

Bellman, R. (1952). On the theory of dynamic programming. *Proceedings of the National Academy of Sciences of the United States of America*, *38*(8), 716–719. doi:10.1073/pnas.38.8.716 PMID:16589166

Ben-Gal, I., Shani, A., Gohr, A., Grau, J., Arviv, S., Shmilovici, A., & Grosse, I. et al. (2005). Identification of transcription factor binding sites with variable-order Bayesian networks. *Bioinformatics (Oxford, England)*, *21*(11), 2657–2666. doi:10.1093/bioinformatics/bti410 PMID:15797905

Benson, D. A., Karsch Mizrachi, I., Lipman, D. J., Ostell, J., & Wheeler, L. (2008). GenBank. *Nucleic Acids Research*, *36*(suppl 1), D25–D30. doi:10.1093/nar/gkm929 PMID:18073190

Berg, O. G., & von Hippel, P. H. (1987). Selection of DNA binding sites by regulatory proteins. Statistical-mechanical theory and application to operators and promoters. *Journal of Molecular Biology*, *193*(4), 723–750. doi:10.1016/0022-2836(87)90354-8 PMID:3612791

Berlow, N., Davis, L., Cantor, E., Seguin, B., Keller, C., & Pal, R. (2013). A new approach for prediction of tumor sensitivity to targeted drugs based on functional data. *BMC Bioinformatics*, *14*(1), 239. doi:10.1186/1471-2105-14-239 PMID:23890326

Bertsekas, D. P. (2005). *Dynamic Programming and Optimal Control*. Belmont, MA: Athena Scientific.

Bhatia, S., Bengani, H., Fish, M., Brown, A., Divizia, M. T., de Marco, R., & Kleinjan, D. A. et al. (2013). Disruption of Autoregulatory Feedback by a Mutation in a Remote, Ultraconserved PAX6 Enhancer Causes Aniridia. *American Journal of Human Genetics*, *93*(6), 1126–1134. doi:10.1016/j.ajhg.2013.10.028 PMID:24290376

Biggar, S., & Crabtree, G. (2001). Cell signaling can direct either binary or graded transcriptional responses. *The EMBO Journal*, *20*(12), 3167–3176. doi:10.1093/emboj/20.12.3167 PMID:11406593

Bishop, L. M., & Qian, H. (2010). Stochastic Bistability and Bifurcation in a Mesoscopic Signaling System with Autocatalytic Kinase. *Biophysical Journal*, *98*(1), 1–11. doi:10.1016/j.bpj.2009.09.055 PMID:20074511

Bittner, M., Meltzer, P., Chen, Y., Jiang, Y., Seftor, E., Hendrix, M., & Trent, J. et al. (2000). Molecular classification of cutaneous malignant melanoma by gene expression profiling. *Nature*, *406*(6795), 536–540. PMID:10952317

Blake, W. J., KÆrn, M., Cantor, C. R., & Collins, J. J. (2003). Noise in eukaryotic gene expression. *Nature*, *422*(6932), 633–637. doi:10.1038/nature01546 PMID:12687005

Bleicher, K. H., Böhm, H.-J., Müller, K., & Alanine, A. I. (2003). Hit and lead generation: Beyond high-throughput screening. *Nature Reviews. Drug Discovery*, *2*(5), 369–378. doi:10.1038/nrd1086 PMID:12750740

Boccaletti, S., Latora, V., Moreno, Y., Chavez, M., & Hwang, D. U. (2006). Complex networks: Structure and dynamics. *Physics Reports*, *424*(4-5), 4–5, 175–308. doi:10.1016/j.physrep.2005.10.009

Borelli, F. F., Camargo, R. Y., Martins, D. C. Jr, & Rozante, L. C. S. (2013). Gene regulatory networks inference using a multi-GPU exhaustive search algorithm. *BMC Bioinformatics*, *14*(Suppl 18), S5. doi:10.1186/1471-2105-14-S18-S5 PMID:24564268

Bowman, A., & Azzalini, A. (1997). *Applied smoothing techniques for data analysis the kernel approach with S-Plus illustrations*. Oxford, UK: Clarendon Press.

Brazma, A., Vilo, J., Ukkonen, E., & Valtonen, K. (1997). Data mining for regulatory elements in yeast genome. *Proceedings of the International Conference on Intelligent Systems for Molecular Biology*, *5*, 65–74. PMID:9322017

Brightman, F. A., & Fell, D. A. (2000). Differential feedback regulation of the MAPK cascade underlies the quantitative differences in EGF and NGF signalling in PC12 cells. *FEBS Letters*, *482*(3), 169–174. doi:10.1016/S0014-5793(00)02037-8 PMID:11024454

Brough, P. A., Barril, X., Borgognoni, J., Chene, P., Davies, N. G. M., Davis, B., & Wright, L. et al. (2009). Combining hit identification strategies: Fragment-based and in silico approaches to orally active 2-aminothieno[2,3-d]pyrimidine inhibitors of the Hsp90 molecular chaperone. *Journal of Medicinal Chemistry*, *52*(15), 4794–4809. doi:10.1021/jm900357y PMID:19610616

Brown, J. A., Sherlock, G., Myers, C. L., Burrows, N. M., Deng, C., Wu, H. I., & Brown, J. M. et al. (2006). Global analysis of gene function in yeast by quantitative phenotypic proling. *Molecular Systems Biology*, *2*, 1–9. doi:10.1038/msb4100043 PMID:16738548

Brunet-Simon, A., Henrion, G., Renard, J. P., & Duranthon, V. (2001). Onset of zygotic transcription and maternal transcript legacy in the rabbit embryo. *Molecular Reproduction and Development*, *58*(2), 127–136. doi:10.1002/1098-2795(200102)58:2<127::AID-MRD1>3.0.CO;2-A PMID:11139224

Brun, M., Dougherty, E. R., & Shmulevich, I. (2005). Steady-state probabilities for attractors in probabilistic Boolean networks. *Signal Processing*, *85*(10), 1993–2013. doi:10.1016/j.sigpro.2005.02.016

Brunner, A. M., Nanni, P., & Mansuy, I. M. (2014). Epigenetic marking of sperm by post-translational modification of histones and protamines. *Epigenetics Chromatin*, *7*(1), 2. doi:10.1186/1756-8935-7-2 PMID:24443974

Brunton, L. L., Chabner, B., & Knollmann, B. C. (2010). *Goodman & Gilman's The Pharmacological Basis of Therapeutics, 12e*. New York, NY: McGraw-Hill Professional.

Budden, D. M., Hurley, D. G., & Crampin, E. J. (2014). Predictive modelling of gene expression from transcriptional regulatory elements. *Briefings in Bioinformatics*. doi:10.1093/bib/bbu034 PMID:25231769

Bui, H. T., Wakayama, S., Mizutani, E., Park, K. K., Kim, J. H., Van Thuan, N., & Wakayama, T. (2011). Essential role of paternal chromatin in the regulation of transcriptional activity during mouse preimplantation development. *Reproduction (Cambridge, England)*, *141*(1), 67–77. doi:10.1530/REP-10-0109 PMID:20974742

Bu, P., Chen, K. Y., Chen, J. H., Wang, L., Walters, J., Shin, Y. J., & Shen, X. et al. (2013). A microRNA miR-34a-regulated bimodal switch targets Notch in colon cancer stem cells. *Cell Stem Cell*, *12*(5), 602–615. doi:10.1016/j.stem.2013.03.002 PMID:23642368

Burger, L., Gaidatzis, D., Schubeler, D., & Stadler, M. B. (2013). Identification of active regulatory regions from DNA methylation data. *Nucleic Acids Research*, *41*(16), e155. doi:10.1093/nar/gkt599 PMID:23828043

Burkart, A. D., Xiong, B., Baibakov, B., Jimenez-Movilla, M., & Dean, J. (2012). Ovastacin, a cortical granule protease, cleaves ZP2 in the zona pellucida to prevent polyspermy. *The Journal of Cell Biology*, *197*(1), 37–44. doi:10.1083/jcb.201112094 PMID:22472438

Bussemaker, H. J., Foat, B. C., & Ward, L. D. (2007). Predictive modeling of genome-wide mRNA expression: From modules to molecules. *Annual Review of Biophysics and Biomolecular Structure*, *36*(1), 329–347. doi:10.1146/annurev.biophys.36.040306.132725 PMID:17311525

Bussemaker, H. J., Li, H., & Siggia, E. D. (2000). Building a dictionary for genomes: Identification of presumptive regulatory sites by statistical analysis. *Proceedings of the National Academy of Sciences of the United States of America*, *97*(18), 10096–10100. doi:10.1073/pnas.180265397 PMID:10944202

Bussemaker, H. J., Li, H., & Siggia, E. D. (2001). Regulatory element detection using correlation with expression. *Nature Genetics*, *27*(2), 167–171. doi:10.1038/84792 PMID:11175784

Butte, A., & Kohane, I. (2000). Mutual information relevance networks: functional genomic clustering using pairwise entropy measurements. In *Pacific Symposium on Biocomputing*, (pp. 418-429).

Caglar, M. U., & Pal, R. (2013). Stochastic Model Simulation Using Kronecker Product Analysis and Zassenhaus Formula Approximation. *IEEE/ACM Transactions on Computational Biology and Bioinformatics*, *10*(5), 1125–1136. doi:10.1109/TCBB.2013.34 PMID:24384703

Cahan, P., Li, H., Morris, S. A., Lummertz da Rocha, E., Daley, G. Q., & Collins, J. J. (2014). CellNet: Network biology applied to stem cell engineering. *Cell*, *158*(4), 903–915. doi:10.1016/j.cell.2014.07.020 PMID:25126793

Cai, X., & Wang, X. (2007). Stochastic Modeling and Simulation of Gene Networks. *IEEE Signal Processing Magazine*, *24*(1), 27–36. doi:10.1109/MSP.2007.273051

Cantone, I., & Fisher, A. G. (2013). Epigenetic programming and reprogramming during development. *Nature Structural & Molecular Biology*, *20*(3), 282–289. doi:10.1038/nsmb.2489 PMID:23463313

Cao, S., Han, J., Wu, J., Li, Q., Liu, S., Zhang, W., & Li, N. et al. (2014). Specific gene-regulation networks during the pre-implantation development of the pig embryo as revealed by deep sequencing. *BMC Genomics*, *15*(1), 4. doi:10.1186/1471-2164-15-4 PMID:24383959

Cao, Y., Gillespie, D. T., & Petzold, L. R. (2006). Efficient step size selection for the tau-leaping simulation method. *The Journal of Chemical Physics*, *124*(4), 044109. doi:10.1063/1.2159468 PMID:16460151

Cao, Y., & Liang, J. (2008). Optimal enumeration of state space of finitely buffered stochastic molecular networks and exact computation of steady state landscape probability. *BMC Systems Biology*, *2*(30). doi:10.1186/1752-0509-2-30 PMID:18373871

Cao, Y., & Liang, J. (2013). Adaptively biased sequential importance sampling for rare events in reaction networks with comparison to exact solutions from finite buffer dCME method. *The Journal of Chemical Physics*, *139*(2), 025101. doi:10.1063/1.4811286 PMID:23862966

Cao, Y., Lu, H. M., & Liang, J. (2010). Probability landscape of heritable and robust epigenetic state of lysogeny in phage lambda. *Proceedings of the National Academy of Sciences of the United States of America*, *107*(43), 18445–18450. doi:10.1073/pnas.1001455107 PMID:20937911

Cao, Y., & Samuels, D. C. (2009). Discrete stochastic simulation methods for chemically reacting systems. *Methods in Enzymology*, *454*, 115–140. doi:10.1016/S0076-6879(08)03805-6 PMID:19216925

Carpenter, A. E., Jones, T. R., Lamprecht, M. R., Clarke, C., Kang, I. H., Friman, O., & Sabatini, D. M. et al. (2006). CellProfiler: Image analysis software for identifying and quantifying cell phenotypes. *Genome Biology*, *7*(10), R100. doi:10.1186/gb-2006-7-10-r100 PMID:17076895

Chai, L. E., Loh, S. K., Low, S. T., Mohamad, M. S., Deris, S., & Zakaria, Z. (2014). A review on the computational approaches for gene regulatory network construction. *Computers in Biology and Medicine*, *48*, 55–65. doi:10.1016/j.compbiomed.2014.02.011 PMID:24637147

Chakraborty, M., Sengupta, A., Bhattacharya, D., Banerjee, S., & Chakrabarti, A. (2010). DNA binding domain of RFX5: Interactions with X-box DNA and RFXANK. *Biochimica et Biophysica Acta*, *1804*(10), 2016–2024. doi:10.1016/j.bbapap.2010.07.009 PMID:20637319

Chalfie, M., Tu, Y., Euskirchen, G., Ward, W., & Prasher, D. (1994). Green fluorescent protein as a marker for gene expression. *Science*, *263*(5148), 802–805. doi:10.1126/science.8303295 PMID:8303295

Chandrasekaran, S., & Price, N. D. (2010). Probabilistic integrative modeling of genome-scale metabolic and regulatory networks in Escherichia coli and Mycobacterium tuberculosis. *Proceedings of the National Academy of Sciences of the United States of America*, *107*(41), 17845–17850. doi:10.1073/pnas.1005139107 PMID:20876091

Chang, C. C., Ma, Y., Jacobs, S., Tian, X. C., Yang, X., & Rasmussen, T. P. (2005). A maternal store of macroH2A is removed from pronuclei prior to onset of somatic macroH2A expression in preimplantation embryos. *Developmental Biology*, *278*(2), 367–380. doi:10.1016/j.ydbio.2004.11.032 PMID:15680357

Chaouiya, C., Remy, E., & Thieffry, D. (2008). Petri net modelling of biological regulatory networks. *Journal of Discrete Algorithms*, *6*(2), 165–177. doi:10.1016/j.jda.2007.06.003

Charbonnier, C., Chiquet, J., & Ambroise, C. (2010). Weighted-lasso for structured network inference from time course data. *Statistical Applications in Genetics and Molecular Biology*, *9*(1), 15.

Chen, L., Arbieva, Z. H., Guo, S. J., Marucha, P. T., Mustoe, T. A., & DiPietro, L. A. (2010). Positional differences in the wound transcriptome of skin and oral mucosa. *BMC Genomics*, *11*, 15. doi: 10.1186/1471-2164-11-471

Chen, C. C., & Zhong, S. (2008). Inferring gene regulatory networks by thermodynamic modeling. *BMC Genomics*, *9*(Suppl 2), S19. doi:10.1186/1471-2164-9-S2-S19 PMID:18831784

Chen, J., Melton, C., Suh, N., Oh, J. S., Horner, K., Xie, F., & Conti, M. et al. (2011). Genome-wide analysis of translation reveals a critical role for deleted in azoospermia-like (Dazl) at the oocyte-to-zygote transition. *Genes & Development*, *25*(7), 755–766. doi:10.1101/gad.2028911 PMID:21460039

Chen, K.-C., Wang, T.-Y., Tseng, H.-H., Huang, C.-Y. F., & Kao, C.-Y. (2005). A stochastic differential equation model for quantifying transcriptional regulatory network in Saccharomyces cerevisiae. *Bioinformatics (Oxford, England)*, *21*(12), 2883–2890. doi:10.1093/bioinformatics/bti415 PMID:15802287

Chen, L., Kostadima, M., Martens, J. H., Canu, G., Garcia, S. P., Turro, E., & Rendon, A. et al. (2014). Transcriptional diversity during lineage commitment of human blood progenitors. *Science*, *345*(6204), 1251033. doi:10.1126/science.1251033 PMID:25258084

Chen, S. J., Zhu, Y. J., Tong, J. H., Dong, S., Huang, W., Chen, Y., & Qian, G. Q. et al. (1991). Rearrangements in the second intron of the RARA gene are present in a large majority of patients with acute promyelocytic leukemia and are used as molecular marker for retinoic acid-induced leukemic cell differentiation. *Blood*, *78*(10), 2696–2701. PMID:1668609

Chen, T., He, H. L., & Church, G. M. (1999). Modeling gene expression with differential equations.*Pacific Symposium on Biocomputing*, (pp. 29-40).

Ching, W.-K., Zhang, S., Jiao, Y., Akutsu, T., Tsing, N., & Wong, A. (2009). Optimal control policy for probabilistic Boolean networks with hard constraints. *IET Systems Biology*, *3*(2), 90–99. doi:10.1049/iet-syb.2008.0120 PMID:19292563

Chorev, M., & Carmel, L. (2012). The function of introns. *Frontiers in Genetics, 3*, 1–15. doi:10.3389/fgene.2012.00055 PMID:22518112

Choudhary, A., Datta, A., Bittner, M. L., & Dougherty, E. R. (2006). Intervention in a family of Boolean networks. *Bioinformatics (Oxford, England), 22*(2), 226–232. doi:10.1093/bioinformatics/bti765 PMID:16286362

Christley, S., Nie, Q., & Xie, X. (2009). Incorporating existing network information into gene network inference. *PLoS ONE, 4*(8), e6799.

Cinquemani, E., Milias-Argeitis, A., Summers, S., & Lygeros, J. (2008). Stochastic dynamics of genetic networks: Modelling and parameter identification. *Bioinformatics (Oxford, England), 24*(23), 2748–2754. doi:10.1093/bioinformatics/btn527 PMID:18845579

Clift, D., & Schuh, M. (2013). Restarting life: Fertilization and the transition from meiosis to mitosis. *Nature Reviews. Molecular Cell Biology, 14*(9), 549–562. doi:10.1038/nrm3643 PMID:23942453

Cochella, L., & Green, R. (2005). Fidelity in protein synthesis. *Current Biology, 15*(14), R536–R540. doi:10.1016/j.cub.2005.07.018 PMID:16051156

Cojocaru, L., & Agur, Z. (1992). A theoretical analysis of interval drug dosing for cell-cycle-phase-specific drugs. *Mathematical Biosciences, 109*(1), 85–97. doi:10.1016/0025-5564(92)90053-Y PMID:1591451

Collins, F. S., Green, E. D., Guttmacher, A. E., & Guyer, M. S. (2003). A vision for the future of genomics research. *Nature, 422*(6934), 835–847. doi:10.1038/nature01626 PMID:12695777

Condic, M. L. (2014). Totipotency: What it is and what it is not. *Stem Cells and Development, 23*(8), 796–812. doi:10.1089/scd.2013.0364 PMID:24368070

Cong, Y., Ching, W.-K., Tsing, N., & Leung, H. (2010). On finite-horizon control of genetic regulatory networks with multiple hard-constraints. *BMC Systems Biology, 4*(Suppl 2), S14. doi:10.1186/1752-0509-4-S2-S14 PMID:20840728

Conlon, E. M., Liu, X. S., Lieb, J. D., & Liu, J. S. (2003). Integrating regulatory motif discovery and genome-wide expression analysis. *Proceedings of the National Academy of Sciences of the United States of America, 100*(6), 3339–3344. doi:10.1073/pnas.0630591100 PMID:12626739

Cook, R. S., Garrett, J. T., Sanchez, V., Stanford, J. C., Young, C., Chakrabarty, A., & Arteaga, C. L. et al. (2011). ErbB3 ablation impairs PI3K/Akt-dependent mammary tumorigenesis. *Cancer Research, 71*(11), 3941–3951. doi:10.1158/0008-5472.CAN-10-3775 PMID:21482676

Cormen, T. H., Leiserson, C. E., & Rivest, R. L. (1990). *Introduction to algorithms*. Cambridge, MA: The MIT Press.

Costa, L. F., Rodrigues, F. A., & Cristino, A. S. (2008). Complex networks: The key to systems biology. *Genetics and Molecular Biology, 31*(3), 591–601. doi:10.1590/S1415-47572008000400001

Covert, M. W., Xiao, N., Chen, T. J., & Karr, J. R. (2008). Integrating metabolic, transcriptional regulatory and signal transduction models in Escherichia coli. *Bioinformatics (Oxford, England)*, *24*(18), 2044–2050. doi:10.1093/bioinformatics/btn352 PMID:18621757

Cowell, I. G., Aucott, R., Mahadevaiah, S. K., Burgoyne, P. S., Huskisson, N., Bongiorni, S., & Singh, P. B. et al. (2002). Heterochromatin, HP1 and methylation at lysine 9 of histone H3 in animals. *Chromosoma*, *111*(1), 22–36. doi:10.1007/s00412-002-0182-8 PMID:12068920

Crick, F. (1970). Central dogma of molecular biology. *Nature*, *227*(5258), 561–563. doi:10.1038/227561a0 PMID:4913914

Cui, X., Wang, T., Chen, H. S., Busov, V., & Wei, H. (2010). Tf-finder: A software package for identifying transcription factors involved in biological processes using microarray data and existing knowledge base. *BMC Bioinformatics*, *11*(1), 425. doi:10.1186/1471-2105-11-425 PMID:20704747

Da Silva, E., & Stumpf, P. (2005). Complex networks and simple models in biology. *Journal of the Royal Society*, *2*(5), 419-430.

Daenen, F., van Roy, F., & De Bleser, P. J. (2008). Low nucleosome occupancy is encoded around functional human transcription factor binding sites. *BMC Genomics*, *9*(1), 332. doi:10.1186/1471-2164-9-332 PMID:18627598

Dahlin, J. L., Inglese, J., & Walters, M. A. (2015). Mitigating risk in academic preclinical drug discovery. *Nature Reviews. Drug Discovery*, *14*(4), 279–294. doi:10.1038/nrd4578 PMID:25829283

Daigle, B. J. Jr, Roh, M. K., Gillespie, D. T., & Petzold, L. R. (2011). Automated estimation of rare event probabilities in biochemical systems. *The Journal of Chemical Physics*, *134*(4), 044110. doi:10.1063/1.3522769 PMID:21280690

D'Angelo, C., Kilian, J., Kudla, J., Batistic, O., & Weinl, S. (2009). *The Arabidopsis Information Resource*. TAIR.

Das, D., Nahle, Z., & Zhang, M. Q. (2006). Adaptively inferring human transcriptional subnetworks. *Molecular Systems Biology*, *2*. doi: 10.1038/msb4100067

Das, D., Pellegrini, M., & Gray, J. W. (2009). A primer on regression methods for decoding cis-regulatory logic. *PLoS Computational Biology*, *5*(1), e1000269. doi:10.1371/journal.pcbi.1000269 PMID:19180174

Das, R., & Gerstein, M. (2004). A method using active-site sequence conservation to find functional shifts in protein families: Application to the enzymes of central metabolism, leading to the identification of an anomalous isocitrate dehydrogenase in pathogens. *Proteins-Structure Function and Bioinformatics*, *55*(2), 455–463. doi:10.1002/prot.10639 PMID:15048835

Datta, A., Choudhary, A., Bittner, M. L., & Dougherty, E. R. (2003). External Control in Markovian Genetic Regulatory Networks. *Machine Learning*, *52*(1), 169–191. doi:10.1023/A:1023909812213

Datta, A., Choudhary, A., Bittner, M. L., & Dougherty, E. R. (2004). External control in Markovian genetic regulatory networks: The imperfect information case. *Bioinformatics (Oxford, England)*, *20*(6), 924–930. doi:10.1093/bioinformatics/bth008 PMID:14751971

Datta, A., & Dougherty, E. R. (2006). *Introduction to Genomic Signal Processing with Control*. Boca Raton, FL: CRC. doi:10.1201/9781420006674

Datta, A., Pal, R., Choudhary, A., & Dougherty, E. R. (2007). Control approaches for probabilistic gene regulatory networks. *IEEE Signal Processing Magazine*, *24*(1), 54–63. doi:10.1109/MSP.2007.273057

Datta, D., & Zhao, H. (2008). Statistical methods to infer cooperative binding among transcription factors in Saccharomyces cerevisiae. *Bioinformatics (Oxford, England)*, *24*(4), 545–552. doi:10.1093/bioinformatics/btm523 PMID:17989095

Davidich, M. I., & Bornholdt, S. (2008). Boolean Network Model Predicts Cell Cycle Sequence of Fission Yeast. *PLoS ONE, 3*(2), e1672.

Davidson, C. (2010). Identifying gene regulatory networks using evolutionary algorithms. *Journal of Computing Sciences in Colleges*, *25*(5), 231–237.

Dayneka, N. L., Garg, V., & Jusko, W. J. (1993). Comparison of four basic models of indirect pharmacodynamic responses. *Journal of Pharmacokinetics and Pharmacodynamics*, *21*(4), 457–478. doi:10.1007/BF01061691 PMID:8133465

De Haan, J., Piek, E., van Schaik, R., de Vlieg, J., Bauerschmidt, S., Buydens, L., & Wehrens, R. (2010). Integrating gene expression and GO classification for PCA by preclustering. *BMC Bioinformatics*, *11*(1), 158. doi:10.1186/1471-2105-11-158 PMID:20346140

De Jong, H. (2002). Modeling and Simulation of Genetic Regulatory Systems: A Literature Review. *Journal of Computational Biology*, *9*(1), 67–103. doi:10.1089/10665270252833208 PMID:11911796

de Smet, R., & Marchal, K. (2010). Advantages and limitations of current network inference methods. *Nature Reviews. Microbiology*, *8*(10), 717–729. PMID:20805835

de Vladar, H. P., & González, J. A. (2004). Dynamic response of cancer under the influence of immunological activity and therapy. *Journal ofTheoretical Biology*, *227*(3), 335–348. doi:10.1016/j.jtbi.2003.11.012 PMID:15019501

Dean, J. (2004). Reassessing the molecular biology of sperm-egg recognition with mouse genetics. *BioEssays*, *26*(1), 29–38. doi:10.1002/bies.10412 PMID:14696038

Deng, Q., Ramskold, D., Reinius, B., & Sandberg, R. (2014). Single-cell RNA-seq reveals dynamic, random monoallelic gene expression in mammalian cells. *Science*, *343*(6167), 193–196. doi:10.1126/science.1245316 PMID:24408435

Derman, C. (1970). *Finite State Markovian Decision Processes* (Vol. 107). Orlando, FL: Academic Press, Inc.

D'haeseleer, P., Liang, S., & Somogyi, R. (2000). Genetic network inference: From co-expression clustering to reverse engineering. *Bioinformatics (Oxford, England)*, *16*(8), 707–726. doi:10.1093/bioinformatics/16.8.707 PMID:11099257

Dias, B. G., & Ressler, K. J. (2014). Parental olfactory experience influences behavior and neural structure in subsequent generations. *Nature Neuroscience*, *17*(1), 89–96. doi:10.1038/nn.3594 PMID:24292232

Dissanayake, S. K., Olkhanud, P. B., O'Connell, M. P., Carter, A., French, A. D., Camilli, T. C., & Weeraratna, A. T. et al. (2008). Wnt5A regulates expression of tumor-associated antigens in melanoma via changes in signal transducers and activators of transcription 3 phosphorylation. *Cancer Research*, *68*(24), 10205–10214. doi:10.1158/0008-5472.CAN-08-2149 PMID:19074888

Dong, H., Kung, J., Bhardwaj, V., & Mcgill, J. (2010). Flow cytometry rapidly identifies all acute promyelocytic leukemias with high specificity independent of underlying cytogenetic abnormalities. *American Journal of Clinical Pathology*, *135*(1), 76–84. doi:10.1309/AJCPW9T-SLQNCZAVT PMID:21173127

Dougherty, E. R., & Braga-Neto, U. (2006). Epistemology of computational biology: Mathematical models and experimental prediction as the basis of their validity. *Journal of Computational Biology*, *14*(1), 65–90.

Dougherty, E. R., Brun, M., Trent, J., & Bittner, M. L. (2009). A Conditioning-Based Model of Contextual Regulation. *IEEE/ACM Transactions on Computational Biology and Bioinformatics*, *6*(2), 310–320. doi:10.1109/TCBB.2007.70247 PMID:19407354

Dougherty, E. R., & Datta, A. (2005). Genomic signal processing: Diagnosis and therapy. *IEEE Signal Processing Magazine*, *22*(1), 107–112. doi:10.1109/MSP.2005.1407722

Dougherty, E. R., Pal, R., Qian, X., Bittner, M. L., & Datta, A. (2010). Stationary and structural control in gene regulatory networks: Basic concepts. *International Journal of Systems Science*, *41*(1), 5–16. doi:10.1080/00207720903144560

Dougherty, E. R., & Shmulevich, I. (2003). Mappings between probabilistic Boolean networks. *Signal Processing*, *83*(4), 799–809. doi:10.1016/S0165-1684(02)00480-2

Dougherty, E. R., & Shmulevich, I. (2012). On the Limitations of Biological Knowledge. *Current Genomics*, *13*(7), 574–587. doi:10.2174/138920212803251445 PMID:23633917

Dougherty, E., & Lotufo, R. (2003). *Hands-on morphological image processing*. Bellingham, WA: SPIE. doi:10.1117/3.501104

Druker, B. J., Sawyers, C. L., Kantarjian, H., Resta, D. J., Reese, S. F., Ford, J. M., & Talpaz, M. et al. (2001). Activity of a specific inhibitor of the BCR-ABL tyrosine kinase in the blast crisis of chronic myeloid leukemia and acute lymphoblastic leukemia with the philadelphia chromosome. *The New England Journal of Medicine*, *344*(14), 1038–1042. doi:10.1056/NEJM200104053441402 PMID:11287973

Dunham, I., Kundaje, A., Aldred, S. F., Collins, P. J., Davis, C. A., & Doyle, F.,…Birney, E., ENCODE Project Consortium. (2012). An integrated encyclopedia of DNA elements in the human genome. *Nature*, *489*, 57–74. PMID:22955616

Dwight, S. S., Harris, M. A., Dolinski, K., Ball, C. A., Binkley, G., Christie, K. R., & Cherry, J. M. et al. (2002). Saccharomyces genome database (SGD) provides secondary gene annotation using the gene ontology (GO). *Nucleic Acids Research*, *30*(1), 69–72. doi:10.1093/nar/30.1.69 PMID:11752257

Edwards, D. (2000). *Introduction to Graphical Modelling*. Springer. doi:10.1007/978-1-4612-0493-0

Egli, D., Chen, A. E., Saphier, G., Ichida, J., Fitzgerald, C., Go, K. J., & Eggan, K. et al. (2011). Reprogramming within hours following nuclear transfer into mouse but not human zygotes. *Nature Communications*, *2*, 488. doi:10.1038/ncomms1503 PMID:21971503

Egozcue, J., Sarrate, Z., Codina-Pascual, M., Egozcue, S., Oliver-Bonet, M., Blanco, J., & Vidal, F. et al. (2005). Meiotic abnormalities in infertile males. *Cytogenetic and Genome Research*, *111*(3-4), 337–342. doi:10.1159/000086907 PMID:16192712

Eisen, M. B., Spellman, P. T., Brown, P. O., & Botstein, D. (1998). Cluster analysis and display of genome-wide expression patterns. *Proceedings of the National Academy of Sciences of the United States of America*, *95*(25), 14863–14868. doi:10.1073/pnas.95.25.14863 PMID:9843981

El Samad, H., Khammash, M., Petzold, L., & Gillespie, D. (2005). Stochastic modelling of gene regulatory networks. *International Journal of Robust and Nonlinear Control*, *15*(15), 691–711. doi:10.1002/rnc.1018

Erkek, S., Hisano, M., Liang, C. Y., Gill, M., Murr, R., Dieker, J., & Peters, A. H. et al. (2013). Molecular determinants of nucleosome retention at CpG-rich sequences in mouse spermatozoa. *Nature Structural & Molecular Biology*, *20*(7), 868–875. doi:10.1038/nsmb.2599 PMID:23770822

Ernst, J., Beg, Q. K., Kay, K. A., Balazsi, G., Oltvai, Z. N., & Bar-Joseph, Z. (2008). A semi-supervised method for predicting transcription factor-gene interactions in *Escherichia coli*. *PLoS Comput Biol*, *4*(3), e1000044.

Evans, J. P., Hardy, D. M., & Storey, B. T. (2001). *Sperm–Egg Interactions: Sperm–Egg Binding in Mammals eLS*. John Wiley & Sons, Ltd.

Faith, J., Hayete, B., Thaden, J., Mogno, I., Wierzbowski, J., Cottarel, G., & Gardner, T. S. et al. (2007). Large-scale mapping and validation of escherichia coli transcriptional regulation from a compendium of expression proles. *PLoS Biology*, *5*(1), 259–265. doi:10.1371/journal.pbio.0050008 PMID:17214507

Fall, C., Marland, E., Wagner, J., & Tyson, J. (Eds.). (2002). *Computational Cell Biology*. New York: Springer.

Farkas, I. J., Jeong, H., Vicsek, T., Barabási, A.-L., & Oltvai, Z. N. (2003). The topology of the transcription regulatory network in the yeast, saccharomyces cerevisiae. *Physica A: Statistical Mechanics and its Applications*, *318*(3-4), 601–612.

Faryabi, B., Chamberland, J.-F., Vahedi, G., Datta, A., & Dougherty, E. R. (2008). Optimal intervention in asynchronous genetic regulatory networks. *IEEE Journal of Selected Topics in Signal Processing*, *2*(3), 412–423. doi:10.1109/JSTSP.2008.923853

Faryabi, B., Datta, A., & Dougherty, E. R. (2007). On approximate stochastic control in genetic regulatory networks. *IET Systems Biology*, *1*(6), 361–368. doi:10.1049/iet-syb:20070015 PMID:18203582

Faryabi, B., Vahedi, G., Chamberland, J.-F., Datta, A., & Dougherty, E. R. (2008). Optimal Constrained Stationary Intervention in Gene Regulatory Networks. *EURASIP Journal on Bioinformatics & Systems Biology*, (1): 620767. PMID:18516276

Fatica, A., & Bozzoni, I. (2014). Long non-coding RNAs: New players in cell differentiation and development. *Nature Reviews. Genetics*, *15*(1), 7–21. doi:10.1038/nrg3606 PMID:24296535

Faure, A., Naldi, A., Chaouiya, C., & Thieffry, D. (2006). Dynamical analysis of a generic Boolean model for the control of the mammalian cell cycle. *Bioinformatics (Oxford, England)*, *22*(14), e124–e131. doi:10.1093/bioinformatics/btl210 PMID:16873462

Felder, S., LaVin, J., Ullrich, A., & Schlessinger, J. (1992). Kinetics of binding, endocytosis, and recycling of EGF receptor mutants. *The Journal of Biological Chemistry*, *117*, 203–212. PMID:1556153

Ferramosca, A., & Zara, V. (2014). Bioenergetics of mammalian sperm capacitation. *Biomed Res Int*, *902953*. doi:10.1155/2014/902953 PMID:24791005

Frech, K., & Werner, T. (1997). Specific modelling of regulatory units in DNA sequences.*Pacific Symposium on Biocomputing*, (pp. 151-162).

Friedman, J., Hastie, T., & Tibshirani, R. (2010). Regularization Paths for Generalized Linear Models via Coordinate Descent. *Journal of Statistical Software*, *33*(1), 1–22. doi:10.18637/jss.v033.i01 PMID:20808728

Friedman, N. (2004). Inferring cellular networks using probabilistic graphical models. *Science Signaling*, *303*(5659), 799–805. PMID:14764868

Friedman, N. (2004). Inferring cellular networks using probabilistic graphical models. *Science, 303*(5659), 799–805. doi:10.1126/science.1094068 PMID:14764868

Friedman, N., Linial, M., Nachman, I., & Pe'er, D. (2000). Using Bayesian networks to analyze expression data. *Journal of Computational Biology, 7*(3-4), 601–620. doi:10.1089/106652700750050961 PMID:11108481

Gamble, M. J., Frizzell, K. M., Yang, C., Krishnakumar, R., & Kraus, W. L. (2010). The histone variant macroH2A1 marks repressed autosomal chromatin, but protects a subset of its target genes from silencing. *Genes & Development, 24*(1), 21–32. doi:10.1101/gad.1876110 PMID:20008927

Gao, F., Foat, B. C., & Bussemaker, H. J. (2004). Defining transcriptional networks through integrative modeling of mRNA expression and transcription factor binding data. *BMC Bioinformatics, 5*(1), 31. doi:10.1186/1471-2105-5-31 PMID:15113405

Garey, M. R., & Johnson, D. S. (1979). *Computers and intractability: a guide to the theory of NP-completeness*. San Francisco: WH Freeman & Co.

Gaspar-Maia, A., Qadeer, Z. A., Hasson, D., Ratnakumar, K., Leu, N. A., Leroy, G., & Bernstein, E. et al. (2013). MacroH2A histone variants act as a barrier upon reprogramming towards pluripotency. *Nature Communications, 4*, 1565. doi:10.1038/ncomms2582 PMID:23463008

Gatewood, J. M., Cook, G. R., Balhorn, R., Schmid, C. W., & Bradbury, E. M. (1990). Isolation of four core histones from human sperm chromatin representing a minor subset of somatic histones. *The Journal of Biological Chemistry, 265*(33), 20662–20666. Retrieved from http://www.ncbi.nlm.nih.gov/pubmed/2243112 PMID:2243112

Gazdag, E., Rajkovic, A., Torres-Padilla, M. E., & Tora, L. (2007). Analysis of TATA-binding protein 2 (TBP2) and TBP expression suggests different roles for the two proteins in regulation of gene expression during oogenesis and early mouse development. *Reproduction (Cambridge, England), 134*(1), 51–62. doi:10.1530/REP-06-0337 PMID:17641088

Geisler, C., Gaisa, N. T., Pfister, D., Fuessel, S., Kristiansen, G., Braunschweig, T., & Henkel, C. et al. (2015). Identification and validation of potential new biomarkers for prostate cancer diagnosis and prognosis using 2D-DIGE and MS. *BioMed Research International, 2015*, 454256. doi:10.1155/2015/454256 PMID:25667921

Gerdes, J. (1990). Ki-67 and other proliferation markers useful for immunohistological diagnostic and prognostic evaluations in human malignancies. *Seminars in Cancer Biology, 1*(3), 199–206. PMID:2103495

Gerlinger, M., Rowan, A. J., Horswell, S., Larkin, J., Endesfelder, D., Gronroos, E., & Swanton, C. et al. (2012). Intratumor heterogeneity and branched evolution revealed by multiregion sequencing. *The New England Journal of Medicine, 366*(10), 883–892. doi:10.1056/NEJMoa1113205 PMID:22397650

Gertz, J., Siggia, E. D., & Cohen, B. A. (2009). Analysis of combinatorial cis-regulation in synthetic and genomic promoters. *Nature*, *457*(7226), 215–218. doi:10.1038/nature07521 PMID:19029883

Ghaffari, N., Ivanov, I., Qian, X., & Dougherty, E. R. (2010). A CoD-based reduction algorithm for designing stationary control policies on Boolean networks. *Bioinformatics (Oxford, England)*, *26*(12), 1556–1563. doi:10.1093/bioinformatics/btq225 PMID:20421196

Giacinti, C., & Giordano, A. (2006). RB and cell cycle progression. *Oncogene*, *25*(38), 5220–5227. doi:10.1038/sj.onc.1209615 PMID:16936740

Gillespie, D. T. (1977). Exact Stochastic Simulation of Coupled Chemical Reactions. *Journal of Physical Chemistry*, *81*(25), 2340–2361. doi:10.1021/j100540a008

Gillespie, D. T. (2000). The chemical Langevin equation. *The Journal of Chemical Physics*, *113*(1), 297–306. doi:10.1063/1.481811

Gillespie, D. T. (2001). Approximate accelerated stochastic simulation of chemically reacting systems. *The Journal of Chemical Physics*, *115*(4), 1716–1733. doi:10.1063/1.1378322

Gillespie, D. T. (2007). Stochastic simulation of chemical kinetics. *Annual Review of Physical Chemistry*, *58*(1), 35–55. doi:10.1146/annurev.physchem.58.032806.104637 PMID:17037977

Goemann, B., Wingender, E., & Potapov, A. (2009). An approach to evaluate the topological significance of motifs and other patterns in regulatory networks. *BMC Systems Biology*, *3*(1), 53. doi:10.1186/1752-0509-3-53 PMID:19454001

Golub, G. H., & Van Loan, C. F. (1996). *Matrix computations* (3rd ed.). Baltimore, MD: Johns Hopkins University Press.

Good, L., Dimri, G. P., Campisi, J., & Chen, K. Y. (1996). Regulation of dihydrofolate reductase gene expression and E2F components in human diploid fibroblasts during growth and senescence. *Journal of Cellular Physiology*, *168*(3), 580–588. doi:10.1002/(SICI)1097-4652(199609)168:3<580::AID-JCP10>3.0.CO;2-3 PMID:8816912

Goss, P. J., & Peccoud, J. (1998). Quantitative modeling of stochastic systems in molecular biology by using stochastic Petri nets. *Proceedings of the National Academy of Sciences of the United States of America*, *95*(12), 6750–6755. doi:10.1073/pnas.95.12.6750 PMID:9618484

Goutsias, J., & Jenkinson, G. (2013). Markovian dynamics on complex reaction networks. *Physics Reports*, *529*(2), 199–264. doi:10.1016/j.physrep.2013.03.004

Grasso, C., Tang, Y., Truffaux, N., Berlow, N. E., Liu, L., Debily, M. A., & Monje, M. et al. (2015). Functionally defined therapeutic targets in diffuse intrinsic pontine glioma. *Nature Medicine*, *21*(6), 555–559. doi:10.1038/nm.3855 PMID:25939062

Greller, L. D., & Somogyi, R. (2002). Reverse engineers map the molecular switching yards. *Trends in Biotechnology*, *20*(11), 445–447. doi:10.1016/S0167-7799(02)02051-6 PMID:12413811

Grima, R., Thomas, P., & Straube, A. V. (2011). How accurate are the nonlinear chemical Fokker-Planck and chemical Langevin equations? *The Journal of Chemical Physics*, *135*(8), 084103. doi:10.1063/1.3625958 PMID:21895155

Guan, Y., Boavida, L., & McCormick, S. (2001). *Gametogenesis eLS*. John Wiley & Sons, Ltd.

Guelzim, N., Bottani, S., Bourgine, P., & Képès, F. (2002). Topological and causal structure of the yeast transcriptional regulatory network. *Nature Genetics*, *31*(1), 60–63. doi:10.1038/ng873 PMID:11967534

GuhaThakurta, D., & Stormo, G. D. (2001). Identifying target sites for cooperatively binding factors. *Bioinformatics (Oxford, England)*, *17*(7), 608–621. doi:10.1093/bioinformatics/17.7.608 PMID:11448879

Gurdon, J. B. (1962). The developmental capacity of nuclei taken from intestinal epithelium cells of feeding tadpoles. *Journal of Embryology and Experimental Morphology*, *10*, 622–640. Retrieved from http://www.ncbi.nlm.nih.gov/pubmed/13951335 PMID:13951335

Haberle, V., Li, N., Hadzhiev, Y., Plessy, C., Previti, C., Nepal, C., & Lenhard, B. et al. (2014). Two independent transcription initiation codes overlap on vertebrate core promoters. *Nature*, *507*(7492), 381–385. doi:10.1038/nature12974 PMID:24531765

Hamatani, T., Carter, M. G., Sharov, A. A., & Ko, M. S. (2004). Dynamics of global gene expression changes during mouse preimplantation development. *Developmental Cell*, *6*(1), 117–131. doi:10.1016/S1534-5807(03)00373-3 PMID:14723852

Hamazaki, N., Uesaka, M., Nakashima, K., Agata, K., & Imamura, T. (2015). Gene activation-associated long noncoding RNAs function in mouse preimplantation development. *Development*, *142*(5), 910–920. doi:10.1242/dev.116996 PMID:25633350

Hammoud, S. S., Nix, D. A., Zhang, H., Purwar, J., Carrell, D. T., & Cairns, B. R. (2009). Distinctive chromatin in human sperm packages genes for embryo development. *Nature*, *460*(7254), 473–478. doi:10.1038/nature08162 PMID:19525931

Handel, M. A., & Schimenti, J. C. (2010). Genetics of mammalian meiosis: Regulation, dynamics and impact on fertility. *Nature Reviews. Genetics*, *11*(2), 124–136. doi:10.1038/nrg2723 PMID:20051984

Hao, L., Johnsen, R., Lauter, G., Baillie, D., & Bürglin, T. R. (2006). Comprehensive analysis of gene expression patterns of hedgehog-related genes. *BMC Genomics*, *7*(1), 280. doi:10.1186/1471-2164-7-280 PMID:17076889

Harbison, C. T., Gordon, D. B., Lee, T. I., Rinaldi, N. J., Macisaac, K. D., Danford, T. W., & Young, R. A. et al. (2004). Transcriptional regulatory code of a eukaryotic genome. *Nature*, *431*(7004), 99–104. doi:10.1038/nature02800 PMID:15343339

Hardy, S., & Robillard, P. N. (2004). Modeling and simulation of molecular biology systems using petri nets: Modeling goals of various approaches. *Journal of Bioinformatics and Computational Biology*, *2*(4), 595–613. doi:10.1142/S0219720004000764 PMID:15617157

Hassa, P. O., Covic, M., Bedford, M. T., & Hottiger, M. O. (2008). Protein arginine methyltransferase 1 coactivates NF-kappaB-dependent gene expression synergistically with CARM1 and PARP1. *Journal of Molecular Biology*, *377*(3), 668–678. doi:10.1016/j.jmb.2008.01.044 PMID:18280497

Hassold, T., & Hunt, P. (2001). To err (meiotically) is human: The genesis of human aneuploidy. *Nature Reviews. Genetics*, *2*(4), 280–291. doi:10.1038/35066065 PMID:11283700

Hastie, T., Tibshirani, R., & Friedman, J. H. (2009). *The elements of statistical learning: data mining, inference, and prediction* (2nd ed.). New York, NY: Springer. doi:10.1007/978-0-387-84858-7

Hasty, J., Pradines, J., Dolnik, M., & Collins, J. J. (2000). Noise-based switches and amplifiers for gene expression. *Proceedings of the National Academy of Sciences of the United States of America*, *97*(5), 2075–2080. doi:10.1073/pnas.040411297 PMID:10681449

Heard, E., & Martienssen, R. A. (2014). Transgenerational epigenetic inheritance: Myths and mechanisms. *Cell*, *157*(1), 95–109. doi:10.1016/j.cell.2014.02.045 PMID:24679529

Hecker, M., Lambeck, S., Toepfer, S., van Someren, E., & Guthke, R. (2009). Gene regulatory network inference: Data integration in dynamic models-a review. *Bio Systems*, *96*(1), 86–103. doi:10.1016/j.biosystems.2008.12.004 PMID:19150482

Hegland, M., Burden, C., Santoso, L., MacNamara, S., & Booth, H. (2007). A solver for the stochastic master equation applied to gene regulatory networks. *Journal of Computational and Applied Mathematics*, *205*(2), 708–724. doi:10.1016/j.cam.2006.02.053

Heinz, S., Romanoski, C. E., Benner, C., & Glass, C. K. (2015). The selection and function of cell type-specific enhancers. *Nature Reviews. Molecular Cell Biology*, *16*(3), 144–154. doi:10.1038/nrm3949 PMID:25650801

Hellyer, N. J., Kim, M. S., & Koland, J. G. (2001). Heregulin-dependent Activation of Phosphoinositide 3-Kinase and Akt via the ErbB2/ErbB3 Co-receptor. *The Journal of Biological Chemistry*, *276*(45), 42153–42161. doi:10.1074/jbc.M102079200 PMID:11546794

Hemmings, B. A., & Restuccia, D. F. (2012). PI3K-PKB/Akt pathway. *Cold Spring Harbor Perspectives in Biology*, *4*(9), a011189. doi:10.1101/cshperspect.a011189 PMID:22952397

Heppner, G. H. (1984). Tumor Heterogeneity. *Cancer Research*, *44*(6), 2259–2265. PMID:6372991

Heydtmann, M. (2002). The nature of truth: Simpson's Paradox and the limits of statistical data. *QJM*, *95*(4), 247–249. doi:10.1093/qjmed/95.4.247 PMID:11937652

Higham, D. J. (2008). Modeling and Simulating Chemical Reactions. *SIAM Review*, *50*(2), 347–368. doi:10.1137/060666457

Hilton, T. L., & Wang, E. H. (2003). Transcription factor IID recruitment and Sp1 activation. Dual function of TAF1 in cyclin D1 transcription. *The Journal of Biological Chemistry*, *278*(15), 12992–13002. doi:10.1074/jbc.M300412200 PMID:12569092

Himmelstein, K. J., & Bischoff, K. B. (1973). Mathematical representations of cancer chemotherapy effects. *Journal of Pharmacokinetics and Pharmacodynamics*, *1*(1), 51–68. doi:10.1007/BF01060027

Hirose, O., Yoshida, R., Imoto, S., Yamaguchi, R., Higuchi, T., Charnock-Jones, D. S., & Miyano, S. et al. (2008). Statistical inference of transcriptional module-based gene networks from time course gene expression profiles by using state space models. *Bioinformatics (Oxford, England)*, *24*(7), 932–942. doi:10.1093/bioinformatics/btm639 PMID:18292116

Hoelder, S., Clarke, P. A., & Workman, P. (2012). Discovery of small molecule cancer drugs: Successes, challenges and opportunities. *Molecular Oncology*, *6*(2), 155–176. doi:10.1016/j.molonc.2012.02.004 PMID:22440008

Holland, J. H. (1992). *Adaptation in Natural and Artificial Systems*. MIT Press.

Holter, N. S., Maritan, A., Cieplak, M., Fedoroff, N. V., & Banavar, J. R. (2001). Dynamic modeling of gene expression data. *Proceedings of the National Academy of Sciences of the United States of America*, *98*(4), 1693–1698. doi:10.1073/pnas.98.4.1693 PMID:11172013

Hoodbhoy, T., Joshi, S., Boja, E. S., Williams, S. A., Stanley, P., & Dean, J. (2005). Human sperm do not bind to rat zonae pellucidae despite the presence of four homologous glycoproteins. *The Journal of Biological Chemistry*, *280*(13), 12721–12731. doi:10.1074/jbc.M413569200 PMID:15677449

Hoops, S., Sahle, S., Gauges, R., Lee, C., Pahle, J., Simus, N., & Kummer, U. et al. (2006). COPASI—a COmplex PAthway SImulator. *Bioinformatics (Oxford, England)*, *22*(24), 3067–3074. doi:10.1093/bioinformatics/btl485 PMID:17032683

Hoskins, R. A., Landolin, J. M., Brown, J. B., Sandler, J. E., Takahashi, H., Lassmann, T., & Celniker, S. E. et al. (2011). Genome-wide analysis of promoter architecture in Drosophila melanogaster. *Genome Research*, *21*(2), 182–192. doi:10.1101/gr.112466.110 PMID:21177961

Hou, Y., Fan, W., Yan, L., Li, R., Lian, Y., Huang, J., & Qiao, J. et al. (2013). Genome analyses of single human oocytes. *Cell*, *155*(7), 1492–1506. doi:10.1016/j.cell.2013.11.040 PMID:24360273

Hovatta, I., Kimppa, K., Lehmussola, A., Pasanen, T., Saarela, J., Saarikko, I., … Laine, M. (2005). DNA microarray data analysis (2nd ed.). CSC - Scientific Computing Ltd.

Hsu, C. H., Shen, Y. C., Lin, Z. Z., Chen, P. J., Shao, Y. Y., Ding, Y. H., & Cheng, A. L. (2010). Phase II study of combining sorafenib with metronomic tegafur/uracil for advanced hepatocellular carcinoma. *Journal of Hepatology*, *53*(1), 126–131. doi:10.1016/j.jhep.2010.01.035 PMID:20416968

Huang, J., Brieba, L. G., & Sousa, R. (2000). Misincorporation by wild-type and mutant T7 RNA polymerases: Identification of interactions that reduce misincorporation rates by stabilizing the catalytically incompetent open conformation. *Biochemistry*, *39*(38), 11571–11580. doi:10.1021/bi000579d PMID:10995224

Huang, S. (1999). Gene expression profiling, genetic networks and cellular states: An integrating concept for tumorigenesis and drug discovery. *Journal of Molecular Medicine*, *77*(6), 469–480. doi:10.1007/s001099900023 PMID:10475062

Huang, S. (1999). Gene expression profiling, genetic networks and cellular states: An integrating concept for tumorigenesis and drug discovery. *Molecular Medicine (Cambridge, Mass.)*, *77*(6), 469–480. PMID:10475062

Huang, S. (2001). Genomics, complexity and drug discovery: Insights from Boolean network models of cellular regulation. *Pharmacomgenomics*, *2*(3), 203–222. doi:10.1517/14622416.2.3.203 PMID:11535110

Huang, Y., Tienda-Luna, I., & Wang, Y. (2009). Reverse engineering gene regulatory networks. *IEEE Signal Processing Magazine*, *26*(1), 76–97. doi:10.1109/MSP.2008.930647 PMID:20046885

Huber, B. R., & Bulyk, M. L. (2006). Meta-analysis discovery of tissue-specific DNA sequence motifs from mammalian gene expression data. *BMC Bioinformatics*, *7*(1), 229. doi:10.1186/1471-2105-7-229 PMID:16643658

Hu, H., & Dai, Y. (2014). A Model-based approach to transcription regulatory network reconstruction from time-course gene expression data.*Conference Proceedings of the Annual International Conference of the IEEE Engineering in Medicine and Biology Society*. doi:10.1109/EMBC.2014.6944690

Human Genome Sequencing Consortium, I.International Human Genome Sequencing Consortium. (2004). Finishing the euchromatic sequence of the human genome. *Nature*, *431*(7011), 931–945. doi:10.1038/nature03001 PMID:15496913

Hunter, J. J. (1986). Stationary distributions of perturbed Markov-chains. *Linear Algebra and Its Applications*, *82*, 201–214. doi:10.1016/0024-3795(86)90152-7

Hunter, J. J. (2005). Stationary distributions and mean first passage times of perturbed Markov chains. *Linear Algebra and Its Applications*, *410*, 217–243. doi:10.1016/j.laa.2005.08.005

Hu, Y. J. (2003). Finding subtle motifs with variable gaps in unaligned DNA sequences. *Computer Methods and Programs in Biomedicine*, *70*(1), 11–20. doi:10.1016/S0169-2607(01)00198-5 PMID:12468123

Ideker, T. E., Thorsson, V., & Karp, R. M. (2000). Discovery of regulatory interactions through perturbation: inference and experimental design.*Pacific Symposium on Biocomputing*, (vol. 5, pp. 302-313).

Iglesias, P. A., & Ingalls, B. P. (2010). *Control Theory and Systems Biology*. MIT Press.

Ioannou, P. A., & Sun, J. (1996). *Robust Adaptive Control*. Upper Saddle River, NJ: PTR Prentice-Hall.

Irie, T., Park, S. J., Yamashita, R., Seki, M., Yada, T., Sugano, S., & Suzuki, Y. et al. (2011). Predicting promoter activities of primary human DNA sequences. *Nucleic Acids Research*, *39*(11), e75. doi:10.1093/nar/gkr173 PMID:21486745

Ivanov, I., & Dougherty, E. R. (2004). Reduction mappings between probabilistic Boolean networks. *EURASIP J. Acoust. Signal Process.*, *1*(1), 125–131. doi:10.1155/S1110865704309182

Ivanov, I., Pal, R., & Dougherty, E. R. (2007). Dynamics preserving size reduction mappings for probabilistic Boolean networks. *IEEE Transactions on Signal Processing*, *55*(5), 2310–2322. doi:10.1109/TSP.2006.890929

Ivanov, I., Simeonov, P., Ghaffari, N., Qian, X., & Dougherty, E. R. (2010). Selection policy-induced reduction mappings for Boolean networks. *IEEE Transactions on Signal Processing*, *58*(9), 4871–4882. doi:10.1109/TSP.2010.2050314

Iwafuchi-Doi, M., & Zaret, K. S. (2014). Pioneer transcription factors in cell reprogramming. *Genes & Development*, *28*(24), 2679–2692. doi:10.1101/gad.253443.114 PMID:25512556

Iyengar, G. N. (2005). Robust dynamic programming. *Mathematics of Operations Research*, *30*(2), 257–280. doi:10.1287/moor.1040.0129

Iyer, V. R., Horak, C. E., Scafe, C. S., Botstein, D., Snyder, M., & Brown, P. O. (2001). Genomic binding sites of the yeast cell-cycle transcription factors SBF and MBF. *Nature*, *409*(6819), 533–538. doi:10.1038/35054095 PMID:11206552

Jacob, F., & Monod, J. (1961). Genetic regulatory mechanisms in the synthesis of proteins+. *Journal of Molecular Biology*, *3*(3), 318–356. doi:10.1016/S0022-2836(61)80072-7 PMID:13718526

Jain, A. K., Duin, R. P. W., & Mao, J. (2000). Statistical Pattern Recognition: A Review. *IEEE Transactions on Pattern Analysis and Machine Intelligence*, *22*(1), 4–37. doi:10.1109/34.824819

Jensen, L. J., & Knudsen, S. (2000). Automatic discovery of regulatory patterns in promoter regions based on whole cell expression data and functional annotation. *Bioinformatics (Oxford, England)*, *16*(4), 326–333. doi:10.1093/bioinformatics/16.4.326 PMID:10869030

Jensen, S. T., Liu, X. S., Zhou, Q., & Liu, J. S. (2004). Computational discovery of gene regulatory binding motifs: A Bayesian perspective. *Statistical Science*, 188–204.

Jeong, H., Mason, S. P., Barabási, A. L., & Oltvai, Z. N. (2001). Lethality and centrality in protein networks. *Nature*, *411*(6833), 41–42. doi:10.1038/35075138 PMID:11333967

Jeong, H., Tombor, B., Albert, R., Oltvai, Z. N., & Barabási, A.-L. (2000). The large-scale organization of metabolic networks. *Nature*, *407*(6804), 651–654. doi:10.1038/35036627 PMID:11034217

Jimenez, R. D., Martins, D. C. Jr, & Santos, C. S. (2014). Gene networks inference through one genetic algorithm per gene. In *14th IEEE International Conference on Bioinformatics and Bioengineering Conference (BIBE)*. Boca Raton, FL: IEEE. doi:10.1109/BIBE.2014.9

Johnson, A. D., Poteete, A. R., Lauer, G., Sauer, R. T., Ackers, G. K., & Ptashne, M. (1981). lambda Repressor and cro--components of an efficient molecular switch. *Nature*, *294*(5838), 217–223. doi:10.1038/294217a0 PMID:6457992

Johnson, D. S., Mortazavi, A., Myers, R. M., & Wold, B. (2007). Genome-wide mapping of in vivo protein-DNA interactions. *Science*, *316*(5830), 1497–1502. doi:10.1126/science.1141319 PMID:17540862

Jonassen, I. (1997). Efficient discovery of conserved patterns using a pattern graph. *Computer Applications in the Biosciences*, *13*(5), 509–522. PMID:9367124

Julius, A. A., Halasz, A., Sakar, M. S., Rubin, H., Kumar, V., & Pappas, G. J. (2008). Stochastic Modeling and Control of Biological Systems: The Lactose Regulation System of Escherichia Coli. *IEEE Transactions on Automatic Control*, *53*(Special Issue), 51–65. doi:10.1109/TAC.2007.911346

Jullien, J., Miyamoto, K., Pasque, V., Allen, G. E., Bradshaw, C. R., Garrett, N. J., & Gurdon, J. B. et al. (2014). Hierarchical molecular events driven by oocyte-specific factors lead to rapid and extensive reprogramming. *Molecular Cell*, *55*(4), 524–536. doi:10.1016/j.molcel.2014.06.024 PMID:25066233

Jullien, J., Pasque, V., Halley-Stott, R. P., Miyamoto, K., & Gurdon, J. B. (2011). Mechanisms of nuclear reprogramming by eggs and oocytes: A deterministic process? *Nature Reviews. Molecular Cell Biology*, *12*(7), 453–459. doi:10.1038/nrm3140 PMID:21697902

Kachalo, S., Lu, H., & Liang, J. (2006). Protein folding dynamics via quantification of kinematic energy landscape. *Physical Review Letters*, *96*(5), 058106. doi:10.1103/PhysRevLett.96.058106 PMID:16487000

Kærn, M., Elston, T. C., Blake, W. J., & Collins, J. J. (2005). Stochasticity in gene expression: From theories to phenotypes. *Nature Reviews. Genetics*, *6*(6), 451–464. doi:10.1038/nrg1615 PMID:15883588

Kampen, N. G. (1991). *Stochastic processes in physics and chemistry*. North Holland.

Kandoth, C., McLellan, M. D., Vandin, F., Ye, K., Niu, B., Lu, C., & Ding, L. et al. (2013). Mutational landscape and significance across 12 major cancer types. *Nature*, *502*(7471), 333–339. doi:10.1038/nature12634 PMID:24132290

Kanehisa, M., Goto, S., Furumichi, M., Tanabe, M., & Hirakawa, M. (2010). KEGG for representation and analysis of molecular networks involving diseases and drugs. *Nucleic Acids Research*, *38*(suppl 1), D355–D360. doi:10.1093/nar/gkp896 PMID:19880382

Karlebach, G., & Shamir, R. (2008). Modelling and analysis of gene regulatory networks. *Nature Reviews. Molecular Cell Biology*, *9*(10), 770–780. doi:10.1038/nrm2503 PMID:18797474

Karr, J. R., Sanghvi, J. C., Macklin, D. N., Gutschow, M. V., Jacobs, J. M. Jr, Bolival, B. Jr, & Covert, M. W. et al. (2012). A Whole-Cell Computational Model Predicts Phenotype from Genotype. *Cell*, *150*(2), 389–401. doi:10.1016/j.cell.2012.05.044 PMID:22817898

Kauffman, S. (1993). *The origins of order: Self-organization and selection in evolution*. Oxford University Press.

Kauffman, S. (1993). *The Origins of Order: Self-organization and Selection in Evolution*. Oxford University Press.

Kauffman, S. A. (1969). Homeostasis and differentiation in random genetic control networks. *Nature*, *224*(215), 177–178. doi:10.1038/224177a0 PMID:5343519

Kauffman, S. A. (1969). Metabolic stability and epigenesis in randomly constructed genetic nets. *Journal of Theoretical Biology*, *22*(3), 437–467. doi:10.1016/0022-5193(69)90015-0 PMID:5803332

Kauffman, S. A. (1993). *The Origins of Order*. Oxford University Press.

Kazeev, V., Khammash, M., Nip, M., & Schwab, C. (2014). Direct solution of the Chemical Master Equation using quantized tensor trains. *PLoS Computational Biology*, *10*(3), e1003359. doi:10.1371/journal.pcbi.1003359 PMID:24626049

Kechris, K. J., van Zwet, E., Bickel, P. J., & Eisen, M. B. (2004). Detecting DNA regulatory motifs by incorporating positional trends in information content. *Genome Biology*, *5*(7), R50. doi:10.1186/gb-2004-5-7-r50 PMID:15239835

Keizer, J. (1977). On the macroscopic equivalence of descriptions of fluctuations for chemical reactions. *Journal of Mathematical Physics*, *18*(7), 1316–1321. doi:10.1063/1.523422

Kel, A. E., Gossling, E., Reuter, I., Cheremushkin, E., Kel-Margoulis, O. V., & Wingender, E. (2003). MATCH: A tool for searching transcription factor binding sites in DNA sequences. *Nucleic Acids Research*, *31*(13), 3576–3579. doi:10.1093/nar/gkg585 PMID:12824369

Kelemen, A., Abraham, A., & Chen, Y. (2008). *Computational Intelligence in Bioinformatics*. Springer. doi:10.1007/978-3-540-76803-6

Keles, S., van der Laan, M., & Eisen, M. B. (2002). Identification of regulatory elements using a feature selection method. *Bioinformatics (Oxford, England)*, *18*(9), 1167–1175. doi:10.1093/bioinformatics/18.9.1167 PMID:12217908

Keméy, J. G., & Snell, J. L. (1983). *Finite Markov Chains*. New York: Springer-Verlag.

Kesanakurti, D., Chetty, C., Rajasekhar Maddirela, D., Gujrati, M., & Rao, J. S. (2013). Essential Role of Cooperative NF-κB and Stat3 Recruitment to ICAM-1 Intronic Consensus Elements in the Regulation of Radiation-induced Invasion and Migration in Glioma. *Oncogene*, *32*(43), 5144–5155. doi:10.1038/onc.2012.546 PMID:23178493

Kholodenko, B. N. (2006). Cell-signalling dynamics in time and space. *Nature Reviews. Molecular Cell Biology*, *7*(3), 165–176. doi:10.1038/nrm1838 PMID:16482094

Kierzek, A. M. (2002). STOCKS: STOChastic Kinetic Simulations of biochemical systems with Gillespie algorithm. *Bioinformatics (Oxford, England)*, *18*(3), 470–481. doi:10.1093/bioinformatics/18.3.470 PMID:11934747

Kim, K. Y., & Wang, J. (2007). Potential energy landscape and robustness of a gene regulatory network: Toggle switch. *PLoS Computational Biology*, *3*(3), e60. doi:10.1371/journal.pcbi.0030060 PMID:17397255

Kim, S. J., Onwuta, U. S., Lee, Y. I., Li, R., Botchan, M. R., & Robbins, P. D. (1992). The retinoblastoma gene product regulates Sp1-mediated transcription. *Molecular and Cellular Biology*, *12*(6), 2455–2463. doi:10.1128/MCB.12.6.2455 PMID:1588949

Kim, S. Y., Imoto, S., & Miyano, S. (2003). Inferring gene networks from time series microarray data using dynamic Bayesian networks. *Briefings in Bioinformatics*, *4*(3), 228–235. doi:10.1093/bib/4.3.228 PMID:14582517

Kim, S., Li, H., Dougherty, E. R., Cao, N., Chen, Y., Bittner, M., & Suh, E. B. (2002). Can Markov chain models mimic biological regulation? *Journal of Biological System*, *10*(04), 337–357. doi:10.1142/S0218339002000676

Klamt, S., Saez Rodriguez, J., & Gilles, E. (2007). Structural and functional analysis of cellular networks with cellnetanalyzer. *BMC Systems Biology*, *1*(1), 2.

Knabe, J. F., Wegner, K., Nehaniv, C. L., & Schilstra, M. J. (2010). Genetic algorithms and their application to in silico evolution of genetic regulatory networks. *Methods in Molecular Biology (Clifton, N.J.)*, *673*, 297–321. doi:10.1007/978-1-60761-842-3_19 PMID:20835807

Kobayashi, H., Azuma, R., & Konagaya, A. (2007a). Kinetic Analysis of Ligand-Receptor Complex Formation with the Aid of Computer Simulation. In *AINAW, 2007, Advanced Information Networking and Applications Workshops, International Conference on, Advanced Information Networking and Applications Workshops, International Conference on 2007*, (vol. 1, pp. 727-732). The Institute of Electrical and Electronics Engineers.

Kobayashi, H. (2003). Computer Simulation of Cytoplasmic pH Regulation Mediated by the F-type H^+-ATPase. *Biochimica et Biophysica Acta*, *1607*(2-3), 211–216. doi:10.1016/j.bbabio.2003.10.001 PMID:14670611

Kobayashi, H., Azuma, R., & Konagaya, A. (2007b). Clustering of membrane proteins in the pre-stimulation stage is required for signal transduction: A computer analysis. *Signal Transduction*, *7*(4), 329–339. doi:10.1002/sita.200600126

Kobayashi, H., Azuma, R., & Yasunaga, T. (2009). Expression of excess receptors and negative feedback control of signal pathways are required for rapid activation and prompt cessation of signal transduction. *Cell Communication and Signaling*, *7*(1), 3. doi:10.1186/1478-811X-7-3 PMID:19254388

Kobayashi, H., Sakurai, T., Imai, M., Takahashi, N., Fukuda, A., Yayoi, O., & Kono, T. et al. (2012). Contribution of intragenic DNA methylation in mouse gametic DNA methylomes to establish oocyte-specific heritable marks. *PLOS Genetics*, *8*(1), e1002440. doi:10.1371/journal.pgen.1002440 PMID:22242016

Kohanoff, J. (2006). *Electronic structure calculations for solids and molecules: theory and computational methods*. Cambridge Univ Pr. doi:10.1017/CBO9780511755613

Kohn, W., & Sham, L. et al. (1965). Self-consistent equations including exchange and correlation effects. *Physical Review*, *140*(4A), A1133–A1138. doi:10.1103/PhysRev.140.A1133

Ko, M. S., Kitchen, J. R., Wang, X., Threat, T. A., Hasegawa, A., Sun, T., & Doi, H. et al. (2000). Large-scale cDNA analysis reveals phased gene expression patterns during preimplantation mouse development. *Development*, *127*(8), 1737–1749. Retrieved from http://www.ncbi.nlm.nih.gov/entrez/query.fcgi?cmd=Retrieve&db=PubMed&dopt=Citation&list_uids=10725249 PMID:10725249

Krawetz, S. A. (2005). Paternal contribution: New insights and future challenges. *Nature Reviews. Genetics*, *6*(8), 633–642. doi:10.1038/nrg1654 PMID:16136654

Krishnapuram, B., Carin, L., Figueiredo, M. A., & Hartemink, A. J. (2005). Sparse multinomial logistic regression: Fast algorithms and generalization bounds. *IEEE Transactions on Pattern Analysis and Machine Intelligence*, *27*(6), 957–968. doi:10.1109/TPAMI.2005.127 PMID:15943426

Kuchaiev, O., Milenkovic, T., Memisevic, V., Hayes, W., & Przulj, N. (2010). Topological network alignment uncovers biological function and phylogeny. *Journal of the Royal Society, Interface*, *7*(50), 1341–1354. doi:10.1098/rsif.2010.0063 PMID:20236959

Kumar, P. R., & Varaiya, P. (1986). *Stochastic Systems: Estimation, Identification and Adaptive Control*. Prentice-Hall.

Kumar, P. R., & Woei, L. (1982). Optimal adaptive controllers for unknown Markov chains. *IEEE Transactions on Automatic Control*, *27*(4), 765–774. doi:10.1109/TAC.1982.1103017

Kurtz, T. G. (1972). The Relationship between Stochastic and Deterministic Models for Chemical Reactions. *The Journal of Chemical Physics*, *57*(7), 2976–2978. doi:10.1063/1.1678692

Kuwahara, H., & Mura, I. (2008). An efficient and exact stochastic simulation method to analyze rare events in biochemical systems. *The Journal of Chemical Physics*, *129*(16), 165101. doi:10.1063/1.2987701 PMID:19045316

Kyoda, K., Baba, K., Onami, S., & Kitano, H. (2004). DBRF–MEGN method: An algorithm for deducing minimum equivalent gene networks from large-scale gene expression profiles of gene deletion mutants. *Bioinformatics (Oxford, England)*, *20*(16), 2662–2675. doi:10.1093/bioinformatics/bth306 PMID:15166016

Lacroix, V., Cottret, L., Thébault, P., & Sagot, M.-F. (2008). An introduction to metabolic networks and their structural analysis. *IEEE/ACM Transactions on Computational Biology and Bioinformatics*, *5*(4), 594–617. doi:10.1109/TCBB.2008.79 PMID:18989046

Lähdesmäki, H., Hautaniemi, S., Shmulevich, I., & Yli-Harjaa, O. (2006). Relationships between probabilistic Boolean networks and dynamic Bayesian networks as models of gene regulatory networks. *Signal Processing*, *86*(4), 814–834. doi:10.1016/j.sigpro.2005.06.008 PMID:17415411

Lamprecht, M. R., Sabatini, D. M., & Carpenter, A. E. (2007). CellProfiler™: Free, versatile software for automated biological image analysis. *BioTechniques*, *42*(1), 71–75. doi:10.2144/000112257 PMID:17269487

Lander, E. S., Linton, L. M., Birren, B., Nusbaum, C., Zody, M. C., Baldwin, J., & Morgan, M. J. et al. (2001). Initial sequencing and analysis of the human genome. *Nature*, *409*(6822), 860–921. doi:10.1038/35057062 PMID:11237011

Landt, S. G., Marinov, G. K., Kundaje, A., Kheradpour, P., Pauli, F., Batzoglou, S., & Snyder, M. et al. (2012). ChIP-seq guidelines and practices of the ENCODE and modENCODE consortia. *Genome Research*, *22*(9), 1813–1831. doi:10.1101/gr.136184.111 PMID:22955991

Larrañaga, P., Poza, M., Yurramendi, Y., Murga, R. H., & Kuijpers, C. M. H. (1996). Structure learning of bayesian networks by genetic algorithms: A performance analysis of control parameters. *IEEE Transactions on Pattern Analysis and Machine Intelligence*, *18*(9), 912–926. doi:10.1109/34.537345

Latchman, D. S. (1997). Transcription factors: An overview. *International Journal of Biochemistry & Cell Biology*, *29*(12), 1305-1312. doi: 10.1016/s1357-2725(97)00085-x

Laubenbacher, R., & Stigler, B. (2004). A computational algebra approach to the reverse engineering of gene regulatory networks. *Journal of Theoretical Biology*, *229*(4), 523–537. doi:10.1016/j.jtbi.2004.04.037 PMID:15246788

Lau, K., Ganguli, S., & Tang, C. (2007). Function constrains network architecture and dynamics: A case study on the yeast cell cycle Boolean network. *Physical Review E: Statistical, Nonlinear, and Soft Matter Physics*, *75*(5), 051907. doi:10.1103/PhysRevE.75.051907 PMID:17677098

Lauritzen, S. (1996). *Graphical Models (Oxford Statistical Science Series)*. Oxford University Press.

Layek, R., Datta, A., & Dougherty, E. (2011). From biological pathways to regulatory networks. *Molecular BioSystems*, *7*(3), 843–851. doi:10.1039/C0MB00263A PMID:21161088

Layek, R., Datta, A., Pal, R., & Dougherty, E. R. (2009). Adaptive intervention in probabilistic Boolean networks. *Bioinformatics (Oxford, England)*, *25*(16), 2042–2048. doi:10.1093/bioinformatics/btp349 PMID:19505946

Lee, D., Yu, M., Lee, E., Kim, H., Yang, Y., Kim, K., & Threadgill, D. et al. (2009). Tumor-specific apoptosis caused by deletion of the ERBB3 pseudo-kinase in mouse intestinal epithelium. *The Journal of Clinical Investigation*, *119*(9), 2702–2713. doi:10.1172/JCI36435 PMID:19690388

Lee, M. T., Bonneau, A. R., & Giraldez, A. J. (2014). Zygotic Genome Activation During the Maternal-to-Zygotic Transition. *Annual Review of Cell and Developmental Biology*, *30*(1), 581–613. doi:10.1146/annurev-cellbio-100913-013027 PMID:25150012

Lee, M. T., Bonneau, A. R., Takacs, C. M., Bazzini, A. A., DiVito, K. R., Fleming, E. S., & Giraldez, A. J. (2013). Nanog, Pou5f1 and SoxB1 activate zygotic gene expression during the maternal-to-zygotic transition. *Nature*, *503*(7476), 360–364. doi:10.1038/nature12632 PMID:24056933

Lee, M., Choi, Y., Kim, K., Jin, H., Lim, J., Nguyen, T. A., & Kim, V. N. et al. (2014). Adenylation of maternally inherited microRNAs by Wispy. *Molecular Cell*, *56*(5), 696–707. doi:10.1016/j.molcel.2014.10.011 PMID:25454948

Lee, S., Lin, X., Nam, N. H., Parang, K., & Sun, G. (2003). Determination of the substrate-docking site of protein tyrosine kinase C-terminal Src kinase. *Proceedings of the National Academy of Sciences of the United States of America*, *100*(25), 14707–14712. doi:10.1073/pnas.2534493100 PMID:14657361

Lee, T. I., Rinaldi, N. J., Robert, F., Odom, D. T., Bar-Joseph, Z., Gerber, G. K., & Young, R. A. et al. (2002). Transcriptional regulatory networks in Saccharomyces cerevisiae. *Science*, *298*(5594), 799–804. doi:10.1126/science.1075090 PMID:12399584

Lee, W. P., & Tzou, W. S. (2009). Computational methods for discovering gene networks from expression data. *Briefings in Bioinformatics*, *10*(4), 408–423. doi:10.1093/bib/bbp028 PMID:19505889

Lee, W.-P., & Yang, K.-C. (2008). A clustering-based approach for inferring recurrent neural networks as gene regulatory networks. *Neurocomputing*, *71*(4), 600–610. doi:10.1016/j.neucom.2007.07.023

Lehoucq, R., Sorensen, D., & Yang, C. (1998). *Arpack users' guide: Solution of large scale eigenvalue problems with implicitly restarted Arnoldi methods*. Philadelphia: SIAM. doi:10.1137/1.9780898719628

Lein, E. S., Hawrylycz, M. J., Ao, N., Ayrės, M., Bensinger, A., Bernard, A., & Jones, A. R. et al. (2007). Genome-wide atlas of gene expression in the adult mouse brain. *Nature*, *445*(7124), 168–176. doi:10.1038/nature05453 PMID:17151600

Lenas, P., Moos, M. Jr, & Luyten, F. P. (2009). Developmental engineering: A new paradigm for the design and manufacturing of cell-based products. part ii. from genes to networks: Tissue engineering from the viewpoint of systems biology and network science. *Tissue Engineering. Part B, Reviews*, *15*(4), 395–422. doi:10.1089/ten.teb.2009.0461 PMID:19589040

Lesch, B. J., & Page, D. C. (2012). Genetics of germ cell development. *Nature Reviews. Genetics*, *13*(11), 781–794. doi:10.1038/nrg3294 PMID:23044825

Levine, M., & Tjian, R. (2003). Transcription regulation and animal diversity. *Nature*, *424*(6945), 147–151. doi:10.1038/nature01763 PMID:12853946

Levin, M. D. (2003). Noise in gene expression as the source of non-genetic individuality in the chemotactic response of Escherichia coli. *FEBS Letters*, *550*(1-3), 135–138. doi:10.1016/S0014-5793(03)00857-3 PMID:12935899

Liang, H. L., Nien, C. Y., Liu, H. Y., Metzstein, M. M., Kirov, N., & Rushlow, C. (2008). The zinc-finger protein Zelda is a key activator of the early zygotic genome in Drosophila. *Nature*, *456*(7220), 400–403. doi:10.1038/nature07388 PMID:18931655

Liang, J., Zhang, J., & Chen, R. (2002). Statistical geometry of packing defects of lattice chain polymer from enumeration and sequential Monte Carlo method. *The Journal of Chemical Physics*, *117*(7), 3511–3521. doi:10.1063/1.1493772

Liang, S., Fuhrman, S., & Somogyi, R. (1998). Reveal: a general reverse engineering algorithm for inference of genetic network architectures. In *Proceedings of Pacific Symposium on Biocomputing*, (pp. 18-29).

Licata, L., Briganti, L., Peluso, D., Perfetto, L., Iannuccelli, M., Galeota, E., & Cesareni, G. et al. (2012). MINT, the molecular interaction database: 2012 update. *Nucleic Acids Research*, *40*(D1), D857–D861. doi:10.1093/nar/gkr930 PMID:22096227

Li, F., Long, T., Lu, Y., Ouyang, Q., & Tang, C. (2004). The yeast cell-cycle network is robustly designed. *Proceedings of the National Academy of Sciences of the United States of America*, *101*(14), 4781–4786. doi:10.1073/pnas.0305937101 PMID:15037758

Li, L., Lu, X., & Dean, J. (2013). The maternal to zygotic transition in mammals. *Molecular Aspects of Medicine*, *34*(5), 919–938. doi:10.1016/j.mam.2013.01.003 PMID:23352575

Lim, S. P., & Garzino-Demo, A. (2000). The human immunodeficiency virus type 1 Tat protein up-regulates the promoter activity of the beta-chemokine monocyte chemoattractant protein 1 in the human astrocytoma cell line U-87 MG: Role of SP-1, AP-1, and NF-kappaB consensus sites. *Journal of Virology*, *74*(4), 1632–1640. doi:10.1128/JVI.74.4.1632-1640.2000 PMID:10644332

Lin, M., Chen, R., & Liu, J. S. (2009). *Lookahead Strategies for Sequential Monte Carlo*. Retrieved from Little, J. W., Shepley, D. P., & Wert, D. W. (1999). Robustness of a gene regulatory circuit. *The EMBO Journal*, *18*(15), 4299–4307.

Li, R., & Albertini, D. F. (2013). The road to maturation: Somatic cell interaction and self-organization of the mammalian oocyte. *Nature Reviews. Molecular Cell Biology*, *14*(3), 141–152. doi:10.1038/nrm3531 PMID:23429793

Liu, J. S., Chen, R., & Logvinenko, T. (2001). A theoretical framework for sequential importance sampling and resampling. *Sequential Monte Carlo Methods in Practice*, 225-246.

Liu, X., Brutlag, D. L., & Liu, J. S. (2001). BioProspector: discovering conserved DNA motifs in upstream regulatory regions of co-expressed genes.*Pacific Symposium on Biocomputing*, (pp. 127-138).

Liu, Y., Jiang, B., & Zhang, X. (2009). Gene-set analysis identifies master transcription factors in developmental courses. *Genomics*, *94*(1), 1–10. doi:10.1016/j.ygeno.2009.02.005 PMID:19272436

Ljung, L. (1987). *System identification: Theory for the user* (Vol. 198). Academic Press.

Lopes, F. M., Martins, D. C., Jr., Barrera, J., & Cesar, R. M., Jr. (2010). SFFS-MR: a floating search strategy for grns inference. In Pattern Recognition in Bioinformatics, Proceedings, (LNCS), (vol. 6282, pp. 407-418). Springer. doi:10.1007/978-3-642-16001-1_35

Lopes, F. M., Cesar, R. M. Jr, & Costa, L. F. (2011a). Gene expression complex networks: Synthesis, identification and analysis. *Journal of Computational Biology*, *18*(10), 1353–1367. doi:10.1089/cmb.2010.0118 PMID:21548810

Lopes, F. M., de Oliveira, E. A., & Cesar, R. M. Jr. (2011b). Inference of gene regulatory networks from time series by Tsallis entropy. *BMC Systems Biology*, *5*(1), 61. doi:10.1186/1752-0509-5-61 PMID:21545720

Lopes, F. M., Martins, D. C. Jr, & Barrera, J. (2008). Feature selection environment for genomic applications. *BMC Bioinformatics*, *9*(1), 451. doi:10.1186/1471-2105-9-451 PMID:18945362

Lopes, F. M., Martins, D. C. Jr, Barrera, J., & Cesar, R. M. Jr. (2014a). A feature selection technique for inference of graphs from their known topological properties: Revealing scale-free gene regulatory networks. *Information Sciences*, *272*, 1–15. doi:10.1016/j.ins.2014.02.096

Lopes, F. M., Ray, S. S., Hashimoto, R. F., & Cesar, R. M. Jr. (2014b). Entropic biological score: A cell cycle investigation for GRNs inference. *Gene*, *541*(2), 129–137. doi:10.1016/j.gene.2014.03.010 PMID:24631265

Lou, H., Yang, Y., Cote, G. J., Berget, S. M., & Gagel, R. F. (1995). An intron enhancer containing a 5' splice site sequence in the human calcitonin/calcitonin gene-related peptide gene. *Molecular and Cellular Biology*, *15*(12), 7135–7142. doi:10.1128/MCB.15.12.7135 PMID:8524281

Louis, M. & Becskei, A. (2002). Binary and graded responses in gene networks. *Science's Signal Transduction Knowledge Environment*, *2002*(143), PE33.

Lu, L. and Zhou, T. (2011). Link prediction in complex networks: a survey. *Physica A: Statistical Mechanics and its Applications*, *390*(6), 1150-1170.

Macfarlan, T. S., Gifford, W. D., Driscoll, S., Lettieri, K., Rowe, H. M., Bonanomi, D., & Pfaff, S. L. et al. (2012). Embryonic stem cell potency fluctuates with endogenous retrovirus activity. *Nature*, *487*(7405), 57–63. doi:10.1038/nature11244 PMID:22722858

Macintyre, G., Bailey, J., Gustafsson, D., Haviv, I., & Kowalczyk, A. (2010). Using gene ontology annotations in exploratory microarray clustering to understand cancer etiology. *Pattern Recognition Letters*, *31*(14), 2138–2146. doi:10.1016/j.patrec.2010.01.006

MacIsaac, K. D., Wang, T., Gordon, D. B., Gifford, D. K., Stormo, G. D., & Fraenkel, E. (2006). An improved map of conserved regulatory sites for Saccharomyces cerevisiae. *BMC Bioinformatics*, *7*(1), 113. doi:10.1186/1471-2105-7-113 PMID:16522208

Maekawa, M., Yamaguchi, K., Nakamura, T., Shibukawa, R., Kodanaka, I., Ichisaka, T., & Yamanaka, S. et al. (2011). Direct reprogramming of somatic cells is promoted by maternal transcription factor Glis1. *Nature*, *474*(7350), 225–229. doi:10.1038/nature10106 PMID:21654807

Magni, P., Germani, M., De Nicolao, G., Bianchini, G., Simeoni, M., Poggesi, I., & Rocchetti, M. (2008). A Minimal Model of Tumor Growth Inhibition. *IEEE Transactions on Bio-Medical Engineering*, *55*(12), 2683–2690. doi:10.1109/TBME.2008.913420 PMID:19126447

Magni, P., Simeoni, M., Poggesi, I., Rocchetti, M., & De Nicolao, G. (2006). A mathematical model to study the effects of drugs administration on tumor growth dynamics. *Mathematical Biosciences*, *200*(2), 127–151. doi:10.1016/j.mbs.2005.12.028 PMID:16516246

Ma, H., Morey, R., O'Neil, R. C., He, Y., Daughtry, B., Schultz, M. D., & Mitalipov, S. et al. (2014). Abnormalities in human pluripotent cells due to reprogramming mechanisms. *Nature*, *511*(7508), 177–183. doi:10.1038/nature13551 PMID:25008523

Maki, Y., Tominaga, D., Okamoto, M., Watanabe, S., & Eguchi, Y. (2001). Development of a system for the inference of large scale genetic networks.*Pacific Symposium on Biocomputing*, (pp. 446-458).

Mamakou, M. E., Sirakoulis, G. C., Andreadis, I., & Karafyllidis, I. (2005). Adaptive reverse engineering of gene regulatory networks using genetic algorithms. In *IEEE International Conference on EUROCON*.Belgrade: IEEE. doi:10.1109/EURCON.2005.1629947

Manninen, T., Linne, M. L., & Ruohonen, K. (2006). Developing Itô stochastic differential equation models for neuronal signal transduction pathways. *Computational Biology and Chemistry*, *30*(4), 280–291. doi:10.1016/j.compbiolchem.2006.04.002 PMID:16880117

Ma, Q., Chirn, G. W., Cai, R., Szustakowski, J. D., & Nirmala, N. (2005). Clustering protein sequences with a novel metric transformed from sequence similarity scores and sequence alignments with neural networks. *BMC Bioinformatics*, 6. PMID:16202129

Marbach, D., Costello, J. C., Küffner, R., Vega, N. M., Prill, R. J., Camacho, D. M., & Zimmer, R. et al. (2012). Wisdom of crowds for robust gene network inference. *Nature Methods*, *9*(8), 796–804. doi:10.1038/nmeth.2016 PMID:22796662

Marbach, D., Prill, R. J., Schaffter, T., Mattiussi, C., Floreano, D., & Stolovitzky, G. (2010). Revealing strengths and weaknesses of methods for gene network inference. *Proceedings of the National Academy of Sciences of the United States of America*, *107*(14), 6286–6291. doi:10.1073/pnas.0913357107 PMID:20308593

Marbach, D., Schaffter, T., Mattiussi, C., & Floreano, D. (2009). Generating realistic in silico gene networks for performance assessment of reverse engineering methods. *Journal of Computational Biology*, *16*(2), 229–239. doi:10.1089/cmb.2008.09TT PMID:19183003

Margolin, A. A., Nemenman, I., Basso, K., Wiggins, C., Stolovitzky, G., Dalla Favera, R., & Calífano, A. (2006). Aracne: An algorithm for the reconstruction of gene regulatory networks in a mammalian cellular context. *BMC Bioinformatics*, *7*(Suppl 1).

Markowetz, F., & Spang, R. (2007). Inferring cellular networks - a review. *BMC Bioinformatics*, *8*(Suppl 6), S5. doi:10.1186/1471-2105-8-S6-S5 PMID:17903286

Marron, J. S. (2004). *Smoothing, functional data analysis, and distance weighted discrimination software*. Retrieved from http://www.unc.edu/~marron/marron_software.html

Marsman, J., & Horsfield, J. A. (2012). Long distance relationships: Enhancer-promoter communication and dynamic gene transcription. *Biochimica et Biophysica Acta (BBA) - Bioenergetics*, *1819*(11-12), 1217–1227. doi:10.1016/j.bbagrm.2012.10.008

Martin, S., Zhang, Z., Martino, A., & Faulon, J. L. (2007). Boolean dynamics of genetic regulatory networks inferred from microarray time series data. *Bioinformatics (Oxford, England)*, *23*(7), 866–874. doi:10.1093/bioinformatics/btm021 PMID:17267426

Martins, D. C. Jr, Braga Neto, U. M., Hashimoto, R. F., Dougherty, E. R., & Bittner, M. L. (2008). Intrinsically Multivariate Predictive Genes. *IEEE Journal of Selected Topics in Signal Processing*, *2*(3), 424–439. doi:10.1109/JSTSP.2008.923841

Martins, D. C. Jr, Cesar, R. M. Jr, & Barrera, J. (2006). W-operator window design by minimization of mean conditional entropy. *Pattern Analysis & Applications*, *9*(2-3), 139–153. doi:10.1007/s10044-006-0031-0

Martins, D. C. Jr, Oliveira, E. A., Braga Neto, U. M., Hashimoto, R. F., & Cesar, R. M. Jr. (2013). Signal propagation in Bayesian networks and its relationship with intrinsically multivariate predictive variables. *Information Sciences*, *225*, 18–34. doi:10.1016/j.ins.2012.10.027

Martone, M. E., Gupta, A., Wong, M., Qian, X., Sosinsky, G., Ludascher, B., & Ellisman, M. H. (2002). A cell-centered database for electron tomographic data. *Journal of Structural Biology*, *138*(1-2), 1–2, 145–155. doi:10.1016/S1047-8477(02)00006-0 PMID:12160711

Maston, G. A., Evans, S. K., & Green, M. R. (2006). Transcriptional Regulatory Elements in the Human Genome. *Annual Review of Genomics and Human Genetics*, *7*(1), 29–59. doi:10.1146/annurev.genom.7.080505.115623 PMID:16719718

Mathelier, A., Zhao, X., Zhang, A. W., Parcy, F., Worsley-Hunt, R., Arenillas, D. J., & Wasserman, W. W. et al. (2014). JASPAR 2014: An extensively expanded and updated open-access database of transcription factor binding profiles. *Nucleic Acids Research*, *42*, D142–D147. doi:10.1093/nar/gkt997 PMID:24194598

Matsumoto, M., & Nishimura, T. (1998). Mersenne twister: A 623-dimensionally equidistributed uniform pseudo-random number generator. *ACM Transactions on Modeling and Computer Simulation*, *8*(1), 3–30. doi:10.1145/272991.272995

Matsuno, H., Doi, A., Nagasaki, M., & Miyano, S. (2000). Hybrid Petri net representation of gene regulatory network.*Pacific Symposium on Biocomputing. Pacific Symposium on Biocomputing*, (pp. 341–352).

Matsuno, H., Inouye, S. T., Okitsu, Y., Fujii, Y., & Miyano, S. (2006). A new regulatory interaction suggested by simulations for circadian genetic control mechanism in mammals. *Journal of Bioinformatics and Computational Biology*, *4*(1), 139–153. doi:10.1142/S021972000600176X PMID:16568547

Matys, V., Fricke, E., Geffers, R., Gossling, E., Haubrock, M., Hehl, R., & Wingender, E. et al. (2003). TRANSFAC (R): Transcriptional regulation, from patterns to profiles. *Nucleic Acids Research*, *31*(1), 374–378. doi:10.1093/nar/gkg108 PMID:12520026

McAdams, H. H., & Arkin, A. (1997). Stochastic mechanisms in gene expression. *Proceedings of the National Academy of Sciences of the United States of America*, *94*(3), 814–819. doi:10.1073/pnas.94.3.814 PMID:9023339

McCue, L., Thompson, W., Carmack, C., Ryan, M. P., Liu, J. S., Derbyshire, V., & Lawrence, C. E. (2001). Phylogenetic footprinting of transcription factor binding sites in proteobacterial genomes. *Nucleic Acids Research*, *29*(3), 774–782. doi:10.1093/nar/29.3.774 PMID:11160901

McCullagh, E., Farlow, J., Fuller, C., Girard, J., Lipinski-Kruszka, J., Lu, D., & El-Samad, H. et al. (2009). Not all quiet on the noise front. *Nature Chemical Biology*, *5*(10), 699–704. doi:10.1038/nchembio.222 PMID:19763097

Meirovitch, H. (1988). Statistical properties of the scanning simulation method for polymer chains. *J. Chemical Phisics, 89*.

Meirovitch, H. (1982). A new method of simulation of real chains: Scanning future steps. *Journal of Physics. A, Mathematical and General*, 15.

Melykuti, B., Burrage, K., & Zygalakis, K. C. (2010). Fast stochastic simulation of biochemical reaction systems by alternative formulations of the chemical Langevin equation. *The Journal of Chemical Physics*, *132*(16), 164109. doi:10.1063/1.3380661 PMID:20441260

Mendes, P., Sha, W., & Ye, K. (2003). Artificial gene networks for objective comparison of analysis algorithms. *Bioinformatics (Oxford, England)*, *19*(Suppl 2), 122–129. doi:10.1093/bioinformatics/btg1069 PMID:14534181

Mendoza, M. R., Lopes, F. M., & Bazzan, A. L. C. (2012). Reverse engineering of grns: An evolutionary approach based on the tsallis entropy. In *Proceedings of the 14th international conference on Genetic and evolutionary computation (GECCO)*. doi:10.1145/2330163.2330190

Messerschmidt, D. M., Knowles, B. B., & Solter, D. (2014). DNA methylation dynamics during epigenetic reprogramming in the germline and preimplantation embryos. *Genes & Development*, *28*(8), 812–828. doi:10.1101/gad.234294.113 PMID:24736841

Mettetal, J. T., Muzzey, D., Pedraza, J. M., Ozbudak, E. M., & van Oudenaarden, A. (2006). Predicting stochastic gene expression dynamics in single cells. *Proceedings of the National Academy of Sciences of the United States of America*, *103*(19), 7304–7309. doi:10.1073/pnas.0509874103 PMID:16648266

Mewes, H. W., Ruepp, A., Theis, F., Rattei, T., Walter, M., Frishman, D., & Antonov, A. et al. (2011). MIPS: Curated databases and comprehensive secondary data resources in 2010. *Nucleic Acids Research*, *39*, D220–D224. doi:10.1093/nar/gkq1157 PMID:21109531

Mews, P., Zee, B. M., Liu, S., Donahue, G., Garcia, B. A., & Berger, S. L. (2014). Histone methylation has dynamics distinct from those of histone acetylation in cell cycle reentry from quiescence. *Molecular and Cellular Biology*, *34*(21), 3968–3980. doi:10.1128/MCB.00763-14 PMID:25154414

Michalewicz, Z. (1996). Genetic algorithms + data structures = evolution programs (3rd rev. ed.). Berlin: Springer-Verlag.

Miller, D., Brinkworth, M., & Iles, D. (2010). Paternal DNA packaging in spermatozoa: More than the sum of its parts? DNA, histones, protamines and epigenetics. *Reproduction (Cambridge, England)*, *139*(2), 287–301. doi:10.1530/REP-09-0281 PMID:19759174

Milo, R., Shann-Orr, S., Itzkovitz, S., Kashtan, N., Chklovskii, D., & Alon, U. (2002). Network motifs: Simple building blocks of complex networks. *Science*, *298*(5594), 824–827. doi:10.1126/science.298.5594.824 PMID:12399590

Mitchell, M. (1996). *An introduction to Genetic Algorithms*. Cambridge, MA: MIT Press.

Mizusawa, Y., Kuji, N., Tanaka, Y., Tanaka, M., Ikeda, E., Komatsu, S., & Yoshimura, Y. et al. (2010). Expression of human oocyte-specific linker histone protein and its incorporation into sperm chromatin during fertilization. *Fertility and Sterility*, *93*(4), 1134–1141. doi:10.1016/j.fertnstert.2008.11.028 PMID:19147139

Montoya Cubas, C. F., Martins, D. C. Jr, Santos, C. S., & Barrera, J. (2014). Gene networks inference through linear grouping of variables. In *14th IEEE International Conference on Bioinformatics and Bioengineering Conference (BIBE)*. Boca Raton, FL: IEEE. doi:10.1109/BIBE.2014.10

Morita, M., Nakamura, M., Hamada, M., & Takahashi, S. (2011). Combinatorial motif analysis of regulatory gene expression in Mafb deficient macrophages. *BMC Systems Biology*, *5*(Suppl 2), S7. doi:10.1186/1752-0509-5-S2-S7 PMID:22784578

Morrison, D. K. (2012). MAP kinase pathways. *Cold Spring Harbor Perspectives in Biology*, *4*(11), a011254. doi:10.1101/cshperspect.a011254 PMID:23125017

Munsky, B., & Khammash, M. (2006). The finite state projection algorithm for the solution of the chemical master equation. *The Journal of Chemical Physics*, *124*(4), 044104. doi:10.1063/1.2145882 PMID:16460146

Munsky, B., & Khammash, M. (2007). A multiple time interval finite state projection algorithm for the solution to the chemical master equation. *Journal of Computational Physics*, *226*(1), 818–835. doi:10.1016/j.jcp.2007.05.016

Murphy, K. (2002). *Dynamic Bayesian Networks: Representation, Inference and Learning*. Berkeley, CA: University of California.

Nair, N. U., Lin, Y., Manasovska, A., Antic, J., Grnarova, P., Sahu, A. D., & Moret, B. M. E. et al. (2014). Study of cell differentiation by phylogenetic analysis using histone modification data. *BMC Bioinformatics*, *15*(1), 269. doi:10.1186/1471-2105-15-269 PMID:25104072

Nair, P., Muthukkumar, S., Sells, S. F., Han, S. S., Sukhatme, V. P., & Rangnekar, V. M. (1997). Early growth response-1-dependent apoptosis is mediated by p53. *The Journal of Biological Chemistry*, *272*(32), 20131–20138. doi:10.1074/jbc.272.32.20131 PMID:9242687

Narasimhan, S., Rengaswamy, R., & Vadigepalli, R. (2009). Structural properties of gene regulatory networks: Definitions and connections. *IEEE/ACM Transactions on Computational Biology and Bioinformatics*, *6*(1), 158–170. doi:10.1109/TCBB.2007.70231 PMID:19179709

Natarajan, A., Yardimci, G. G., Sheffield, N. C., Crawford, G. E., & Ohler, U. (2012). Predicting cell-type-specific gene expression from regions of open chromatin. *Genome Research*, *22*(9), 1711–1722. doi:10.1101/gr.135129.111 PMID:22955983

Necsulea, A., Soumillon, M., Warnefors, M., Liechti, A., Daish, T., Zeller, U., & Kaessmann, H. et al. (2014). The evolution of lncRNA repertoires and expression patterns in tetrapods. *Nature*, *505*(7485), 635–640. doi:10.1038/nature12943 PMID:24463510

Network, T. C. G. A. (2008). Comprehensive genomic characterization defines human glioblastoma genes and core pathways. *Nature*, *455*, 1061–1068. doi:10.1038/nature07385 PMID:18772890

Netzer, N., Goodenbour, J. M., David, A., Kimberly, A., Dittmar, K. A., & Jones, R. B. (2009). Innate Immune and Chemically Triggered Oxidative Stress Modifies Translational Fidelity. *Nature*, *462*(7272), 522–526. doi:10.1038/nature08576 PMID:19940929

Newman, M. E. J. (2003). The structure and function of complex networks. *SIAM Review*, *45*(2), 167–256. doi:10.1137/S003614450342480

Ng, M., Zhang, S.-Q., Ching, W.-K., & Akutsu, T. (2006). A Control Model for Markovian Genetic Regulatory Networks. *Transactions on Computational Systems Biology V*, *4070*, 36–48. doi:10.1007/11790105_4

Nishida, K., Frith, M. C., & Nakai, K. (2009). Pseudocounts for transcription factor binding sites. *Nucleic Acids Research*, *37*(3), 939–944. doi:10.1093/nar/gkn1019 PMID:19106141

Nolan, G. P., Ghosh, S., Liou, H. C., Tempst, P., & Baltimore, D. (1991). DNA binding and I kappa B inhibition of the cloned p65 subunit of NF-kappa B, a rel-related polypeptide. *Cell*, *64*(5), 961–969. doi:10.1016/0092-8674(91)90320-X PMID:2001591

Nothias, J. Y., Miranda, M., & DePamphilis, M. L. (1996). Uncoupling of transcription and translation during zygotic gene activation in the mouse. *The EMBO Journal*, *15*(20), 5715–5725. Retrieved from http://www.ncbi.nlm.nih.gov/pubmed/8896464 PMID:8896464

Nowell, P. C., & Hungerford, D. A. (1960). A minute chromosome in human chronic granulocytic leukemia. *Science*, *132*, 1497.

Nowicki, M., Pawlowski, P., Fischer, T., Hess, G., Pawlowski, T., & Skorski, T. (2003). Chronic myelogenous leukemia molecular signature. *Oncogene*, *22*(25), 3952–3963. doi:10.1038/sj.onc.1206620 PMID:12813469

O'Connor, T. R., & Bailey, T. L. (2014). Creating and validating cis-regulatory maps of tissue-specific gene expression regulation. *Nucleic Acids Research*, *42*(17), 11000–11010. doi:10.1093/nar/gku801 PMID:25200088

O'Flanagan, R. A., Paillard, G., Lavery, R., & Sengupta, A. M. (2005). Non-additivity in protein-DNA binding. *Bioinformatics (Oxford, England)*, *21*(10), 2254–2263. doi:10.1093/bioinformatics/bti361 PMID:15746285

Ogawa, M. (1989). Hemopoietic stem cells: Stochastic differentiation and humoral control of proliferation. *Environmental Health Perspectives*, *80*, 199–207. doi:10.1289/ehp.8980199 PMID:2647480

Ohlsson, H., Karlsson, O., & Edlund, T. (1988). A beta-cell-specific protein binds to the two major regulatory sequences of the insulin gene enhancer. *Proceedings of the National Academy of Sciences of the United States of America*, *85*(12), 4228–4231. doi:10.1073/pnas.85.12.4228 PMID:3288987

O'Neill, L. A., Golenbock, D., & Bowie, A. G. (2013). The history of Toll-like receptors - re-defining innate immunity. *Nature Reviews. Immunology*, *13*(6), 453–460. doi:10.1038/nri3446 PMID:23681101

Orchard, S., Ammari, M., Aranda, B., Breuza, L., Briganti, L., Broackes Carter, F., & Hermjakob, H. et al. (2014). The MIntAct project - IntAct as a common curation platform for 11 molecular interaction databases. *Nucleic Acids Research*, *42*(D1), D358–D363. doi:10.1093/nar/gkt1115 PMID:24234451

Orth, J. D., Tang, Y., Shi, J., Loy, C. T., Amendt, C., Wilm, C., & Mitchison, T. J. et al. (2008). Quantitative live imaging of cancer and normal cells treated with Kinesin-5 inhibitors indicates significant differences in phenotypic responses and cell fate. *Molecular Cancer Therapeutics*, *7*(11), 3480–3489. doi:10.1158/1535-7163.MCT-08-0684 PMID:18974392

Ott, S., Hansen, A., Kim, S.-Y., & Miyano, S. (2005). Superiority of network motifs over optimal networks and an application to the revelation of gene network evolution. *Bioinformatics (Oxford, England)*, *21*(2), 227–238. doi:10.1093/bioinformatics/bth484 PMID:15377501

Ouyang, Z., Zhou, Q., & Wong, W. H. (2009). ChIP-Seq of transcription factors predicts absolute and differential gene expression in embryonic stem cells. *Proceedings of the National Academy of Sciences of the United States of America*, *106*(51), 21521–21526. doi:10.1073/pnas.0904863106 PMID:19995984

Ozbudak, E. M., Thattai, M., Kurtser, I., Grossman, A. D., & van Oudenaarden, A. (2002). Regulation of noise in the expression of a single gene. *Nature Genetics*, *31*(1), 69–73. doi:10.1038/ng869 PMID:11967532

Pal, R. (2010). Context-sensitive probabilistic Boolean networks: Steady-state properties, reduction, and steady-state approximation. *IEEE Transactions on Signal Processing*, *58*(2), 879–890. doi:10.1109/TSP.2009.2030832

Pal, R., & Berlow, N. (2012). *A kinase inhibition map approach for tumor sensitivity prediction and combination therapy design for targeted drugs. In Pac Symp Biocomput* (pp. 351–362). Kohala Coast: World Scientific.

Pal, R., & Bhattacharya, S. (2010). Characterizing the Effect of Coarse-Scale PBN Modeling on Dynamics and Intervention Performance of Genetic Regulatory Networks Represented by Stochastic Master Equation Models. *IEEE Transactions on Signal Processing*, *58*(6), 3341–3351. doi:10.1109/TSP.2010.2043970

Pal, R., & Bhattacharya, S. (2012). Transient dynamics of reduced-order models of genetic regulatory networks. *IEEE/ACM Transactions on Computational Biology and Bioinformatics*, *9*(4), 1230–1244. doi:10.1109/TCBB.2012.37 PMID:22411891

Pal, R., Bhattacharya, S., & Caglar, M. (2012). Robust approaches for genetic regulatory network modeling and intervention: A review of recent advances. *IEEE Signal Processing Magazine*, *1*(29), 66–76. doi:10.1109/MSP.2011.942818

Pal, R., Datta, A., Bittner, M. L., & Dougherty, E. R. (2005). Intervention in context-sensitive probabilistic Boolean networks. *Bioinformatics (Oxford, England)*, *21*(7), 1211–1218. doi:10.1093/bioinformatics/bti131 PMID:15531600

Pal, R., Datta, A., & Dougherty, E. R. (2006). Optimal infinite-horizon control for probabilistic Boolean networks. *IEEE Transactions on Signal Processing*, *54*(6), 2375–2387. doi:10.1109/TSP.2006.873740

Pal, R., Datta, A., & Dougherty, E. R. (2008). Robust intervention in probabilistic Boolean networks. *IEEE Transactions on Signal Processing*, *56*(3), 1280–1294. doi:10.1109/TSP.2007.908964

Pal, R., Datta, A., & Dougherty, E. R. (2009). Bayesian Robustness in the Control of Gene Regulatory Networks. *IEEE Transactions on Signal Processing*, *57*(9), 3667–3678. doi:10.1109/TSP.2009.2022872

Pal, R., Ivanov, I., Datta, A., Bittner, M. L., & Dougherty, E. R. (2005). Generating Boolean networks with a prescribed attractor structure. *Bioinformatics (Oxford, England)*, *21*(21), 4021–4025. doi:10.1093/bioinformatics/bti664 PMID:16150807

Pan, X. (2014). Single Cell Analysis: From Technology to Biology and Medicine. *Single Cell Biol*, *3*(1). doi:10.4172/2168-9431.1000106 PMID:25177539

Park, S. J., Ichinose, N., & Yada, T. (2008). Probabilistic Graphical Modeling for Large-scale Combinatorial Regulation of Transcription Factors. In *Proc. the workshop on Knowledge, Language, and Learning in Bioinformatics* (KLLBI) (pp. 72-86). Retrieved from http://www.hgc.jp/~park/research/data/kllbi2008/park_kllbi2008.pdf

Park, J., Karplus, K., Barrett, C., Hughey, R., Haussler, D., Hubbard, T., & Chothia, C. (1998). Sequence comparisons using multiple sequences detect three times as many remote homologues as pairwise methods. *Journal of Molecular Biology*, *284*(4), 1201–1210. doi:10.1006/jmbi.1998.2221 PMID:9837738

Park, S. J., Komata, M., Inoue, F., Yamada, K., Nakai, K., Ohsugi, M., & Shirahige, K. (2013). Inferring the choreography of parental genomes during fertilization from ultralarge-scale whole-transcriptome analysis. *Genes & Development*, *27*(24), 2736–2748. doi:10.1101/gad.227926.113 PMID:24352427

Park, S. J., & Nakai, K. (2011). A regression analysis of gene expression in ES cells reveals two gene classes that are significantly different in epigenetic patterns. *BMC Bioinformatics*, *12*(Suppl 1), S50. doi:10.1186/1471-2105-12-S1-S50 PMID:21342583

Park, S. J., Shirahige, K., Ohsugi, M., & Nakai, K. (2015). DBTMEE: A database of transcriptome in mouse early embryos. *Nucleic Acids Research*, *43*, D771–D776. doi:10.1093/nar/gku1001 PMID:25336621

Park, S. J., Umemoto, T., Saito-Adachi, M., Shiratsuchi, Y., Yamato, M., & Nakai, K. (2014). Computational promoter modeling identifies the modes of transcriptional regulation in hematopoietic stem cells. *PLoS ONE*, *9*(4), e93853. doi:10.1371/journal.pone.0093853 PMID:24710559

Parzen, E. (1962). On Estimation of a Probability Density Function and Mode. *Annals of Mathematical Statistics*, *33*(3), 1065–1076. doi:10.1214/aoms/1177704472

Patil, A., Kumagai, Y., Liang, K. C., Suzuki, Y., & Nakai, K. (2013). Linking transcriptional changes over time in stimulated dendritic cells to identify gene networks activated during the innate immune response. *PLoS Computational Biology*, *9*(11), e1003323. doi:10.1371/journal.pcbi.1003323 PMID:24244133

Paulsson, J., & Ehrenberg, M. (2000). Random signal fluctuations can reduce random fluctuations in regulated components of chemical regulatory networks. *Physical Review Letters*, *84*(23), 5447–5450. doi:10.1103/PhysRevLett.84.5447 PMID:10990965

Pawson, T., & Nash, P. (2003). Assembly of cell regulatory systems through protein interaction domains. *Science*, *300*(5618), 445–452. doi:10.1126/science.1083653 PMID:12702867

Pe'er, D., & Hacohen, N. (2011). Principles and Strategies for Developing Network Models in Cancer. *Cell*, *144*(6), 864–873. doi:10.1016/j.cell.2011.03.001 PMID:21414479

Pear, W. S., Miller, J. P., Xu, L., Pui, J. C., Soffer, B., Quackenbush, R. C., & Baltimore, D. et al. (1998). Efficient and rapid induction of a chronic myelogenous leukemia-like myeloproliferative disease in mice receiving P210 bcr/abl-transduced bone marrow. *Blood*, *92*(10), 3780–3792. PMID:9808572

Peat, J. R., Dean, W., Clark, S. J., Krueger, F., Smallwood, S. A., Ficz, G., & Reik, W. et al. (2014). Genome-wide Bisulfite Sequencing in Zygotes Identifies Demethylation Targets and Maps the Contribution of TET3 Oxidation. *Cell Reports*, *9*(6), 1990–2000. doi:10.1016/j.celrep.2014.11.034 PMID:25497087

Pe'er, D., Regev, A., Elidan, G., & Friedman, N. (2001). Inferring subnetworks from perturbed expression profiles. *Bioinformatics (Oxford, England)*, *17*(Suppl 1), S215–S224. doi:10.1093/bioinformatics/17.suppl_1.S215 PMID:11473012

Penfold, C. A., Wild, D. L. (2011). How to infer gene networks from expression profiles, revisited. *Interface Focus*, 1.

Peng, H., Long, F., & Ding, C. (2005). Feature selection based on mutual information: Criteria of max-dependency, max-relevance, and min-redundancy. *IEEE Transactions on Pattern Analysis and Machine Intelligence*, *27*(8), 1226–1238. doi:10.1109/TPAMI.2005.159 PMID:16119262

Perou, C. M., Sørlie, T., Eisen, M. B., van de Rijn, M., Jeffrey, S. S., Rees, C. A., & Botstein, D. et al. (2000). Molecular portraits of human breast tumours. *Nature*, *406*(6797), 747–752. doi:10.1038/35021093 PMID:10963602

Perrin, B. E., Ralaivola, L., Mazurie, A., Bottani, S., Mallet, J., & d'Alché Buc, F. (2003). Gene networks inference using dynamic Bayesian networks. *Bioinformatics (Oxford, England)*, *19*(Suppl 2), 138–148. doi:10.1093/bioinformatics/btg1071

Piechota, M., Korostynski, M., & Przewlocki, R. (2010). Identification of cis-regulatory elements in the mammalian genome: The cREMaG database. *PLoS ONE*, *5*(8), e12465. doi:10.1371/journal.pone.0012465 PMID:20824209

Piko, L., & Clegg, K. B. (1982). Quantitative changes in total RNA, total poly(A), and ribosomes in early mouse embryos. *Dev Biol*, *89*(2), 362-378. doi:0012-1606(82)90325-6

Pinney, J. W., Westhead, D. R., & McConkey, G. A. (2003). Petri Net representations in systems biology. *Biochemical Society Transactions, 31*(6), 1513–1515. doi:10.1042/bst0311513 PMID:14641101

Pournara, I., & Wernisch, L. (2004). Reconstruction of gene networks using Bayesian learning and manipulation experiments. *Bioinformatics (Oxford, England), 20*(17), 2934–2942. doi:10.1093/bioinformatics/bth337 PMID:15180938

Poustelnikova, E., Pisarev, A., Blagov, M., Samsonova, M., & Reinitz, J. (2004). A database for management of gene expression data in situ. *Bioinformatics (Oxford, England), 20*(14), 2212–2221. doi:10.1093/bioinformatics/bth222 PMID:15059825

Prasad, T. S. K., Goel, R., Kandasamy, K., Keerthikumar, S., Kumar, S., Mathivanan, S., & Pandey, A. (2009). Human Protein Reference Database - 2009 update. *Nucleic Acids Research, 37*(Database), D767–D772. doi:10.1093/nar/gkn892 PMID:18988627

Prifti, E., Zucker, J., Clement, K., & Henegar, C. (2008). Funnet: An integrative tool for exploring transcriptional interactions. *Bioinformatics (Oxford, England), 24*(22), 2636–2638. doi:10.1093/bioinformatics/btn492 PMID:18799481

Przytycka, T., & Kim, Y. A. (2010). Network integration meets network dynamics. *BMC Biology, 8*(1), 48. doi:10.1186/1741-7007-8-48 PMID:20513250

Ptashne, M. (2004). A Genetic Switch: Phage Lambda Revisited (3rd ed.). Cold Spring Harbor Laboratory Press.

Ptashne, M. (2007). On the use of the word 'epigenetic'. *Current Biology, 17*(7), R233–R236. doi:10.1016/j.cub.2007.02.030 PMID:17407749

Ptashne, M., Backman, K., Humayun, M., Jeffrey, A., Maurer, R., Meyer, B., & Sauer, R. (1976). Autoregulation and function of a repressor in bacteriophage lambda. *Science, 194*(4261), 156–161. doi:10.1126/science.959843 PMID:959843

Pudil, P., Novovicová, J., & Kittler, J. (1994). Floating search methods in feature-selection. *Pattern Recognition Letters, 15*(11), 1119–1125. doi:10.1016/0167-8655(94)90127-9

Pudimat, R., Schukat-Talamazzini, E. G., & Backofen, R. (2004). *Feature Based Representation and Detection of Transcription Factor Binding Sites.* Paper presented at the German Conference on Bioinformatics.

Qian, J., Esumi, N., Chen, Y., Wang, Q., Chowers, I., & Zack, D. J. (2005). Identification of regulatory targets of tissue-specific transcription factors: Application to retina-specific gene regulation. *Nucleic Acids Research, 33*(11), 3479–3491. doi:10.1093/nar/gki658 PMID:15967807

Qian, X. (2012). A new phenotypically constrained control policy in Boolean networks based on basins of attraction. In *Proceedings of the IEEE International Workshop on Genomic Signal Processing and Statistics (GENSIPS)*. Washington, DC: IEEE. doi:10.1109/GENSIPS.2012.6507725

Qian, X., & Dougherty, E. R. (2008). Effect of Function Perturbation on the Steady-State Distribution of Genetic Regulatory Networks: Optimal Structural Intervention. *IEEE Transactions on Signal Processing*, *56*(10), 4966–4976. doi:10.1109/TSP.2008.928089

Qian, X., & Dougherty, E. R. (2012). Intervention in Gene Regulatory Networks via Phenotypically Constrained Control Policies Based on Long-Run Behavior. *IEEE/ACM Transactions on Computational Biology and Bioinformatics*, *9*(1), 123–136. doi:10.1109/TCBB.2011.107 PMID:21788674

Qian, X., Ghaffari, N., Ivanov, I., & Dougherty, E. R. (2010). State reduction for network intervention in probabilistic Boolean networks. *Bioinformatics (Oxford, England)*, *26*(24), 3098–3104. doi:10.1093/bioinformatics/btq575 PMID:20956246

Qian, X., Ivanov, I., Ghaffari, N., & Dougherty, E. R. (2009). Intervention in gene regulatory networks via greedy control policies based on long-run behavior. *BMC Systems Biology*, *3*(1), 61. doi:10.1186/1752-0509-3-61 PMID:19527511

Qian, X., Yoon, B.-J., & Dougherty, E. R. (2012). Structural intervention of gene regulatory networks by general rank-k matrix perturbation. In *Proceedings of the 37th IEEE International Conference on Acoustics, Speech, and Signal Processing (ICASSP)*. Kyoto, Japan: IEEE. doi:10.1109/ICASSP.2012.6287987

Queinnec, I., Garcia, G., Niculescu, S.-I., & Tarbouriech, S. (2007). *Biology and control theory: current challenges* (Vol. 357). Springer. doi:10.1007/978-3-540-71988-5

Ramadan, W. M., Kashir, J., Jones, C., & Coward, K. (2012). Oocyte activation and phospholipase C zeta (PLCzeta): Diagnostic and therapeutic implications for assisted reproductive technology. *Cell Communication and Signaling*, *10*(1), 12. doi:10.1186/1478-811X-10-12 PMID:22591604

Raser, J. M., & O'Shea, E. K. (2005). Noise in gene expression: Origins, consequences, control. *Science*, *309*(5743), 2010–2013. doi:10.1126/science.1105891 PMID:16179466

Ray, S. S., Bandyopadhyay, S., & Pal, S. (2009). Combining multisource information through functional-annotation-based weighting: Gene function prediction in yeast. *Biomedical Engineering. IEEE Transactions on*, *56*(2), 229–236.

Ray, S. S., Bandyopadhyay, S., & Pal, S. (2012). A Weighted power framework for integrating multi-source information: Gene function prediction in yeast. *IEEE Transactions on Bio-Medical Engineering*, *59*(4), 1162–1168. doi:10.1109/TBME.2012.2186689 PMID:22318478

Rebeiz, M., Reeves, N. L., & Posakony, J. W. (2002). SCORE: A computational approach to the identification of cis-regulatory modules and target genes in whole-genome sequence data. Site clustering over random expectation. *Proceedings of the National Academy of Sciences of the United States of America*, *99*(15), 9888–9893. doi:10.1073/pnas.152320899 PMID:12107285

Ren, B., Robert, F., Wyrick, J. J., Aparicio, O., Jennings, E. G., Simon, I., & Young, R. A. et al. (2000). Genome-wide location and function of DNA binding proteins. *Science, 290*(5500), 2306–2309. doi:10.1126/science.290.5500.2306 PMID:11125145

Rigoutsos, I., & Floratos, A. (1998). Combinatorial pattern discovery in biological sequences: The TEIRESIAS algorithm. *Bioinformatics (Oxford, England), 14*(1), 55–67. doi:10.1093/bioinformatics/14.1.55 PMID:9520502

Ristevski, B. (2013). A survey of models for inference of gene regulatory networks. *Nonlinear Analysis: Modelling and Control, 18*(4), 444–465.

Robertson, G., Hirst, M., Bainbridge, M., Bilenky, M., Zhao, Y. J., Zeng, T., & Jones, S. et al. (2007). Genome-wide profiles of STAT1 DNA association using chromatin immunoprecipitation and massively parallel sequencing. *Nature Methods, 4*(8), 651–657. doi:10.1038/nmeth1068 PMID:17558387

Roh, M. K., Daigle, B. J. Jr, Gillespie, D. T., & Petzold, L. R. (2011). State-dependent doubly weighted stochastic simulation algorithm for automatic characterization of stochastic biochemical rare events. *The Journal of Chemical Physics, 135*(23), 234108. doi:10.1063/1.3668100 PMID:22191865

Roh, M. K., Gillespie, D. T., & Petzold, L. R. (2010). State-dependent biasing method for importance sampling in the weighted stochastic simulation algorithm. *The Journal of Chemical Physics, 133*(17), 174106. doi:10.1063/1.3493460 PMID:21054005

Rosenberg, M., Fan, A. X., Lin, I. J., Liang, S. Y., & Bungert, J. (2013). Cell-cycle specific association of transcription factors and RNA polymerase ii with the human β-globin gene locus. *Journal of Cellular Biochemistry, 114*(9), 1997–2006. doi:10.1002/jcb.24542 PMID:23519692

Ross, K. W. (1989). Randomized and past-dependent policies for Markov decision processes with multiple constraints. *Operations Research, 37*(3), 474–477. doi:10.1287/opre.37.3.474

Rowley, J. D., Golomb, H. M., & Dougherty, C. (1977). 15/17 translocation, a consistent chromosomal change in acute promyelocytic leukaemia. *Lancet, 1*(8010), 549–550. doi:10.1016/S0140-6736(77)91415-5 PMID:65649

Ruklisa, D., Brazma, A., & Viksna, J. (2005). Reconstruction of gene regulatory networks under the Finite State Linear Model. *Genome Informatics, 16*(1), 225–236. PMID:16901105

Saad, Y. (2003). *Iterative methods for sparse linear systems*. Siam. doi:10.1137/1.9780898718003

Samoilov, M., Plyasunov, S., & Arkin, A. (2005). Stochastic amplification and signaling in enzymatic futile cycles through noise-induced bistability with oscillations. *Proceedings of the National Academy of Sciences of the United States of America, 102*(7), 2310–2315. doi:10.1073/pnas.0406841102 PMID:15701703

Sánchez-Moreno, I., Coral-Vázquez, R., Méndez, J. P., & Canto, P. (2008). Full-length SRY protein is essential for DNA binding. *Molecular Human Reproduction*, *14*(6), 325–330. doi:10.1093/molehr/gan021 PMID:18453550

Sandelin, A., Alkema, W., Engstrom, P., Wasserman, W. W., & Lenhard, B. (2004). JASPAR: An open-access database for eukaryotic transcription factor binding profiles. *Nucleic Acids Research*, *32*(90001), D91–D94. doi:10.1093/nar/gkh012 PMID:14681366

Sandve, G. K., & Drablos, F. (2006). A survey of motif discovery methods in an integrated framework. *Biology Direct*, *1*(1), 11. doi:10.1186/1745-6150-1-11 PMID:16600018

Sanft, K. R., Wu, S., Roh, M., Fu, J., & Petzold, L. R. (2011). StochKit2: Software for discrete stochastic simulation of biochemical systems with events. *Bioinformatics (Oxford, England)*, *27*(17), 2457–2458. doi:10.1093/bioinformatics/btr401 PMID:21727139

Santos, D. C., Camargo, R. Y., Martins, D. C., Jr., Song, S. W., Rozante, L. C. S., & Borelli, F. F. (2015). A multi-GPU hitting set algorithm for GRNs inference. *15th IEEE/ACM International Symposium on Cluster, Cloud and Grid Computing (CCGrid)*. Shenzhen, China: IEEE/ACM.

Santra, T. (2014). A bayesian framework that integrates heterogeneous data for inferring gene regulatory networks. *Front Bioeng Biotechnol*, *2*, 13. doi:10.3389/fbioe.2014.00013 PMID:25152886

Sarrate, Z., Vidal, F., & Blanco, J. (2014). Meiotic abnormalities in metaphase I human spermatocytes from infertile males: Frequencies, chromosomes involved, and the relationships with polymorphic karyotype and seminal parameters. *Asian Journal of Andrology*, *16*(6), 838–844. doi:10.4103/1008-682X.135126 PMID:25080930

Sato, K. (1979). Polyspermy-preventing mechanisms in mouse eggs fertilized in vitro. *The Journal of Experimental Zoology*, *210*(2), 353–359. doi:10.1002/jez.1402100219 PMID:541603

Savinkova, L. K., Ponomarenko, M. P., Ponomarenko, P. M., Drachkova, I. A., Lysova, M. V., Arshinova, T. V., & Kolchanov, N. A. (2009). TATA box polymorphisms in human gene promoters and associated hereditary pathologies. *Biochemistry (Mosc)*, *74*(2), 117–129. doi:10.1134/S0006297909020011 PMID:19267666

Schena, M., Shalon, D., Davis, R. W., & Brown, P. O. (1995). Quantitative monitoring of gene expression patterns with a complementary DNA microarray. *Science*, *270*(5235), 467–470. doi:10.1126/science.270.5235.467 PMID:7569999

Scherf, M., Klingenhoff, A., & Werner, T. (2000). Highly specific localization of promoter regions in large genomic sequences by PromoterInspector: A novel context analysis approach. *Journal of Molecular Biology*, *297*(3), 599–606. doi:10.1006/jmbi.2000.3589 PMID:10731414

Schlitt, T., & Brazma, A. (2007). Current approaches to gene regulatory network modelling. *BMC Bioinformatics*, *59*(Suppl 6). PMID:17903290

Scholzen, T., & Gerdes, J. (2000). The Ki-67 protein: From the known and the unknown. *Journal of Cellular Physiology*, *182*(3), 311–322. doi:10.1002/(SICI)1097-4652(200003)182:3<311::AID-JCP1>3.0.CO;2-9 PMID:10653597

Schultz, D., Onuchic, J. N., & Wolynes, P. G. (2007). Understanding stochastic simulations of the smallest genetic networks. *The Journal of Chemical Physics*, *126*(24), 245102. doi:10.1063/1.2741544 PMID:17614590

Schultz, R. M. (2002). The molecular foundations of the maternal to zygotic transition in the preimplantation embryo. *Human Reproduction Update*, *8*(4), 323–331. doi:10.1093/humupd/8.4.323 PMID:12206467

Schweitzer, P. J. (1968). Perturbation theory and finite Markov chains. *Journal of Applied Probability*, *5*(2), 401–413. doi:10.2307/3212261

Seok, J., Kaushal, A., Davis, R., & Xiao, W. (2010). Knowledge-based analysis of microarrays for the discovery of transcriptional regulation relationships. *BMC Bioinformatics*, *11*(Suppl 1), S8. doi:10.1186/1471-2105-11-S1-S8 PMID:20122245

Shah, N. P., Kasap, C., Weier, C., Balbas, M., Nicoll, J. M., Bleickardt, E., & Sawyers, C. L. et al. (2008). Transient potent BCR-ABL inhibition is sufficient to commit chronic myeloid leukemia cells irreversibly to apoptosis. *Cancer Cell*, *14*(6), 485–493. doi:10.1016/j.ccr.2008.11.001 PMID:19061839

Shav Tal, Y., Singer, R. H., & Darzacq, X. (2004). Imaging gene expression in single living cells. *Nature Reviews. Molecular Cell Biology*, *5*(10), 855–862. doi:10.1038/nrm1494 PMID:15459666

Shea, M., & Ackers, G. (1985). The OR control system of bacteriophage lambda: A physical-chemical model for gene regulation. *Journal of Molecular Biology*, *181*(2), 211–230. doi:10.1016/0022-2836(85)90086-5 PMID:3157005

Sheather, S. J., & Jones, M. C. (1991). A reliable data-based bandwidth selection method for Kernel density estimation. *Journal of the Royal Statistical Society. Series B. Methodological*, *53*(3), 683–690.

Sherman, J., & Morrison, W. J. (1950). Adjustment of an inverse matrix corresponding to a change in one element of a given matrix. *Annals of Mathematical Statistics*, *21*(1), 124–127. doi:10.1214/aoms/1177729893

Shiio, Y., Yamamoto, T., & Yamaguchi, N. (1992). Negative regulation of Rb expression by the p53 gene product. *Proceedings of the National Academy of Sciences of the United States of America*, *89*(12), 5206–5210. doi:10.1073/pnas.89.12.5206 PMID:1608930

Shin, A., & Iba, H. (2003). Construction of genetic network using evolutionary algorithm and combined fitness function. *Genome Informatics*, *14*, 94–103. PMID:15706524

Shinagawa, T., Takagi, T., Tsukamoto, D., Tomaru, C., Huynh, L. M., Sivaraman, P., & Ishii, S. et al. (2014). Histone variants enriched in oocytes enhance reprogramming to induced pluripotent stem cells. *Cell Stem Cell, 14*(2), 217–227. doi:10.1016/j.stem.2013.12.015 PMID:24506885

Shiraishi, T., Matsuyama, S., & Kitano, H. (2010). Large-scale analysis of network bistability for human cancers. *PLoS Computational Biology, 6*(7), e1000851. doi:10.1371/journal.pcbi.1000851 PMID:20628618

Shmulevich, I., & Aitchison, J. D. (2009). Deterministic and stochastic models of gene regulatory networks. *Methods in Enzymology, 467*, 335–356. doi:10.1016/S0076-6879(09)67013-0 PMID:19897099

Shmulevich, I., & Dougherty, E. R. (2007). *Genomic Signal Processing*. Princeton University Press. doi:10.1515/9781400865260

Shmulevich, I., & Dougherty, E. R. (2010). *Probabilistic Boolean Networks: The Modeling and Control of Gene Regulatory Networks*. New York: SIAM Press. doi:10.1137/1.9780898717631

Shmulevich, I., Dougherty, E. R., Kim, S., & Zhang, W. (2002). Probabilistic Boolean networks: A rule-based uncertainty model for gene regulatory networks. *Bioinformatics (Oxford, England), 18*(2), 261–274. doi:10.1093/bioinformatics/18.2.261 PMID:11847074

Shmulevich, I., Dougherty, E. R., & Zhang, W. (2002a). Control of stationary behavior in probabilistic Boolean networks by means of structural intervention. *Journal of Biological System, 10*(04), 431–445. doi:10.1142/S0218339002000706

Shmulevich, I., Dougherty, E. R., & Zhang, W. (2002b). Gene perturbation and intervention in probabilistic Boolean networks. *Bioinformatics (Oxford, England), 18*(10), 1319–1331. doi:10.1093/bioinformatics/18.10.1319 PMID:12376376

Sidje, R. B. (1998). Expokit: A software package for computing matrix exponentials. *ACM Transactions on Mathematical Software, 24*(1), 130–156. doi:10.1145/285861.285868

Simcha, D., Price, N. D., & Geman, D. (2012). The limits of de novo DNA motif discovery. *PLoS ONE, 7*(11), e47836. doi:10.1371/journal.pone.0047836 PMID:23144830

Simon, R., & Norton, L. (2006). The Norton-Simon hypothesis: Designing more effective and less toxic chemotherapeutic regimens. *Nature Clinical Practice. Oncology, 3*(8), 406–407. doi:10.1038/ncponc0560 PMID:16894366

Simpson, A. J., Caballero, O. L., Jungbluth, A., Chen, Y. T., & Old, L. J. (2005). Cancer/testis antigens, gametogenesis and cancer. *Nature Reviews. Cancer, 5*(8), 615–625. doi:10.1038/nrc1669 PMID:16034368

Sinha, S. (2006). On counting position weight matrix matches in a sequence, with application to discriminative motif finding. *Bioinformatics (Oxford, England), 22*(14), e454–e463. doi:10.1093/bioinformatics/btl227 PMID:16873507

Sirota, M., Dudley, J. T., Kim, J., Chiang, A. P., Morgan, A. A., Sweet-Cordero, A., & Butte, A. J. et al. (2011). Discovery and preclinical validation of drug indications using compendia of public gene expression data. *Science Translational Medicine*, *3*(96), 96ra77. doi:10.1126/scitranslmed.3001318 PMID:21849665

Sloop, K. W., Showalter, A. D., Von Kap-Herr, C., Pettenati, M. J., & Rhodes, S. J. (2000). Analysis of the human LHX3 neuroendocrine transcription factor gene and mapping to the subtelomeric region of chromosome 9. *Gene*, *245*(2), 237–243. doi:10.1016/S0378-1119(00)00025-1 PMID:10717474

Smadbeck, P., & Kaznessis, Y. N. (2014). Solution of Chemical Master Equations for Nonlinear Stochastic Reaction Networks. *Curr Opin Chem Eng*, *5*, 90–95. doi:10.1016/j.coche.2014.07.001 PMID:25215268

Smith, A. D., Sumazin, P., Xuan, Z., & Zhang, M. Q. (2006). DNA motifs in human and mouse proximal promoters predict tissue-specific expression. *Proceedings of the National Academy of Sciences of the United States of America*, *103*(16), 6275–6280. doi:10.1073/pnas.0508169103 PMID:16606849

Smith, L., Watson, M. B., O'Kane, S. L., Drew, P. J., Lind, M. J., & Cawkwell, L. (2006). The analysis of doxorubicin resistance in human breast cancer cells using antibody microarrays. *Molecular Cancer Therapeutics*, *5*(8), 2115–2120. doi:10.1158/1535-7163.MCT-06-0190 PMID:16928833

Smith, Z. D., Chan, M. M., Mikkelsen, T. S., Gu, H., Gnirke, A., Regev, A., & Meissner, A. (2012). A unique regulatory phase of DNA methylation in the early mammalian embryo. *Nature*, *484*(7394), 339–344. doi:10.1038/nature10960 PMID:22456710

Snoep, J. L., & Westerhoff, H. V. (2005). From isolation to integration, a systems biology approach for building the Silicon Cell. *Topics in Current Genetics*, *13*, 13–30. doi:10.1007/b106456

Soinov, L. A., Krestyaninova, M. A., & Brazma, A. (2003). Towards reconstruction of gene networks from expression data by supervised learning. *Genome Biology*, *4*(1), R6. doi:10.1186/gb-2003-4-1-r6 PMID:12540298

Soltoff, S. P., Carraway, K. L. III, Prigent, S. A., Gullick, W. G., & Cantley, L. C. (1994). ErbB3 is involved in activation of phosphatidylinositol 3-kinase by epidermal growth factor. *Molecular and Cellular Biology*, *14*(6), 3550–3558. doi:10.1128/MCB.14.6.3550 PMID:7515147

Spellman, P. T., Sherlock, G., Zhang, M. Q., Iyer, V. R., Anders, K., Eisen, M. B., & Futcher, B. et al. (1998). Comprehensive identification of cell cycle-regulated genes of the yeast Saccharomyces cerevisiae by microarray hybridization. *Molecular Biology of the Cell*, *9*(12), 3273–3297. doi:10.1091/mbc.9.12.3273 PMID:9843569

Spiro, C., Bazett-Jones, D. P., Wu, X., & McMurray, C. T. (1995). DNA structure determines protein binding and transcriptional efficiency of the proenkephalin cAMP-responsive enhancer. *The Journal of Biological Chemistry*, *270*(46), 27702–27710. doi:10.1074/jbc.270.46.27702 PMID:7499237

Spyratos, F., Ferrero-Poüs, M., Trassard, M., Hacène, K., Phillips, E., Tubiana-Hulin, M., & Le Doussal, V. (2002). Correlation between MIB-1 and other proliferation markers: Clinical implications of the MIB-1 cutoff value. *Cancer*, *94*(8), 2151–2159. doi:10.1002/cncr.10458 PMID:12001111

Stark, C., Breitkreutz, B., Reguly, T., Boucher, L., Breitkreutz, A., & Tyers, M. (2006). BioGRID: A general repository for interaction datasets. *Nucleic Acids Research*, *34*(90001), D535–D539. doi:10.1093/nar/gkj109 PMID:16381927

Steinbach, B., & Posthoff, C. (2012). Sources and Obstacles for Parallelization-a Comprehensive Exploration of the Unate Covering Problem Using Both CPU and GPU. In J. Astola, M. Kameiama, M. Lukac, & M. Stankovic (Eds.), GPU Computing with Applications in Digital Logic. Tampere International Center for Signal Processing.

Stephen, T. S., & Kadonaga, J. T. (2003). The RNA polymerase II core promoter. *Annual Review of Biochemistry*, *72*(1), 449–479. doi:10.1146/annurev.biochem.72.121801.161520 PMID:12651739

Stergachis, A. B., Neph, S., Sandstrom, R., Haugen, E., Reynolds, A. P., Zhang, M., & Stamatoyannopoulos, J. A. et al. (2014). Conservation of trans-acting circuitry during mammalian regulatory evolution. *Nature*, *515*(7527), 365–370. doi:10.1038/nature13972 PMID:25409825

Stewart, W. J. (1994). *Introduction to the numerical solution of Markov chains* (Vol. 41). Princeton University Press Princeton.

Storey, J. D., Xiao, W., Leek, J. T., Tompkins, R. G., & Davis, R. W. (2005). Significance analysis of time course microarray experiments. *Proceedings of the National Academy of Sciences of the United States of America*, *102*(36), 12837–12842. doi:10.1073/pnas.0504609102 PMID:16141318

Stormo, G. D. (2000). DNA binding sites: Representation and discovery. *Bioinformatics (Oxford, England)*, *16*(1), 16–23. doi:10.1093/bioinformatics/16.1.16 PMID:10812473

Stormo, G. D., Schneider, T. D., Gold, L., & Ehrenfeucht, A. (1982). Use of the 'Perceptron' algorithm to distinguish translational initiation sites in E. coli. *Nucleic Acids Research*, *10*(9), 2997–3011. doi:10.1093/nar/10.9.2997 PMID:7048259

Strogatz, S. H. (2001). Exploring complex networks. *Nature*, *410*(6825), 268–276. doi:10.1038/35065725 PMID:11258382

Styczynski, M. P., & Stephanopoulos, G. (2005). Overview of computational methods for the inference of gene regulatory networks. *Engineering*, *29*(3), 519–534.

Suzuki, H., Forrest, A. R., van Nimwegen, E., Daub, C. O., Balwierz, P. J., Irvine, K. M., & Hayashizaki, Y. et al. (2009). The transcriptional network that controls growth arrest and differentiation in a human myeloid leukemia cell line. *Nature Genetics*, *41*(5), 553–562. doi:10.1038/ng.375 PMID:19377474

Svoboda, P., & Flemr, M. (2010). The role of miRNAs and endogenous siRNAs in maternal-to-zygotic reprogramming and the establishment of pluripotency. *EMBO Reports*, *11*(8), 590–597. doi:10.1038/embor.2010.102 PMID:20651740

Swain, M., Hunniford, T., Dubitsky, W., Mandel, J., & Palfreyman, N. (2005). Reverse-engineering gene-regulatory networks using evolutionary algorithms and grid computing. *Journal of Clinical Monitoring and Computing*, *19*(4-5), 329–337. doi:10.1007/s10877-005-0678-x PMID:16328947

Szallasi, Z., Stelling, J., & Periwal, V. (2006). *System Modeling in Cellular Biology: From Concepts to Nuts and Bolts*. Cambridge, MA: The MIT Press. doi:10.7551/mitpress/9780262195485.001.0001

Székely, T. Jr, & Burragea, K. (2014). Stochastic simulation in systems biology. *Computational and Structural Biotechnology Journal*, *12*(20-21), 14–25. doi:10.1016/j.csbj.2014.10.003 PMID:25505503

Tachibana-Konwalski, K., Godwin, J., van der Weyden, L., Champion, L., Kudo, N. R., Adams, D. J., & Nasmyth, K. (2010). Rec8-containing cohesin maintains bivalents without turnover during the growing phase of mouse oocytes. *Genes & Development*, *24*(22), 2505–2516. doi:10.1101/gad.605910 PMID:20971813

Tahiliani, M., Koh, K. P., Shen, Y., Pastor, W. A., Bandukwala, H., Brudno, Y., & Rao, A. et al. (2009). Conversion of 5-methylcytosine to 5-hydroxymethylcytosine in mammalian DNA by MLL partner TET1. *Science*, *324*(5929), 930–935. doi:10.1126/science.1170116 PMID:19372391

Takahashi, K., Tanabe, K., Ohnuki, M., Narita, M., Ichisaka, T., Tomoda, K., & Yamanaka, S. (2007). Induction of pluripotent stem cells from adult human fibroblasts by defined factors. *Cell*, *131*(5), 861–872. doi:10.1016/j.cell.2007.11.019 PMID:18035408

Talbert, P. B., & Henikoff, S. (2010). Histone variants--ancient wrap artists of the epigenome. *Nature Reviews. Molecular Cell Biology*, *11*(4), 264–275. doi:10.1038/nrm2861 PMID:20197778

Tang, F., Barbacioru, C., Wang, Y., Nordman, E., Lee, C., Xu, N., & Surani, M. A. et al. (2009). mRNA-Seq whole-transcriptome analysis of a single cell. *Nature Methods*, *6*(5), 377–382. doi:10.1038/nmeth.1315 PMID:19349980

Tang, F., Kaneda, M., O'Carroll, D., Hajkova, P., Barton, S. C., Sun, Y. A., & Surani, M. A. et al. (2007). Maternal microRNAs are essential for mouse zygotic development. *Genes & Development*, *21*(6), 644–648. doi:10.1101/gad.418707 PMID:17369397

Teramura, Y., Ichinose, J., Takagi, H., Nishida, K., Yanagida, T., & Sako, Y. (2006). Single-molecule analysis of epidermal growth factor binding on the surface of living cells. *The EMBO Journal*, *25*(18), 4215–4222. doi:10.1038/sj.emboj.7601308 PMID:16946702

Thackray, V. G. (2014). Fox tales: Regulation of gonadotropin gene expression by forkhead transcription factors. *Molecular and Cellular Endocrinology*, *385*(1-2), 62–70. doi:10.1016/j.mce.2013.09.034 PMID:24099863

Theodoridis, S., & Koutroumbas, K. (2008). *Pattern Recognition* (4th ed.). Academic Press.

Thomas, R. (1998). Laws for the dynamics of regulatory networks. *The International Journal of Developmental Biology*, *42*(3), 479–485. PMID:9654035

Thomas, R., Thieffry, D., & Kaufman, M. (1995). Dynamical behaviour of biological regulatory networks-I. Biological role of feedback loops and practical use of the concept of the loop-characteristic states. *Bulletin of Mathematical Biology*, *57*(2), 247–276. doi:10.1007/BF02460618 PMID:7703920

Tibshirani, R. (1996). Regression shrinkage and selection via the Lasso. *Journal of the Royal Statistical Society. Series B. Methodological*, *58*(1), 267–288.

Torres-Padilla, M. E., Bannister, A. J., Hurd, P. J., Kouzarides, T., & Zernicka-Goetz, M. (2006). Dynamic distribution of the replacement histone variant H3.3 in the mouse oocyte and preimplantation embryos. *The International Journal of Developmental Biology*, *50*(5), 455–461. doi:10.1387/ijdb.052073mt PMID:16586346

Torrie, G. M., & Valleau, J. P. (1977). Nonphysical sampling distributions in Monte Carlo free-energy estimation: Umbrella sampling. *Journal of Computational Physics*, *23*(2), 187–199. doi:10.1016/0021-9991(77)90121-8

Troyanskaya, O. G. (2005). Putting microarrays in a context: Integrated analysis of diverse biological data. *Briefings in Bioinformatics*, *6*(1), 34–43. doi:10.1093/bib/6.1.34 PMID:15826355

Ulitsky, I., & Shamir, R. (2007). Identification of functional modules using network topology and high-throughput data. *BMC Systems Biology*, *1*(1), 8. doi:10.1186/1752-0509-1-8 PMID:17408515

Vahedi, G., Faryabi, B., Chamberland, J.-F., Datta, A., & Dougherty, E. R. (2008). Intervention in gene regulatory networks via a stationary mean-first-passage-time control policy. *IEEE Transactions on Bio-Medical Engineering*, *55*(10), 2319–2331. doi:10.1109/TBME.2008.925677 PMID:18838357

Vahedi, G., Faryabi, B., Chamberland, J.-F., Datta, A., & Dougherty, E. R. (2009). Optimal intervention strategies for cyclic therapeutic methods. *IEEE Transactions on Bio-Medical Engineering*, *56*(2), 281–291. doi:10.1109/TBME.2008.2003092 PMID:19272947

van der Heijden, G. W., Dieker, J. W., Derijck, A. A., Muller, S., Berden, J. H., Braat, D. D., & de Boer, P. et al. (2005). Asymmetry in histone H3 variants and lysine methylation between paternal and maternal chromatin of the early mouse zygote. *Mechanisms of Development*, *122*(9), 1008–1022. doi:10.1016/j.mod.2005.04.009 PMID:15922569

van der Heijden, G. W., Ramos, L., Baart, E. B., van den Berg, I. M., Derijck, A. A., van der Vlag, J., & de Boer, P. et al. (2008). Sperm-derived histones contribute to zygotic chromatin in humans. *BMC Developmental Biology*, *8*(1), 34. doi:10.1186/1471-213X-8-34 PMID:18377649 van Helden, J., Andre, B., & Collado-Vides, J. (1998). Extracting regulatory sites from the upstream region of yeast genes by computational analysis of oligonucleotide frequencies. *Journal of Molecular Biology*, *281*(5), 827–842. doi:10.1006/jmbi.1998.1947 PMID:9719638

Van Kampen, N. G. (2007). Stochastic processes in physics and chemistry (3rd ed.). Elsevier Science and Technology.

Vassena, R., Boue, S., Gonzalez-Roca, E., Aran, B., Auer, H., Veiga, A., & Izpisua Belmonte, J. C. (2011). Waves of early transcriptional activation and pluripotency program initiation during human preimplantation development. *Development*, *138*(17), 3699–3709. doi:10.1242/dev.064741 PMID:21775417

Veena, M. S., Wilken, R., Zheng, J. Y., Gholkar, A., Venkatesan, N., Vira, D., & Srivatsan, E. S. et al. (2014). p16 Protein and gigaxonin are associated with the ubiquitination of NFkappaB in cisplatin-induced senescence of cancer cells. *The Journal of Biological Chemistry*, *289*(50), 34921–34937. doi:10.1074/jbc.M114.568543 PMID:25331947

Velculescu, V. E., Zhang, L., Vogelstein, B., & Kinzler, K. W. (1995). Serial Analysis of Gene Expression. *Science*, *270*(5235), 484–487. doi:10.1126/science.270.5235.484 PMID:7570003

Vellela, M., & Qian, H. (2007). A Quasistationary Analysis of a Stochastic Chemical Reaction: Keizer's Paradox. *Bulletin of Mathematical Biology*, *69*(5), 1727–1746. doi:10.1007/s11538-006-9188-3 PMID:17318672

Vellela, M., & Qian, H. (2009). Stochastic dynamics and non-equilibrium thermodynamics of a bistable chemical system: The Schlogl model revisited. *Journal of the Royal Society, Interface*, *6*(39), 925–940. doi:10.1098/rsif.2008.0476 PMID:19095615

Venkateswarlu, S., Dawson, D. M., St Clair, P., Gupta, A., Willson, J. K., & Brattain, M. G. (2002). Autocrine heregulin generates growth factor independence and blocks apoptosis in colon cancer cells. *Oncogene*, *21*(1), 78–86. doi:10.1038/sj.onc.1205011 PMID:11791178

Venter, J. C., Adams, M. D., Myers, E. W., Li, P. W., Mural, R. J., Sutton, G. G., & Zhu, X. et al. (2001). The Sequence of the Human Genome. *Science*, *291*(5507), 1304–1351. doi:10.1126/science.1058040 PMID:11181995

Vicente, F. F. R., & Lopes, F. M. (2014). SFFS-WS: A feature selection algorithm exploring the smallworld properties of GNs. In Pattern Recognition in Bioinformatics, Proceedings, (LNCS), (vol. 8626, pp. 60-71). Berlin: Springer.

Vidal, M. (2005). Interactome modeling. *FEBS Letters*, *579*(8), 1834–1838. doi:10.1016/j.febslet.2005.02.030 PMID:15763560

Vincent, L. (1993). Grayscale area opening and closing, their efficient implementation and applications. In *Proceedings in Eurasip Workshop on Mathematical Morphology and its Applications to Signal Processing*. Barcelona, Spain: Academic Press.

Vincent, L., & Soille, P. (1991). Watersheds in digital spaces: An efficient algorithm based on immersion simulations. *IEEE Transactions on Pattern Analysis and Machine Intelligence*, *13*(6), 583–598. doi:10.1109/34.87344

Vlieghe, D., Sandelin, A., De Bleser, P. J., Vleminckx, K., Wasserman, W. W., van Roy, F., & Lenhard, B. (2006). A new generation of JASPAR, the open-access repository for transcription factor binding site profiles. *Nucleic Acids Research*, *34*, D95–D97. doi:10.1093/nar/gkj115 PMID:16381983

Voet, D., Voet, J., & Pratt, C. W. (2005). *Fundamentals of Biochemistry: Life at the Molecular Level* (2nd ed.). John Wiley & Sons.

Vohradsky, J. (2001). Neural network model of gene expression. *The FASEB Journal*, *15*(3), 846–854. doi:10.1096/fj.00-0361com PMID:11259403

Volfson, D., Marciniak, J., Blake, W. J., Ostroff, N., Tsimring, L. S., & Hasty, J. (2006). Origins of extrinsic variability in eukaryotic gene expression. *Nature*, *439*(7078), 861–864. doi:10.1038/nature04281 PMID:16372021

von Hippel, P. H., Johnson, N. P., & Marcus, A. H. (2013). Fifty years of DNA "Breathing": Reflections on old and new approaches. *Biopolymers*, *99*, 923–954. PMID:23840028

Wagner, A. (1999). Genes regulated cooperatively by one or more transcription factors and their identification in whole eukaryotic genomes. *Bioinformatics (Oxford, England)*, *15*(10), 776–784. doi:10.1093/bioinformatics/15.10.776 PMID:10705431

Wahde, M., & Hertz, J. (2000). Coarse-grained reverse engineering of genetic regulatory networks. *Bio Systems*, *55*(1-3), 129–136. doi:10.1016/S0303-2647(99)00090-8 PMID:10745116

Wang, Q. T., Piotrowska, K., Ciemerych, M. A., Milenkovic, L., Scott, M. P., Davis, R. W., & Zernicka-Goetz, M. (2004). A genome-wide study of gene activity reveals developmental signaling pathways in the preimplantation mouse embryo. *Dev Cell*, *6*(1), 133-144. doi:S1534580703004040

Wang, S., Kou, Z., Jing, Z., Zhang, Y., Guo, X., Dong, M., . . . Gao, S. (2010). Proteome of mouse oocytes at different developmental stages. *Proc Natl Acad Sci U S A*, *107*(41), 17639-17644. doi: 10.1073/pnas.1013185107

Wang, H., & Dey, S. K. (2006). Roadmap to embryo implantation: Clues from mouse models. *Nature Reviews. Genetics*, *7*(3), 185–199. doi:10.1038/nrg1808 PMID:16485018

Wang, T., & Stormo, G. D. (2003). Combining phylogenetic data with co-regulated genes to identify regulatory motifs. *Bioinformatics (Oxford, England)*, *19*(18), 2369–2380. doi:10.1093/bioinformatics/btg329 PMID:14668220

Wang, Z., Gerstein, M., & Snyder, M. (2009). RNA-Seq: A revolutionary tool for transcriptomics. *Nature Reviews. Genetics*, *10*(1), 57–63. doi:10.1038/nrg2484 PMID:19015660

Wang, Z., Xu, W., San Lucas, F. A., & Liu, Y. (2013). Incorporating prior knowledge into Gene Network Study. *Bioinformatics (Oxford, England)*, *29*(20), 2633–2640. doi:10.1093/bioinformatics/btt443 PMID:23956306

Ward, W. S., & Coffey, D. S. (1991). DNA packaging and organization in mammalian spermatozoa: Comparison with somatic cells. *Biology of Reproduction*, *44*(4), 569–574. doi:10.1095/biolreprod44.4.569 PMID:2043729

Wassarman, P. M. (1999). Mammalian fertilization: Molecular aspects of gamete adhesion, exocytosis, and fusion. *Cell*, *96*(2), 175–183. doi:10.1016/S0092-8674(00)80558-9 PMID:9988213

Wassarman, P., Chen, J., Cohen, N., Litscher, E., Liu, C., Qi, H., & Williams, Z. (1999). Structure and function of the mammalian egg zona pellucida. *The Journal of Experimental Zoology*, *285*(3), 251–258. doi:10.1002/(SICI)1097-010X(19991015)285:3<251::AID-JEZ8>3.0.CO;2-1 PMID:10497324

Wasserman, W. W., & Sandelin, A. (2004). Applied bioinformatics for the identification of regulatory elements. *Nature Reviews. Genetics*, *5*(4), 276–287. doi:10.1038/nrg1315 PMID:15131651

Weeraratna, A. T., Jiang, Y., Hostetter, G., Rosenblatt, K., Duray, P., Bittner, M., & Trent, J. M. (2002). Wnt5a signaling directly affects cell motility and invasion of metastatic melanoma. *Cancer Cell*, *1*(3), 279–288. doi:10.1016/S1535-6108(02)00045-4 PMID:12086864

Wen, D., Banaszynski, L. A., Liu, Y., Geng, F., Noh, K. M., Xiang, J., & Rafii, S. et al. (2014). Histone variant H3.3 is an essential maternal factor for oocyte reprogramming. *Proceedings of the National Academy of Sciences of the United States of America*, *111*(20), 7325–7330. doi:10.1073/pnas.1406389111 PMID:24799717

Weng, K. A., Jeffreys, C. A., & Bickel, S. E. (2014). Rejuvenation of meiotic cohesion in oocytes during prophase I is required for chiasma maintenance and accurate chromosome segregation. *PLOS Genetics*, *10*(9), e1004607. doi:10.1371/journal.pgen.1004607 PMID:25211017

Werhli, A. V., Grzegorczyk, M., & Husmeier, D. (2006). Comparative evaluation of reverse engineering gene regulatory networks with relevance networks, graphical gaussian models and bayesian networks. *Bioinformatics (Oxford, England)*, *22*(20), 2523–2531. doi:10.1093/bioinformatics/btl391 PMID:16844710

Werhli, A. V., & Husmeier, D. (2008). Gene regulatory network reconstruction by bayesian integration of prior knowledge and/or different experimental conditions. *Journal of Bioinformatics and Computational Biology*, *6*(3), 543–572. doi:10.1142/S0219720008003539 PMID:18574862

Werner, T. (1999). Models for prediction and recognition of eukaryotic promoters. *Mammalian Genome*, *10*(2), 168–175. doi:10.1007/s003359900963 PMID:9922398

White, D. J. (1978). *Finite Dynamic Programming: An Approach to Finite Markov Decision Processes*. New York, NY: John Wiley & Sons, Inc.

Whitfield, T. W., Wang, J., Collins, P. J., Partridge, E. C., Aldred, S. F., Trinklein, N. D., & Weng, Z. et al. (2012). Functional analysis of transcription factor binding sites in human promoters. *Genome Biology*, *13*, R50. PMID:22951020

Wilkinson, J. C., Stein, R. A., Guyer, C. A., Beechem, J. M., & Staros, J. V. (2001). Realtime kinetics of ligand/cell surface receptor interactions in living cells: Binding of epidermal growth factor to the epidermal growth factor receptor. *Biochemistry*, *40*(34), 10230–10242. doi:10.1021/bi010705t PMID:11513601

Wingender, E. (2008). The TRANSFAC project as an example of framework technology that supports the analysis of genomic regulation. *Briefings in Bioinformatics*, *9*(4), 326–332. doi:10.1093/bib/bbn016 PMID:18436575

Wingender, E., Chen, X., Hehl, R., Karas, H., Liebich, I., Matys, V., & Schacherer, F. et al. (2000). TRANSFAC: An integrated system for gene expression regulation. *Nucleic Acids Research*, *28*(1), 316–319. doi:10.1093/nar/28.1.316 PMID:10592259

Worrad, D. M., Ram, P. T., & Schultz, R. M. (1994). Regulation of gene expression in the mouse oocyte and early preimplantation embryo: Developmental changes in Sp1 and TATA box-binding protein, TBP. *Development*, *120*(8), 2347–2357. Retrieved from http://www.ncbi.nlm.nih.gov/pubmed/7925035 PMID:7925035

Xiao, Y. (2009). A tutorial on analysis and simulation of boolean gene regulatory network models. *Current Genomics*, *10*(7), 511–525. doi:10.2174/138920209789208237 PMID:20436877

Xiao, Y., & Dougherty, E. R. (2007). The impact of function perturbations in Boolean networks. *Bioinformatics (Oxford, England)*, *23*(10), 1265–1273. doi:10.1093/bioinformatics/btm093 PMID:17379691

Xing, E. P., Wu, W., Jordan, M. I., & Karp, R. M. (2004). Logos: A modular bayesian model for de novo motif detection. *Journal of Bioinformatics and Computational Biology*, *2*(1), 127–154. doi:10.1142/S0219720004000508 PMID:15272436

Xue, W., Wang, J., Shen, Z., & Zhu, H. (2004). Enrichment of transcriptional regulatory sites in non-coding genomic region. *Bioinformatics (Oxford, England)*, *20*(4), 569–575. doi:10.1093/bioinformatics/btg450 PMID:14990453

Xue, Z., Huang, K., Cai, C., Cai, L., Jiang, C. Y., Feng, Y., & Fan, G. et al. (2013). Genetic programs in human and mouse early embryos revealed by single-cell RNA sequencing. *Nature*, *500*(7464), 593–597. doi:10.1038/nature12364 PMID:23892778

Xu, X., Smorag, L., Nakamura, T., Kimura, T., Dressel, R., Fitzner, A., & Krishna Pantakani, D. V. et al. (2015). Dppa3 expression is critical for generation of fully reprogrammed iPS cells and maintenance of Dlk1-Dio3 imprinting. *Nature Communications*, *6*, 6008. doi:10.1038/ncomms7008 PMID:25613421

Yamada, S., Shiono, S., Joo, A., & Yoshimura, A. (2003). Control mechanism of JAK/STAT signal transduction pathway. *FEBS Letters*, *534*(1-3), 190–196. doi:10.1016/S0014-5793(02)03842-5 PMID:12527385

Yang, J. P., Hori, M., Sanda, T., & Okamoto, T. (1999). Identification of a novel inhibitor of nuclear factor-kappaB, RelA-associated inhibitor. *The Journal of Biological Chemistry*, *274*(22), 15662–15670. doi:10.1074/jbc.274.22.15662 PMID:10336463

Yeung, M. K., Tegner, J., & Collins, J. J. (2002). Reverse engineering gene networks using singular value decomposition and robust regression. *Proceedings of the National Academy of Sciences of the United States of America*, *99*(9), 6163–6168. doi:10.1073/pnas.092576199 PMID:11983907

Yoon, B.-J., Qian, X., & Dougherty, E. R. (2013). Quantifying the objective cost of uncertainty in complex dynamical systems. *IEEE Transactions on Signal Processing*, *61*(9), 2256–2266. doi:10.1109/TSP.2013.2251336

Yousefi, M. R., Datta, A., & Dougherty, E. R. (2012). Optimal Intervention Strategies for Therapeutic Methods With Fixed-Length Duration of Drug Effectiveness. *IEEE Transactions on Signal Processing*, *60*(9), 4930–4944. doi:10.1109/TSP.2012.2202114

Yousefi, M. R., Datta, A., & Dougherty, E. R. (2013). Optimal Intervention in Markovian Gene Regulatory Networks With Random-Length Therapeutic Response to Antitumor Drug. *IEEE Transactions on Bio-Medical Engineering*, *60*(12), 3542–3552. doi:10.1109/TBME.2013.2272891 PMID:23864151

Yousefi, M. R., & Dougherty, E. R. (2013). Intervention in gene regulatory networks with maximal phenotype alteration. *Bioinformatics (Oxford, England)*, *29*(14), 1758–1767. doi:10.1093/bioinformatics/btt242 PMID:23630177

Yousefi, M. R., & Dougherty, E. R. (2014). A comparison study of optimal and suboptimal intervention policies for gene regulatory networks in the presence of uncertainty. *EURASIP Journal on Bioinformatics & Systems Biology*, (6): 1–13. PMID:24708650

Yousefi, M. R., & Ivanov, I. (2013). Optimal control of gene regulatory networks with uncertain intervention effects. In *Global Conference on Signal and Information Processing (GlobalSIP)* (pp. 105–108). IEEE. doi:10.1109/GlobalSIP.2013.6736826

Yuan, M., & Lin, Y. (2006). Model selection and estimation in regression with grouped variables. *Journal of the Royal Statistical Society. Series B, Statistical Methodology*, *68*(1), 49–67. doi:10.1111/j.1467-9868.2005.00532.x

Yu, H., Paccanaro, A., Trifonov, V., & Gerstein, M. (2006). Predicting interactions in protein networks by completing defective cliques. *Bioinformatics (Oxford, England)*, *22*(7), 823–829. doi:10.1093/bioinformatics/btl014 PMID:16455753

Yurttas, P., Morency, E., & Coonrod, S. A. (2010). Use of proteomics to identify highly abundant maternal factors that drive the egg-to-embryo transition. *Reproduction (Cambridge, England)*, *139*(5), 809–823. doi:10.1530/REP-09-0538 PMID:20106898

Yu, X., Lin, J., Zack, D. J., & Qian, J. (2006). Computational analysis of tissue-specific combinatorial gene regulation: Predicting interaction between transcription factors in human tissues. *Nucleic Acids Research*, *34*(17), 4925–4936. doi:10.1093/nar/gkl595 PMID:16982645

Zanotto, E., Häkkinen, A., Teku, G., Shen, B., Ribeiro, A. S., & Jacobs, H. T. (2009). NF-Y influences directionality of transcription from the bidirectional Mrps12/Sarsm promoter in both mouse and human cells. *Biochimica et Biophysica Acta*, *1789*(5), 432–442. doi:10.1016/j.bbagrm.2009.05.001 PMID:19439209

Zañudo, J. G. T., & Albert, R. (2015). Cell Fate Reprogramming by Control of Intracellular Network Dynamics. *PLoS Computational Biology*, *11*(4), e1004193. doi:10.1371/journal.pcbi.1004193 PMID:25849586

Zappavigna, V., Renucci, A., Izpisúa-Belmonte, J. C., Urier, G., Peschle, C., & Duboule, D. (1991). HOX4 genes encode transcription factors with potential auto- and cross-regulatory capacities. *The EMBO Journal*, *10*, 4177–4187. PMID:1756725

Zeng, F., Baldwin, D. A., & Schultz, R. M. (2004). Transcript profiling during preimplantation mouse development. *Developmental Biology*, *272*(2), 483–496. doi:10.1016/j.ydbio.2004.05.018 PMID:15282163

Zhang, S. Q., Ching, W. K., Tsing, N. K., Leung, H. Y., & Guo, D. (2010). A new multiple regression approach for the construction of genetic regulatory networks. *Artificial Intelligence in Medicine*, *48*(2-3), 153–160. doi:10.1016/j.artmed.2009.11.001 PMID:19963359

Zhang, X., Jin, J. Y., Wu, J., Qin, X., Streilein, R., Hall, R. P., & Zhang, J. Y. (2015). RNA-Seq and ChIP-Seq reveals SQSTM1/p62 as a key mediator of JunB suppression of NF-κB-dependent inflammation. *The Journal of Investigative Dermatology*, *135*(4), 1016–1024. doi:10.1038/jid.2014.519 PMID:25501661

Zhang, Y., Larsen, C. A., Stadler, H. S., & Ames, J. B. (2011). Structural basis for sequence specific DNA binding and protein dimerization of HOXA13. *PLoS ONE*, *6*(8), e23069. doi:10.1371/journal.pone.0023069 PMID:21829694

Zhang, Y., Quian, M., Ouyang, Q., Deng, M., Li, F., & Tang, C. (2006, July). Stochastic Model of Yeast Cell-Cycle Network. *Physica D. Nonlinear Phenomena*, *219*(1), 35–39. doi:10.1016/j.physd.2006.05.009

Zhao, X., Huang, H., & Speed, T. P. (2005). Finding short DNA motifs using permuted Markov models. *Journal of Computational Biology*, *12*(6), 894–906. doi:10.1089/cmb.2005.12.894 PMID:16108724

Zhou, H., Zarubin, T., Ji, Z., Min, Z., Zhu, W., Downey, J. S., & Han, J. (2005). Frequency and Distribution of AP-1 Sites in the Human Genome. *DNA Research*, *12*(2), 139–150. doi:10.1093/dnares/12.2.139 PMID:16303745

Zhou, T., Chen, L., & Aihara, K. (2005). Molecular communication through stochastic synchronization induced by extracellular fluctuations. *Physical Review Letters*, *95*(17), 178103. doi:10.1103/PhysRevLett.95.178103 PMID:16383875

Zhou, X., Wang, X., Pal, R., Ivanov, I., Bittner, M. L., & Dougherty, E. R. (2004). A Bayesian connectivity-based approach to constructing probabilistic gene regulatory networks. *Bioinformatics (Oxford, England)*, *20*(17), 2918–2927. doi:10.1093/bioinformatics/bth318 PMID:15145802

Zhou, Y., Li, S., Hu, Y. P., Wang, J., Hauser, J., Conway, A. N., & Brattain, M. G. et al. (2006). Blockade of EGFR and ErbB2 by the novel dual EGFR and ErbB2 tyrosine kinase inhibitor GW572016 sensitizes human colon carcinoma GEO cells to apoptosis. *Cancer Research*, *66*(1), 404–411. doi:10.1158/0008-5472.CAN-05-2506 PMID:16397255

Zhu, X. M., Yin, L., Hood, L., & Ao, P. (2004). Robustness, stability and efficiency of phage lambda genetic switch: Dynamical structure analysis. *Journal of Bioinformatics and Computational Biology*, *2*(4), 785–817. doi:10.1142/S0219720004000946 PMID:15617166

Zhu, Z., Pilpel, Y., & Church, G. M. (2002). Computational identification of transcription factor binding sites via a transcription-factor-centric clustering (TFCC) algorithm. *Journal of Molecular Biology*, *318*(1), 71–81. doi:10.1016/S0022-2836(02)00026-8 PMID:12054769

Zou, H., & Hastie, T. (2005). Regularization and variable selection via the elastic net. *Journal of the Royal Statistical Society. Series B, Statistical Methodology*, *67*(2), 301–320. doi:10.1111/j.1467-9868.2005.00503.x

Zou, M., & Conzen, S. D. (2005). A new dynamic Bayesian network (DBN) approach for identifying gene regulatory networks from time course microarray data. *Bioinformatics (Oxford, England)*, *21*(1), 71–79. doi:10.1093/bioinformatics/bth463 PMID:15308537

About the Contributors

Xiaoning Qian received the Ph.D. degree in Electrical Engineering from Yale University, New Haven, CT, in 2005. Currently, he is an assistant professor with the Department of Electrical & Computer Engineering, Texas A&M University, College Station, TX. He is affiliated with the Center for Bioinformatics & Genomic Systems Engineering (CBGSE) and the Center for Translational Environmental Health Research (CTEHR) at Texas A\&M. His current research interests include computational network biology, genomic signal processing, and biomedical image analysis.

Ranadip Pal received the BTech degree in electronics and electrical communication engineering from the Indian Institute of Technology, Kharagpur, India, in 2002, and the MS and PhD degrees in electrical engineering from Texas A & M University, College Station, in 2004 and 2007, respectively. From August 2007, he has been with Texas Tech University where he is currently an associate professor at Electrical and Computer Engineering Department. His research areas are stochastic modeling and control, genomic signal processing, and computational biology. He received the Chancellor's Council Distinguished Research Award, 2016; Whitacre Research Award, 2014; President's excellence in Teaching Award, 2012, and US NSF CAREER Award, 2010.

* * *

Youfang Cao is a passionate early career scientist in the field of computational systems biology and mathematical biology. He has extensive expertise in theories and applications of stochastic modeling in complex biological networks. He received his PhD degree in Biomedical Engineering from the Shanghai Center for Systems Biomedicine in Shanghai Jiao Tong University in 2011. He was awarded the prestigious Anthony Leggett Award, named after the Nobel laureate Anthony James Leggett, for the achievements during his PhD thesis research. He was a Research Assistant Professor from 2012 to 2015 in University of Illinois at Chicago, and now a postdoctoral research associate in the Theoretical Biology and Biophysics (T-6) group, and in the Center for Nonlinear Studies (CNLS) in Los Alamos National Laboratory. He has led the development of the theoretical framework for the accurate direct solution method for the chemical master equation (ACME). He also designed the algorithms and developed the software package for ACME (http://tanto.bioe.uic.edu/dcme). The ACME method is currently the only method able to solve both the steady state and time-evolving probability landscapes of realistic biological networks with controlled truncation errors. The method can be applied to study broad problems in systems biology. His major research interest is to attack complex realistic biological problems using mathematical models. His current research includes multi-scale stochastic models in HIV viral infections and immunology. More details can be found from his website: http://www.youfangcao.com/.

Yang Dai received Ph.D degree in Management Science and Engineering from University of Tsukuba, Japan, in 1991. From 1991 to 1997 she was a research associate, an assistant professor at the Department of Management Science of Kobe University of Commerce. From 1997-2001 she was an assistant professor of Department of Mathematical and Computing Sciences, Tokyo Institute of Technology, Japan. Since 2001 she is on faculty of Department of Bioengineering, the university of Illinois at Chicago. Her recent research interests include the analysis of gene expression and pathways, inference of biological network and data analysis on large scale functional genomic data. Professor Dai has published more than 90 peer reviewed papers in journals and conference proceedings.

Hong Hu is a Ph.D. candidate in Bioinformatics program in Department of Bioengineering at University of Illinois at Chicago. His research interest includes next generation sequencing, transcriptional regulation study and machine learning.

Jianping Hua received the B.S. and M.S. degrees in Electrical Engineering from the Tsinghua University, Beijing, China, in 1998 and 2000, respectively. He received the Ph.D. degree in Eletectrical Engineering from Texas A&M University in 2004. Currently he is the research Assistant Professor at the Center for Bioinformatics and Genomics Systems Engineering (CBGSE) at Texas A&M University. His main research interest lies in genomic signal processing, signal and image processing, and high-throughput content analysis.

Hiroshi Kobayashi, earned his Ph.D in biochemistry from Tokyo University, Japan in 1974. He is currently Professor Emeritus at the Graduate School of Pharmaceutical Sciences, Chiba University in Japan. His research interests are adaptation strategies of living cells to acidic environments and computer simulation analysis of cellular functions.

Jie Liang is the Richard and Loan Hill Professor of Bioengineering and Computer Science in the Richard and Loan Hill Dept of Bioengineering at the University of Illinois at Chicago. He received his B.S. degree from Fudan University (1986), MCS and Ph.D. from the University of Illinois at Urbana-Champaign (1994). He was an NSF CISE postdoctoral research associate (1994-1996) at the Beckman Institute and National Center for Supercomputing and its Applications (NCSA) in Urbana. He was a visiting fellow at the Institute of Mathematics and Applications at Minneapolis in 1996, and an Investigator at SmithKline Beecham Pharmaceuticals from 1997 to 1999. He was a recipient of the NSF CAREER award in 2003. He is an AIMBE fellow, and was a university scholar (19992012). His research interests are in computational proteomics, stochastic networks, cellular pattern formation, and structural bioinformatics. His recent work can be accessed at http://gila.bioe.uic.edu/liang/liang_pub.html.

Fabricio M. Lopes received the PhD degree in Bioinformatics from University of Sao Paulo (USP), Brazil in 2011. His Ph.D. thesis received an honorable mention from USP Thesis Award 2011. Since 2003, he has been professor at the Computer Science Department at Federal University of Technology - Paraná (UTFPR), Brazil. His main research topics are in systems biology and pattern recognition for bioinformatics and computer vision, including but not limited to, gene network identification and data integration. His recent work can be accessed at http://pessoal.utfpr.edu.br/fabricio/.

David C. Martins Jr. received the B.Sc. (2001), M.Sc. (2004) and Ph.D. (2008) degrees in computer science from the Institute of Mathematics and Statistics, University of Sao Paulo (USP), Brazil. In 2007 he did a research stage (doctorate sandwich) at the Genomic Signal Processing Laboratory - Texas A\&M University - USA. His Ph.D. thesis received an honorable mention from CAPES Thesis Award 2009. Since 2009, he has been an assistant professor at the Center of Mathematics, Computing and Cognition of Federal University of ABC (UFABC), Brazil. In 2015 he received the Academic Excellence Award from the UFABC computer science graduation program. His main research topics are in pattern recognition for computer vision, bioinformatics and systems biology, including, but not limited to, gene network identification and genes/biomarkers prioritization associated to complex diseases.

Kenta Nakai was born in Osaka, Japan, in 1963. He received the PhD degree on the prediction of subcellular localization sites of proteins from Kyoto University, Kyoto, Japan, in 1992. From 1989, he has worked at several institutions, including Kyoto University, National Institute of Basic Biology, and Osaka University. From 1999 to 2003, he was an Associate Professor at Human Genome Center, the Institute of Medical Science, the University of Tokyo, Japan. Since 2003, he has been a full Professor at the same institute. He was elected as the president of the Japanese Society for Bioinformatics in 2006 and 2007. His main research interest is to develop computational ways for interpreting biological information, especially, that of transcriptional regulation.

Sung-Joon Park received the PhD degree in Engineering from Tokyo Institute of Technology in 2005. He has worked as a researcher in Kobe University and Kyoto University. He is currently a project senior assistant professor in Human Genome Center, the Institute of Medical Science, the University of Tokyo. His current research interest includes genome and epigenome information science using high-throughput sequencing technologies. He is a member of Information Processing Society of Japan, The Japan Society for Artificial Intelligence, and Japanese Society for Bioinformatics.

Shubhra Sankar Ray is an Associate Professor in the Machine Intelligence Unit, Indian Statistical Institute, Kolkata. He is also associated with the Center for Soft Computing research, funded by the Department of Science and Technology, Govt. of India, at the same institute. He received the B. Sc degree with Physics Honours, M. Sc degree in Electronic Science and the M. Tech degree in Radio Physics and Electronics in 1998, 2000 and 2002, respectively. He received the Ph.D. degree in Engineering in 2008 from Jadavpur University. He worked as a CSIR Senior Research Fellow at Machine Intelligence Unit, Indian Statistical Institute in 2003-2006, Visiting Research Fellow at Center for Soft Computing Research, Indian Statistical Institute in 2006-2008, and Post Doctoral Fellow at the Biophysics Division, Saha Institute of Nuclear Physics, India, in 2008-2009. His current research activities are on Bioinformatics, Neural Networks, Genetic Algorithms, Data mining and Soft Computing. Three of his journal publications are now curated paper in Saccharomyces Genome Database, Department of Genetics, School of Medicine, Stanford University, California, USA. His name has been included in Marquis Who's Who in the World, 2007 Edition and 30th Pearl Anniversary Edition, 2012. He received the Microsoft Young Faculty Award 2010-2011 from Microsoft Research Laboratory, India and Indian Statistical Institute.

Chao Sima's research interest is in eliciting novel biological knowledge through mathematical and engineering approach using various genomic data, and the application of pattern recognition and statistical techniques in biological data. By characterizing and profiling drugs responses of key cell components over time for live cells, much insight can be gained toward understanding their roles in cell dynamics under stress and cell's key decisionmaking in the direction of apoptosis or survival. Also the well-developed methodologies in existing literature on classification, feature selection and error estimation can be misleading due to the uniqueness of the genomic data, which is often characterized by limited sample size and large feature dimension. Establishing these potential misuses and adapting them to applications like cancer classification, biomarker discovery and high throughput screening target identification have helped tremendously in propelling us to new discoveries.

Anna Terebus is currently a Ph.D. student in Bioinformatics at University of Illinois in Chicago. She graduated with a Master's Degree in Applied Mathematics from Rivne State Humanitarian University, where she worked as a Research Assistant in Applied Mathematics and Informatics Department (Rivne, Ukraine). Anna is currently a Research Assistant in Bioengineering department at University of Illinois in Chicago working with Professor Jie Liang. Her research interests are solution of discrete Chemical Master Equation and stochastic modeling of genetic and biochemical networks and network motifs.

Jeffrey M. Trent is President and Research Director of the Translational Genomics Research Institute (TGen) in Phoenix, Arizona. TGen's main goal as a non-profit medical research institute is focused on applying advances from the Human Genome Project to discover the molecular basis of disease, identifying factors contributing to disease progression and severity, and accelerating new drug development to advance clinical treatments for patients. TGen's research focuses on a variety of diseases, including Alzheimer's, autism, diabetes and multiple subtypes of cancer. TGen also engages in bio-defense research involving deadly pathogens, and has developed (under TGen's world-recognized Physician-in-Chief, Dr. Daniel Von Hoff) a clinical research service in oncology, which annually touches the lives of hundreds of patients with cancer. Prior to forming TGen in 2002, Dr. Trent served for 10 years as the Scientific Director of the National Human Genome Research Institute at the National Institutes of Health in Bethesda, Maryland. Under his guidance, NHGRI's Division of Intramural Research became an internationally recognized research center in human genetics. Dr. Trent's research has provided important insights into the genetic basis of cancer. He is the author of more than 300 manuscripts in the scientific literature, numerous book chapters, invited reviews, and has given hundreds of invited lectures. He has received numerous honors and awards, and has sat on the editorial boards of a dozen scientific publications. He specializes in developing and integrating novel "omic" technologies, supporting studies of molecular changes related to cancer risk and progression. He continues to participate in studies of other complex diseases in humans, as well as leading TGen's canine hereditary cancer program. Dr. Trent's previous faculty positions included: The University of Arizona, where he was Deputy Director and Director for Basic Science of the Arizona Comprehensive Cancer Center; the University of Michigan, where he held the E. Maisel Endowed Professorship in Cancer Genetics, Professor of Human Genetics and Radiation Oncology, Head of the Cancer Biology Division of the Department of Radiation Oncology, and Deputy Director and Director of Basic Research for the Michigan Comprehensive Cancer Center. He also is a Diplomat of the American College of Medical Genetics. Work in Dr. Trent's laboratory focuses on the study of genetic changes related to cancer predisposition and progression. He has worked the majority of his career on melanoma, most recently serving as the Co-Principal Investigator of the Stand Up to Cancer Melanoma Dream Team. The focus on that project is using molecularly-guided therapy for patients with BRAF wild-type (BRAFwt) metastatic melanoma. In addition to continuing work on germline genetic alterations associated with melanoma risk, his laboratory has been among the most active in identifying and understanding the somatic changes associated with canine melanoma. The canine is a critically important model of human disease, and in the case of melanoma the clear clinical association to the human is for the largely understudied mucosal melanomas. Other work in his laboratory has been focused upon relating the recent advances in both molecular biology and cancer genetics of ovarian cancer. Specifically, he was one of the leaders of an international consortium which recently identified that Small Cell Carcinoma of the Ovary, hypercalcemic type, (SCCOHT) displays frequent inactivating germline and somatic mutations in SMARCA4 {http://www.nature.com/ng/journal/v46/n5/full/ng.2928.html}. SCCOHT is an extremely rare, aggressive cancer affecting children and young women (average age of diagnosis 23yo compared to 63yo for the common epithelial ovarian cancers). Working with investigators headed by Dr. David Huntsman of the British Columbia Cancer Agency they identified germline and somatic inactivating mutations in this SWI/SNF chromatin-remodeling gene in nearly all SCCOHT. The genetic changes lead to SMARCA4 protein loss in >95% of SCCOHT tumors but in only 0.4% (2/485) of other primary ovarian tumors. These data implicate SMARCA4 in SCCOHT oncogenesis.

M. R. Yousefi received his master's degree in electrical engineering from the University of Tehran, Iran, in 2008 and his Ph.D. in electrical engineering from Texas A&M University, College Station, Texas in 2013. His Ph.D. research focused on designing optimal control strategies, taking into account probabilistic variabilities in tumor response to external interventions, for Markovian gene regulatory networks. He joined the Department of Electrical and Computer Engineering at The Ohio State University in June 2013 as a Research Assistant Professor. His current research interests include genomic signal processing and studying optimal intervention problems in gene regulatory networks from a probabilistic perspective.

Index

S

T

Z

CPSIA information can be obtained at www.ICGtesting.com
Printed in the USA
BVOW04*0802070616

450937BV00007B/20/P

9 781522 503538